Experimental Physics of Gravitational Waves

Institute of Physics
Urbino University
Italy

INFN – Italian Institute
for Nuclear Physics
Florence Section

Proceedings of the International Summer School on

Experimental Physics of Gravitational Waves

Urbino, Italy September 6–18, 1999

Editors

M. Barone
Urbino University, Italy

G. Calamai
Astrophysical Observatory, Arcetri, Italy

M. Mazzoni & R. Stanga
Florence University, Italy

F. Vetrano
Urbino University, Italy

World Scientific
Singapore • New Jersey • London • Hong Kong

Published by

World Scientific Publishing Co. Pte. Ltd.

5 Toh Tuck Link, Singapore 596224

USA office: 27 Warren Street, Suite 401-402, Hackensack, NJ 07601

UK office: 57 Shelton Street, Covent Garden, London WC2H 9HE

British Library Cataloguing-in-Publication Data
A catalogue record for this book is available from the British Library.

PROCEEDINGS OF THE INTERNATIONAL SUMMER SCHOOL ON EXPERIMENTAL PHYSICS OF GRAVITATIONAL WAVES

ISBN-13 978-981-02-4306-7
ISBN-10 981-02-4306-5

International Advisory Committee:

C. Boccara, ESPCI, *Paris*
A. Brillet, Observatoire de la Cote d'Azur, *Nice*
G. Conforto, Physics Institute, University, *Urbino*
M. Davier, LAL, *Orsay*
A. de Rujula, CERN, *Geneva*
A. Giazotto, INFN, *Pisa*
M. Inguscio, LENS & University, *Florence*
F. Pacini, Astrophysical Observatory of Arcetri, *Florence*

Local Organizing Committee:

M. Barone, Institute of Physics, University, *Urbino*
P. Dominici, Institute of Physics, University, *Urbino*
F. Frasconi, INFN, *Pisa*
C. Grimani, Institute of Physics, University, *Urbino*
M.A. Piergiovanni, Institute of Physics, University, *Urbino*

Scientific Secretariat:

R. Persi, Institute of Physics, University, *Urbino*

Council of the School:

G. Calamai, Astrophysical Observatory of Arcetri, *Florence*
F. Fidecaro, Department of Physics, University, *Pisa*
M. Mazzoni, Dpt. of Astronomy and Space Science, University, *Florence*
R. Stanga, Department of Astronomy and Space Science, University, *Florence*
F. Vetrano, Institute of Physics, University, *Urbino* (Director)

Supported by:

Institute of Physics, Urbino University
INFN - Italian Institute for Nuclear Physics, Florence Section
Astrophysics Observatory of Arcetri, Florence
Department of Astronomy and Space Science, Florence University
Fondazione Cassa di Risparmio di Pesaro

Preface

The idea of a School on "Experimental Physics of Gravitational Waves" arose at one of the numerous meetings on the VIRGO Project during a coffee-break. A group of Florence-Urbino people was discussing about many (too many indeed!) technical and scientific topics related to build up the antenna. Suddenly, old (or, let's say, not too young) people were put in trouble by young fellows asking for wide references on these arguments. In fact, due to the continuous production of ideas concerning physical concepts and technical solutions, it became hard and harder to find exhaustive documentation on this topic.

Thanks to the encouragement received from many colleagues, especially those members of the VIRGO Project, and to the availability of Urbino University structures, the idea has finally been pursued. A two weeks School was held in Urbino, Italy, on September 1999. It has been addressed to graduate students and recently post-graduate students. After the introduction of needed theoretical instruments, lectures offered up-to-date insights in fundamental physics and experimental tools used and developed for construction and exploitation of gravitational wave antennae. Problems in the fields of optics, signal processing, control and feedback in active mechanical filtering are also deeply analyzed, with reference to recent solutions adopted in building up present gravitational detectors.

It is a pleasure to acknowledge all the Lecturers for their participation in the School, for their work in preparing valuable lectures and in providing the manuscripts. Thanks to their excellent contributions, these proceedings give a precious and up-to-date outline on gravitational antennae and related physics and technology. We apologize to the reader for the three missing papers: unfortunately, the manuscripts were provided too late.

We are very grateful to many colleagues members of VIRGO (especially the people of the Executive Committee) for encouragement and suggestions. We owe thanks to all people of the Institute of Physics of Urbino University for providing total and reliable support during the School.

Finally, we express our gratitude to the Head of Urbino University, Carlo Bo, to the Directors of the Institute of Physics of Urbino University, Gianni Conforto, of the INFN Florence Section, Paolo Maurenzig, of the Astrophysical Observatory in Arcetri, Franco Pacini, of the Department of Astronomy and Space Science of Florence University, Claudio Chiuderi, and to the President of "Fondazione Cassa di Risparmio di Pesaro", Bruno Brusciotti, for their support in realizing the School.

The Editors

Participants

Amato Francesco, University, *Bari*

Amico Paolo, University, *Perugia*

Arnaud Nicolas, University, *Paris*

Bompiani Edoardo, University, *Roma*

Cesaretti Paolo, University, *Bologna*

Cinelli Alessandro, University, *Roma*

Fabbroni Leonardo, University, *Firenze*

Grote Hartmut, University, *MPI- Hannover*

Guidi Gianluca, University, *Bruxelles*

Mantovani Roberto, University, *Urbino*

Marin Andrea, INFN, *Pisa*

Martelli Filippo, University, *Urbino*

Moriondo Giovanni, University, *Firenze*

Nucita Achille A., INFN, *Pisa*

Pinter Luca, CNR, *Bologna*

Pradier Thierry, University, *Paris*

Scheidecker Jean-Paul, O.C.A., *Nice*

Spani Molella Luca, University, *Pavia*

Varvella Monica, University, *Napoli*

Vinante Andrea, University, *Trento*

Vocca Helios, University, *Perugia*

Program and Lecturers:

Courses:

General Relativity and Gravitational Waves,
P. Tourrenc, P. et M. Curie University, Paris

Astrophysical Sources for Gravitational Waves,
S. Bonazzola, Meudon, Paris

Signal Processing and Data Analysis,
M.G. Di Benedetto, La Sapienza University, Roma;
A. Vannucci, University, Parma;
A. Vicerè, INFN, Pisa

Issues on Mechanics,
L. Benvenuti, La Sapienza University, Roma;
M.D. Di Benedetto, University, L'Aquila;
G. Losurdo, INFN, Pisa;
A. Vicerè, INFN, Pisa;
S. Vitale, University, Trento

Optics for Gravitational Antennae,
V. Loriette, ESPCI, Paris;
J.Y. Vinet, Observatoire de la Cote d'Azur, Nice

Short Talks on Selected Topics:

VIRGO R&D, **A. Brillet**, Observatoire de la Cote d'Azur, Nice
Time-Frequency Analysis, **P. Flandrin**, ENS, Lyon
Monolithic Suspensions, **L. Gammaitoni**, University, Perugia
Interferometric Gravitational Antennae, **A. Giazotto**, INFN, Pisa
An Optical Transducer for Bar Detectors, **F. Marin**, University, Florence
The Mystery of UHE Cosmic Rays, **A. Olinto**, University, Chicago
Supernovae, **N. Panagia**, ESA-STScI, Baltimore
Seismic and Newtonian Noise, **R. Stanga**, University, Florence
Gamma-Ray Bursts, **M. Vietri**, University, Roma
The LISA Project, **S. Vitale**, University, Trento
Bar Detectors,
M. Visco, CNR, Frascati,
L. Votano, INFN, Frascati

Contents

GENERAL RELATIVITY AND GRAVITATIONAL WAVES

Philippe TOURRENC
Université Pierre et Marie Curie (LGCR, Case 142))
4, place Jussieu - F-75252 Paris cedex 05
E-mail : PhT@ccr.jussieu.fr
et
Observatoire de la Côte d'Azur (ILGA)
B.P. 4229 - F-06304 Nice cedex 4

This course is an introduction to experimental gravitation and gravitational waves. It is an approach to the theoretical bases of experimental gravitation, especially in mechanics and optics. After a general presentation of relativistic metric theories of gravitation, including General Relativity, we consider the case of the quasi Newtonian approximation. We study the emission of gravitational radiations and the elementary theories of mechanics, elasticity and optics which are relevant in the laboratory (especially for the interferometric and mechanical detectors of gravitational waves on Earth). Our aim is to discuss the conceptual link between the abstract theoretical constructions and the more concrete experimental schemes. We hope that restricting our developments to the minimum theoretical knowledge will help to discover experimental gravitation and the gravitational wave detection programs.

1 Introduction

1.1 The nature of these lectures

The direct detection of gravitational waves is a huge experimental challenge which requires the best capabilities in many domains such as optics and mechanics. Joe Weber pioneered this research forty years ago with mechanical detectors. Until now a generation of physicists has already worked in the field and many years are still necessary to achieve the final goal i.e. the birth of a new astronomy. Because of the long time scale involved, the rate of renewal of the physicists should be maintained in this particular domain and more generally in experimental gravitation. Because of the difficulties, the participation of experts specialized in different fields is necessary. However General Relativity and experimental gravitation do not form a part of the training of most of physicists.

Many books and advanced schools have been dedicated to relativistic physics and gravitational waves. Many articles and reports have been written. Thus a large amount of informations can be presently found in the literature. However it remains easier to go thoroughly into a specialized problem than to start in the field. It is not straightforward to find the precise relation between current experimental works (with lasers, mirrors, servo loops, suspensions, etc.) and the theory of General Relativity or the behavior of astrophysical objects. In this context we will summarize what can be found in any text book but we will discuss with more details the conceptual links between the abstract theoretical constructions and the more concrete experimental schemes.

This course is an introduction to relativistic experimental gravitation and gravitational waves. Our aim is to present the theoretical basis of experimental gravitation rather than technical and specialized materials which are exposed in the lectures and the talks on astrophysical sources, mechanics, optics, etc.. We restrict our developments to the *minimum* knowledge necessary to any physicist involved in a gravitational wave detection program. Obviously the distinction between fundamental and technical materials and the choice of the "minimum necessary knowledge" contains a part of arbitrary choices.

As a starting point we make the trivial observation that Newtonian physics is involved in the daily work of any experimental physicist. Newtonian physics rely on pre-relativistic concepts whereas the theory of Relativity denies some of them any meaning. This is an actual difficulty when beginning in the field of experimental gravitation. Fortunately current laboratory experiments take place in the quasi Newtonian framework (weak field and low velocities). In such a case it is possible to elaborate a *Newtonian language.* Several possibilities however exist where the same words do not have exactly the same meanings it is therefore necessary to remain very cautious. As an example let us consider a free falling mass point. A Newtonian physicist would claim that it is accelerated towards the center of the Earth. On the other hand one can consider that an acceleration is what an accelerometer measures. From this point of view a free falling mass is not accelerated.

Expressions such as "acceleration", "velocity of the light", "distance", "displacement noise", etc. are currently used [a]. They all refer to some Galilean frame of the pre-relativistic physics : most often, the necessity of such a frame is underlying although not explicitly mentioned. However there exist no Galilean coordinates in General Relativity and the very concept of distance has no meaning independently of the experiment designed for its measurement (proper distance, luminosity distance, red shift distance, etc.).

In General Relativity (in the quasi Newtonian case) one can build coordinate systems where the physical laws have nearly the same expression than in pre-relativistic physics in a Galilean coordinate system. These quasi Galilean coordinates are interpreted as Galilean coordinates within the framework of Newtonian physics whereas the physical laws must be slightly modified. This

[a]For instance a frequently asked question is "Does a high Q pendulum goes on moving after it has been hit by gravitational wave pulse?".

neo-Newtonian conception robs the theory of one of its characteristics but it is a simple way to use the physical intuition based on the pre-relativistic physics. However in a given situation several quasi Galilean coordinates can be found; therefore several interpretations are available and a great care is necessary to give a coherent description of any experiment.

1.2 Contents

The next three sections deal with the general classical (non quantum) framework. The quasi Newtonian case is introduced in section 5 with a special consideration for the emission of the gravitational radiations. In the two last sections we consider mechanics and optics in the laboratory.

1 Introduction

This section with the presentation above, the contents, some notations and conventions, and orders of magnitude.

2 Fundamental Concepts

Space, time, space-time and mathematical tools.

3 Equivalence principles

Equivalence principles, the free fall of a test particle, local "operational" coordinates and discussion of experiments.

4 Relativistic theories of gravitation

The metric theories, the General Relativity (including the energy-momentum tensor, the Einstein equations, the Schwarzschild solution and the Birkhoff theorem, the gravitational waves).

5 General Relativity in the quasi Newtonian case

The general framework and the multipolar expansion at the lowest order, a parallel between electromagnetic and gravitational waves.

6 Mechanics and elasticity

The pre-relativistic equations and their generalization, the action of gravitational waves on a material medium (Einstein and Fermi-like coordinates), an acceleration free pendulum.

7 Interferometric detection of gravitational waves

The propagation of the light and Michelson-like interferometers.

Appendix

The gravitational waves emitted by a rotating ellipsoid and a binary system.

A bibliography as short as possible is given at the end. It is limited to physically or more mathematically oriented text books. Few basic articles are quoted when they constitute a necessary or useful complement on the topics of the present course.

Several easily tractable exercises should help the progression along the following developments.

1.3 Notations and Conventions

- $A := B$ means that this equality is just the definition of a quantity included in A or B or the definition of some notation.

- *Unit:*

c is the speed of the light in the vacuum ($3 \cdot 10^8 \mathrm{ms}^{-1}$).
G is the Newtonian gravitational constant ($6.67 \cdot 10^{-11} \mathrm{m}^3\mathrm{kg}^{-1}\mathrm{s}^{-2}$).
The units are chosen in such a way that $G = c = 1$ except when these quantities are explicitly mentioned. In that case the time, t, and the mass, m, are expressed in meters.

$$t_{meter} := c_{\mathrm{ms}^{-1}} \cdot t_{second} \qquad \text{and} \qquad m_{meter} = \left(\tfrac{G}{c^2}\right)_{\mathrm{mkg}^{-1}} \cdot m_{\mathrm{kg}}.$$

- *Indices*

  Greek $(\alpha, \beta, etc) = 0, 1, 2$ or 3.

  Latin $(i, j, k, \text{etc}) = 1, 2$ or 3.

- *Einstein summation rule of repeated indices*

  The so-called **contraction of indices** concern two (and only two) indices with the same name, one up and the other down :

$$Q^{\alpha\beta\ldots}_{\beta k..} := \sum_{\beta=0}^{\beta=3} Q^{\alpha\beta\ldots}_{\beta k..} \quad \text{and} \quad S^{\alpha km..}_{im\beta\ldots} := \sum_{m=1}^{m=3} S^{\alpha km..}_{im\beta\ldots}.$$

- *Partial derivatives* of a function $f(x^\alpha)$

$$\frac{\partial}{\partial x^\alpha} f(x) := \partial_\alpha f := f_{,\,\alpha}.$$

- *Covariant derivatives* : a generalization of partial derivatives (see below Fundamental concepts/Covariant derivative and curvature tensor). Its notation is ∇_α

$$\nabla_\alpha (\,) := (\,)_{;\,\alpha} := \frac{\nabla}{\nabla_\alpha} (\,).$$

- *Change of coordinates*
  - We define new coordinates $y^{\alpha'}$:
$$x^{\alpha} = X^{\alpha}(y^{\sigma'}) \Leftrightarrow y^{\rho'} = Y^{\rho'}(x^{\beta})$$
$$a^{\alpha}_{\sigma'} := \frac{\partial X^{\alpha}}{\partial y^{\sigma'}} \ , \ a^{\rho'}_{\beta} := \frac{\partial Y^{\rho'}}{\partial x^{\beta}} \Leftrightarrow a^{\alpha}_{\sigma'} a^{\sigma'}_{\beta} = \delta^{\alpha}_{\beta} \text{ and } a^{\alpha}_{\sigma'} a^{\rho'}_{\alpha} = \delta^{\rho'}_{\sigma'}.$$
  - A quantity F is considered as a function of x^{α} : $F := \Phi(x^{\alpha})$. It can be considered as a function of $y^{\sigma'}$ as well : $F = \Psi\left(y^{\sigma'}\right) = \Phi\left[X^{\alpha}\left(y^{\sigma'}\right)\right]$ then
$$\partial_{\sigma'} F = a^{\alpha}_{\sigma'} \cdot \partial_{\alpha} F \text{ and } \quad \partial_{\alpha} F = a^{\sigma'}_{\alpha} \cdot \partial_{\sigma'} F.$$
- *Miscellaneous*
  - $\overrightarrow{a}$ is a vectorial notation for $\{a^1, a^2, a^3\} := \{a^k\}$.
    We define also $\overrightarrow{a} \cdot \overrightarrow{b} := \sum_k a^k b^k$.
  - The Cartesian-like coordinates : $\{x^0, x^k\} := \{ct, x, y, z\} := \{x^{\alpha}\}$ (and $\{x\}$ in short).
  - The **Kronecker symbol** : $\delta^{\alpha}_{\beta} := (1 \text{ when } \alpha = \beta \text{ and } 0 \text{ when } \alpha \neq \beta)$.
    N.B. The 3-dimensional Kronecker symbol will be written δ^j_k, δ_{jk} or δ^{jk}.
  - The **Minkowski tensor** : $\eta_{\alpha\beta} := diag\,(1, -1, -1, -1)$; its inverse is $\eta^{\alpha\beta} = \eta^{\beta\alpha}$: $\eta_{\alpha\beta}\eta^{\alpha\gamma} := \delta^{\gamma}_{\beta} \Leftrightarrow \eta^{\alpha\beta} := diag\,(1, -1, -1, -1)$.
  - The d'Alembertian is the differential operator $\Box$:
$$\Box F := \frac{\partial^2 F}{c^2 \partial t^2} - \Delta F := \eta^{\alpha\beta}\partial_{\alpha\beta} F$$
  - "iff" means "if and only if".
- *Other conventions* : In the literature one finds various conventions of sign. Here we define the metric tensor $g_{\alpha\beta} \sim \eta_{\alpha\beta} = diag(1, -1, -1, -1)$. The spatial metric is $\sim -\delta_{jk} = \eta_{jk}$. A different convention is often chosen. One defines the "flat space-time metric" $f_{\alpha\beta} := diag\,(-1, 1, 1, 1) = -\eta_{\alpha\beta}$. Then one chooses the metric $g_{\alpha\beta} \sim f_{\alpha\beta}$. In such a case the spatial metric is $\sim \delta_{jk}$. The "$-$" sign has disappeared. The various formulae are obtained from our formulae under the substitutions $\eta_{\alpha\beta} \to -\eta_{\alpha\beta} := f_{\alpha\beta}$ and $g_{\alpha\beta} \to -g_{\alpha\beta}$. For instance, $d\tau^2 = g_{\alpha\beta} dx^{\alpha} dx^{\beta}$ (here)$\to$ $d\tau^2 = -g_{\alpha\beta} dx^{\alpha} dx^{\beta}$ where $d\tau$ is the proper time.

1.4 Orders of magnitude

In Minkowskian coordinates, the **fundamental element** of special Relativity reads

$$d\tau^2 = dt^2 - dx^2 - dy^2 - dz^2.$$

Where $d\tau$ is the proper time (and $c = 1$).

If the coordinates are not Minkowskian (in accelerated frame or if the gravitation is taken into account) some extra quadratic terms appear :

$$d\tau^2 = g_{00}dt^2 + 2g_{0k}dtdx^k + g_{jk}dx^j dx^k.$$

With suitable coordinates the fundamental element is very similar to the special relativistic one in usual situations.

- Newtonian gravitation

  At the distance r from a mass M with spherical symmetry g_{00} is of order of $g_{00} \sim 1 - \frac{2GM}{rc^2}$ with $\frac{2GM}{rc^2} \lesssim 2 \cdot 10^{-6}$ in the Solar system.

- Accelerated frame

  The reference frame is accelerated, its acceleration is $\overrightarrow{a}$. At point $\overrightarrow{x}$, fixed relatively to the accelerated frame, the inertial force is similar to a gravitational force in the sense that it produces the same acceleration on any mass point. Its description is included in $g_{\alpha\beta}$: $g_{00} \sim 1 + 2\overrightarrow{a} \cdot \overrightarrow{x}/c^2$. For $\|\overrightarrow{a}\| \sim 10\text{ms}^{-2}$ one finds $|g_{00} - 1|/x \sim 10^{-16}\text{m}^{-1}$.

- Rotating frame

  - If the reference frame rotates around a fixed axis (relatively to a Galilean frame) with the angular velocity ω, a term g_{0k} is present. At the distance r from the axis $g_{0k} \sim \omega r/c$. For $\omega \sim 6 \cdot 10^3\text{rad/mn} \simeq 100\text{s}^{-1}$ and $r \sim 1\text{m}$ one finds $g_{0k} \sim 3 \cdot 10^{-7}$.
  - The Lense-Thirring effect is a gravitational coupling between the spin of the Earth and the spin of a gyroscope in its neighborhood. It is described in a geocentric frame by terms g_{0k} which does not exceed $7 \cdot 10^{-16}$. They can be cancelled in a locally rotating frame.

- Gravitational waves

  The gravitational waves produce perturbations of the g_{jk}-terms depending on the time. In the suitable quasi Minkowskian coordinates $g_{xx} = 1 + h_+$, $g_{yy} = 1 - h_+$ and $g_{xy} = g_{yx} = h_\times$. For a coalescing binary system in the Virgo cluster one can expect that $h_{+,\times}$ will not exceed 10^{-21} on Earth.

2 Fundamental concepts

2.1 Space and time

a- Pre-relativistic kinematics

In Newtonian physics an event is defined by its nature (a lamp is switched on or off, a nucleus disintegrates, etc). Three coordinates (x, y, z) give the position of the event (in space) and one coordinate (t) specifies the time.

The motion of a small object, P, is described by a curve in a four dimensional diagram (t, x, y, z) : $x = X(t)$, $y = Y(t)$, $z = Z(t)$.

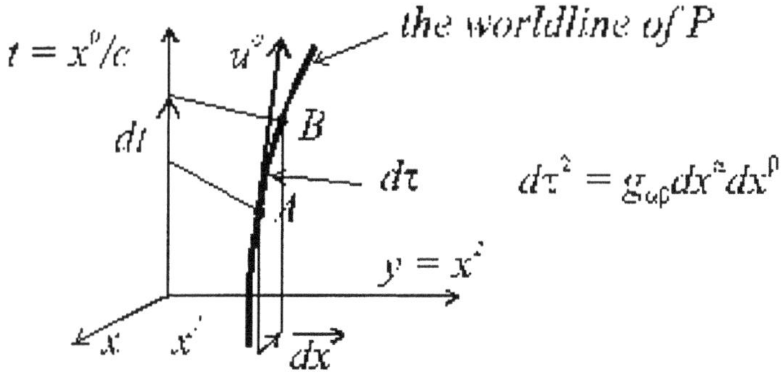

Figure 1: Representation of the motion of a point P.

The distance, $d\ell$, between two neighboring points fixed in space is a physical quantity which can be measured with standards rods. The coordinates can be chosen in such a way that $d\ell = \sqrt{dx^2 + dy^2 + dz^2} = \sqrt{d\vec{x}^2}$ where dx, dy, dz are the difference of the space coordinates of the two events.

We assume that the small object is a clock (a biological clock such as a human being or a physical clock like an atomic clock). We consider two arbitrary events, A and B, which happen to this object. It is always possible to decide that A happens prior to B (we note $A \preccurlyeq B$). The relation "$\preccurlyeq$" is an ordering relation. Moreover if two different clocks meet at A and B, the same relation $A \preccurlyeq B$ holds for both of them. This property defines a natural chronology. Between A and B, the clock "ages", i.e. it is submitted to irreversible transformations. The ageing of the clock is a physical process which is measured by a number : the **proper time**. The time coordinate can be chosen in such a way that $d\tau = dt$ where dt is the difference of the time-coordinates of the two neighboring events A and B.

We will call Newtonian coordinates these very special space and time coordinates that we have mentioned above.

The coordinates can be defined by the means of theoretical properties (e.g. the Newtonian coordinates above), but they need also to be experimentally defined. A difficult problem consists in defining an operational coordinate system which is at the same time a Newtonian system. The same difficulty remains in relativistic physics.

b- Relativistic kinematics

Within the framework of relativistic physics the concept of event remains unchanged. The events are still associated with four coordinates (t, x, y, z) and the operational coordinates used by the "pre-relativistic" physicists are still valid. Their physical interpretation is however deeply modified.

From the relativistic point of view, an event happens on a four dimensional manifold (this is the reason why four coordinates are necessary). The concept of absolute space or time does not remain any more. The distinction between space and time becomes a matter of interpretation which depends on the observer. As a consequence the notion of distance is always related to an operational procedure or a mathematical definition (optical distance and proper distance for instance).

The only concepts which remain are the concept of ageing and proper time, $d\tau$, altogether with the existence of a natural chronology.

Special Relativity is the relativistic theory which describes a world without gravitation. The space-time manifold is very special in that case. It is called the Minkowski space-time. In Minkowski space-time there exist special coordinates (i.e. Minkowskian coordinates) where the expression for the proper time between two neighboring events is $d\tau$:

$$d\tau^2 = c^2 dt^2 - dx^2 - dy^2 - dz^2. \tag{1}$$

In eq.1 the units are : t in *seconds*, x, y, z in *meters* and c a constant (ms^{-1}).

One can define the notations $\{x^\alpha\} = \{x^0 = ct, x^1 = x, x^2 = y, x^3 = z\}$. Then $d\tau^2 = \eta_{\alpha\beta} dx^\alpha dx^\beta$ where $\eta_{\alpha\beta} = diag(1, -1, -1, -1)$.

c- Free mass points and velocities

In both Newtonian and special relativistic physics, the concept of "free particle" is introduced. In Newtonian physics one expects that there exists Newtonian Galilean coordinates where a free particle keeps a constant velocity, i.e.

$$\frac{d^2x}{dt^2} = 0 = \frac{d^2y}{dt^2} = \frac{d^2z}{dt^2}. \tag{2}$$

Let us assume that we have defined an experimental protocol for the determination of such coordinates. As Special Relativity is the actual correct theory, these operational coordinates are not Galilean coordinates but rather Minkowskian coordinates.

Using Minkowskian coordinates (t, x, y, z), the motion of a free particle in Special Relativity is

$$\frac{d^2x}{dt^2} = 0 = \frac{d^2y}{dt^2} = \frac{d^2z}{dt^2}. \tag{3}$$

It is useful to define two different velocities : the **four-velocity** (see Fig. 1 above), $u^\alpha = dx^\alpha/d\tau$ and the velocity relatively to the coordinate system, $v^k = dx^k/dt$. The Michelson and Morley experiment give the numerical value of c which is interpreted as the velocity of the light in an empty space relatively to a Minkowskian observer (the accuracy is better than 10^{-12} and the isotropy of c better than 10^{-14}).

Newtonian physics (Galilean coord.)	Special Relativity (Minkowskian coord.)	Metric Theories (No special coord.)
$u^0 = c = 1,$ $u^k = v^k$	$u^0 = \gamma = \frac{1}{\sqrt{1-\vec{v}^2/c^2}},$ $u^k = \gamma v^k$	$u^\alpha := \frac{dx^\alpha}{d\tau}$

d- The gravitation

We can now change the coordinates and use for instance spherical coordinates (t, r, θ, φ)

$$x = r\sin\theta\cos\varphi,\ y = r\sin\theta\sin\varphi,\ z = r\cos\theta. \tag{4}$$

The proper time reads

$$d\tau^2 = dt^2 - \left(dr^2 + r^2 d\theta^2 + r^2\sin^2\theta\, d\varphi^2\right) \tag{5}$$

where $c = 1$.

More generally one can use any coordinates $\{x^0, x^1, x^2, x^3\} = \{x^\alpha\}$

$$t = T(x^\alpha),\ x = X(x^\alpha),\ y = Y(x^\alpha),\ z = Z(x^\alpha) \tag{6}$$

where T, X, Y, Z are arbitrary function (regular enough). Then one finds

$$d\tau^2 = g_{\alpha\beta} dx^\alpha dx^\beta \quad \text{with} \qquad (7)$$
$$g_{\alpha\beta} = \partial_\alpha T \partial_\beta T - \partial_\alpha X \partial_\beta X - \partial_\alpha Y \partial_\beta Y - \partial_\alpha Z \partial_\beta Z.$$

Relativistic gravitation can be introduced within the framework of the so-called metric theories. In such a case the proper time reads

$$d\tau^2 = g_{\alpha\beta} dx^\alpha dx^\beta \qquad (8)$$

where $g_{\alpha\beta}$ cannot be equal to $\eta_{\alpha\beta}$ everywhere (whatever the coordinates).

The $g_{\alpha\beta}$ are called the coefficients of **the metric**; they are symmetric functions of the coordinates : $g_{\alpha\beta} = g_{\beta\alpha}$.

We can summarize the differences between the various theories by the various expression of the proper time :

Newtonian physics (Newtonian coord.)	Special Relativity (Minkowskian coord.)	Metric Theories (No special coord.)
$d\tau = dt$	$d\tau^2 = c^2dt^2 - dx^2 - dy^2 - dz^2$ $d\tau^2 = \eta_{\alpha\beta} dx^\alpha dx^\beta$	$d\tau^2 = g_{\alpha\beta} dx^\alpha dx^\beta$ $g_{\alpha\beta} \neq \eta_{\alpha\beta}$

Exercises : 1) Using eq.4 calculate dx, dy and dz and check the validity of eq.5.

2) Prove that eq.5 is also a consequence of eq.7.

2.2 Scalars, vectors and tensors

Let us perform some chemical analysis (e.g. we measure the proportion, p, of oxygen in the air). This analysis is an event which takes place somewhere on space-time. The result is a function of the position on space-time : $p = p(x^\alpha)$. The function $x^\alpha \mapsto p$ depends on the coordinate system but the result (the numerical value of p at some given event) is coordinate independent : p is a **scalar**. The mass of a particle, the proper time between two neighboring events, the result of a given experiment are scalars.

Now we consider a four component quantity e.g. the coordinate separation between two neighboring events in a given coordinate system : $\{dx^\alpha\}$. We change the coordinates : $x^\alpha \to x^{\beta'} = x^{\beta'}(x^\alpha)$ and $x^\alpha = x^\alpha(x^{\beta'})$. Then

$$\{dx^\alpha\} \leftrightarrow \{dx^{\beta'}\} \text{ with } dx^\alpha = \frac{\partial x^\alpha}{\partial x^{\beta'}} dx^{\beta'} \text{ and } dx^{\beta'} = \frac{\partial x^{\beta'}}{\partial x^\alpha} dx^\alpha. \qquad (9)$$

The quantity $\{dx^\alpha\}$ is called a **contravariant** vector.

More generally we define

$$a^\alpha_{\beta'} := \frac{\partial x^\alpha}{\partial x^{\beta'}} := \partial_{\beta'}(x^\alpha) \text{ and } a^{\beta'}_\alpha := \frac{\partial x^{\beta'}}{\partial x^\alpha} := \partial_\alpha \left(x^{\beta'}\right). \qquad (10)$$

Any four component quantity $A^\alpha(x)$ which becomes $A^{\beta'}(x') = \left(a^{\beta'}_\alpha A^\alpha\right)_{(x)}$ in the change of coordinates is a contravariant vector.

Exercises : 1) Consider that the proper time between two neighboring events is a physical quantity whose value must not depend on the coordinates and prove that u^α is a contravariant vector .

2) Prove that $a^\alpha_{\beta'} a^{\beta'}_\gamma = \delta^\alpha_\gamma$ and $a^{\beta'}_\alpha a^\alpha_{\gamma'} = \delta^{\alpha'}_{\gamma'}$ and check that $A^{\beta'} = a^{\beta'}_\alpha A^\alpha \Leftrightarrow A^\beta = a^\beta_{\alpha'} A^{\alpha'}$.

One defines also **covariant vectors**. They are four components quantities A_α such as

$$\{x^\alpha\} \leftrightarrow \{x^{\beta'}\} \Rightarrow A_{\beta'} = a^\alpha_{\beta'} A_\alpha \Leftrightarrow A_\alpha = a^{\beta'}_\alpha A_{\beta'}. \qquad (11)$$

Exercise : Prove that $A_\alpha B^\alpha$ is a scalar.

One can consider more complicated quantities which are called **tensors**. A tensor $A^{\alpha\beta}{}_\gamma$ displays 64 components ($64 = 4 \times 4 \times 4$). It is a mathematical object which satisfies the transformation law

$$\{x^\alpha\} \leftrightarrow \{x^{\beta'}\} \Rightarrow A^{\sigma'\mu'}{}_{\rho'} = a^{\sigma'}_\alpha a^{\mu'}_\beta a^\gamma_{\rho'} A^{\alpha\beta}{}_\gamma \Leftrightarrow A^{\alpha\beta}{}_\gamma = a^\alpha_{\sigma'} a^\beta_{\mu'} a^{\rho'}_\gamma A^{\sigma'\mu'}{}_{\rho'}. \qquad (12)$$

This definition is immediately extended to an arbitrary number of indices.

Exercises : 1) Prove that δ^α_β is a tensor.

2) Prove that $g_{\alpha\beta}$ and its inverse $g^{\alpha\beta}$ (defined by the means of the relations $g_{\alpha\beta} g^{\alpha\gamma} := \delta^\gamma_\beta$) are two tensors.

The metric tensor $g_{\alpha\beta}$ and its inverse $g^{\alpha\beta}$ provide **a fundamental one to one correspondence** between covariant and contravariant quantities. For instance

$$\begin{aligned} &A_\alpha \;\rightarrow\; A^\alpha := g^{\alpha\beta} A_\beta \rightarrow A_\beta := g_{\alpha\beta} A^\alpha \; ; \\ &T^\alpha{}_\beta \longleftrightarrow T^{\alpha\gamma} = g^{\gamma\beta} T^\alpha{}_\beta \leftrightarrow T_{\gamma\beta} = g_{\gamma\alpha} T^\alpha{}_\beta = g_{\gamma\alpha} g_{\beta\sigma} T^{\alpha\sigma}. \end{aligned} \qquad (13)$$

Notice that the position of the indices (up or down, left or right) plays an important role in the transformation formulae above (except for symmetric tensors).

Because of this correspondence we consider that covariant and contravariant vectors are just the covariant and contravariant components of the same vector.

The tensorial nature of a physical quantity is a physical property and not only a matter of mathematical definition. Let us consider an example.

A Minkowskian observer O_1 measures the energy of a particle of mass m; he finds E_1. A second Minkowskian observer, O_2, measures the energy of the same particle; he finds E_2. Both observers perform two similar experiments : they use reference frames pointing toward the same far away stars, they use similar clocks and similar apparatus with some relative velocity. The two observers being Minkowskian observers, the correspondence between their coordinates is a Lorentz transformation, their relative velocity is a constant.

If O_2 describes the first experiment in his own coordinate system, he has to predict the same result E_1 which has been observed by O_1 (the mathematical construction and the interpretation of the theory must imply that the result of a given experiment does not depend on the observer). This does not tell anything on the nature of the energy.

Now if we consider the two different experiments performed on the same physical system by the two different observers we do know that $E_1 \neq E_2$. In order to find some physical quantities tied together, we must consider the 4-momentum $\{P^\alpha = mu^\alpha\}$. The first observer measures the 4-momentum; he finds P_1^α. The second observer perform similar experiments, he finds $P_2^{\beta'}$. One can check experimentally that the correspondence between the 4 components P_1^α and $P_2^{\beta'}$ is just the Lorentz transformation which relates two contravariant vectors in the change of coordinates from O_1 to O_2 (the energy is cP^0). We claim that the 4-momentum is a contravariant vector. It is a physical property which can be experimentally checked (it implies $E_1 \neq E_2$ for instance) and not only the consequence of some mathematical definition.

2.3 Covariant derivative and curvature tensor

From the metric tensor one defines the **Christoffel symbols**

$$\left\{{}^{\alpha}_{\beta\gamma}\right\} := \frac{1}{2} g^{\alpha\sigma} \left\{g_{\sigma\beta,\gamma} + g_{\sigma\gamma,\beta} - g_{\beta\gamma,\sigma}\right\} = \left\{{}^{\alpha}_{\gamma\beta}\right\} \tag{14}$$

The Christoffel symbols are used to define the covariant derivative, ∇_α, of a scalar, a vector or a tensor field. These fields are tensor fields but the Christoffel symbols are not the components of a tensor.

$$
\begin{array}{lll}
\text{Scalar} & f(x): & \nabla_\alpha f := \partial_\alpha f \\
\text{Vector} & \left\{ \begin{array}{l} A^\alpha(x): \\ A_\alpha(x): \end{array} \right. & \begin{array}{l} \nabla_\sigma A^\alpha := \partial_\sigma A^\alpha + \left\{ {\alpha \atop \beta\sigma} \right\} A^\beta \\ \nabla_\sigma A_\alpha := \partial_\sigma A_\alpha - \left\{ {\beta \atop \alpha\sigma} \right\} A_\beta \end{array} \\
\text{Tensor} & T^{\alpha\dots}_{\beta\dots}(x): & \nabla_\sigma T^{\alpha\dots}_{\beta\dots} := \partial_\sigma T^{\alpha\dots}_{\beta\dots} + \left\{ {\alpha \atop \rho\sigma} \right\} T^{\rho\dots}_{\beta\dots} + \dots - \left\{ {\rho \atop \beta\sigma} \right\} T^{\alpha\dots}_{\rho\dots} - \dots
\end{array}
$$

The covariant derivative is an extension of the ordinary partial derivative in Special Relativity. It fulfils the Leibnitz rule

$$\nabla_\sigma \left(A^{\alpha\dots}_{\beta\dots} B^{\gamma\dots}_{\delta\dots} \right) = \left(\nabla_\sigma A^{\alpha\dots}_{\beta\dots} \right) B^{\gamma\dots}_{\delta\dots} + A^{\alpha\dots}_{\beta\dots} \left(\nabla_\sigma B^{\gamma\dots}_{\delta\dots} \right). \tag{15}$$

One can check that

$$\nabla_\sigma g_{\alpha\beta} = 0 = \nabla_\sigma g^{\alpha\beta} = \nabla_\sigma \delta^\alpha_\beta. \tag{16}$$

Exercises : 1) Check that one can perform any contraction either before or after a covariant derivation (for instance $\nabla_\beta A_\alpha^{\ \alpha} = g_{\rho\sigma} \nabla_\beta A^{\rho\sigma}$).

2) What are in Special Relativity the quantities $\nabla_\alpha A^\alpha$ and $\nabla_\alpha \nabla^\alpha f$ where A^α is a vector and f a scalar? (make use of Minkowski coordinates).

Partial derivatives commute, $[\partial_\alpha, \partial_\beta] := \partial_{\alpha\beta} - \partial_{\beta\alpha} = 0$ but this is not generally the case for covariant derivative :

$$[\nabla_\rho, \nabla_\sigma] A^{\nu\dots}_{\mu\dots} = R^\nu{}_{\alpha\rho\sigma} A^{\alpha\dots}_{\mu\dots} + \dots - R^\alpha{}_{\mu\rho\sigma} A^{\nu\dots}_{\alpha\dots} - \dots \tag{17}$$

where $R^\alpha{}_{\beta\mu\nu}$ is the **Riemann-Christoffel** tensor (curvature tensor) :

$$R^\alpha{}_{\beta\mu\nu} := \partial_\mu \left\{ {\alpha \atop \beta\nu} \right\} - \partial_\nu \left\{ {\alpha \atop \beta\mu} \right\} + \left\{ {\alpha \atop \varepsilon\mu} \right\} \left\{ {\varepsilon \atop \beta\nu} \right\} - \left\{ {\alpha \atop \varepsilon\nu} \right\} \left\{ {\varepsilon \atop \beta\mu} \right\}. \tag{18}$$

The curvature tensor displays the following symmetries:

$$R_{\alpha\beta\mu\nu} = -R_{\alpha\beta\nu\mu} = R_{\mu\nu\alpha\beta}.$$

It satisfies the identities (whatever the metric $g_{\alpha\beta}$)

$$R_{\alpha\beta\mu\nu} + R_{\alpha\nu\beta\mu} + R_{\alpha\mu\nu\beta} = 0$$

and the **Bianchi identities**

$$\nabla_\sigma R_{\alpha\beta\mu\nu} + \nabla_\mu R_{\alpha\beta\nu\sigma} + \nabla_\nu R_{\alpha\beta\sigma\mu} = 0.$$

In Special Relativity, one can check that $R^{\alpha}_{\ \beta\mu\nu} \equiv 0$. One can actually show that $R^{\alpha}_{\ \beta\mu\nu} \equiv 0$ iff there is no gravitation (i.e. iff there exists a Minkowskian coordinate system where $g_{\alpha\beta} \equiv \eta_{\alpha\beta}$).

In the sequel it will be useful to introduce the **Ricci tensor** : $R_{\beta\nu} := R^{\alpha}_{\ \beta\alpha\nu}$ and the **Einstein tensor** $S_{\alpha\beta} := R_{\alpha\beta} - \frac{1}{2}g_{\alpha\beta}R$ where the curvature scalar is $R := R^{\alpha}_{\alpha}$.

Exercise : 1) Using the symmetries of the curvature tensor, prove that $R_{\alpha\beta} = R_{\beta\alpha}$.

2) From the Bianchi identities derive the following **contracted Bianchi identities** : $\nabla_{\alpha}S^{\alpha\beta} \equiv 0$.

2.4 Covariant equations

The physical contents of the Universe is described by the means of tensor fields, Φ, (momentum vector, electromagnetic tensor, etc.). The physical laws are expressed as differential equations involving these quantities. A coordinate system is required to express these equations. To guarantee the coordinate independence of the physical laws one generally assumes the mathematical form

$$E\left[\Phi\right] = 0 \tag{19}$$

where E is the set of the components of a tensor considered as a function of the fields Φ and their derivatives.

Due to the definition of a tensor, eq.19 guarantees that it is satisfied in every coordinate systems iff it is satisfied in any special coordinate system.

The **minimum coupling** consists in the substitutions $\eta_{\alpha\beta} \rightarrow g_{\alpha\beta}$ and $\partial_{\alpha} \rightarrow \nabla_{\alpha}$ in first order special relativistic equations. For instance a scalar field $\psi(x)$ which propagates at the velocity $c = 1$ in a Minkowskian coordinate system satisfies the equation $\Box\psi := \eta^{\alpha\beta}\partial_{\alpha}\partial_{\beta}\psi = 0$. This second order equation is split into a set of two first order equations : $\psi^{\alpha} := \eta^{\alpha\beta}\partial_{\beta}\psi$ and $\partial_{\alpha}\psi^{\alpha} = 0$. The minimum coupling with gravitation give $\psi^{\alpha} = g^{\alpha\beta}\nabla_{\beta}\psi$ and $\nabla_{\alpha}\psi^{\alpha} = 0 \Rightarrow g^{\alpha\beta}\nabla_{\alpha}\nabla_{\beta}\psi = 0$. The left hand side of this equation is obviously a tensor (here it is a scalar). Such an equation is a so-called manifestly covariant equation.

Exercise : The magnetic field, B^k, and the electric field, E^k, measured by a Minkowskian observer are the components of the antisymmetric Faraday tensor $F^{\alpha\beta}$: $E^k = F^{k0}$, $B^1 = F^{23}, B^2 = F^{31}$, $B^3 = F^{12}$ in suitable units. The Maxwell equations in an empty space read $F_{\alpha\beta} = A_{\alpha,\beta} - A_{\beta,\alpha}$ and $F^{\alpha\beta}{}_{,\beta} = 0$

where A_α is the 4-vector potential. Prove that the generalization of Maxwell equations read $F_{\alpha\beta} = A_{\alpha;\beta} - A_{\beta;\alpha}$ with $\nabla^\beta \nabla_\beta A_\alpha - \nabla_\alpha \left(\nabla_\beta A^\beta\right) - R_{\beta\alpha} A^\beta = 0$.

3 Equivalence principles

3.1 Newtonian equivalence principle

In Newtonian physics the **inertial mass**, m, is a coefficient which appears in the second Newton law : $Force = m \times acceleration$. In Special Relativity the inertial mass m of a body appears to be proportional to the energy of the body at rest : $E = mc^2$.

When considering the Newtonian gravitational attraction, one introduces the gravitational mass M. The gravitational attraction force between two bodies of **gravitational mass** M and M' is $F = GMM'/r^2$ where G is the Newtonian gravitational constant and r the distance between the two bodies.

Various experiments demonstrate that (in suitable units) $m = M$. The present accuracy is of order of 10^{-13}. Spatial projects are currently designed to achieve 10^{-15} (μSCOPE) and 10^{-18} (miniSTEP).

One can assume that the equality $m = M$ holds true exactly : this is the Newtonian equivalence principle[b].

We consider two free falling particles in a gravitational field; the equivalence principle implies that at any given position they take the same acceleration independently of their internal structure. Thus they will have the same motion if they have the same initial conditions. This is actually true only for test particles which are not dressed with any long range non negligible field. It would not be true if the mutual interaction between the particles were preponderant or if one of the particle were a charged particle (its electric field would interact with far away gravitational fields and modify its motion). Such hypothetical particles are known as "test particles".

Let us consider test particles which are initially in the same small free falling laboratory in the neighborhood of the Earth. We can give two descriptions of the situation. An observer tied to the Earth claims that the particles and the laboratory are free falling with the same acceleration. His conclusion is that no particle is accelerated relatively to the laboratory.

On the other hand an observer tied to the walls of the laboratory does not feel any gravitation; he claims that the particles are free particles and that their velocities remain constant.

[b] When one tries to build a quantum theory of gravitation, it seems that extra fields are necessary. They would produce a breakdown of the equivalence principle. For this reason μScope and STEP are important experiments.

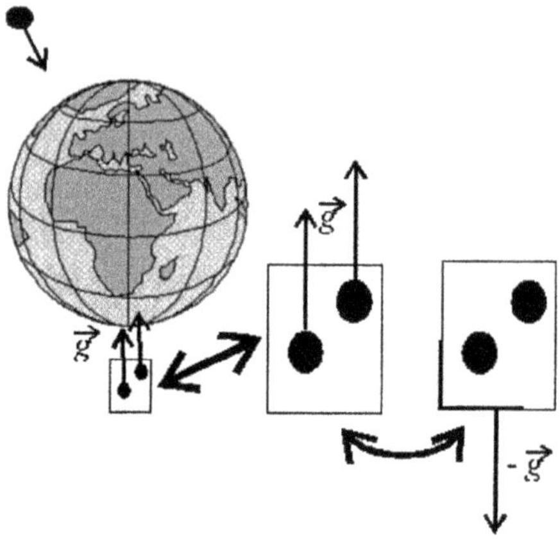

Figure 2: Gravity compared to inertia.

The second observer ignores the gravitational field. He can study the various laws of physics and discover Special Relativity (Relativity without gravitation). He can use gyroscopes and rockets in order to cancel any rotation of the laboratory and he can build a Minkowskian operational coordinate system comoving with the laboratory.

However two neighboring particles do not have exactly the same acceleration. Their acceleration points toward the center of the Earth and therefore they cannot be exactly parallel. Moreover if the particles have different celerities, their distance increases and the difference between their accelerations becomes observable. However these effects are inefficient if the particles remain near enough. For these reasons one can consider that Special Relativity is valid locally (i.e. in a small enough laboratory for a lapse of time short enough).

3.2 Einstein equivalence principle

The claim that Special Relativity is locally valid is known as Einstein's Equivalence Principle (EEP).

EEP holds for test system. If one considers the mutual gravitational interaction between particles, one can never cancel the gravitational effects by a suitable motion of the laboratory. In that sense, the Einstein equivalence principle is more a clue to formulate the laws of physics in a gravitational field than an actual principle.

EEP has several consequences.

The space-time manifold is endowed with the coordinate system $\{x^\alpha\}$. The proper time between two neighboring events is $d\tau$ with $d\tau^2 = g_{\alpha\beta}dx^\alpha dx^\beta$. First we consider an event e. EEP implies that there exists a Minkowskian coordinate system, $\{x^{\alpha'}\}$ in the neighborhood of e (for instance the coordinate system above, which is comoving with a free falling observer at e). The existence of locally Minkowskian coordinates implies that there exists a change of coordinates depending on e, $x^\alpha \to x^{\alpha'} = f^{\alpha'}_{(e)}(x^\rho)$, such that $g_{\alpha\beta} \to a^\alpha_{(e)\mu'} a^\beta_{(e)\nu'} g_{\alpha\beta} = g_{\mu'\nu'}$ with $(g_{\mu'\nu'})_{(e)} = \eta_{\mu'\nu'}$. In mathematics this property means that the signature of the metric is -2. In the special case where the previous change of coordinates is independent of the event e there exists a global Minkowskian coordinate system and the space-time manifold is the **flat space-time** of Special Relativity (no gravitation).

In the locally Minkowskian coordinate system the metric must remain constant in the neighborhood of e. In general it cannot strictly be a constant (otherwise space-time would be Minkowskian, without gravitation, in the neighborhood of e). However one can always choose the coordinates in such a way that $(g_{\mu'\nu',\sigma'})_{(e)} = 0 \Leftrightarrow \left\{{}^{\alpha'}_{\beta'\mu'}\right\}_{(e)} = 0$. Thus in the locally Minkowskian coordinate system in the neighborhood of e the following relations hold true

$$(g_{\mu'\nu'})_{(e)} = \eta_{\mu'\nu'}\ , \quad \left\{{}^{\alpha'}_{\beta'\mu'}\right\}_{(e)} = 0\ , \quad (\nabla_{\alpha'}[K])_{(e)} = (\partial_{\alpha'}[K])_{(e)} \tag{20}$$

where K is an arbitrary tensor field.

Eq.20 gives a clue to generalize special relativistic laws to General Relativity. We give now two examples.

a- Example 1

Let us notice that $\nabla_{\alpha'}\nabla_{\beta'} \neq \partial_{\alpha'}\partial_{\beta'}$. As a consequence it is necessary to deal with first order derivatives only. To see how to proceed, let us consider a scalar field[c] ψ whose law of propagation is $\Box\psi = 0$ in a Minkowskian coordinate system of Special Relativity.

We define $\psi^\alpha := \eta^{\alpha\beta}\partial_\alpha\psi$, thus $\Box\psi := \partial_\alpha\psi^\alpha = 0$.

In a gravitational field, in the neighborhood of any event, e, one can use the locally Minkowskian coordinates. We define $\left(\psi^{\alpha'}\right)_{(e)} := \left(\eta^{\alpha'\beta'}\partial_{\beta'}\psi\right)_{(e)}$, EEP suggests that the propagation law reads $\left(\partial_{\alpha'}\psi^{\alpha'}\right)_{(e)} = 0$.

---

[c] In that special case $\nabla_{\alpha'}\nabla_{\beta'}\psi = \partial_{\alpha'}\partial_{\beta'}\psi$ but the method that we describe is more general.

From eq.20 we obtain $\psi^{\alpha'}_{(e)} = g^{\alpha'\beta'}_{(e)} \nabla_{\beta'}\psi_{(e)}$ and $\partial_{\alpha'}\psi^{\alpha'}_{(e)} = \nabla_{\alpha'}\psi^{\alpha'}_{(e)} = 0$. Using the eq.15 and eq.16 we find

$$\left(\nabla_{\alpha'}\left(g^{\alpha'\beta'}\nabla_{\beta'}\psi\right)\right)_{(e)} = \left(g^{\alpha'\beta'}\nabla_{\alpha'}\nabla_{\beta'}\psi\right)_{(e)} = 0. \tag{21}$$

The left hand side of this equation can be considered as the component of a tensor (here it is a scalar) which value is zero. Thus the components of this tensor are zero in any coordinate system. We use a global coordinate system, $\{x^\alpha\}$, which does not depend on the event e. The equation reads $\left(g^{\alpha\beta}\nabla_\alpha\nabla_\beta\psi\right)_{(x_e)} = 0$. It is valid at any event (for any $x_{(e)}$). Therefore we find the generalization of the propagation equation :

$$\Box\psi = 0 \to g^{\alpha\beta}\nabla_\alpha\nabla_\beta\psi = 0. \tag{22}$$

We already derived this equation with the simplest formal generalization (see the "minimum coupling" in "Fundamental concepts / Covariant equations"); however the physical reasons of such a generalization (based on EEP) are more transparent here.

b- Example 2

The second example deals with null accelerations.

One can always define the acceleration of a particle relatively to any coordinate system but this is not a physical quantity. The physical acceleration is what an accelerometer measures. Following EEP the physical acceleration is the acceleration relatively to a locally Minkowskian frame.

The coordinates of a particle is $x^\alpha(\tau)$ in general and $x^{\alpha'}(\tau)$ in the locally Minkowskian coordinates in the neighborhood of the event e chosen on the worldline of the particle. The velocity relatively to the coordinate system is $v^{k'} := dx^{k'}/dt'$. We consider the event E in the neighborhood of e on the worldline of the particle. We define $\tau_{(E)} - \tau_{(e)} := d\tau \simeq dt'$ (because $v^{k'}_{(e)} = 0$). The 4-velocity is $u^{\alpha'} = dx^{\alpha'}/d\tau \simeq \left\{1, v^{k'}\right\}$. The variation of the 4-velocity is $du^{\alpha'} := u^{\alpha'}_{(E)} - u^{\alpha'}_{(e)} \simeq \left\{0, v^{k'}_{(E)} - v^{k'}_{(e)}\right\} := \left\{0, a^{k'}dt'\right\}$ where $a^{k'}$ is the physical acceleration.

Now we define $\nabla u^\alpha := \left(du^\alpha/d\tau + \left\{{}^{\alpha}_{\rho\sigma}\right\} u^\rho u^\sigma\right) d\tau$ and $\gamma^\alpha := \nabla u^\alpha/d\tau$. One can check that γ^α is a 4-vector it is called **the 4-acceleration.** One can also check that $\gamma^{\alpha'} := \nabla u^{\alpha'}/d\tau = \left\{0, a^{k'}\right\}$ because $dt' \simeq d\tau$ and $\left\{{}^{\alpha'}_{\rho'\sigma'}\right\}_{(e)} = 0$.

N.B. The quantity $\gamma^\alpha u_\alpha$ is a scalar; its value is calculated in the locally Minkowskian system : $\gamma^\alpha u_\alpha = 0$ thus the 4-acceleration is orthogonal to the 4-velocity.

When the particle does not feel any acceleration $a^{k'} = 0$ thus $\gamma^{\alpha'} = 0 \Leftrightarrow \gamma^{\alpha} = 0$.

Exercise : Prove that a free falling particle does not feel any acceleration.

3.3 Free fall of a test particle

Another consequence of the Einstein equivalence principle is the equations of the motion of a free falling particle. These equations can be derived from the analysis of the preceding section :

$$\gamma^{\alpha} := \frac{du^{\alpha}}{d\tau} + \left\{ {\alpha \atop \rho\sigma} \right\} u^{\rho} u^{\sigma} = 0 \text{ with } u^{\alpha} := \frac{dx^{\alpha}}{d\tau}. \tag{23}$$

Here, we will consider another method.

In Special Relativity a free particle has a constant velocity. Now one can consider the worldline of a free particle, $\mathcal{C}_0$ between two events A and B and the worldline, $\mathcal{C}$, of another particle between these two events. One can calculate the proper time between A and B along $\mathcal{C}_0$ and $\mathcal{C}$. One finds τ_0 and τ with $\tau_0 > \tau$. Thus the proper time along the worldline of a free particle is an extremum

$$\tau = \int_A^B \sqrt{\dot{t}^2 - \dot{x}^2 - \dot{y}^2 - \dot{z}^2}\, du \qquad \text{is an extremum} \tag{24}$$

where t, x, y, z are the Minkowskian coordinates considered as four functions of a parameter u and where $\dot{(\ )} = d(\ )/du$.

τ being an extremum, the equations of $\mathcal{C}_0$ are the Lagrange equations[d] derived from the Lagrangian $\mathcal{L} = \sqrt{\dot{t}^2 - \dot{x}^2 - \dot{y}^2 - \dot{z}^2}$.

Now we consider a free falling particle in a gravitational field, and a locally Minkowskian observer O_e in the neighborhood of an event e on the trajectory of the particle. We choose two events on the trajectory in the vicinity of e. Following EEP, the free falling particle is considered as a free particle by O_e thus its proper time is an extremum. The proper time can be expressed in any coordinate system :

$$\tau := \int_A^B \sqrt{g_{\alpha\beta}\, \dot{x}^{\alpha} \dot{x}^{\beta}}\, du. \tag{25}$$

[d] If a Lagrangian L is a function of x^{α} and $\dot{x}^{\alpha}$ the Lagrange equations read
$$\frac{d}{du}\left(\frac{\partial L}{\partial \dot{x}^{\alpha}}\right) - \frac{\partial L}{\partial x^{\alpha}} = 0.$$

Therefore the equations are derived from the Lagrangian $\mathcal{L} = \sqrt{g_{\alpha\beta}\, \dot{x}^\alpha \dot{x}^\beta}$. These equations are local equations, they are independent of the eventual vicinity of A and B. It is easy to verify that the Lagrange equations are precisely eq.23 above.

The Lagrangian $\mathcal{L}$ has the nice property that the parameter, u, is arbitrary. It can be chosen in the Lagrangian itself, before calculating the Lagrange equations. We define $u := x^0 := t \Leftrightarrow \dot{x}^0 = 1$ and $v^k := dx^k/dt = \dot{x}^k$. Then

$$\mathcal{L} = \sqrt{g_{00} + 2g_{0k}v^k + g_{kj}v^k v^j}. \tag{26}$$

The space-time curves where the proper time is extremum are called (time like) **geodesics**. One can prove that the geodesics can also be derived from the Lagrangian $L = g_{\alpha\beta}\, \dot{x}^\alpha \dot{x}^\beta$ but in that case the parameter is fixed (up to an affine transformation). Such parameters are called affine parameters. From the Lagrange equations one can prove that the numerical value of L remains a constant along a geodesic and that the proper time itself is an affine parameter associated with the numerical value $L = 1 = \mathcal{L}$.

In Special Relativity a massless particle (e.g. a photon) has the celerity c in a Minkowskian coordinate system. This is equivalent to the condition $d\tau = 0$ along the worldline of the particle. Once more EEP implies in that case $g_{\alpha\beta}\, \dot{x}^\alpha \dot{x}^\beta = L = 0$. The results are summarize below :

$$\text{Lagrangians} : \Lambda = \text{ either } \mathcal{L} = \sqrt{g_{\alpha\beta}\, \dot{x}^\alpha \dot{x}^\beta} \text{ or } L = g_{\alpha\beta}\, \dot{x}^\alpha \dot{x}^\beta$$

$$\text{Lagrange equations} : \frac{d}{du}\frac{\partial \Lambda}{\partial \dot{x}^\alpha} - \frac{\partial \Lambda}{\partial x^\alpha} = 0 \Leftrightarrow \frac{d^2 x^\alpha}{d\tau^2} + \left\{ {\alpha \atop \mu\nu} \right\} \frac{dx^\mu}{d\tau}\frac{dx^\nu}{d\tau} = 0$$

$$\text{First integral } (u = \tau) : L = 1 \text{ (for } m \neq 0) \ , \ \ L = 0 \text{ (for } m = 0)$$

3.4 Local physical coordinates

We consider now a so-called "**fiducial observer**", O, whose world line is not a geodesic. The observer builds an orthogonal frame in the usual way (he welds bars and uses triangles). He measures the components of its acceleration $\vec{a}$ with accelerometers and the components of its angular velocity $\vec{\omega}$ with gyroscopes.

Exercise : Calculate $\vec{a}$ and $\vec{\omega}$ for an observer at rest on the Earth (consider that an accelerometer is a mass fixed at the extremity of a spring).

The fiducial observer needs a coordinate system in his neighborhood. If he was in free fall he could use the locally Minkowskian coordinates above where $g_{\alpha\beta} \simeq \eta_{\alpha\beta}$. Here there are several problems with these coordinates : i) they are not comoving with the fiducial observer, ii) the metric is known up to first order only relatively to $||\vec{x}||$ and we look for a second order formula.

A possible answer based on geometrical considerations gives the following metric coefficients :

$$g_{00} = \left(1 + 2\vec{a} \cdot \vec{x} + (\vec{a} \cdot \vec{x})^2 - (\vec{\omega} \wedge \vec{x})^2\right) + R_{0jk0} x^j x^k, \tag{27}$$

$$g_{0k} = -(\vec{\omega} \wedge \vec{x})^k + \frac{2}{3} R_{0jki} x^j x^k \text{ and } g_{jk} = \eta_{jk} + \frac{1}{3} R_{imkj} x^m x^k.$$

where $R_{\alpha\beta\nu\sigma}$ is the Riemann-Christoffel tensor[e].

In any gravitational field, one can choose the fiducial observer and the coordinates in such a way that $\vec{a} = \vec{\omega} = \vec{0}$, then the fiducial observer is free falling, the coordinates are the so-called **Fermi coordinates**, they are locally Minkowskian. The operational coordinates form a "**local inertial frame**".

In the case of a time-independent metric an operational coordinate system associated to the metric eq.27 can be built in the following way.

1. We stretch a non extensible thread between M and O and we measure its length ℓ (with a usual standard of length).
2. At the origin, O, we measure the angles, θ_k, between the direction of the thread and the axes of the orthogonal frame already built above.
3. Points M are said to be at rest iff ℓ and θ_k remain constant. The space coordinates in eq.27 are $x^k = \ell \cos\theta_k$.
4. The previous thread is now considered as a wave guide for high frequency electromagnetic pulses. At the origin, the observer uses an ideal clock to send a pulse to M along the thread at a precise time t_O. The pulse arrives at M at a coordinate time that we define as $t_O + \ell$.
5. By repeating this procedure we define the coordinate time for any event at any point at rest.

[e] We consider here and thereafter that the fiducial observer is at the origin of the coordinate system but this is not necessary in general.

N.B. The proper time at O, between two successive pulses is $(\Delta t)_O$. The coordinate time difference at point M between two successive pulses is also $\Delta t = (\Delta t)_O$ but Δt is not the proper time at M.

Exercise : The propagation of an electromagnetic pulse along the wave guide from O to M is characterized by the equations $\vec{x} = \vec{T} \cdot u$ where $\vec{T}$ is constant and $d\tau = 0$. Calculate the proper time (measured at point O) necessary for the round trip OMO as a function of ℓ. Consider only the case of a stationary metric given by eq.27.

In relativity there is no separation between space and time independently of any observer. Thus the concept of the form of a solid is not valid, neither is the concept of rigidity for a 3-dimensional body. However one can define non extensible thread. This concept is relevant (in physics) when the metric varies slowly with the time. The previous discussion has been performed with the stationary metric eq.27. It remains valid when the coefficients of the metric vary slowly with the time but in the general case the discussion is more complicated[f].

3.5 Discussion of experiments

Let us first consider the Newtonian limit of the theory.

We use the operational coordinates which are called "Galilean coordinates" by the pre-relativistic physicist and we address the question of the metric in such coordinates.

We notice that Special Relativity is a good theory and that these coordinates are quasi Minkowskian coordinates. Due to the presence of the gravitational field, some discrepancy between the theory of Special Relativity and the experimental results can however be observed. Thus one can assume that $g_{\alpha\beta} = \eta_{\alpha\beta} + k_{\alpha\beta}$ where $|k_{\alpha\beta}| << 1$.

Using eq 26 we find $\mathcal{L} = \sqrt{(1 + k_{00}) + 2k_{0k}v^k - \vec{v}^2 + k_{kj}v^k v^j}$.

Now we remind that $c = 1$. Then $v^k = v^k/c$. In the case of low velocities where the Newtonian physics is valid $|v^k| << 1$.

We expand $L : L = 1 + \frac{1}{2}k_{00} - \frac{1}{2}\vec{v}^2 + k_{0k}v^k + \frac{1}{2}k_{kj}v^k v^j - ....$

This Lagrangian gives the same equations than $\Lambda := 1 - L = \frac{1}{2}\vec{v}^2 - \frac{1}{2}k_{00} - k_{0k}v^k - \frac{1}{2}k_{kj}v^k v^j - ...$ Now we compare Λ to the Newtonian Lagrangian $L_N := \frac{1}{2}\vec{v}^2 - \phi$ where ϕ is the Newtonian potential. These two Lagrangians

---

[f] In that case a real thread do not behave as a non extensible thread of the theory.

give the same Lagrange equation iff

$$k_{00} \simeq 2\phi\ , \quad k_{0k}v^k \text{ and } k_{kj}v^k v^j \text{ negligible.} \tag{28}$$

These are the conditions for the Newtonian theory to be valid in the low velocity approximation.

The relation $k_{00} \simeq 2\phi$ is a property of the metric. It is independent of the velocity of the particle that we consider. This gives a possibility to check the theory with high velocity particles such as photons. The corresponding effect is called the **Einstein gravitational frequency shift.**

We consider the case of a stationary metric : $\partial g_{\alpha\beta}/\partial t = 0$. For the sake of simplicity we consider the Schwarzschild solution in standard coordinates (see below) : $d\tau^2 = Bdt^2 - B^{-1}dr^2 - r^2\left(\sin^2\theta\, d\varphi^2 + d\theta^2\right)$ (with $B = 1 + 2\phi$). In the surrounding of the Earth the gravitational potential is $\phi = (GM_\oplus)\,/\,(rc^2)$ where $M_\oplus$ is the mass of the Earth ($GM_\oplus/c^2 \simeq 4,43$mm) and r is the radial coordinate ($r \simeq$ the "distance" to the center of the Earth). The worldline of a light signal is a **null geodesic** ($L = 0$). Along the radial worldlines of the light signals $d\tau^2 = Bdt^2 - B^{-1}dr^2 = 0$. A signal emitted by E at time t_E is received by R at time t_R such as $t_R - t_E = \int_{r_E}^{r_R} \frac{dr}{B}$ for $r_E < r_R$. The same relation holds for a second signal emitted at time t'_E and received at time t'_R. Thus $t_R - t_E = t'_R - t'_E \Leftrightarrow t'_R - t_R = t'_E - t_E\ := \Delta t$.

Now we calculate the proper time measured by E between two successive emissions. Observer E is at rest and the following relations hold true along its worldline : $dr = d\varphi = d\theta = 0$, $d\tau^2 = B_E dt^2$ where B_E remains constant. Thus the time measured by the emitter is $T_E = \sqrt{B_E}\Delta t$; a similar relation holds for R. Therefore $T_R/T_E = \sqrt{B_R/B_E}$. For periodic signals, the period decreases when the observer R moves away from the emitter E towards infinity.

In the neighborhood of the Earth one finds $B_R/B_E \simeq 1 - gz/c^2$ where $g = (GM_\oplus)\,/\,\left(R_\oplus^2 c^2\right) \sim 10\text{ms}^{-2}$ ($R_\oplus$ is the radius of the Earth $\simeq 6400$km).

Exercises : 1) Draw the space-time diagram corresponding to the preceding discussion.

2) Consider a scalar light field ψ which follows the eq.22 in a stationary metric. Show that there exist periodic solutions such as $\psi = e^{-i\omega x^0} \cdot f\left(x^k\right)$ where ω is a constant. Prove that the period for an observer at rest ($x^k = cte$) is $T = \sqrt{g_{00}} \cdot 2\pi/\omega$. Derive from the preceding expression the general formula for Einstein's gravitational frequency shift between two observers $O1$ and $O2$:

$$\frac{\nu_{O2}}{\nu_{O1}} = \frac{T_{O1}}{T_{O2}} = \sqrt{\frac{(g_{00})_{O1}}{(g_{00})_{O2}}}. \tag{29}$$

3) In the preceding Schwarzschild space-time consider i) a satellite S, on a circular orbit ($r_S = cte$, $\theta = \pi/2$), ii) an observer, M, at rest ($r_M = r_S$, $\theta = \pi/2$, $\varphi_M = cte$), iii) An observer M_0 at rest very far away ($r_{M_0} \to \infty$, $\theta = \pi/2$, $\varphi_{M_0} = \varphi_M$). *N.B.* Because of the spherical symmetry the value $\theta = \pi/2$ does not correspond to any particular case.

Figure 3: Times and clocks.

Check that the Lagrange equations for the free falling satellite read $r = r_S$, $\theta = \pi/2$, $\varphi = \Omega\,(t - t_0)$ with $\Omega^2 = GM_\oplus/r^3$.

When the satellite meets M, an electromagnetic signal is emitted. Prove that the period of the signal is $T_S = \left(1 - \frac{3GM}{rc^2}\right) \cdot (2\pi/\Omega)$ for an observer in the satellite, $T_M = \left(1 - \frac{GM}{rc^2}\right) \cdot (2\pi/\Omega)$ for the observer M and $T_{M_0} = 2\pi/\Omega$ for the observer at infinity.

4) Consider eq.27 above in the case when no gravitation is present (or neglect the terms of order $x^j x^k$). A light beam propagates along the circle $\|\vec{x}\| = R$ in the x-y plane orthogonal to $\vec{\omega}$. It is emitted by the source S and it is detected at point D diametrically opposed. Calculate the difference between the two flight coordinate time of a photon depending on which way he takes from O to D. To arrive at the same moment on the detector the two photons must leave S at different times. It results a phase difference, $\delta\Phi$, between the two beams. The phase of the light emitted at point S is assumed to be $\Phi = \omega\tau$ where τ is the proper time at S. Calculate $\delta\Phi$ as a function of the area of the circle and its perimeter. This effect is known as the **Sagnac effect**.

4 The relativistic theories of gravitation

4.1 The metric theories

We have considered theories where space-time is endowed with a metric tensor and where Einstein's equivalence principle is valid. Such theories are called "metric theories". However extra fields might be necessary to describe the gravitation (e.g. scalar fields).

One can distinguish two sets of equations. The first set connects the gravitational fields to the non gravitational contents of the Universe (the matter distribution, the electromagnetic fields, etc..). The second set of equations give the evolution of the non gravitational fields. Within the framework of metric theories, these laws depends only on the metric. This is a consequence of EEP and the minimal coupling.

In most theories, including General Relativity, the second set of equations are derived from the first set which is the only fundamental one. However in many situations the two sets are decoupled. Let us consider as an example the case of gravitational waves (i.e. a special gravitational field that we will study later).

On one hand the gravitational waves are produced by astrophysical sources. The metric is derived from the solution of the first set of equations. These equations are generally non linear differential equations; they give the gravitational fields and the motion of the sources. On the other hand the gravitational waves are detected by a physical device which (at the lowest order of approximation) does not modify the gravitational field. Thus it can be considered as a pure passive test device. Only the second set of equations is relevant in order to describe the detector. The same situation holds also in the solar system when we consider the motion of the planets.

The gravitational field is mainly due to the Sun. The Eddington and Robertson expansion gives the corresponding metric.

In the case of spherical symmetry, for a static gravitational field, one can prove that there exists a coordinate system (so-called isotropic) such as

$$d\tau^2 = B(r)dt^2 - A(r)\left(dr^2 + r^2 \sin^2\theta \, d\varphi^2 + r^2 d\theta^2\right). \qquad (30)$$

The Newtonian gravitational field does not exceed $M_\odot/R_\odot c^2 \sim 2 \cdot 10^{-6}$ ($M_\odot$ is the mass of the Sun and $R_\odot$ its radius). The metric is quasi Minkowskian. A and B are non dimensional quantities, they depend on G, M, c and r. The only pure number that can be built with these four quantities is GM/rc^2.

Eddington and Robertson expansion is a Taylor expansion of A and B

relatively to GM/rc^2 :

$$d\tau^2 = \left(1 - 2\alpha\frac{GM}{rc^2} + 2\beta\left(\frac{GM}{rc^2}\right)^2 + ...\right) dt^2, \tag{31}$$
$$- \left(1 + 2\gamma\frac{GM}{rc^2} + ...\right) \left(dr^2 + r^2\sin^2\theta\, d\varphi^2 + r^2 d\theta^2\right).$$

The coefficients α, β, γ are called post Newtonian parameters; they depend on the theory that we consider. In General Relativity $\alpha = \beta = \gamma = 1$.

The Sun can be considered as the main source of the gravitational field in the Solar system. A light ray, a radar signal or a planet are assumed to behave like pure test systems.

The post Newtonian parameters can be measured (not only in the Solar system but also in relativistic binary neutron stars such as PSR 1916+13). The main results are summarized below

The validity of the Newtonian limit at low speed	$\Rightarrow \alpha = 1$
The Einstein gravitational frequency shift	$\Rightarrow \vert 1-\alpha\vert \lesssim 10^{-4}$
The deflection of a light ray by the Sun The delay of radar echoes from the planets	$\Rightarrow \vert 1-\gamma\vert \lesssim 10^{-3}$
The orbital precession of the planets ($\alpha = \gamma = 1$)	$\Rightarrow \vert 1-\beta\vert \lesssim 10^{-3}$

A generalization of the preceding formalism is known as the PPN formalism (PPN = Parametrized Post Newtonian).

4.2 General Relativity

General Relativity is a metric theory where the metric tensor is the only gravitational field.

Two questions arise. i) How shall we describe the non gravitational contents of the Universe? ii) What are the fields equations which relate the metric to the contents of the Universe?

a- Continuous material medium

Within the framework of Newtonian physics a material continuous medium is described by several classical fields :

$\rho\,(t, \vec{x})$ is the mass density. It depends on the position $\vec{x}$ and on the time t (in a Galilean coordinate system)

$v^k\,(t, \vec{x})$ is the velocity at time t of the material particle at point $\vec{x}$.

$\Theta^{jk}\,(t, \vec{x}) = \Theta^{kj}\,(t, \vec{x})$ is the stress tensor.

The mass conservation and the Newton second law read

$$\partial_t \rho + \partial_k \left(\rho v^k\right) = 0 \;, \quad \rho \partial_t v^k + \rho v^i \partial_i v^k = \partial_i \Theta^{ik}. \tag{32}$$

In order to generalize the preceding equations within the framework of elativistic physics we consider a local Minkowskian frame (coordinates $x^{\alpha'}$) in ie neighborhood of any event e such as the velocity is $v^{k'}_{(e)} = 0$. The velocities $^{k'}(t, \overrightarrow{x})$ remain low in the neighborhood of e, therefore Newtonian equations re valid.

The 4-velocity in the neighborhood of e is $\left\{u^{\alpha'}\right\} := \left\{1, v^{k'}\right\} / \sqrt{1 - \overrightarrow{v}^2/c^2}$. Ve define the ten quantities $T^{\alpha'\beta'} = T^{\beta'\alpha'}$ in such a way that $T^{0'0'} = \rho u^{0'} u^{0'}$, $^{0'k'} = \rho u^{0'} u^{k'}$ and $T^{k'j'} = \rho u^{k'} u^{j'} - \Theta^{k'j'}$.

Now we assume that $T^{\alpha'\beta'}$ are the components of a tensor. This assumpon defines the components $T^{\alpha\beta}$ in any coordinate system. This tensor is the **energy-momentum tensor**".

One can check that the mass conservation and the Newton second law read

$$\partial_{\alpha'} T^{\alpha'0'} = 0 \;, \quad \partial_{\alpha'} T^{\alpha'k'} = 0 \Leftrightarrow \partial_{\alpha'} T^{\alpha'\beta'} = 0. \tag{33}$$

Now we generalize the equations above

$$\nabla_\alpha T^{\alpha\beta} = 0. \tag{34}$$

These equations are called either the conservation laws or the equations of ie motion.

Most of the physical systems (e.g. material systems, electromagnetic fields, c..) are associated with a symmetric energy-momentum tensor which fulfils ie eq.34. Due to the simple structure of the stress tensor, the energy momenım tensor for a perfect fluid is

$$T^{\alpha\beta} = \left(\rho c^2 + P\right) u^\alpha u^\beta - P g^{\alpha\beta}. \tag{35}$$

here u^α is the four-velocity, P the pressure and ρc^2 the energy density for a cally comoving observer.

· The field equations

ı General Relativity the field equations are $G^{\alpha\beta} = \chi T^{\alpha\beta}$ where χ is a constant nd $G^{\alpha\beta}$ a tensor which depends only on the metric and its partial derivative p to second order only.

$G^{\alpha\beta}$ is chosen in such a way that the field equations implies eq.34. The nly possibility is

$$G^{\alpha\beta} := S^{\alpha\beta} + \Lambda g^{\alpha\beta} \tag{36}$$

where $S^{\alpha\beta}$ is the Einstein tensor ($S^{\alpha\beta} := R^{\alpha\beta} - \frac{1}{2}g^{\alpha\beta}R$) and Λ the cosmological constant.

The cosmological constant (if any) is very small. It plays a role at the cosmological level but it can be neglected at the scale of the astronomical sources of gravitational waves. In the sequel we assume $\Lambda = 0$. Thus the field equations read

$$R^{\alpha\beta} - \frac{1}{2}g^{\alpha\beta}R = \chi T^{\alpha\beta} \Leftrightarrow R^{\alpha\beta} = \chi\left(T^{\alpha\beta} - \frac{1}{2}g^{\alpha\beta}T\right) \tag{37}$$

where $T := T^{\alpha}_{\alpha} = -R^{\alpha}_{\alpha} := -R$. These equations are the **Einstein equations.**

Exercises : 1) Use the contracted Bianchi identities above to derive eq 34 from the Einstein equations.

2) Prove that $u^{\beta}u_{\beta} = cte \Rightarrow u_{\beta}\nabla_{\alpha}u^{\beta} = 0$. Consider an incoherent fluid ($T^{\alpha\beta} = \rho u^{\alpha}u^{\beta}$), develop the eq.34, contract them with u_{β} and derive the conservation law $\nabla_{\alpha}(\rho u^{\alpha}) = 0$. Prove the relation $u^{\alpha}\nabla_{\alpha}u^{\beta} = 0$ and show that this relation means that the free falling particles of the incoherent fluid have a geodesic motion.

c- *The Schwarzschild solution and the Birkhoff theorem*

We assume the spherical symmetry and we consider the solutions of the vacuum Einstein's equations ($T^{\alpha\beta} = 0$).

The spherical symmetry implies that there exists **standard coordinates** such that

$$d\tau^2 = e^{\nu}dt^2 - e^{\lambda}dr^2 - r^2\left(\sin^2\theta\, d\varphi^2 + d\theta^2\right) \tag{38}$$

where ν and λ are two functions of t and r only.

The solution of the vacuum Einstein's equations, $S^{\alpha\beta} = 0$, is

$$e^{\nu} = e^{-\lambda}f(t)\ ,\ \ e^{-\lambda} = 1 + a/r \tag{39}$$

where $f(t)$ is an arbitrary (positive) function of t and a a constant. We introduce new coordinates such as $dt_{new} = \sqrt{f(t)}dt$:

$$d\tau^2 = (1 + a/r)\left(dt_{new}\right)^2 - \frac{1}{(1 + a/r)}dr^2 - r^2\left(\sin^2\theta\, d\varphi^2 + d\theta^2\right). \tag{40}$$

The metric is time independent : this is the Birkhoff theorem. It states that the explosion of a supernova does not produce any time dependant modification if it is a spherically symmetric explosion (i.e. no production of gravitational waves[g]).

We consider that the source of the field above is a mass at the origin $r = 0$. In order to recover the Newtonian case far away from the origin one must assume $a/r = 2\phi = -2\,(GM)\,/\,(rc^2)$. The corresponding solution is called the "Schwarzschild solution" (in standard coordinates) :

$$d\tau^2 = B(r)dt^2 - \frac{dr^2}{B(r)} - r^2\left(\sin^2\theta\, d\varphi^2 + d\theta^2\right)\ ,\quad B(r) := 1 - \frac{2GM}{rc^2} \qquad (41)$$

where M is the mass of the central body.

Exercise : Introduce a new radial coordinate $\overline{r}$: $r = \overline{r}\left(1 + (GM)\,/\,(2\overline{r}c^2)\right)^2$. Prove that the proper time takes an isotropic form (cf. eq.30). Calculate α, β and γ (you should find $\alpha = \beta = \gamma = 1$).

Generally the **Schwarzschild radius** of the central mass ($R_S := 2GM/c^2$) is much smaller than the radius R_0 of the body itself. Therefore the solution above is valid in a region where $r \geq R_0 > R_S$. However one can consider a compact object with $R_0 < R_S$. In that case the surface $r = R_S$ separates the space-time into two regions.

In the exterior where $r > R_S$ the coordinate t and r are respectively time-like and space-like coordinates ($g_{tt} = B > 0$ and $g_{rr} = -1/B < 0$). When a light source approaches the surface $r = R_S$, the Einstein gravitational shift becomes so huge that the frequency of the light becomes zero everywhere; the photons loose their energy and cannot be seen any more. The surface $r = R_S$ is the last surface that can be seen from outside : it is called **an horizon**. Such a compact object is called a **Schwarzschild black hole**.

In the interior where $r < R_S$ one has $g_{tt} = B < 0$ and $g_{rr} = -1/B > 0$. It means that t is a space-like coordinate while r is a time like-coordinate.

Exercise : Consider a spherically symmetric distribution of matter with a cavity around the origin ($r = 0$). Prove that in the cavity space-time is flat. To make the demonstration use the fact that the metric and its derivatives are regular at the origin.

---

[g]This is a handicap for the production of gravitational waves because a strong anisotropy is rather rare.

d- Gravitational waves

We consider a locally Minkowskian observer. In his neighborhood the metric is quasi Minkowskian, $g_{\alpha\beta} = \eta_{\alpha\beta} + k_{\alpha\beta}$ where $|k_{\alpha\beta}| \sim k << 1$. This observer can consider that the basic concepts of Special Relativity are valid and that $k_{\alpha\beta}$ is a field which perturbs the laws of physics (the propagation of the light, the behavior of the clocks, etc...). When the motions are slow, the quasi Minkowskian coordinate system can even be assimilated to a Galilean system. This academic neo-Newtonian interpretation distorts the fundamental ideas of General Relativity; it is however useful because our intuition is mostly based on Newtonian physics.

We write down the Einstein equations in an empty space and we neglect the terms of order of k^2 :

$$\Box k_{\alpha\beta} - k_{\alpha},_{\beta} - k_{\beta},_{\alpha} + k,_{\alpha\beta} = 0 \text{ with } k_{\alpha} := \eta^{\sigma\rho} k_{\alpha\rho,\sigma} \text{ and } k := \eta^{\sigma\rho} k_{\sigma\rho}. \quad (42)$$

N.B. k is used twice; it is the order of magnitude of $k_{\alpha\beta}$ and it is precisely $\eta^{\sigma\rho} k_{\sigma\rho}$ in 42 however no confusion is possible.

Now we proceed to a first order change of the coordinates (first order means $\sim k$) : $g_{\alpha\beta} = \eta_{\alpha\beta} + k_{\alpha\beta} \to \eta_{\alpha\beta} + h_{\alpha\beta}$. We choose the transformation in such a way that the so-called harmonic condition is fulfilled : $h_{\alpha} - h_{,\alpha}/2 = 0$ (where $h_{\alpha} := \partial_{\rho} h^{\rho}_{\alpha}$, $h := h^{\rho}_{\rho}$, $h^{\rho}_{\alpha} := \eta^{\rho\beta} h_{\alpha\beta}$ This is always possible. Therefore eq.42 becomes

$$\Box h_{\alpha\beta} = 0. \quad (43)$$

Such an equation is valid in the "neighborhood" of the observer. However when the source of the gravitational field is far away at several tens of megaparsec (in Virgo cluster for instance) the neighborhood of a terrestrial observer can be as large as the solar system. Within the framework of the neo-Newtonian interpretation, the coordinates are considered as Galilean coordinates therefore Eq.43 is the equation of a field which propagates at the speed of the light, the so-called "**gravitational waves**".

When the source of the gravitational field points towards $z \to -\infty$. The local solution is a plane wave :

$$h_{\alpha\beta} = h_{\alpha\beta}(t - z). \quad (44)$$

It is still possible to choose the harmonic coordinates in such a way that $h_{0\alpha} = 0$. The harmonicity condition gives $h_{z\alpha} = cte$ and $h_{xx} + h_{yy} = cte$. Therefore the only quantities which depend on t are $h_{xx} - h_{yy}$ and $h_{xy} = h_{yx}$. Thus the time dependant components of the gravitational wave are

$$\overset{w}{h}_{xx} = -\overset{w}{h}_{yy} = h_{+}(t - z)\ , \text{ and } \quad \overset{w}{h}_{xy} = \overset{w}{h}_{yx} = h_{\times}(t - z). \quad (45)$$

One can also consider the question of the gravitational waves from a phenomenological point of view.

First we use a coordinate system called Gauss coordinates where $k_{0\alpha} = 0$. We assume that the terms of order k^2 are negligible. Therefore the metric is characterized by the 3×3 symmetric matrix $[h] = [h_{jk}]$.

We consider the following basis of the 3×3 matrices

$$[U_+] = \begin{bmatrix} 1 & 0 & 0 \\ 0 & 1 & 0 \\ 0 & 0 & 0 \end{bmatrix} \quad [U_\times] = \begin{bmatrix} 0 & 1 & 0 \\ 1 & 0 & 0 \\ 0 & 0 & 0 \end{bmatrix} \quad [T] = \begin{bmatrix} 1 & 0 & 0 \\ 0 & 1 & 0 \\ 0 & 0 & 1 \end{bmatrix}$$

$$[U_0] = \begin{bmatrix} 1 & 0 & 0 \\ 0 & 1 & 0 \\ 0 & 0 & -2 \end{bmatrix} \quad [U_1] = \begin{bmatrix} 0 & 0 & 1 \\ 0 & 0 & 0 \\ 1 & 0 & 0 \end{bmatrix} \quad [U_2] = \begin{bmatrix} 0 & 0 & 0 \\ 0 & 0 & 1 \\ 0 & 1 & 0 \end{bmatrix}$$

$$[V_0] = \begin{bmatrix} 0 & -1 & 0 \\ 1 & 0 & 0 \\ 0 & 0 & 0 \end{bmatrix} \quad [V_1] = \begin{bmatrix} 0 & 0 & 0 \\ 0 & 0 & -1 \\ 0 & 1 & 0 \end{bmatrix} \quad [V_2] = \begin{bmatrix} 0 & 0 & 1 \\ 0 & 0 & 0 \\ -1 & 0 & 0 \end{bmatrix}$$

We make the decomposition

$$\begin{aligned} [h] = h_+ [U_+] + h_\times [U_+] + h_1 [U_1] + h_2 [U_2] + h_0 [U_0] + +h_T [T] \\ +f_1 [V_1] + f_2 [V_2] + f_0 [V_0] \,. \end{aligned} \tag{46}$$

We assume that the propagation takes place along the z-axis. The decomposition eq.46 is a spin[h] decomposition of the most general 3×3 matrix.

The matrices $[U]$ represent spin 2 states, $|U\rangle$, with $S_z^2 |U\rangle = 4 |U\rangle$ for $[U_+]$ and $[U_\times]$, $S_z^2 |U\rangle = |U\rangle$ for $[U_1]$ and $[U_2]$ and $S_z^2 |U_0\rangle = 0$. The matrix $[T]$ represents the spin zero state while the antisymmetric $[V]$ matrices represent the spin 1 states.

Within the framework of metric theories, $[h]$ is a symmetric matrix therefore $f_0 = f_1 = f_2 = 0$.

General Relativity corresponds to the case $h_0 = h_1 = h_2 = 0 = h_T$ with $\Box h_+ = 0 = \Box h_\times$ (in the empty space). The corresponding field quantum is **the graviton**. One thus considers that the graviton has a mass equal to zero and an helicity 2.

Within the framework of more general theories the gravitation can be described as a combination of states with spin 2 and spin zero (then it displays

[h] Under an infinitesimal rotation the matrix $[P]$ becomes $[P] + \delta [P]$ with $\delta [P] = -i\delta\varphi^k [[S_k], [P]]$ where $\{S_k\} = \{S_x, S_y, S_z\}$ are the generators of the rotations relatively to Ox, Oy and Oz e.g. $S_z = \begin{bmatrix} 0 & -i & 0 \\ i & 0 & 0 \\ 0 & 0 & 0 \end{bmatrix}$

6 polarizations) or as a particle with spin maximum 2 (in such a case the states with spin 1 exist but they do not couple to ordinary matter in the framework of the metric theories above).

One can also imagine that the mass of the graviton is $m \neq 0$ or that there exists a mass spectrum.

At the present moment there is no observational reason to be in doubt about General Relativity (and the present observational limits on the mass of the graviton is severe). However it will be the task of the detectors of gravitational waves to observe directly the properties of the graviton.

Exercise : The binary pulsar PSR 1913+16 emits gravitational waves whose effect on its orbital motion has been observed. The angular frequency, ω, of the expected gravitational waves is of order of $7 \cdot 10^{-5} \mathrm{s}^{-1}$ (twice the orbital frequency see the appendix below). The emission of a gravitational wave is possible if $\hbar\omega \gtrsim mc^2$ where $\hbar \simeq 10^{-34}$Js is Planck's constant and m the mass of the graviton). Calculate an upper value of the mass, m, of the graviton.

5 General Relativity : The quasi Newtonian case

5.1 The general framework

We consider the weak field approximation : $g_{\alpha\beta} = \eta_{\alpha\beta} + k_{\alpha\beta}$. We assume the order of magnitude $|k_{\alpha\beta}| \sim k << 1$ everywhere. If the gravitational constant, χ, were equal to zero, no gravitation would be present and the space-time would be flat. Therefore we consider that the gravitational constant is of order of k (times a dimensional quantity).

As we neglect k^2 terms the indices are lowered (or raised) by the means of $\eta_{\alpha\beta}$ (or $\eta^{\alpha\beta}$) rather than $g_{\alpha\beta}$ (or $g^{\alpha\beta}$) in any first order quantity.

Exercise : Check that $g^{\alpha\beta} = \eta^{\alpha\beta} - k^{\alpha\beta} + O(k^2)$.

The Ricci tensor and the Einstein equations read

$$\begin{aligned} R_{\alpha\beta} \ :&= \frac{1}{2}\left(k_{\alpha},_{\beta} + k_{\beta},_{\alpha} - k,_{\alpha\beta} - \Box k_{\alpha\beta}\right) \\ &= \chi \cdot j_{\alpha\beta} := \chi\left(T_{\alpha\beta} - \frac{1}{2}\eta_{\alpha\beta}T\right). \end{aligned} \tag{47}$$

where k and k_α have been defined above (eq.42)

One can check the weak field expression of the Bianchi identities i.e. $S^{\alpha}_{\beta,\alpha} := R^{\alpha}_{\beta},_{\alpha} - R,_{\beta}/2 \equiv 0$. Therefore integration conditions of eq.47 are

$$\partial_\alpha T^{\alpha}_{\beta} = 0 + O(k) \Leftrightarrow \partial_\alpha j^{\alpha}_{\beta} - \frac{1}{2} j^{\alpha}_{\alpha,\beta} = 0 + O(k). \tag{48}$$

The eq.48 are the equations of the motions in Special Relativity.

Within the neo-Newtonian interpretation one consider that Special Relativity is valid (as a first approximation). The non gravitational fields are described by their energy-momentum tensor $T^{\alpha\beta}$. They create a gravitational perturbation of the metric which is given by eq.47. Then the metric $g_{\alpha\beta} := \eta_{\alpha\beta} + k_{\alpha\beta}$ is used to calculate the perturbations to special relativistic laws (e.g. the propagation of the light or the proper times, etc..).

Now we consider a first order change of the coordinates :

$$\underset{old}{x}^{\alpha} := \underset{new}{x}^{\alpha} + \varepsilon^{\alpha}(\underset{new}{x}) \tag{49}$$

where ε^{α} are four arbitrary regular functions of order $\sim k$.

One can check that (up to first order) any first order function, $f(x)$ and any component of a first order tensor, $A(x)$, fulfils the relations

$$f(\underset{new}{x}) = f(\underset{old}{x}), \quad \overset{new}{A}(\underset{new}{x}) = \overset{new}{A}(\underset{old}{x}) = \overset{old}{A}(\underset{new}{x}) = \overset{old}{A}(\underset{old}{x}) \qquad [k^2]. \tag{50}$$

We define the functions $\overset{new}{k}_{\alpha\beta}$:

$$\overset{new}{k}_{\alpha\beta}(x) := \left(\overset{old}{k}_{\alpha\beta} + \varepsilon_{\alpha,\beta} + \varepsilon_{\beta,\alpha}\right)_{(x)}. \tag{51}$$

We calculate the metric in the new coordinate system :

$$\overset{new}{g}_{\alpha\beta}\left(\underset{new}{x}\right) = \eta_{\alpha\beta} + \overset{new}{k}_{\alpha\beta}\left(\underset{new}{x}\right) = \eta_{\alpha\beta} + \overset{new}{k}_{\alpha\beta}\left(\underset{old}{x}\right) \qquad [k^2]. \tag{52}$$

We notice that $\overset{new}{k}$ is different from $\overset{old}{k}$ but that there is no necessity to distinguish $\underset{new}{x}$ from $\underset{old}{x}$. Thus one can consider that we have not changed the coordinates but rather the metric by adding a so-called gauge term $\varepsilon_{\alpha,\beta} + \varepsilon_{\beta,\alpha}$. The quantity $\varepsilon_{\alpha,\beta} + \varepsilon_{\beta,\alpha}$ originates from the transformation of the quantity $\eta_{\alpha\beta}$ which was present in the old expression of the metric. The transformation

$$\overset{old}{k}_{\alpha\beta}(x) \to \overset{new}{k}_{\alpha\beta}(x) := \left(\overset{old}{k}_{\alpha\beta} + \varepsilon_{\alpha,\beta} + \varepsilon_{\beta,\alpha}\right)_{(x)}. \tag{53}$$

is called a gauge transformation. It does not correspond to a change of the gravitational field but rather to a change in its representation.

One can easily check that $R_{\alpha\beta}[k_{\rho\sigma}] \equiv R_{\alpha\beta}[k_{\rho\sigma} + \varepsilon_{\alpha,\beta} + \varepsilon_{\beta,\alpha}] + O(k^2)$. Therefore the Einstein equations, eq.47, can be solved up to a gauge transformation only. On the other hand one can choose the gauge in order to simplify the Einstein equations. In the de Donder gauge the supplementary conditions $k_\alpha - k_{,\alpha}/2 = 0$ hold true (cf. harmonicity conditions above). The Einstein equations and their retarded solutions read

$$\Box k_{\alpha\beta} = -2\chi \cdot j_{\alpha\beta} \Longrightarrow \tag{54}$$
$$k_{\alpha\beta}(t, \vec{r}) = \frac{-\chi}{2\pi} \iiint \frac{j_{\alpha\beta}(t-R, \vec{x},)}{R} d^3x \,, \quad \text{with} \quad R := \|\vec{x} - \vec{r}\| \,.$$

The integration is performed over the space, more precisely over the regions where the sources are present ($\vec{x}$ such that $j_{\alpha\beta}(\vec{x}, t-R) \neq 0$).

We consider the static case in the non relativistic limit. We have seen that the Newtonian theory is recovered for $k_{00} = 2\phi$ where ϕ is the Newtonian potential :

$$k_{00} = 2\phi := -2\frac{G}{c^2} \iiint \frac{\rho(\vec{x})}{R} d^3x. \tag{55}$$

ρ is the mass density.

In the Newtonian approximation $T^{\alpha\beta} \simeq \rho u^\alpha u^\beta \Rightarrow j_{00} \simeq \rho c^2/2$. The solution (eq.54) reads

$$k_{00} = -\frac{\chi}{4\pi} \iiint \frac{\rho c^2}{R} d^3x. \tag{56}$$

The coordinates being interpreted as Galilean coordinates of the pre-relativistic theory, the Einstein equations give the Newtonian theory in the non relativistic limit when

$$\chi = \frac{8\pi G}{c^4}. \tag{57}$$

Let us notice that the eq.54 are linear equations. Therefore one can calculate the gravitational fields, $k_{(S)\alpha\beta}$, from the various sources (S) independently and perform a summation over the sources S in order to find the total gravitational field (up to a gauge transformation).

5.2 Multipolar expansion

a- The approximations and the method

We consider a gravitational source in the neighborhood of the origin. More precisely we assume $T^{\alpha\beta}(t, \vec{x}) \neq 0$ only if $\|\vec{x}\| < a$. The observer is far from the source, $r := \|\vec{r}\| >> a$. We neglect any quantity of order of $(1/r)^2$.

The solution (eq.54) reads

$$k_{\alpha\beta}(t, \vec{r}) \simeq \frac{-\chi}{2\pi} \cdot \frac{1}{r} \iiint j_{\alpha\beta}(t-R, \vec{x}) \cdot d^3x \,, \quad \text{with} \quad R \simeq r - \vec{n} \cdot \vec{x} \tag{58}$$

where $\vec{n} = \vec{r}/r$.

Now we assume the slow motion approximation. We consider $\left|\partial_t j^{\alpha\beta}\right| \sim \omega_{grav}\left|j^{\alpha\beta}\right|$ where $T_{grav} = 2\pi/\omega_{grav}$ is the order of the time necessary for a non negligible modification of the source (T_{grav} is of the order of *half the period* for a periodic source. *N.B. half the period* $\sim$ *the period!*). The speed of the matter, v^k, does not exceed the order of magnitude $a\omega_{grav}$. We assume $\varepsilon := a\omega_{grav}/c << 1 \Longleftrightarrow a << \lambda_{grav}$ where $\lambda_{grav} := cT_{grav}$. In such a case the following expansion is valid.

$$j^{\alpha\beta}_{(t-R,\vec{x})} = \left[j^{\alpha\beta} + \frac{(\vec{n}\cdot\vec{x})}{c}\partial_t j^{\alpha\beta} + \frac{(\vec{n}\cdot\vec{x})^2}{2!\cdot c^2}\partial_{tt} j^{\alpha\beta} + ...\right]_{(t-r,\vec{x})} . \quad (59)$$

The successive terms do not exceed the order of magnitude $\left|j^{\alpha\beta}\right|$, $\varepsilon\left|j^{\alpha\beta}\right|$, $\varepsilon^2\left|j^{\alpha\beta}\right|$, etc.

We consider now the energy-momentum tensor in the low velocity case :

$$T^{00} \sim \rho c^2,\ T^{0k} \sim \rho c^2\frac{v^k}{c},\quad T^{jk} \sim \rho c^2\frac{v^j v^k}{c^2} \text{ with } \quad \left|\frac{v^i}{c}\right| \lesssim \varepsilon << 1. \quad (60)$$

Considering the expansion of eq.58 at order ε^2, one notices that it is not necessary to know precisely the ten functions $T^{\alpha\beta}(t,\vec{x})$ but only the multipolar description of the source in order to calculate $k_{\alpha\beta}$.

b- Multipolar expansion

We define the following multipoles :

$$\text{Monopole} : M(t) := \frac{1}{c^2}\iiint T^{00}(t,\vec{x})d^3x$$
$$\text{Dipole} : M^k(t) := \frac{1}{c^2}\iiint x^k T^{00}(t,\vec{x})d^3x$$
$$\text{Quadrupole} : M^{jk}(t) := \frac{1}{c^2}\iiint x^j x^k T^{00}(t,\vec{x})d^3x$$

From the equation of the motions eq.48, $\partial_\alpha T^{\alpha\beta} = 0$, one obtains

$$M = cte$$
$$\frac{d}{dt}M^k = P^k := \frac{1}{c}\iiint T^{0k}(t,\vec{x})d^3x = cte'$$
$$\frac{1}{2}\frac{d^2}{dt^2}M^{jk}(t) = \iiint T^{jk}(t,\vec{x})d^3x$$

Considering eq.60 M appears as the inertial mass of the gravitational source (it is the sum of the elementary inertial masses of the source). It is a constant.

P^k is the momentum it remains constant too.

<u>*Exercise*</u> : Using similar methods as in the footnote[i], prove the preceding properties.

c- The quadrupole formula

Using the eq.59 to expand the eq.58 one finds the solution $k_{\alpha\beta}$

$$k_{\alpha\beta} = h_{\alpha\beta} + \textit{arbitrary first order gauge transformation} \tag{61}$$

with

$$h_{00} = -2\frac{GM}{rc^2}\ ,\quad k_{0k} = h_{k0} = 0\ ,\quad h_{jk} = h_{kj} = \frac{2GM}{rc^2}\eta_{jk} + \overset{w}{h}_{jk} \tag{62}$$

We use Cartesian coordinates $\{x, y, z\}$ centered on the observer, N, and such as the *z-axis* points along the line of sight (see Fig. 4)

Figure 4: Polarizations $h_+ = h_{xx} = -h_{yy}$ and $h_\times = h_{xy} = h_{yx}$

One finds (in the neighborhood of N)

$$\overset{w}{h}_{zk} = 0\ ,\quad \overset{w}{h}_{xx} = -\overset{w}{h}_{yy} = h_+(t-z)\ ,\quad \overset{w}{h}_{xy} = \overset{w}{h}_{yx} = h_\times(t-z) \tag{63}$$

[i]Consider $dM^k/dt = \iiint x^k \partial_0 T^{00} d^3x$ (here c = 1). The relation $\partial_\alpha T^{\alpha 0} = 0$ reads $dM^k/dt = -\iiint x^k \partial_j T^{j0} d^3x = -\iiint \partial_j \left(x^k T^{j0}\right) d^3x + \iiint \delta^k_j T^{j0} d^3x$. The first integral is the integral of a divergence over the source; it is equal to the surface integral $-\iint x^k T^{j0} d\sigma_j$ on any surface surrounding the source (where $T^{0j} = 0$), therefore it is equal to zero. Thus $dM^k/dt = \iiint T^{k0} d^3x := P^k$.

$$h_+(t-z) = \frac{-G}{c^4 r_N}\left(\overset{..xx}{I} - \overset{..yy}{I}\right)_{ret.} \quad , \quad h_\times(t-z) = \left(\frac{-2G}{c^4 r_N}\overset{..xy}{I}\right)_{ret} \tag{64}$$

where $(F)_{ret} := F(t-z-r_N)$ and $\ddot{(\ )} := \frac{d^2}{dt^2}(\ )$. The quantities I^{jk} are the reduced quadrupoles.

$$I^{jk} := M^{jk} - \frac{1}{3}\delta^{jk}\sum_i M^{ii} := \frac{1}{c^2}\iiint T^{00}(t,\vec{x})\left(x^j x^k - \frac{1}{3}\delta^{jk}\vec{x}^2\right)d^3x. \tag{65}$$

The only terms which depend on t are the $\overset{w}{h}_{jk}$. They describe the gravitational radiation emitted by the source. One can check that the gravitational polarizations h_+ and $h_\times$ do not depend on the origin but only on the orientation of the axes.

Considering the case where $h_{\alpha\beta}$ displays the spherical symmetry, one finds $\overset{w}{h}_{jk} = 0$ (cf. Birkhoff theorem). Moreover in h_{00} the inertial mass M appears as the gravitational mass (cf. equivalence principle).

The gravitational radiation depends only on the quadrupole M^{jk}. A similar calculation can be performed for electromagnetic waves. In the Lorentz gauge the equations are $\Box A^\alpha = J^\alpha$ where A^α is the 4-potential vector and J^α the 4-current (J^0 is the charge density and J^k the current in suitable units). The monopole is the total charge $Q = \iiint J^0 d^3x$, the dipole is $Q^k = \iiint x^k J^0 d^3x$ and the quadrupole $Q^{jk} = \iiint x^j x^k J^0 d^3x$.

From the conservation laws $\partial_\alpha J^\alpha = 0$ one obtains $Q = cte$. In general the dipole is a function of the time and it results in a dipolar emission of electromagnetic waves. However Q^k is a constant when the current originates from the motion of particles with identical charge to mass ratio, then the emission is a quadrupolar emission.

Exercise : Consider N punctual bodies with mass $m_{(n)}$ and coordinates $x^k_{(n)}(t)$. Prove that $I^{jk} = \sum_{n=1}^{N} m_{(n)}\left(x^j_{(n)}x^k_{(n)} - \frac{1}{3}\delta^{jk}\vec{x}^{\,2}_{(n)}\right)$.

The amount of the Newtonian gravitational potential energy and the kinetic energy is E. Due to the emission of gravitational radiations E decreases. For a Newtonian gravitational quasi equilibrium the virial theorem[j] is valid and the kinetic energy of the source increases while the potential gravitational energy decreases (the source becomes more compact).

[j] The virial theorem states that in a quasi Newtonian equilibrium $2E = V$ where V is the gravitational potential energy.

This loss of energy corresponds to a gravitational power L_{grav} radiated by the gravitational waves. One can prove that

$$
\begin{aligned}
L_{grav}(t) \; &:= \frac{1}{32\pi} \iint_{\Sigma} \left(\left(\dot{h}_{+} \right)^2 + \left(\dot{h}_{\times} \right)^2 \right)_{(t)} d\sigma \\
&= \frac{G}{5c^5} \sum_{j,k} \left(\dddot{I}^{jk} \right)^2_{(t - r_N/c)}
\end{aligned}
\qquad (66)
$$

We can consider the question of the production of gravitational waves on Earth. Let us consider the oscillation of a mass m with an angular frequency ω and an amplitude a thus $I \sim ma^2$, $\dot{I} \sim I\omega$ and the kinetic energy stored $E \sim I\omega^2$. Eq.64 and eq.66 give $h_{+,\times} \sim \left(GI\omega^2\right) / \left(c^4 r\right)$, $L_{grav} \sim \left(GI^2\omega^6\right) / \left(5c^5\right)$. In the wave zone where Eq.64 is valid $r > 2\pi c/\omega$. Therefore the power to supply is $P > L_{grav} > \left(0.8 \cdot \pi^2 c^5 / G\right) \left(h_{+,\times}\right)^2$ ($\sim 0.3 \cdot 10^6$MW for $h_{+,\times} \sim 10^{-21}$) and the energy stored is $E \sim \left(c^4/G\right) r h_{+,\times}$ ($\sim 10^{23}$J for $r \sim 1m$ and $h_{+,\times} \sim 10^{-21}$). These orders of magnitude show that the astrophysical sources are the only interesting sources of gravitational radiations.

It was not easy to demonstrate rigorously that eq.66 is a consequence of the full non linear Einstein equations.

The quantities I^{jk} concern the sources while $h_{+,\times}$ describes the gravitational field far away from the sources.

In order to prove eq.66 it is necessary to know the equations of the motion and the gravitational field up to the radiation reaction term and to match the two regions together[k].

5.3 Electromagnetic and Gravitational waves

Some analogy exists between the electromagnetic waves and the gravitational waves at the weak field approximation. We summarize the analogy below.

---

[k] See in the bibliography given at the end the course by T. Damour in "Les Houches" (1983).

$EM - Waves$	$GR - Waves$
4 potentials : $A^\alpha(x)$ or $A_\beta := \eta_{\beta\sigma}A^\sigma$	10 potentials : $k_{\alpha\beta}(x) = k_{\beta\alpha}$ or $k^{\alpha\beta} := \eta^{\alpha\sigma}\eta^{\beta\rho}k_{\sigma\rho}$
The field : $F_{\alpha\beta} := A_{\alpha,\beta} - A_{\beta,\alpha}$	The weak field : $R^\alpha_{\ \beta\rho\sigma} := \left\{ {\alpha \atop \beta\sigma} \right\}_{,\rho} - \left\{ {\alpha \atop \beta\rho} \right\}_{,\sigma}$
Gauge transformation : $A_\alpha \to A_\alpha + \varepsilon_{,\alpha}$ Lorentz gauge : $A^\alpha_{\ ,\alpha} = 0 \Downarrow$	Gauge transformation : $k_{\alpha\beta} \to k_{\alpha\beta} + \varepsilon_{\alpha,\beta} + \varepsilon_{\beta,\alpha}$ deDonder gauge : $2k^\beta_{\alpha,\beta} - h^\beta_{\beta,\alpha} = 0 \Downarrow$
Field equations : $\Box A^\alpha = j^\alpha$ current $= j^\alpha$ $\Downarrow$ Conservation (charge, current) : $j^\alpha{}_{,\alpha} = 0$	Field equations : $\Box k^{\alpha\beta} = j^{\alpha\beta}$ momentum, energy $= T^{\alpha\beta} \to j^{\alpha\beta}$ $\Downarrow$ Conservation (energy, momentum) + eq. of the motion : $T^{\alpha\beta}_{\ \ ,\beta} = 0$
Waves : celerity c, helicity 1 Two polarizations	Waves : celerity c, helicity 2 Two polarizations

6 Mechanics and Elasticity

In this section we consider the case of an extended body in a weak gravitational field.

The matter is initially at rest relatively to a Galilean (Minkowskian) coordinate system. The material points are labelled by their initial space coordinates r^k. At some time, the body is mechanically excited and moreover the coordinates become quasi Galilean coordinates (e.g. a gravitational radiation strikes the body). The space coordinates of the material point becomes $x^k(t, r^j)$ (with $x^k = r^k$ when $t \mapsto -\infty$). Each material point is represented by its world line and the body by a **world tube**.

We define the displacement (relatively to the coordinate system), S^k, and the deformation, S_{jk} :

$$S^k(t, r^j) := x^k(t, r^j) - r^k \ , \ S_{jk} := \left(S^k_{\ ,j} + S^j_{\ ,k}\right)/2 = S_{kj}.$$

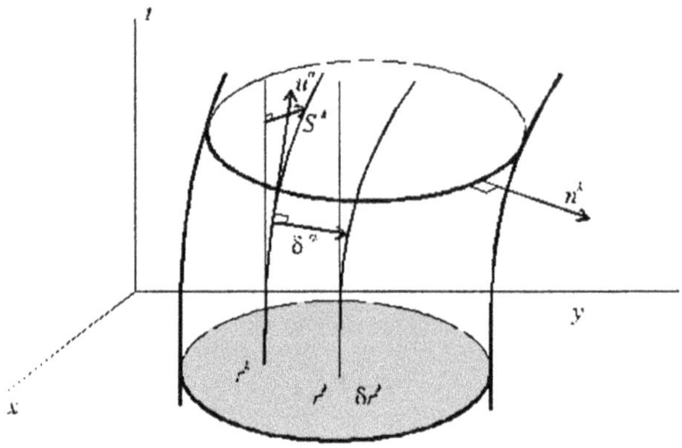

Figure 5: Space-time representation of a deformable body.

The quantities S^k and S_{jk} are formally similar to the pre-relativistic quantities.

6.1 The Newtonian equations

For small displacements, one considers that any term proportional to S^k or its derivative is a first order term.

The velocity of the mass point " r^k " is $v^k = \partial_t S^k$ and the mass density is $\rho = \rho_0 + \delta\rho$ where ρ_0 is the initial mass density and $\delta\rho$ a first order term. The mass conservation reads

$$\partial_t \rho + \partial_k \left(\rho v^k\right) = 0. \tag{67}$$

Within the framework of the theory of elasticity the deformation tensor, S_{jk} and the stress tensor Θ_{jk} are proportional. For an isotropic medium

$$\Theta_{jk} = 2\mu S_{jk} + \lambda \delta_{ik} S. \tag{68}$$

where μ and λ are the Lamé coefficients and $S := \sum_k S_{kk}$. Thus the elastic medium is characterized by its mass density at rest ρ_0 and the two Lamé coefficients λ and μ (or equivalently by the Young modulus Y and the Poisson coefficient σ).

From the stress tensor one calculates the force density. The Newton second law reads

$$\rho_0 \partial_{tt} S^k = \sum_j \Theta_{kj,\,j}. \tag{69}$$

Considering a free body, the boundary conditions on the surface Σ of body are

$$\left. \sum_j \Theta_{kj} n^j \right]_\Sigma = 0 \tag{70}$$

where n^j is the unitary vector orthogonal to Σ.

The preceding first order equations (from eq.67 to eq.70) are very general iff the displacements are small enough. We want to generalize these equations in the case of a weak gravitational field.

6.2 The generalization to the gravitational case

Several generalizations have been imagined most of them are equivalent in the quasi Newtonian case. However no thoroughly experimental test is presently available. This point is important because the theory of the mechanical detectors of gravitational radiations rely on the validity of the present extension of elasticity.

- First we write the fundamental equations (from eq.68 to eq.70) in a quasi Galilean coordinate system.

  We define

  $$\underset{old}{x}{}^{\alpha} := \{T, X, Y, Z\} = \text{Galilean (Minkowskian) coordinates}$$
  $$\underset{new}{x}{}^{\alpha} := \{t, x, y, z\} := \underset{old}{x}{}^{\alpha} - \varepsilon^\alpha(\underset{old}{x}) = \text{quasi Galilean coordinates}$$

- The functions ε^α are first order quantities which cancel in the past. Thus the mass points are labelled by the same quantity, r^k, in both coordinate system.

  We define the first order quantity $k_{\alpha\beta}$:

  $$k_{ij} := \varepsilon^i_{\ ,j} + \varepsilon^j_{\ ,i}\,, \quad k_{0j} := k_{j0} := \varepsilon^0_{\ ,j} - \varepsilon^j_{\ ,0}\,, \quad k_{00} := 2\varepsilon^0_{\ ,0}. \tag{71}$$

__Exercise__ : f is any first order quantity. Check the following relations (valid at first order only) $f\left(\underset{old}{x}\right) = f\left(\underset{new}{x}\right) = f(t, r^k) = f(T, r^k)$, $\partial f/\partial r^1 = \partial f/\partial X = \partial f/\partial x$ and similar relations for r^2 and r^3.

- Relatively to the new coordinate system $U^k(t, \vec{r}) := \underset{new}{x}{}^k(t, \vec{r}) - r^k$ is the displacement. A straightforward calculation gives

$$U^k(t, \vec{r}) = S^k(t, \vec{r}) - \varepsilon^k(t, \vec{r}), \tag{72}$$
$$S^k_{,00} = U^k_{,00} + \frac{1}{2}k_{00,\,k} - k_{k0,\,0}\,, \quad S_{jk} = U_{jk} - \frac{1}{2}k_{jk}.$$

- Therefore, the basic equations (from eq.68 to eq.70) read

$$\rho_0\left(U^k_{,00} + \frac{1}{2}k_{00,\,k} - k_{k0,\,0}\right) = \sum_j 2\mu\left(U_{jk} - \frac{1}{2}k_{jk}\right)_{,\,j} + \sum_j \lambda\left(U_{jj} - \frac{1}{2}k_{jj}\right)_{,\,k} \tag{73}$$

$$\left.\sum_j 2\mu\left(U_{jk} - \frac{1}{2}k_{jk}\right)n^j + \lambda\left(U_{jj} - \frac{1}{2}k_{jj}\right)n^k\right]_\Sigma = 0 \tag{74}$$

- Now we calculate the metric in the new coordinate system. The metric coefficients are Minkowskian in the old coordinate system, they become quasi Minkowskian in the new coordinate system (however the space-time itself has not changed, it remains Minkowskian)

$$\overset{old}{g}_{\alpha\beta} = \eta_{\alpha\beta} \to \overset{new}{g}_{\alpha\beta} = \eta_{\alpha\beta} + \eta_{\rho\alpha}\varepsilon^\rho_{,\,\beta} + \eta_{\rho\beta}\varepsilon^\rho_{,\,\alpha} = \eta_{\alpha\beta} + k_{\alpha\beta}. \tag{75}$$

  One can check that only the displacement and the perturbation of the metric are involved in eq.73 and eq.74.

- Finally, in the presence of a gravitational field we assume the validity of eq.73 and eq.74 although no Minkowskian coordinates exist in such a case.

Exercise : Check that the eq.73 and eq.74 are gauge invariant (covariant) when $U^k \to U^k + \varepsilon^k$ and $k_{\alpha\beta} \to k_{\alpha\beta} + \varepsilon_{\alpha,\,\beta} + \varepsilon_{\beta,\,\alpha}$).

N.B. These equations are consistent with the equation of the **geodesic deviation** and its generalization which was the early equation used by J. Weber to describe a mechanical detector of gravitational waves :

$$u^\beta \nabla_\beta \left(u^\gamma \nabla_\gamma \delta^\alpha\right) + R^\alpha_{\beta\rho\gamma} u^\beta \delta^\rho u^\gamma = \delta^\beta \nabla_\beta \left(\gamma^\alpha\right) \tag{76}$$

where γ^α is the 4-acceleration field (cf. the definition in the section Equivalence principles/Einstein equivalence principle/The second example) while the other quantities are defined in Figure 5 above.

6.3 Interpretations

We define $M := \iiint \rho_0 d^3r$ and the center of mass G :

$$x_G{}^k(t) := (1/M) \iiint \rho_0 \left(r^k + U^k\right) d^3r.$$

The worldline of G remains inside the world tube of the body (because $\rho_0 > 0$). When the size of the body is small, it can be assimilated to a point and the quantities $k_{\alpha\beta}$ remain nearly constant all over the body. The world-tube becomes the worldline of G. We integrate eq.73 over r^k and we use the boundary conditions eq.74. We obtain the equation of the motion of a mass point in free fall.

$$\iiint \rho_0 \partial_{tt} U^k d^3r + M\left(\frac{1}{2}k_{00,\,k} - k_{k0,\,0}\right) = 0 \Leftrightarrow$$

$$\frac{d^2 x_G{}^k}{dt^2} = -\left\{{k \atop 00}\right\} := f^k_{grav}. \tag{77}$$

<u>*Exercises*</u> : 1) Prove that eq.77 is the equation of the geodesic motion of a free falling test particle (within the present approximations).

2) Calculate the components of the gravitational acceleration f^k_{grav} for the PPN metric eq.31, for the canonical metric eq.27 and for the gravitational wave eq.45.

3) f^k_{grav} is the gravitational acceleration of G relatively to the coordinate system. What is the true physical acceleration?

a- The gravitational radiation in the Einstein coordinates

We consider the metric eq.45 which describes plane gravitational waves far away from the source : $g_{\alpha\beta} = \eta_{\alpha\beta} + h_{\alpha\beta}$ with $h_{\alpha 0} = h_{0\alpha} = h_{\alpha 3} = h_{3\alpha} = 0$, $h_{11} = -h_{22} = h_+(t-z)$, $h_{12} = h_{21} = h_\times(t-z)$. We call the corresponding coordinates "**Einstein coordinates**". From the gauge point of view the corresponding gauge is the Transverse Traceless gauge (**TT gauge**). A straight forward calculation gives $f^k_{grav} = 0$. Thus any mass point initially at rest remains at rest (relatively to the coordinate system).

In the interferometric detectors which are considered in the next section, the light goes forth and back between two mirrors. The mirrors are suspended at rest. When a gravitational radiation hits the detector the mirrors remain at rest. It does not mean that no effect can be observed because a full description of the detector needs a description of the behavior of the light in the same coordinate system.

More generally, considering equations eq.73 one finds

$$\rho_0 U^k_{,00} = \sum_j 2\mu U_{jk,\,j} + \lambda U_{jj,\,k} \text{ with} \tag{78}$$

$$\left.\sum_j 2\mu U_{jk} n^j + \lambda U_{jj} n^k = \mu h_{jk} n^j \right]_\Sigma .$$

One can consider that a gravitational wave does not modify the equations but generates a surface force density.

In a mechanical detector these forces produce mechanical waves, strains and stresses that one tries to observe.

b- The gravitational radiation in Fermi-like coordinates

The detectors of gravitational radiation on Earth are designed for the detection of signals at frequencies lower than a few kilohertz. Their size, d, does not exceed a few kilometers. Therefore we can assume $d/\lambda_{grav} << 1$ where λ_{grav} is the wavelength of the gravitational radiation. We perform the following change from the (old) Einstein coordinates to the Fermi-like coordinates[l] :

$$t = t_{old} - \tfrac{1}{4}\left(x^2 - y^2\right)_{old} \dot{h}_+ - \tfrac{1}{2}\left(xy\right)_{old} \dot{h}_\times, \quad x = x_{old} - \tfrac{1}{2}x_{old}h_+ - \tfrac{1}{2}y_{old}h_\times,$$
$$z = z_{old} - \tfrac{1}{4}\left(x^2 - y^2\right)_{old} \dot{h}_+ - \tfrac{1}{2}\left(xy\right)_{old} \dot{h}_\times, \quad y = y_{old} + \tfrac{1}{2}y_{old}h_+ - \tfrac{1}{2}x_{old}h_\times.$$

Here $h_{+,\times} = h_{+,\times}(t-z) \simeq h_{+,\times}(t)$ over the experimental device (because $d << \lambda_{grav}$) and $\dot{h}_{+,\times} := \partial_t h_{+,\times}$.

We calculate the proper time in the new coordinates

$$d\tau^2 = \left(\eta_{\alpha\beta} + h_{\alpha\beta}\right) dx^\alpha_{old} dx^\beta_{old} =$$
$$= (1+K)\,dt^2 - dx^2 - dy^2 - (1-K)dz^2 - 2K dt dz \text{ with} \tag{79}$$

[l] An extra quadratic gauge transformation is necessary to obtain the canonical form 27. At the current order of approximation, the results that we obtain in the sequel are the same in both coordinate system (the present one and the canonical one).

$$K \; := \frac{1}{2}\left(x^2-y^2\right)\ddot{h}_{+}+(xy)\,\ddot{h}_{\times}\sim h\cdot\left(\frac{d}{\lambda_{grav}}\right)^2 << h << 1$$

where we have used the order of magnitude $h \sim h_{+,\times}$.

The metric differs from Minkowski's metric by a term which is can be neglected. Most of the physical laws will remain the Special Relativistic ones[m].

A free falling particle initially at rest remains at rest in Einstein coordinates ($x^k_{old} = r^k$). However it displays a motion in the new coordinate system.

$$x^k = d\left(\frac{r^k}{d}+\frac{1}{2}h^k_j(t)\frac{r^j}{d}+\frac{d}{\lambda_{grav}}O(h)\right). \tag{80}$$

The gravitational acceleration (relatively to the coordinate system) can be derived either from eq.77 or from eq.80 :

$$\frac{d^2x^k_G}{dt^2} = f^k_{grav} = \frac{1}{2}\ddot{h}^k_j(t)r^j. \tag{81}$$

In the interferometric detectors of gravitational wave, the light behaves as in Special Relativity. The effect arises from the motion of the mirrors due to the gravitational acceleration.

The equations eq.73 and eq.74 read

$$\rho_0\left(U^k_{,00}+\frac{1}{2}K_{,k}\right) = \sum_j 2\mu U_{jk,\,j}+\lambda U_{jj,\,k}\,, \tag{82}$$

$$\left.\sum_j 2\mu U_{jk}n^j+\lambda U_{jj}n^k\right]_{\Sigma} = 0.$$

The boundary conditions are not modified when a gravitational wave hits a mechanical detector but a volume force appears. It derives from the potential $K/2 = \frac{1}{4}\left(x^2-y^2\right)\ddot{h}_{+}(t)+\frac{1}{2}(xy)\,\ddot{h}_{\times}(t)$ which depends on the time t. These forces are tidal forces; they produce mechanical waves, displacements and stresses that are (hopefully) measured in the mechanical detectors.

c- Comparison with electromagnetic waves

Eq.80 yields a direct comparison between the effect of an electromagnetic wave on test charged particles and the effect of a gravitational wave on test masses.

---

[m] It is the same in eq.27 where $\vec{a}$ and $\vec{\omega}$ remain however to describe the rotation and the acceleration of the frame.

In the figure below (Fig.6) we consider three particles.

1) Identical charged particles in the field of a plane electromagnetic wave. We represent their position in Minkowskian coordinates at time t and $t + \Delta t$. We assume that their distance is large enough to neglect their mutual interaction (we neglect the radiation pressure too).

2) Test masses in the field of a plane gravitational wave. We represent their position in Fermi-like coordinates at time t and $t + \Delta t$.

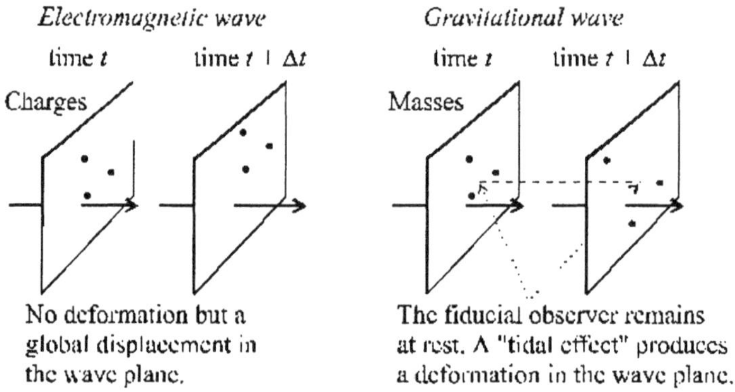

Figure 6: Displacements of identical test charges and test masses.

No longitudinal displacement is observed. The electromagnetic and gravitational waves are both transverse.

6.4 An acceleration-free pendulum

In the interferometric detectors of gravitational waves the mirrors are not in free fall. Thus an extra force appears in eq.74 where the term on the right is not zero but some surface force $f^k_{susp} + \delta f^k_{mech}(t)$ where f_{susp} is a static force mainly due to the suspension. δf^k_{mech} is time dependant, it represents various perturbations including the possible servos. Moreover the metric must include the stationary gravitational field f^k_{stat} (mostly due to the Earth) and the time dependant gravitational fields, $\delta f^k_{grav}(t)$, due to the motion of the various masses in the surrounding of the mirrors, including the planets and the deformation of the Earth (fortunately not all of them are in the detection frequency range). The theory that we consider is a linearized theory, thus using the previous method (integration over the r^k) we find

$$\frac{d^2 x_G^k}{dt^2} = f_{stat}^k + f_{waves}^k + f_{susp}^k + \delta f_{grav}^k + \delta f_{mech}^k \tag{83}$$

where $f_{waves}^k(t)$ is due to the gravitational waves that we want to detect.

Let us assume that no time dependant perturbation is initially present (when $t \to -\infty$) and that the mirrors are at rest. Thus $f_{stat}^k + f_{susp}^k = 0$. Moreover we assume that an ideal accelerometer is fixed on the mirrors. It will measure γ^k (cf. above Equivalence principle/Einstein equivalence principle/the second example)

$$\gamma^k = \frac{d^2 x_G^k}{dt^2} - \left(f_{stat}^k + f_{waves}^k(t) + \delta f_{grav}^k(t)\right) = f_{susp}^k + \delta f_{mech}^k(t). \tag{84}$$

The ideal situation corresponds to $\gamma^k = cte$ (which implies $\delta f_{mech}^k = 0$), with the condition $f_{stat}^k + f_{susp}^k = 0$. Therefore the acceleration relatively to the coordinate system is

$$\frac{d^2 x_G^k}{dt^2} = f_{waves}^k + \delta f_{grav}^k. \tag{85}$$

In the Fermi-like coordinates the interferometric detectors are sensitive to the time dependance of x_G^k (see the next section). Thus eq.85 shows that the detection of the gravitational waves depends on the possibility that δf_{grav}^k remains negligible compared to f_{waves}^k in the detection frequency range[n]. Of course, the real situation is more complicated because no real isolation can give exactly $\gamma^k = cte$ (a residual thermal noise remains) and because the position noise is not the alone mechanical noise; for instance the vibrations of the mirrors themselves have to be taken into account. However usual (non relativistic, non gravitational) technics are relevant to deal with this problem and are therefore out of the scope of the present course.

7 Interferometric detection of gravitational waves

7.1 *The high frequency approximation in optics*

From the components of the electric and magnetic fields, one can build the Faraday tensor $F_{\alpha\beta}$ in Special Relativity (see above the last exercise in Fundamental concepts / Covariant equations). In a Minkowskian coordinate system the Maxwell equations read $F_{\alpha\beta} = A_{\alpha,\,\beta} - A_{\beta,\,\alpha}$, $F^{\alpha\beta}{}_{,\beta} = 0$.

[n] f_{wave}^k is given by 81. The integration of 85 in the case where δf_{grav}^k is negligible shows that a pendulum does not oscillate any more once the gravitational wave has passed over (in the ideal case). The situation is different for high Q mechanical detectors.

The Einstein's equivalence principle gives the Maxwell equations in General Relativity : $F_{\alpha\beta} = A_{\alpha;\,\beta} - A_{\beta;\,\alpha}$, $F^{\alpha\beta}{}_{;\,\beta} = 0 \Rightarrow$

$$F^{\alpha\beta}{}_{;\,\beta} := \nabla^{\rho}\nabla_{\rho}A^{\alpha} - R^{\alpha}_{\rho}A^{\rho} - \nabla^{\alpha}\nabla_{\rho}A^{\rho} = 0. \tag{86}$$

with $R^{\alpha}_{\rho} = 0$ in the empty space.

We notice that $F_{\alpha\beta} = A_{\alpha;\,\beta} - A_{\beta;\,\alpha} \equiv A_{\alpha,\,\beta} - A_{\beta,\,\alpha}$ therefore $F_{\alpha\beta}\left[A_{\rho}\right] \equiv F_{\alpha\beta}\left[A_{\rho} + \partial_{\rho}f\right]$. We choose the function f (i.e. we choose the electromagnetic gauge) in such a way that $\nabla_{\rho}A^{\rho} = 0$. This is the Lorentz gauge in General Relativity. The field equations fulfilled by the 4-potential A^{ρ} are $\nabla^{\rho}\nabla_{\rho}A^{\alpha} = 0$.

We define $A^{\alpha} := a^{\alpha} \cdot e^{-iS}$ where the phase S is a real scalar and the amplitude, a^{ρ}, a real vector. Now we consider the high frequency approximation (i.e. along any elementary space-time displacement a^{α} varies slowly while S changes very quickly). The standard easiest way to proceed is to make the substitution $S \to S/\varepsilon$ and to expand the Maxwell equations. At the order $(1/\varepsilon)^2$ we find the eikonal equation

$$S_{;}{}^{\rho}S_{;\,\rho} := g^{\rho\sigma}S_{,\,\sigma}S_{,\,\rho} = 0. \tag{87}$$

We define now a four dimensional light ray $\mathcal{C} : x^{\alpha} = x^{\alpha}(u)$

$$\frac{dx^{\alpha}}{du} := S^{\alpha} := g^{\alpha\sigma}S_{\sigma}(x^{\rho}) \tag{88}$$

where u is some parameter and $S_{\rho} := S_{;\rho} := S_{,\rho}$. The eikonal equation implies that along a four dimensional light ray the phase remains a constant :

$$\frac{d}{du}S\left(x^{\rho}(u)\right) = \frac{dx^{\rho}}{du} \cdot \partial_{\rho}S := S^{\rho}S_{\rho} = 0. \tag{89}$$

Using eq.87, the phase being a scalar, one checks that

$$S^{\rho}\nabla_{\rho}S_{\alpha} = S^{\rho}\nabla_{\alpha}S_{\rho} = \nabla_{\alpha}\left(S^{\rho}S_{\rho}\right)/2 = 0.$$

Therefore eq.89 means that any four dimensional light ray is a null geodesic i.e. a geodesic along which $d\tau = 0$ (cf. General Relativity/The field equations/the second exercise). It is the worldline of a photon.

We consider now the weak field approximation where $g_{\alpha\beta} = \eta_{\alpha\beta} + k_{\alpha\beta}$ with $|k_{\alpha\beta}| \sim \kappa << 1$ and $S = \overset{0}{S} + \delta S$. The function $\overset{0}{S} = \omega t - \overrightarrow{k} \cdot \overrightarrow{x}$ (where ω and $\overrightarrow{k}$ are constant with $\overrightarrow{k}^2 = \omega^2$) is a solution of the eikonal equation in the Minkowski space time and δS is its perturbation due to the gravitation; it is of order k. At the first order relatively to k, the eikonal equation reads

$$\eta^{\rho\sigma}\,\overset{0}{S}_{,\sigma}\,\delta S_{,\,\rho} = \frac{1}{2}k^{\rho\sigma}\,\overset{0}{S}_{,\rho}\,\overset{0}{S}_{,\sigma}\,. \tag{90}$$

We consider now two events A and B on the same light ray where the phases are S_A and S_B. The light goes from A to B.

First we neglect the gravitational perturbation and we consider the unperturbed four dimensional light ray, $\mathcal{C}_0$

$$t = t_A + \omega u\,, \quad \overrightarrow{x} = \overrightarrow{x}_A + \overrightarrow{k}\,u \;\text{ with} \tag{91}$$
$$\overrightarrow{k} \;: = \omega \cdot \frac{\overrightarrow{x}_B - \overrightarrow{x}_A}{\ell_{AB}}\,, \quad \ell_{AB} := \|\overrightarrow{x}_B - \overrightarrow{x}_A\|\,.$$

The mathematical interpretation of eq.90 is straightforward. It results in the following equation

$$\frac{d}{dt}\delta S = \frac{\omega}{2}\left(k_{00} + 2k_{0j}n_0^j + k_{ij}n_0^i n_0^j\right)\Big]_{\mathcal{C}_0} \quad \text{with } n_0^j := \frac{x_B^j - x_A^j}{\ell_{AB}}. \tag{92}$$

$(\,)]_{\mathcal{C}_0}$ means that $\overrightarrow{x} = \overrightarrow{x}_A + \overrightarrow{n}_0\,(t - t_A) = \overrightarrow{x}_B - \overrightarrow{n}_0\,(t_B - t)$ in $(\,)$.

Therefore we obtain the phase S_B :

$$S_B \;: = S(t_B, \overrightarrow{x}_B) = S\left(t_B - \ell_{AB}, \overrightarrow{x}_A\right) + \delta S_B \tag{93}$$
$$\delta S_B \;: = \int_{t_B - \ell_{AB}}^{t_B} \frac{\omega}{2}\left(k_{00} + 2k_{0j}n_0^j + k_{ij}n_0^i n_0^j\right)\Big]_{\mathcal{C}_0} dt$$

Eq. 93 gives the solution of any problem in optics (at the eikonal approximation). It can be generalize in order to consider the propagation in a medium with an optical index $n_0 \neq 1$ and to the case of matter waves as well.

Now we give an interpretation of the theory based on eq.93. This interpretation is coordinate-dependent.

The quasi Minkowskian coordinates are interpreted as Minkowskian coordinates.

We define the elementary "length" $d\ell$, the "unitary" vector, $\overrightarrow{n} = \{n^k\}$ along the light ray and the optical index n :

$$d\ell^2 \;: = (d\overrightarrow{x})^2 := -\eta_{jk}dx^j dx^k\;,\; n^k := \frac{dx^k}{d\ell} := n_0^k + \delta n^k\Big]_{\mathcal{C}}\,, \tag{94}$$
$$n \;: = \frac{dt}{d\ell} = 1 + \delta n$$

where $\delta\,(\,)$ is the perturbation of the quantity $(\,)$.

Let us notice that $\overrightarrow{n}^2 = \overrightarrow{n}_0^{\,2} = 1 \Rightarrow \overrightarrow{n}_0 \cdot \delta\overrightarrow{n} = 0$.

As a consequence of eq.89 and eq.94 one can prove that

$$n^k g^{0\alpha} S_{,\,\alpha} = n g^{k\alpha} S_{,\,\alpha}. \tag{95}$$

We use this relation to develop the eikonal eq.87. A straightforward calculation gives

$$n = 1 - \frac{1}{2}\left(k_{00} + 2k_{0j}n_0^j + k_{ij}n_0^i n_0^j\right) = 1 + \delta n \tag{96}$$

The optical index depends on the direction of the propagation (n_0^k). The quantity $k_{0j}n_0^j$ produces a non reciprocal effect which is responsible for the Sagnac effect in eq.27 above.

We assume that the index varies slowly with the position and the time (considering a gravitational wave we mean that $\ell_{AB} << \lambda_{grav}$).

We rewrite eq.93 :

$$\begin{aligned} S_B \ : &= S(t_B, \vec{x}_B) = S\left(t_B - \ell_{AB}, \vec{x}_A\right) - \int_{t_B - \ell_{AB}}^{t_B} \omega \cdot \delta n]_{C_0}\, dt, \\ &\simeq S\left(t_B - \ell_{AB}, \vec{x}_A\right) - \omega \cdot \delta n \ell_{AB} \simeq S\left(t_B - n\ell_{AB}, \vec{x}_A\right). \end{aligned} \tag{97}$$

Then, using eq.97 and eq.95 we find that $\delta n^k = 0$.

This relation means that

1. the light propagates along a straight line whatever the quasi Minkowskian coordinates that we consider (the same result holds true if we consider that δn is a function of the time only),
2. the effect is completely described by the phase equation eq.97 where the optical index (eq.96) varies slowly with the time and depends on the direction, $\vec{n}_0$ of the propagation of the light.

Now we can study completely the behavior of an interferometric detector of gravitational waves in the two coordinate systems that we have already considered above (see Mechanic and Elasticity/Interpretations).

7.2 Interferometric detectors of gravitational waves

a- General equations

We consider the simplified model of detector represented in Fig.7 below. In order to obtain the first order terms relatively to $h \sim h_{+,\times}$ we assume an ideal situation where the mirrors are in free fall[o]. Any other time dependent

[o]The gravitational time dependent effects are the same with ideal suspensions in the static gravitational field of the Earth.

perturbation must be added to the phase shift that we calculate here. In the range of frequencies that we consider, the main troublesome contributions arises from the noises and from the Newtonian gravitational tidal effects due to moving masses (e.g. human activity).

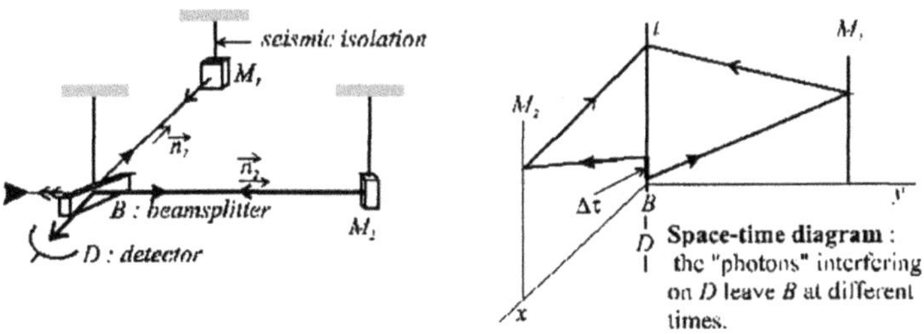

Figure 7: Representations of a simple interferometric detector.

In the previous section we have considered that in any quasi Minkowskian coordinates the light propagates along a straight line. However the expression of the phase depends on the $k_{\alpha\beta}$ and therefore on the coordinates that we are using. The description of the motion of the mirrors depends also on the coordinates it results that the physical effect is independent of the particular description involved.

We choose the origin of the coordinates on the beamsplitter B.

- In the Einstein coordinates $k_{0\alpha} = k_{3\alpha} = 0$, $k_{ij} = h_{ij}$ with

$$h_{11} = -h_{22} = h_+ \text{ and } h_{12} = h_{21} = h_\times .$$

- The mirrors remain at rest, i.e. the beam splitter B is at the origin and any end mirror M has constant coordinates

$$\{r_M^k\} = \overrightarrow{x}_M = L_M \cdot \overrightarrow{n}_M = L_{0M} \cdot \overrightarrow{n_0}_M .$$

- The optical index is $n = 1 - \frac{1}{2} h_{jk} n_M^j n_M^k$ (cf. eq.96).

The duration for a round trip BMB (measured at point B) is

$$\tau_E = 2L_M \cdot \sqrt{g_{00}} \cdot n/c = 2L_{0M}\left(1 - \tfrac{1}{2} h_{jk} n_M^j n_M^k\right) \text{ where } L_{0M}^2 = \left(\overrightarrow{r}_M\right)^2 .$$

- In the Fermi-like coordinates the beamsplitter, B, is chosen as the fiducial observer.

$$k_{1\alpha} = k_{2\alpha} = 0,\ k_{00} = k_{33} = -k_{03} = \tfrac{1}{2}\ \ddot{h}_{jk}\ r_M^j r_M^k := K.$$

- The beamsplitter remains at rest ($\overrightarrow{r}_B = \overrightarrow{0} \Rightarrow \overrightarrow{x}_B = \overrightarrow{0}$) but the end mirrors do not :

$$x^k_M = r^k + \tfrac{1}{2} h^k_j r^j_M \quad \text{(cf. eq.80)}.$$

The distance $L_M := \sqrt{(\overrightarrow{x}_M)^2}$ changes with the gravitational wave :

$$L_M = L_{0M}\left(1 + \sum_k h^k_j n^j_M n^k_M\right) = L_{0M}\left(1 - \tfrac{1}{2} h_{jk} n^j_M n^k_M\right).$$

- The optical index is $n = 1 + \delta n$ with $\delta n = -\frac{1}{2} K \left(1 - n^z_M\right)^2$. The perturbation $\delta n \lesssim h_{+,\times} \cdot \left(L_{0M}/\lambda_{grav}\right)^2$ is negligible because of the condition $\left(L_{0M}/\lambda_{grav}\right) << 1$ therefore $n \simeq 1$ and $g_{00} \simeq 1$.

The duration of a round trip BMB (measured at point B) is

$$\tau_F = 2\sqrt{g_{00}} L_M \cdot n/c = 2L_{0M}\left(1 - \tfrac{1}{2} h_{jk} n^j_M n^k_M\right).$$

The duration of a round trip (measured at point B) is a physical quantity and $\tau_E = \tau_F$ as expected [p].

The Fermi-like coordinates are convenient because the expression of the proper time is the same as in Special Relativity in Minkowskian coordinates up to a very high accuracy ($\left(L_{0M}/\lambda_{grav}\right)^2 \cdot h \lesssim 10^{-4} h \lesssim 10^{-25}$). Therefore one can assume that the description of most of the physical phenomena (including the Maxwell equations) is very similar to the special relativistic one. Thus a gravitational wave does not produce any noticeable effect on the polarization of the light during its flight in the vacuum.

b- The physical effect

It is now straightforward to calculate the phase difference between the two beams. The proper time necessary for a round trip is different in each arm. We use the notation L_a instead of L_{0M} and n^k_a instead of n^k_M where $a = 1$ or 2 depending on the arm that we consider. We find

$$\begin{aligned} \tau_a &= 2L_a - L_a h_{jk} n^j_a n^k_a \Rightarrow \\ \Delta\Phi &= 2\omega \frac{L_2 - L_1}{c} + \delta\Phi \text{ with} \\ \delta\Phi &:= h_{jk}(t)\left(n^j_1 n^k_1 - n^j_2 n^k_2\right) \frac{L_0}{c} \end{aligned} \tag{98}$$

[p]Notice that the global description must be coherent in the sense that the motion of the mirrors in the Fermi-like coordinates must not be consideredas a real motion whose effect could combine with the variation of the optical index which appears in the Einstein coordinates.

where we have assumed $L_1 \simeq L_2$ and $2L_0 := L_1 + L_2$. The quantity $\delta\Phi$ changes with the time. It results in a change of the optical power which will be (hopefully) detected.

The situation is more complicated in real devices (e.g. Virgo, Ligo and similar detectors) where the light is stored in each arm in a Fabry-Perot interferometer. In such a case one can either consider that the mirrors remain at rest and that the optical index, $n = 1-\frac{1}{2}h_{jk}n_a^{\,j}n_a^{\,k}$, is a function of the time and the orientation of the arms or that the mirrors are moving $(x_M^{\,k} = r^k + \frac{1}{2}h^k_j r_M^{\,j})$ while the light propagates in an empty space. Whatever the point of view, the problem is a classical interferometry problem where gravitation does not play any special role any more. Therefore we will not study this question here.

In order to discuss the effect of the orientation of the detector relatively to the gravitational wave we first consider the optimistic case where the plane of the arms is orthogonal to the direction of the gravitational wave.

We consider a circle of free particles in the x-y plane (in the Einstein coordinates for instance) : $x^2 + y^2 = R^2$. Two of these particles are the end mirrors of the interferometer while the center of the circle is the beamsplitter. The basic effect depends on the optical length, nR, between the origin and the different particles. We give below a representation of the optical length for a periodic gravitational wave.

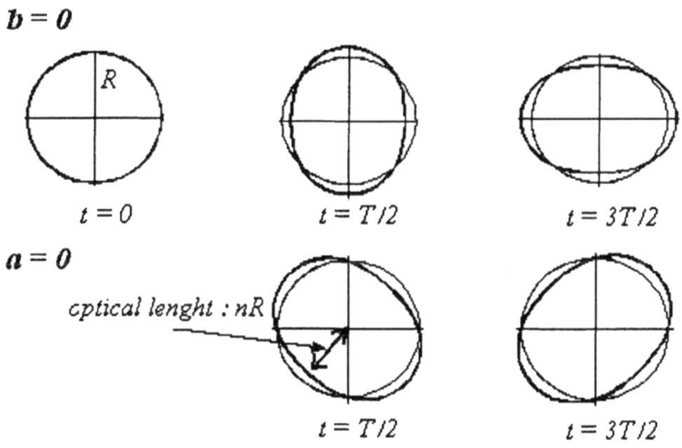

Figure 8: $h_+ = a \sin \omega t$, $h_\times = b \sin \omega t$, $T = 2\pi/\omega$

Here the gravitational perturbations have been amplified because the expected relative variation of nR does not exceed 10^{-21} on Earth.

In the general case the arms are not in the x-y plane orthogonal to the direction of the gravitational wave. To compare the general situation to the optimistic case we introduce the polarization vector $\overrightarrow{h}$:

$$\overrightarrow{h} := h\{\cos\Psi, \sin\Psi\} \quad \text{with} \quad h\sin 2\Psi := h_\times \ , \quad h\cos 2\Psi := h_+. \tag{99}$$

N.B. Here h is precisely defined; it is no longer the previous order of magnitude that we have introduce above.

One can check that $\overrightarrow{h}$ is a vector in the sense that a rotation of the x-y axes modifies its components (in the same way as the components of a vector) but does not change $\overrightarrow{h}$ itself.

We consider the projections, ℓ_1 and ℓ_2 of the two arms of the interferometer on the x-y plane.

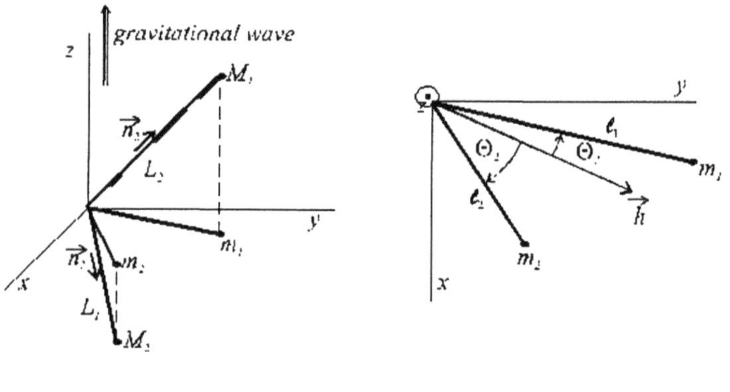

Figure 9: Projection of the interferometer on the wave plane.

The phase difference is given by the relation

$$\delta\Phi = 2\pi \frac{h}{\lambda_{opt}} \left(\cos 2\Theta_1 \cdot \frac{\ell_1^2}{L_1} - \cos 2\Theta_2 \cdot \frac{\ell_2^2}{L_2} \right) \tag{100}$$

where λ_{opt} is the wave length of the light in the interferometer and Θ_1 and Θ_2 are defined above on Figure 9.

One can easily check that $|\delta\Phi|$ varies with the orientation of the detector between zero and the maximum value $2\pi h\,(L_1 + L_2)\,/\lambda_{opt}$ (which does not exceed $4 \cdot 10^{-11}$ for $L_1 \sim L_2 \sim 3$km, $\lambda_{opt} \sim 1\mu$m and $h \lesssim 10^{-21}$).

c- Array of interferometric detectors

At the present moment three large detectors are under construction : the two LIGO detectors in the US ($L_0 \sim 4$km) and the Italian-French detector, Virgo, in Italy ($L_0 \sim 3$km). A smaller detector is under construction in Germany : the German-British detector GEO ($L_0 \sim 0.6$km). Several prototypes and projects are also developed (or considered) in Japan, Australia and Russia. They are based on the basic principle described above. Each detector can be considered as an element of a global Earth bound observatory.

The gravitational signal, $S_{(N)}$, of the detector number N is proportional to the quantity $\delta\Phi_{(N)}$ which can be calculated from eq.100.

One can choose an arbitrary reference frame, $\left\{\overrightarrow{i}, \overrightarrow{j}, \overrightarrow{k}\right\}$ (a Copernican frame for instance). The direction of the gravitational wave is given by the unitary vector $\overrightarrow{u}$. Thus the direction of the source on the sky is $-\overrightarrow{u}$. One can choose the unitary vectors $\overrightarrow{e}_1 = \overrightarrow{j} \wedge \overrightarrow{u} / \left\| \overrightarrow{j} \wedge \overrightarrow{u} \right\|$ and $\overrightarrow{e}_2 = \overrightarrow{u} \wedge \overrightarrow{e}_1$ in order to define the polarizations $h_{11} = -h_{22} = h_+$ and $h_{12} = h_{21} = h_\times$. Thus the gravitational wave is defined by two angles, θ and φ which give the position of the source on the sky and two functions of the time, h_+ and $h_\times$. We define precisely $h_{+,\times}(t)$ as the value of $h_{+,\times}$ at the origin at time t.

The signal reads

$$S_{(N)}(t) = h_+(t - t_N)F_{+(N)}(\theta, \varphi) + h_+(t - t_N)F_{\times(N)}(\theta, \varphi) \qquad (101)$$

where $t_N = \overrightarrow{u} \cdot \overrightarrow{x}_N$ is the delay from the origin to the detector. It depends on the position of the detector along the direction of the gravitational wave. Due to the motion of the Earth, t_N is a slowly varying function of the time which is known once the two angles θ and φ are known. The quantities $F_{+,\times\,(N)}$ depend on the orientation of the detector. Their mathematical dependence on θ and φ can be easily calculated.

In order to find what information can be extracted from a given array of detector let us discuss a simple model where the array is operated with three detectors.

First we consider a short gravitational pulse. It will not be possible to eliminate all the false signals but no false signal will appear on two detectors at the same moment (at least if the detectors are independent and if the false signals are rare enough; this is precisely what is expected). Thus the correlation of the observations between the different detectors will give the gravitational signals.

During a short signal, the quantities $F_{+,\times(N)}$ and t_N remain constant. The arrival times of the signal on the various detectors give the direction of the source on the sky (i.e. θ and φ). Thus the quantities t_N and $F_{+,\times(N)}$ are known. We write eq.101 : $S_{(N)}(t+t_N) = h_+(t)F_{+(N)} + h_+(t)F_{\times(N)}$. Considering the response of only two detectors ($N = 1,2$) one finds the two polarizations h_+ and $h_\times$ as functions of $S_{(1)}(t-t_1)$ and $S_{(2)}(t-t_2)$.

The quantities $\theta, \varphi, h_+(t)$ and $h_\times(t)$ are known without using the full information brought by the detector $N = 3$. We can now calculate the predicted detailed signal, $S_{(3)p}(t)$ of the third detector. The comparison between the functions $S_{(3)p}(t)$ and the function $S_{(3)}(t)$ which is actually observed is a test of the theory.

As a conclusion we put forward that we can check General Relativity and obtain all the information it predicts with three detectors only (two of them would not be enough).

Now we consider a periodic gravitational waves with frequencies suitable for a detection i.e. between a few tens of hertz and a few kilohertz. The period is short; the quantities $F_{+,\times(N)}$ and t_N remain constant during one period. We can consider that the signal during one period is a pulse. Thus the periodic gravitational wave is a repetition of identical short pulses (even if it is a sinusoidal wave).

The position and the orientation of a detector changes during the day and the year. One detector alone works in the same way as an array of many detectors located at different places which receive the same signal at the same moment (except for the delay t_N+ *a multiple of the period*). In that case it is possible to obtain θ, φ, and $h_{+,\times}$ with arguments similar to the previous ones. These previous arguments were based on the independence of the detectors. This might not be the case when the array is formed by the same detector at different times. However the situation is not hopeless because the change of the orientation of the detectors is known and therefore we know how the signal changes with the time. Such a characteristic signature is not expected from local perturbations.

Thus, in principle, one detector alone could observe periodic waves. Periodic gravitational waves are expected at low frequencies and unfortunately the detection is not possible on Earth at frequencies lower than a few hertz. However an interferometer in space could solve the problem (i.e. the LISA project).

Acknowledgments

It is a pleasure to thank the organizers of the 1999 Urbino International Summer School on Experimental physics of Gravitational Waves for this meeting and especially Professor Flavio Vetrano.

Appendix

The preceding theory gives the gravitational waves in two interesting cases : a rotating neutron star which emits periodic waves and a binary system who produces a so-called "chirp signal".

Neutron star

The neutron star is assumed to be an oblate ellipsoid with a cylindrical symmetry. Its principal moment of inertia are I_z and I_Δ. Its angular velocity relatively to the "fixed stars" is Ω. The other quantities involved are defined in the Figure 10 below.

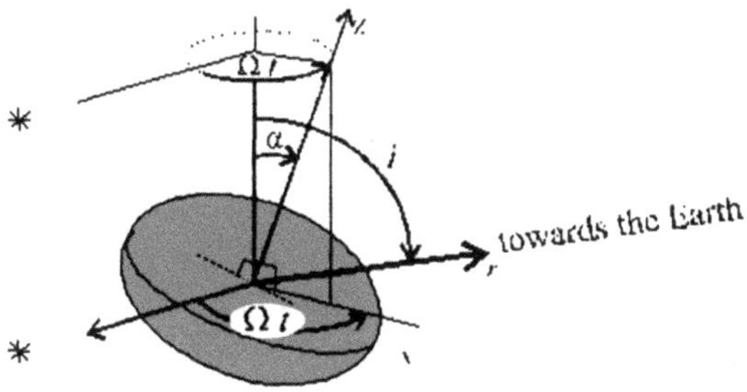

Figure 10: Rotation of a neutron star.

The polarizations are

$$
\begin{aligned}
h_+ &= \frac{1}{2} h_0 \sin\alpha \cdot \{\cos\alpha \cdot \sin i \cdot \cos i \cdot \cos \Omega t \\
&\quad - \sin\alpha \cdot \left(1+\cos^2 i\right) \cdot \cos 2\Omega t\} \\
h_\times &= \frac{1}{2} h_0 \sin\alpha \cdot \{\cos\alpha \cdot \sin i \cdot \sin \Omega t \\
&\quad -2 \sin\alpha \cdot \cos i \cdot \sin 2\Omega t\} \quad \text{where} \\
h_0 &= \varepsilon \frac{4G\Omega^2}{c^4 r} I_z \,, \quad \varepsilon = \frac{I_z - I_\Delta}{I_z}
\end{aligned}
$$

Let us notice the two angular frequencies Ω and 2Ω in the gravitational wave.

Most of the expected signals have an amplitude which does not exceed 10^{-26} at frequencies of order of a few tens of hertz. A long integration time ($\sim$ 1y) will be necessary to achieve the observation with a sensitivity of order of $3 \cdot 10^{-22}/\sqrt{\mathrm{Hz}}$. Some signals are however expected at higher frequencies where the sensitiviy is better with an amplitude $h_0 \sim 10^{-24}$.

Coalescing binaries

Two neutron stars of mass M_1 and M_2 are gravitationally bounded. They turn around their center of mass at a distance a from one another with an angular velocity ω. A loss of energy is due to the emission of gravitational waves. It results in a contraction of the system and an increase of ω. If the quasi Newtonian theory above were completely exact and if the two neutron stars were two mass points, the angular velocity would become infinite and d would reduce to zero in a finite lapse of time, at $t = t_c$.

We define the total mass $M := M_1 + M_2$, the reduced mass $m = M_1 M_2/M$ and the chirp mass $\mathcal{M} = M^{2/5} m^{3/5}$.

The distance between the two stars is $a = (GM)^{1/3} \omega^{-2/3}$. The angular velocity is

$$\omega = \frac{1}{2}\frac{d\varsigma}{dt} := \frac{1}{8\tau}\left(\frac{t_c - t}{5\tau}\right)^{-3/8}$$

where $\tau := G\mathcal{M}/c^3$ is a characteristic time of the system (of order of 6μs for $\mathcal{M} \sim 1.2 M_\odot$).

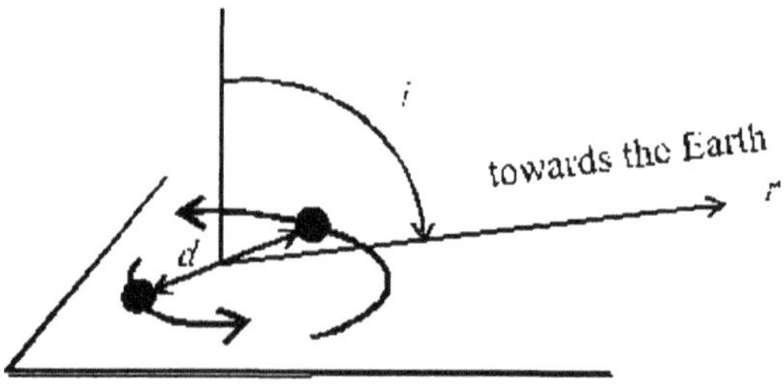

Figure 11: A quasi Newtonian binary system.

The two polarizations are

$$h_+ = h_0 \frac{1+\cos^2 i}{2} \cdot \cos\varsigma\ , \quad h_\times = h_0 \cos i \cdot \sin\varsigma \text{ with}$$

$$h_0\ :=4\frac{c\tau}{r}\left(\omega\tau\right)^{2/3} \text{ and } \varsigma := -2\left(\frac{t_c - t}{5\tau}\right)^{5/8} + cte$$

At time t_0, long before the coalescence ($t_c - t_0 >> 5\tau$), one can give a Taylor expansion of ς and prove that $\varsigma \simeq 2\omega t + cte'$ where ω can be considered as a constant during a time of order of $(t_c - t_0)$: the binary system does not radiate efficiently until the end of the process.

Let us notice that, here, 2ω is the alone angular frequency in the gravitational wave.

Bibliography

In the following bibliography (and the references quoted therein) one finds the possibility to start with and to study more thoroughly General Relativity, Experimental Gravitation and Gravitational Radiation.

Many books are available, among them we suggest

to start with

- Ph. Tourrenc : *Relativity and Gravitation*, Cambridge University Press (1997). 240 p.

  Relativity and Relativistic Gravitation with the minimum necessary mathematics. Several experimental and observational examples are discussed (the classical tests, the interferometric detection of gravitational waves and the Gravity Probe-B experiment).

- B.F. Schutz : *A first course in General Relativity*, Cambridge University Press (1985). 400 p.

  General Relativity with tractable exercises and an annotated bibliography.

then

- S. Weinberg : *Gravitation and Cosmology*, Wiley (1972). 650 p.

  A physical approach to gravitation and astrophysics. Many subjects are presented : Special and General Relativity, experimental tests in the Solar system, relativistic stars and relativistic cosmology. No thorough attention is paid on experimental gravitation in the laboratory.

- C.W. Misner, K. Thorne, J.A. Wheeler : *Gravitation*, W.H. Freeman (1973) 1300 p.

  A reference book for junior as well as senior physicists. Most of the fundamental topics are discussed from various points of view.

- C. Will : *Theory and Experiment in Gravitational Physics*, Cambridge University Press (1993). 350 p.

  Detailed discussion of the main experimental and observational results.

and the proceedings of two summer school at "Les Houches".

- *Gravitational Radiation*, edited by N. Deruelle and T. Piran (NorthHolland 1983). This book deals with experiments, as well as theory and numerical methods.

- *Relativistic Gravitation and Gravitational Radiation*, edited by J-A Marck and J-P Lasota (Cambridge University Press 1997, CD-Rom available).

  An interesting recent theoretical review : General Relativity (physical content, mathematical and numerical methods), the sources of the gravitational waves (astrophysical and cosmological), the expected signals (properties and analysis) ...

Consider also some articles

- K. Thorne in *Three Hundred Years of Gravitation*, edited by S.W. Hawking and W. Israel (Cambridge University Press 1987). A basic paper on gravitational wave detection with theoretical considerations useful for experiment.

- W-T. Ni and M. Zimmermann : *Phys. Rev. D* **17**, p. 1473 (1978). The standard form of the metric eq.27 is established. For an expansion up to cubic terms see W-Q. Li and W-T. Ni *PJ. Math. Phys.* **20**, p. 1473 (1979).

- B. Carter and H. Quintana : *Proc. Roy. Soc.* **A 331,** p. 57 (1972). The theory of elasticity in General Relativity. For the quasi Newtonian case see A. Papapetrou : *Ann. Inst. Henri Poincaré.* **A XVI** p. 63 (1972). For the generalization to visco-elastic media see R. Gambini : *Ann. Inst. Henri Poincaré.* **A XXIII** p. 389 (1975).

- B. Linet and Ph. Tourrenc. *Can. J. Phys.* **54**, p. 1129 (1976). The eq.92 is established and discussed. For a generalization to an isotropic medium see Ph. Tourrenc in the *Third Symposium on Frequency Standards and Metrology* published in *J. Phys. Fr.* **42** C8 p. 441 (1981).

and five books

- L. Landau and E. Lifschitz : *Classical theory of fields*, Pergamon Press (1985).
- R. d'Inverno : *Introducing Einstein's Relativity*, Clarendon Press (1993) and R.M. Wald : *General Relativity*, Univ. of Chicago Press (1984), theoretically and mathematically oriented.
- *Special Theory of Relativity,* Pergamon Press (1970) where a succinct presentation of the subject is given by C.W. Kilminster with a collection of important papers between 1881 and 1937 and *The principle of Relativity*, Dover (1923), important papers on the way to General Relativity.

PHYSICS OF SOURCES OF GRAVITATIONAL WAVES

S. BONAZZOLA AND E. GOURGOULHON
Département d'Astrophysique Relativiste et de Cosmologie,
UMR 8629 C.N.R.S., Observatoire de Paris,
F-92195 Meudon Cedex, France

Foreword

Few years ago, one of us, (S.B.) in collaboration with J.A. Marck, has written the review paper "Astrophysical Sources of Gravitational Waves" [1]. Since this paper a lot of work has been done in modeling the source and in understanding the most pertinent mechanisms of gravitation wave emission. When Prof F. Vetrano asked very kindly one of us (S.B.) to give a course at the Urbino summer school, I (S.B.) have decided, in order to avoid a non interesting repetition of what can be found in the above review paper, to emphasize the physics of astrophysical sources of gravitational radiation. In other words, these lectures should be considered as a complement of the above review paper. I am the only responsible of this choice. The writing of this lecture was done in collaboration with E. Gourgoulhon.

1 Introduction

Soon after publishing the General Theory of Relativity, Einstein showed that in the weak field approximation, its linearized equation are hyperbolical-one, thus implying propagation of the gravitational waves with the speed of the light.

However, because the Einstein Equations involve geometrical objects whose coordinates expressions do not have invariant meaning, it was unclear for several decades whether these solutions revealed physical properties of the gravitational field or merely the choice of coordinates system for the linearized field equations. In spite of a beautiful result obtained by B. Finzi [2], who showed in a covariant way that the characteristic surfaces of the Einstein equations propagate at the speed of the light, there were doubts on the reality of the gravitational waves. The Finzi result was, and it is until now, completely ignored by the scientific community. Thanks the work of several authors, specially that of Pirani [3] , who gave a precise answer of the question "what would

happen if a gravitational wave crossed my laboratory ?", the question is now well settled. The discovery by R. Hulse and J. Taylor of the binary pulsar PSR 1913+16,[4] the measured orbital period of which decreases following the predictions of he General Relativity Theory, is a beautiful observational confirmation of the existence of gravitational waves [a] .
Modern technology allows us to built ground and space experiments sensitive enough to detect gravitational waves radiated by astrophysical objects. In these notes, we shall describe the astrophysical sources that are plausible candidates as efficient gravitational wave emitters and we shall discuss their possible detection by ground based detectors. Because of the compactness of these sources (mainly Neutron Stars (NS) and Black Holes (BH)) General Relativity Theory is the natural frame for their correct modeling. In order to avoid the intricacy and complexity of the theory, we have chosen the strategy of describing and explaining as simple as possible the most important and pertinent concepts. Technical and detailed description can be found on the Appendices. The reader will found useful complements in the articles published by different authors on the Les Houches summer school book (1995) [6].

1.1 General considerations

Detection of Gravitational Waves: Why ?
There are various motivations pushing people to try to detect gravitational waves. A personal poll among our colleagues on the motivations of detection of gravitational waves has given the following answers:

-Because it is a technical challenge
-Best determination of the upper limit of the graviton mass
-Possible detection of a scalar field coupled with the tensorial field.
-Physics of the NSs and physics of dense nuclear matter.
-Possible test of of General Relativity (GR) in the non linear regime.
-Mass function of massive stars.
-Determination of an absolute astronomical distance
-Direct evidence of the existence of BH.
-Because they (GW) are there.
-Other ([1])

[a]The formula giving the decreasing rate of the period of two orbiting objects was obtained by Einstein itself just after his theory of general relativity. This formula, often called Landau formula, was criticized because its deficiency of covariance. Only recently T. Damour and N. Deruelle [5] has confirmed the validity of the Einstein formula.

As can be seen, there are numerous and different motivations to be involved in a so exiting challenge. Our personal preference is the Hymalaian[b] motivation "Because they are there". More seriously, there exists a quite shared credo, that at least at the beginning the detection of GW by ground based experiment (GEO, LIGO, TAMA, VIRGO) should be considered as a physical experiments consisting in the direct detection of GW; only later,when the sensitivity will be improved, the detectors would become a new kind of astronomical observatory. We don't share this point of view: in fact from the first detections of GW emitted by coalescing objects important astrophysical informations on the evolutionary track of massive stars can be obtained. We shall see later that an analogous statement for GW emitted by rotating objects holds.

Some notions on Theory of Gravitation

Following Newton and Poisson,the gravitational potential Φ must satisfy the Poisson equation

$$\Delta\Phi = 4\pi G\,\mathrm{n} \tag{1}$$

where G is the gravitational constant ($G = 6.67\,10^{-8}\,g^{-1}\,cm^3\,s^{-2}$) and n $(\mathrm{g\,cm^{-3}})$ is the baryonic density. This equation is invariant with respect a Galileo transform, but is *non* invariant with respect a Lorentz transform. Nordström tried to generalize the Poisson equation by replacing the Laplacian operator Δ by the d"Alembertian operator $\Box = \Delta - 1/c^2\,\partial^2/\,\partial t^2$ where c is the light velocity. Therefore the Nordström equation reads

$$\Box\,\Phi = 4\pi\,G\,\mathrm{n} \tag{2}$$

It is easy to see that the gravitational potential Φ propagates at finite velocity c. Unfortunately this simple generalization of the Newtonian theory is not in agreement with the observational facts: it predicts a precession of the perihelium of Mercury of -14 arc-seconds per century instead of the observed $+42$ arc-seconds. Conclusion: a scalar theory, or a spin 0 theory of the gravitation does not work and we can hope that a theory based on an higher spin field will describe successfully the gravitation.
The vector field (spin 1 field) is the next candidate to build a relativistic theory of the gravitation (semi-integer spin fields must be ruled out because

[b] A paraphrase of the famous answer "*because it is there*", given by Mallory, who was the first alpinist to try to climb the mount Everest, to a journalist who asked him the motivation of a such hard and dangerous enterprise

the Pauli principle forbids large occupation numbers and the field cannot be treated as a classic one). This field W^α $(\alpha = 0...3)$ would satisfy a vectorial d'Alembert equation

$$\Box W^\alpha = g\, J^\alpha \tag{3}$$

where J^α is a current vector. But this results in a repulsive force between two masses of the same signs, therefore this cannot describe a gravitation theory. Only a spin 2 field or tensorial field $h_{\alpha\beta}$ satisfying d'Alembert equation

$$\Box h_{\alpha\beta} = \frac{8\pi\, G}{c^4}\, T_{\alpha\beta} \tag{4}$$

allows us to build a satisfactory theory of the gravitation (here $T_{\alpha\beta}$ is the energy-momentum tensor of the matter).
Actually for reasons of consistency, a non linear theory is necessary; this theory is the General Theory of Relativity, The equations are the well known Einstein equations:

$$R_{\alpha\beta} - \frac{1}{2} g_{\alpha\beta} R = \frac{8\pi G}{c^4} T_{\alpha\beta} \tag{5}$$

where $R_{\alpha\beta}$ is the Ricci tensor , $R = R^\alpha_\alpha$ and $g_{\alpha\beta}$ is the metric tensor.
The above system of 10 equations is hyperbolic and "actions" propagate at the velocity of the light.
The price to pay for this new theory is quite high: the above equations look quite simple, but when they are written in an explicit way, hundred of terms appear.
As we have said in the spin 2 theory there exist 10 potentials $g_{\alpha\beta}$. It is known that in general, a symmetric tensor $g_{\alpha\beta}$ contains fields of lower spin. The spin 0 components g_{00} represents the Newtonian potential, The 3-objet (or spin 1 component) g_{0i} represents the generalized Coriolis acceleration, and the traceless part of g_{ik} is the genuine spin 2 field describing the gravitational waves.
Gravitational waves carry energy and angular momentum. The most important contribution is given by the quadrupole terms.(Dipole terms vanish because masses of opposite sign do not exist). Here we give the expressions giving the gravitational luminosity L, the angular momentum radiated per second $\dot{\mathcal{L}}$ and the amplitude h of the gravitational wave as function of the variation of the quadrupole moment Q_{ik} of the source [7]

$$L = \frac{1}{5}\frac{G}{c^5} \sum_{ik} \frac{d^3 Q_{ik}}{dt^3} \frac{d^3 Q_{ik}}{dt^3} \quad i, k = 1, 2, 3 \tag{6}$$

$$\dot{\mathcal{L}} = \frac{2}{5}\frac{G}{c^5}\sum_{ik}\frac{d^3Q_{ik}}{dt^3}\frac{d^2Q_{ik}}{dt^2} \tag{7}$$

$$h = \frac{2G}{c^4}\ddot{Q} \tag{8}$$

We recall that , [c]

$$Q_{ik} = \mathcal{I}_{ik} - \frac{1}{3}\mathcal{I}^j_j\,\delta_{ik}\,, \quad \mathcal{I}^{ik} = \int_{Vol} n x^i x^k\, dV \tag{9}$$

We have now all what we need to give an answer to the following question:
What are the conditions to have an efficient source of gravitational waves ? We shall show that we need 3 necessary but not sufficient conditions:

-The source must be in non a steady state configuration and must be as massive as possible.
-The source must be compact.
-The motion of the source must be coherent.

It is easy to understand the first condition: Because of the time derivative of Q_{ik} in the Eq. (6),(7),(8) show that the source must evolve. The numerical factor $1/5\,G\,/c^5$ in c.g.s. units in front of Eq.(6) or Eq.(7) implies that only large masses, or better, large quadrupole momenta and fast motion typical of the star-like objects are necessary in order to fight against a so terribly small factor.

Compactness

Here we show that the source (a star-like object) must be compact. From Eq. (6), we see that the gravitational luminosity L depends on the square of the variations of the quadrupole moment Q. Because Q is proportional to the mass M and to the square of the radius R of the star it seems that for a given mass, stars with large radius are more efficient GW emitter than compact stars. This is a hasty conclusion: In fact a large star cannot "move" fast (its 3rd derivative cannot be very large). It turns out that for the same mass, a compact star is a more efficient GW emitter than a high quadrupole huge star.

---

[c]It exists a linear relationship between h_{ik} and the quadrupole Q_{ik} (See lectures by P. Tourrenc [8]) In Eq.(8) the relationship between h_{ik} and Q_{ik} are written in a symbolic form.

Compactness is a quite intuitive notion but for quantitative estimation, we need a definition. We define a compactness factor $\mathcal{C}$ the quantity

$$\mathcal{C} = \frac{2\,G\,M}{c^2 R} = \frac{R_s}{R} \tag{10}$$

where R_s is the Swarzschild radius of the star. Typical values of compactness factors for different objects are:
Black holes or the Universe: $\mathcal{C} = 1\,, \quad R = 2.96\,10^5\,\frac{M}{M_\odot}$ cm
Neutron stars: $\mathcal{C} \sim 0.3\,, \quad R \sim 10^6\,cm$;
White dwarfs: $\mathcal{C} \sim 10^{-4}\,, \quad R \sim 5\,10^8$ cm
The Sun: $\mathcal{C} = 4.3\,10^{-6} \quad R = 6.96\,10^{10}$ cm
Jupiter: $\mathcal{C} = 4\,10^{-8} \quad R = 7.13\,10^9$ cm
The Earth: $\mathcal{C} \sim 1.4\,10^{-9}\,, \quad R = 6.38\,10^8$ cm
A Weber type antenna: $\mathcal{C} \sim 1.5\,10^{-24}\,, \quad R = 100$ cm
The proton: $\mathcal{C} = 1.9\,10^{-39}\,, \quad R = 1.3\,10^{-13}\,cm$

In what follows, we shall express the GW luminosity L as function of the mass M of the star and of its compactness parameter $\mathcal{C}$.
An isolated star radiates only if oscillations are present or if it rotates having broken its axial symmetry. Mechanisms that break the axial symmetry of a rotating NS will be discussed in a next section.
Consider first an oscillating star. The lowest frequency ν is of the order of the crossing time of the sound waves $\nu \sim c_s/(2R)$ where c_s is the sound velocity. An order of magnitude of c_s is given by $c_s^2 \sim P/\mathrm{n} \sim PV/M$ where P and n are respectively some averaged pressure and the averaged baryonic density and V is the volume of the star. The virial theorem gives a relationship between the pressure and the gravitational energy E_{gr}:

$$E_{gr} = -3 \int P\,dV \sim 3PV \tag{11}$$

in the case of an homogeneous star $E_{gr} \sim 3/5 M^2\,G/R$ from which we obtain the frequency oscillation ν

$$\nu \sim 0.21\alpha \sqrt{\frac{G\,M}{R^3}} \tag{12}$$

where $\alpha \sim 1$ is a numerical constant depending on the equation of the plasma forming the star. We parameterize the variation δQ of the quadrupole moment of the star in the following way: $\delta Q = \epsilon\,M\,R^2$. We have now the expression of $\ddot{Q}$:

$$\ddot{Q} = 4\,\pi^2\,\nu^2\,\epsilon\,M\,R^2 \sim 1.7\alpha^2\,\epsilon\frac{G\,M^2}{R} \tag{13}$$

By taking into account Eq.(10) and (8) we have

$$h \sim 3.4\alpha^2 \frac{\epsilon M G}{c^2 D} \mathcal{C} \tag{14}$$

were D is the distance of the source. The importance of the compactness factor is evident.
In a same way, the GW luminosity can be computed:

$$L \sim 3.3\,10^{-2}\,\alpha^6\,\epsilon^2\,c^5\,\frac{\mathcal{C}^5}{G} \tag{15}$$

Notice that the mass and the radius of the star are included in the compactness parameter $\mathcal{C}$.

We can compute the gravitational damping time τ and the quality factor $\mathcal{Q} = \tau\nu$: $\tau = E_k/L$ where $E_k = 1/2\,MR^2\,\epsilon^2\,(2\pi\,\nu)^2$ is the kinetic energy of the oscillation. From which

$$\tau \sim 13\,\frac{G\,M}{\mathcal{C}^4\,c^3\,\alpha^4}\,s \quad \mathcal{Q} \sim 2.6\,\mathcal{C}^{-\frac{5}{2}} \tag{16}$$

where $\mathcal{Q} = \nu\,\tau$ is the quality factor of the oscillation. An analogous computation can be performed for the rotating stars. An axi-symmetric rotating star does not radiates ($\ddot{Q} = 0$!) consequently also for a rotating star a deformation ϵ which breakes the symmetry of the star is required. Replace the value of ν in the Eq.12, by $\frac{\Omega}{2\,\pi}$ were $\frac{\Omega}{2\,\pi}$ is the frequency of the rotation of the star and introduce a new parameter $\mu = \Omega/\,\Omega_K$ where $\Omega_K = \sqrt{GM/R^3}$ is the Keplerian angular velocity,i.e. the maximum angular velocity at which the star can rotate without loosing matter at the equator.
Conclusions: only compact objets, i.e.black holes and neutron stars are potential sources of GW detectable by ground based experiments, in fact no other star-like objects exist in the gap between the neutron stars and the white dwarf ($6\,10^{-4} \le \mathcal{C} \le .3$). The reason of this gap will be explained in a following section. Note that the lower VIRGO bond ($\nu_l = 10\ Hz$) will exclude the detection of BH with a Schwarzschild radius R_s larger then $\sim\ 3\,10^4$ km ($R_s > c/\nu_l$) or equivalently, with a mass greather than 10^4 solar masses. LISA space experiment will allow to explore lower frequencies.

Coherence of the motion

The motion of the source (oscillation, rotation,...) must be coherent: in fact the variation of the quadrupole $\ddot{Q}^{ik}$ is given by

$$\ddot{Q}^{iK} = \int \ddot{n}(x,t)(x^i\,x^k - \frac{r^2}{3}\delta^{ik})\,dV\,, \quad r^2 = (x^1)^2 + (x^2)^2 + (x^3)^2 \tag{17}$$

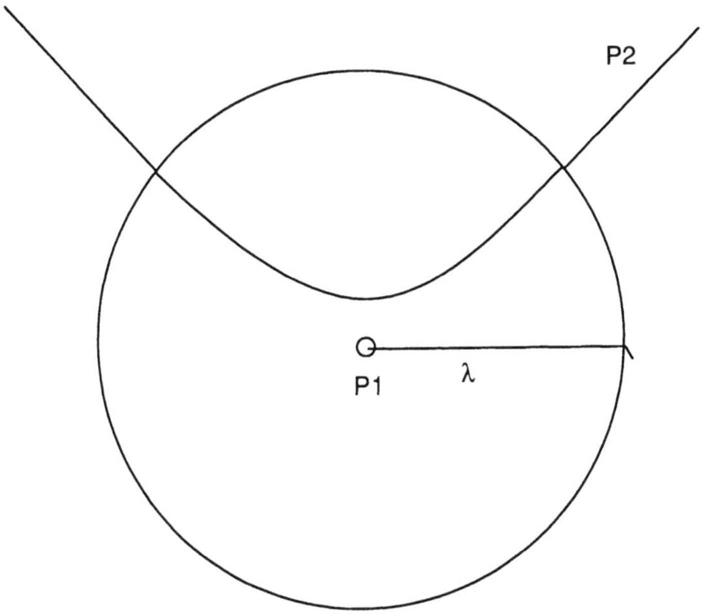

Figure 1. The proton P_2 is scattered by P_1 It radiates only in the Debye sphere the radius of which is λ

in absence of coherence of the motion, this quantity vanishes although $\int \ddot{n}(x,t)^2\, dV$ is a non zero quantity. That does not mean that the gravitational radiation is completely inhibited, it is only reduced by a factor depending on the coherence of the motion.
An example is the best way to clarify this concept.
Consider the Sun, it radiates electro-magnetic energy ($3.82\, 10^{33}\, erg\, s^{-1}$) but it radiates gravitational energy too: in fact during collision of a proton with an other proton their quadrupole moment varies and consequently they radiate GW. In what follows, we shall estimate the total (high frequency) gravitational radiation.

Consider a proton P_1 of the plasma forming the star, colliding with an other proton P_2. P_2 is scattered by the Coulombian field generated by P_1. (see Fig.1)

Because of the screening effect due to the electrons, the interaction potential U is

$$U = \frac{\exp - \frac{r}{\lambda}}{r} \tag{18}$$

where λ is the Debye length

$$\lambda = \sqrt{\frac{k\,T}{4\,\pi\,e^2\,n_e}} \tag{19}$$

and k T e n_e are respectively the Boltzmann constant $k = 1.38\,10^{-16}\,erg\,/K^\circ$, T the temperature of the plasma, the electron charge ($e = 4.8\,10^{-10}$ ESU) and n_e the electronic density.
An order of magnitude of λ in stellar conditions is $\lambda \sim 3\,10^{-8}\sqrt{\frac{T_7}{ne/g\,cm^{-3}}}\;cm$
We shall need also the mean velocity of the protons V_p

$$V_p \sim V_{th} \sim \sqrt{\frac{k\,T}{m_p}} \tag{20}$$

where V_{th} is the thermal velocity.(In stellar conditions $V_{th} \sim 3\,10^7 T^{1/2}\,cm\,s^{-1}$)
Because the proton P_2 experiences the action of the proton P_1 only when it is inside of Debye sphere, it radiates GW only during the time τ spent inside this volume.

$$\tau = \frac{\lambda}{V_{th}} = \sqrt{\frac{m_p}{4\,\pi\,e^2\,n_e}} \tag{21}$$

The quadrupole moment and its variation of the system P_1, P_2 are

$$Q \sim m_p^2\,\lambda^2, \quad \frac{d^3Q}{dt^3} \sim m_p^2\,\frac{\lambda^2}{\tau^3} \tag{22}$$

by taking into account the Eq.(15) we have the GW energy E_s radiated during the scattering of P_2

$$E_s \sim \frac{1}{5}\frac{G}{c^5}\left(\frac{d^3Q}{dt^3}\right)^2 \tau = \frac{1}{5}\,G\,\frac{m_p^2}{\lambda}\,\frac{V_{th}^5}{c^5} \tag{23}$$

For stellar conditions $E_s \sim 10^{-63}$ erg.
The total power radiated by each proton is $L_p = E_s$ times the number of particle impinging the Debye sphere per second:

$$L_p = \frac{dE_p}{dt} \sim E_s\,4\pi\,\lambda^2\,n_p\,V_{th} \tag{24}$$

where n_p is the proton density that we have supposed to be equal to the electron density.
By taking into account Eq.(19),Eq.(20) ,Eq.(23)we have

$$L_p \sim \frac{2}{5}\frac{\pi^{1/2}}{c^5}\frac{G}{m_p\, e}\sqrt{(kT)^7\, n_p} \sim 1.2\,10^{-46}T^{7/2}\left(\frac{n}{g/cm^3}\right)^{1/2} erg\, s^{-1} \quad (25)$$

The total GW luminosity is given $N_p\, L_p$ where N_p is the total number of protons ($N_p = M/m_p \sim 1.2\,10^{57} M_\odot$) from which

$$L \sim \frac{2}{5}\frac{\pi^{1/2}}{c^5}\frac{GM}{m_p^2\, e}\sqrt{(kT)^7\, n_p} \quad (26)$$

or

$$L \sim 1.4\,10^{11}T^{7/2}\left(\frac{n}{g/cm^3}\right)^{1/2} M_\odot\, erg\, s^{-1} \quad (27)$$

the power emitted is about $10kW$! In the case of a very hot NS ($T = 10^{10}K^\circ$) and $n_p \sim 10^{14} g\, cm^{-3}$, the efficiency is higher but still low: $L \sim 4\,10^{28}\, erg\, s$ [d]
It is worth to note that in the expressions computed in the previous section, giving the GW luminosity, the mass M of the star appears with a power of two. (cf.Eq.(6)). This fact is typical of the coherent radiation for which the strain h emitted by each proton adds each other. In the Eq.(26), the luminosity is proportional to the mass of the star. This fact is typical of the random radiation: only the square of h of each proton adds each other.
Finally it is interesting to compute the cooling time τ_c of the star and to compare it with the relaxation time τ_r of the oscillations computed in the previous section. We have

$$\tau_c = \frac{E_{th}}{L} \sim \frac{2}{5}\frac{\pi^{1/2}}{c^5}\frac{GM}{m_p\, e}\sqrt{(kT)^5\, n_p} \quad (28)$$

where E_{th} is the thermal energy of the star. We obtain $\tau_c = 10^{36} s$ for the sun, and $\tau_c = 10^{17}\, s$ for the NS previously considered. (Remember that the age of the Universe is *only* $\sim 510^{17}\, s$

Conclusions

Only B.H. and NS are possible interesting source of gravitation waves under the condition that their motion are coherent. Possible coherent motions are binary systems, oscillating isolated NS and BH,rotating NS.

[d]The value by Eq. (26) must be considered as an upper limit of the gravitational luminosity: in fact at this temperature protons and neutrons are partially degenerated, moreover electrons are relativistic and Debye length is no longer given by Eq.(18)

2 Physics of black holes

The physics of an isolated BH is quite simple: gravitational mass, total angular momentum and total electric charge are the only informations that can go out from the interior of the horizon, therefore an infinity of collapsing systems can be the origin of the same BH. Moreover, a BH is axi- symmetric around its rotation axis and the so called "no hair theorem" states the an uncharged BH have zero magnetic field. The surface of the horizon has an electric resistance equal to the impedance of the vacuum: $R = 377\,\Omega$ (the resistance of a spherical shell does not depend on its radius). When the BH is not alone in space, for example when plasma falls on its horizon, the physics is much more complicated but this case is not interesting for our purposes.

3 Physics of neutron stars

If a BH is a geometrical object, a NS is a physical object, and its physics is much more complicate.
Let us describe briefly how the equilibrium configuration of a star like object can be computed: In order to be in equilibrium, the gradient of the pressure $gradP$ must compensate gravitational force $n\, grad\, U$:

$$-grad\, U = \frac{1}{n}\, grad\, P \tag{29}$$

The Equation of State (EOS) gives the relationship between the pressure P and other parameters characterizing the fluid as the baryonic density n, the temperature T the chemical potentials μ_1, $\mu_2, \ldots$ Among all the EOS the *barotropic* EOS $P = P(n)$ plays a major role in the theory of the equilibrium configurations of a star. By introducing the enthalpy $\epsilon = \int_0^{\mathrm{n}} 1/\mathrm{n}\, dP$ the Eq.(29) can by written as

$$\epsilon + U = Const \tag{30}$$

this equation together the Poisson equation

$$\Delta U = 4\pi\, G\, \mathrm{n}(\epsilon) \tag{31}$$

allows us to compute (numerically) the equilibrium configuration of the star [e].
Barotropic EOS are general enough to describe star equilibrium configurations

[e]It easy to see that in general it is not possible to write the hydrostatic Eqs.(29) in the integral form given by the Eq.(30), in fact the integrability conditions of the system of 3 equations (29) are that the acceleration $1/\mathrm{n}\, gradP(\mathrm{n}, T, \mu_1, \mu_2, \ldots)$ can be written as a gradient of a some thermodynamical potential. As it was shown barotropic EOS satisfies this condition.

in most interesting cases. In what follows we shall describe the behavior of the matter at high density.
Let us consider a gas (for example hydrogen) at temperature $T = 10^5 K$ At this temperature and a low pressure, the gas can be considered mono-atomic. If its density is low enough,its EOS is $P = \mathrm{n}/m_p KT = n\,kT$ where m_p is the proton mass and n is the numerical density of the gas.(we neglect the electron mass). Let us describe the qualitative behavior when the gas is more and more compressed at constant temperature T. From a qualitative point of view noting happens until when the particle density n is lower than $1/R_a^3 \sim 6.7\,10^{24} cm^{-3}$ ($n = 11\,g\,cm^{-3}$) where R_a is the radius of the hydrogen atom. At this density the atoms start touch each other, the EOS becomes stiffer because the pressure is not longer generated by the thermal motion of the atoms, but by the mutual interaction of the electrons of contiguous atoms. The fluid behaves as a liquid, but the distinction between gas and liquid is not pertinent for our purpose. At this point the electrons of hydrogen atoms start to live the nuclei and the vacuum existing among the atoms is filled more and more when the density increases. At the above temperature electrons are completely degenerated. For this density, the electrons behave as perfect degenerated gas: in fact the electrostatic energy due to the mutual interaction of electrons is less than the Fermi energy. In other words,the degenerated gas, becomes more and more "perfect" when the density rises. [f] The most important contribution to the pressure of the gas is given by degenerated non relativistic electrons,therefore the EOS is

$$P = \frac{\hbar^2}{20\,m_e}\left(\frac{3}{\pi}\right)^{2/3} n_e^{5/3} \tag{33}$$

i.e. the pressure depends only on the density. A EOS of this type is called *polytropic* of adiabatic index $\gamma = 5/3$.
When the density reaches a value for which the corresponding Fermi energy $\mu \geq .2 m_e c^2$, ($n_e = 1.5\,10^{29}\,cm^{-3}$ $\mathrm{n} = 9.04\,10^4\,g\,cm^{-3}$) electrons becomes relativistic. It exists a simple asymptotic expression for EOS in the ultra-

---

[f] The electrostatic energy is given by $E_{el} = 1/2e^2/r = e^2\,n_e^{1/3}$ where $e = 4.8\,10^{-10}$ ESU is the electronic charge. The relationship between the Fermi energy μ and the electronic density ne is given by

$$\mu = (3\pi)^{2/3}\hbar^2/(2\,m_e)n_e^{2/3} = 5.8\,10^{-27}\,n_e^{2/3}\,erg \tag{32}$$

(where m_e (electron mass) $=9.1\,10^{-28}\,g$, $2\pi\,\hbar$ is the Planck constant $= 6.62\,10^{-27} erg\,s$. The electron density n_e at which the Fermi energy is equal or larger then to the electrostatic energy is easily computed:$n_{ei} = 6.5\,10^{23}\ cm^{-3}$. Note that under the hypothesis of the neutrality of the plasma, protons and electron density n_p, n_e must be equal.

relativistic case:

$$P = \frac{\pi}{4}\sqrt{3}\,\hbar\, c\, n_e^{4/3} \tag{34}$$

We can say that the EOS is a polytropic with an adiabatic index γ varying between 4/3 and 5/3, although the exact expression is more complicate.
The behavior of EOS changes dramatically when the Fermi energy of the electrons reaches the threshold value $E_t = (m_n - m_p)\,c^2 = 2.07\,10^{-6}\,erg$ or $1.29\,Mev$. (The corresponding electron or proton density is $n_e = n_p = 2.26\,10^{31}\,cm^{-3}$ or n $= 3.77\,10^{7}\,g\,cm^{-3}$) in fact at this energy, the capture of electrons by protons is energetically more convenient, the electrons density falls down brutally, neutrons start to be present in the plasma, pressure stays constant until when the Fermi pressure of the neutrons becomes dominant. Note the typical behavior of a second order phase transition. [g] (Fig.(2), (3))

Actually at this density, the EOS is softer then what is shown in Fig. (3): In fact once neutrons are formed, nuclear reactions form other nuclei (i.e. deuterium). This nuclear soup is at the chemical equilibrium. When the barionic n density reaches the values close the nuclear density ($n_{nuc} \sim 10^{14}\ g\,cm^{-3}$) the nuclei decay to protons and neutrons. At nuclear density that touch each other, strong interactions forces start to play the dominant role in generating the pressure and the EOS can be described roughly by a polytropic of index $\gamma = 2 \div 2.2$
For higher densities, typical of the interior of NS,($n \geq n_{nuc}$), other nuclear reactions produce non symmetric nuclei (stable ordinary nuclei are called "*symmetric*" because the numbers of protons and neutrons are equal), the computation of the EOS becomes less reliable: in fact for symmetric nuclear matter experimental cross sections (neutron-ion, proton-ions) allow us to compute a reliable EOS. The lack of experimental cross section for non-symmetric nuclei requires N-bodies technique in order to obtain "reliable" EOS. This is the main reason why we can find more then twenty EOS in the literature. They all agree at densities lower then the nuclear density, and they start to diverge at higher densities. In spite of this difficulty, progress in the last years were so important that a consensus exists in the nuclear physics community, that the more recent EOS, in which two-bodies interactions and relativistic effects are taken into account, describe correctly the nuclear

[g]This behavior is due to the creation of neutron in the proton-electron plasma. It looks as phase transition but it is not: in fact the plateau showed in the fig. (3) it is not horizontal, and the derivative of the pressure with respect to the density is continuous.

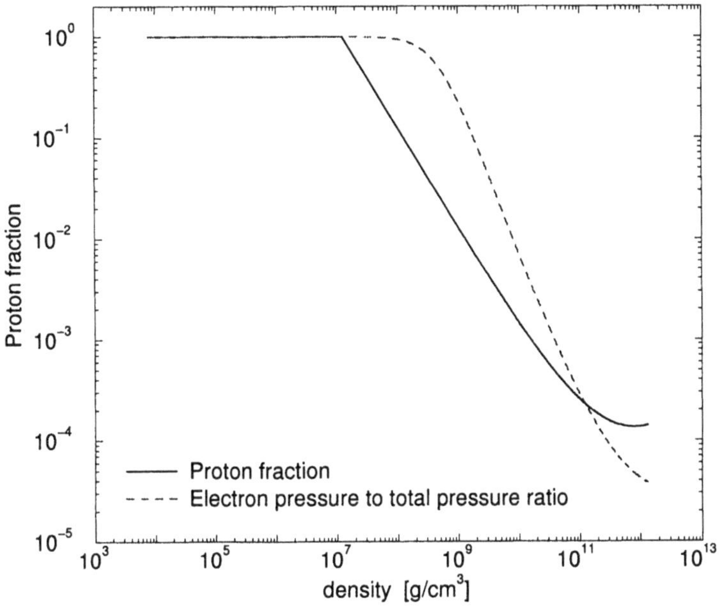

Figure 2. Proton/Baryon and (Electron pressure)/(Total pressure) ratio versus baryonic density

matter behavior at least in the range of density existing in NS ($n \leq 10\, n_{nuc}$). It is worth to note that the above scenario is highly speculative, in real life, the initial plasma is composed by a mixture of ions, (magnesium, silicon, carbon, oxygen, iron) depending on the evolutionary track of the star. Nevertheless the actual behavior of the plasma is very similar to the one described above. At high densities, the EOS depends slightly on the initial composition of the plasma. The main difference is the critical density at which the neutrons form. For example critical densities $4.45\, 10^{10}\, g\, cm^{-3}$ and $1.15\, 10^{9}\, g\, cm^{-3}$ respectively for carbon and iron nuclei [9]. In spite of these differences, the chosen academic example contains all the most pertinent features of the nuclear EOS.

Now it appears clearly why there is a gap in the compactness factor between the white dwarfs and NS: if the central density is larger then the critical density at which the phase transition happens, the EOS is much softer and therefore the star is more compact.

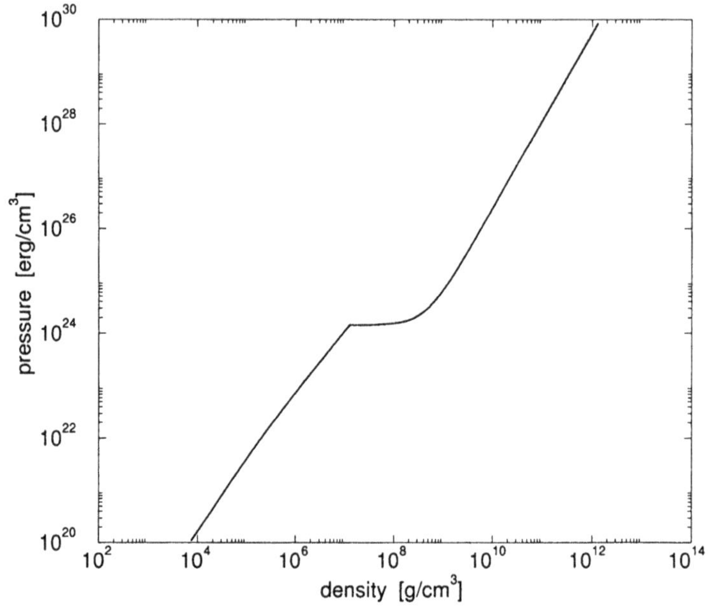

Figure 3. Total pressure versus barionic density. The slope of the plot before the plateau is about 4/3 because electrons are relativistic. At the right of the plateau the pressure is due to the non relativistic neutrons therefore the slope is 5/3.

Chandrasekhar Mass Limit

As it was already said, NS are compact objects, therefore relativistic effects are important. The most dramatic difference between a Newtonian star and a general-relativistic one is that a Newtonian star can have any mass, on the contrary a general-relativistic star has necessarily a mass below a limit: the maximum mass.

There are two relativistic effects that are the origin of the existence of a critical mass: the first one is due to the special relativity that states that actions, therefore acoustic waves can not propagate with velocity c_s larger than the light velocity c. Consequently the EOS can not be indefinitely stiff. The second reason is due to the structure of Einstein equations: the sources of gravitational potentials contain the energy density and the pressure of the plasma (cf. Eq.(5)). There exists a critical density at the center of a NS at which an ulterior compression of the star, increases the gravitation field faster than the pressure: this is due, as it was already said,to the fact the EOS cannot be indefinitely stiff and the pressure acts as source of the gravitational field: in other words, more the pressure increase more the gravitational field becomes strong. The critical mass of a non rotating NS depends slightly on the EOS, and is about $2\,M_{\odot}$. Less relativistic objects, such white dwarfs show the same behavior, their critical mass, or Chandrasekhar mass, is about $1.4\,M_{\odot}$. This value is slightly dependent on EOS.

4 Origin and Structure of Neutron Stars

General considerations

There is a large consensus that NS are formed by the core collapse of a supernova. A massive star ($8M_{\odot} \leq M \leq 25\,M_{\odot}$), during his life, ejects matter from outer layers,via stellar wind. In few millions of years it is out of fuel, the ashes of the nuclear burning consist of various elements whose distribution inside of the star depends on the density at which the combustion takes place. The central region, with a mass of $\sim 1.5\,M_{\odot}$ is filled with the most stable elements, iron and nickel. The so-called iron core (the $1.5\,M_{\odot}$ central core) has a structure very similar to the one of a white dwarf. At the end of the combustion, its mass exceeds the Chandrasekhar mass and it starts to collapse toward a hot NS: the new equilibrium configuration. The cooling time is very short: one minute is enough to cool the NS down to $10^{10}\,K$: a new NS is born.

The structure of a NS is quite complex. Fig. (4) shows a cut of a NS. When the temperature of a NS is low enough ($T \leq 10^{10}\,K^{\circ}$) the surface layers can solidify and a nuclear lattice is formed [10]. In the solid and liquid phase neutrons, thanks to the nuclear forces can form pairs that behave as bosons. Transition from an ordinary liquid to a superfluid phase can happen if the temperature of the star is below a critical temperature T_c, the corresponding energy being equal to the energy gap between the free neutrons and neutron pairs. This gap is quite high ($\sim 1\,MeV$) because strong interactions are the origin of pairing force. Consequently the critical temperature T_c at which a NS can have a superfluid phase is about $T_c \sim 10^{10}\,K$. The superfluidity of the neutron matter is very similar to the ${H_e}^3$ one: The coupling of fermions ${H_e}^3$ atoms, forms a pair of quasi-particles behaving as bosons. The only difference is the strength of the coupling: superfluid ${H_e}^3$ has a critical temperature of $\sim 1.9\,10^{-3}K$ at $10\,bar$ pressure. A so low transition temperature is due to the weakness of the magnetic dipole

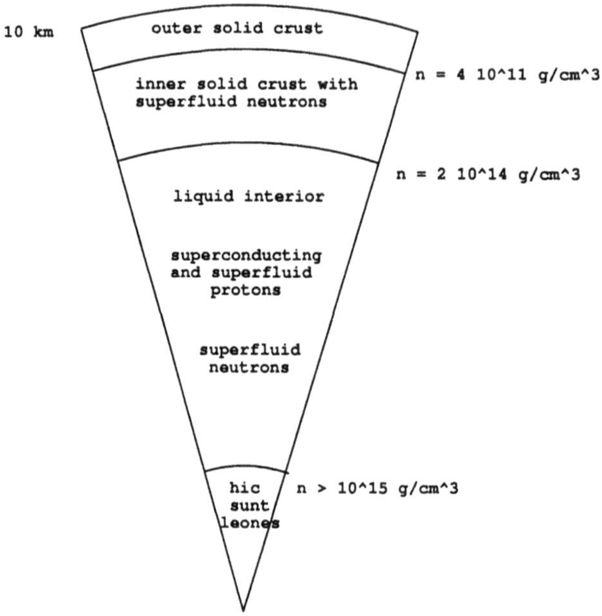

Figure 4. Cross section of a NS. In the solid and liquid core neutrons are superfluid. In the liquid phase protons can become superconductor.

interaction between two $H_e{}^3$ atoms.

Neutron Stars and Pulsar

The discovery of pulsars in 1967, confirmed the theoretical prediction of the existence of NS. Few months after the observation of the first pulsar, it was clear that the only explanation of the behavior of such a peculiar radio-source was a rotating magnetized NS. The strength of the magnetic field $B \sim 10^{12}\, gauss$, was estimated thanks the observed slowing down of the rotation period P. Observed cyclotron lines in the X-rays spectrum of few binary X-rays pulsars were a beautiful confirmation of the existence of such a huge magnetic field. The magnetic field is supposed to be the fossil one generated during the life of the progenitor star and magnified by compression during the collapse [h].

[h]The magnification of the magnetic field is due to the high electrical conductivity of the plasma forming the pre-supernova core and the NS: Under the zero resistivity hypothesis,

Finally, observed suddenly variations of the rotation period, (glitches) were considered as a proof of the existence of a solid (crust) zone in the NS.
The main problems that be solved on the pulsars models are:
- The structure of the magnetosphere.
- The structure of the magnetic field and its stability inside of the NS.
- The rheology of the nuclear matter of rotating NS.

In these lectures we are interested only in the last two points: in fact, as we shall see in a next section, the structure of the magnetic field can play a major role on the deformation of a rotating NS, a necessary condition to have an efficient GW emitter .
We shall see also that possible instabilities of a rotating NS depend on the transport coefficients of nuclear matter.

Rotating NS.
As it was already said, NS are relativistic objects, therefore all computations in modeling their structure or in describing the motion must be done in the General Relativity frame. The structure of an axi-symmetric rotating NS can be easily computed by solving a system of 4 partial differential equations (PDE). Fig.(5) shows the shape and the structure of a rotating $1.4\, M_{\odot}$ NS.

Rigid rotation is a simple and intuitive concept, unfortunately superfluidity requires that $curl\, \vec{v} = 0$ and consequently superfluid state is incompatible with rigid motion because a superfluid flow must derive from a potential ($\vec{v} = grad\, S$). Moreover, if the superfluid can be considered incompressible the potential S must be harmonic ($\Delta S = 0$) [i]

---

the magnetic flux across a surface defined by a closed line co-moving with the plasma is conserved (see Appendix 1). During the life of the star, the equatorial surface S of the core shrinks from, say, $10^6\, km$ to the radius of the NS ($10\, km$), the corresponding surface is reduced by a factor of 10^{10}, and because of the conservation of the magnetic flux $\Phi \sim B\, S$, B is magnified by the same factor.

[i]The superfluidity is characterized by the existence of a macroscopic quantum state due to the Bose condensation: a macroscopic number of spin integer (bosons) particles condensate in the same ground quantum state. In nuclear matter, the attractive part of the strong interaction forces can form pairs of neutrons and pairs of protons. These pairs behave as bosons and can condensate in the space of momenta. The main idea underlying the theory of the supefluidity and the theory of the superconductivity is that this macroscopic quantum state is described by a wave function Ψ. Let be $\Psi = \psi \exp(-iS)$ the wave function, where ψ and S are respectively the amplitude and the phase. The normalization factor ψ is chosen in a such a way that $\psi\psi^*$ is equal to the average density of superfluid nuclei. Both functions ψ and S are supposed to be constant, or vary very slowly. The value of the momentum density $\vec{p}$ is $\vec{p} = -i\hbar\psi\psi^* \vec{\nabla} S$. Moreover, under the approximation that the fluid is incompressible,

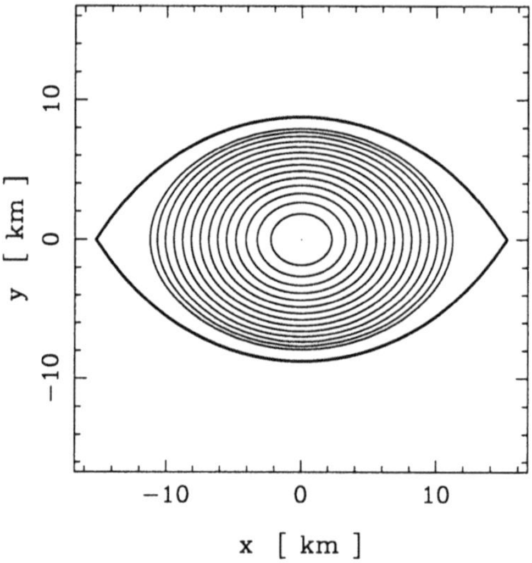

Figure 5. Isodensity lines of a $1.4\,M_{\odot}$ NS rotating rigidly Note that the star is quite homogeneous: the baryonic density decreases quite brutally near the surface. The cusp on the equatorial plane is due to the Keplerian angular velocity Ω_K For this value of Ω_K, the centrifugal forces acting on a matter at the surface of the star on an equatorial plane, compensates exactly the gravitational force.

An interesting harmonic function is (in cylindrical coordinates ρ, θ, z) $S = 2\pi i l\theta$ (l is an integer number in order to have a single-valued wave function), from which we can compute the azimuthal component of the momentum of one pair $p_\theta = hm/\rho$. This solution describes a cylindrical vortex. The path integral along a closed line is quantified and it depends how the integration path encircle the singularity line $\rho = 0$. Let us take a circular path of radius R orthogonal to the z axis and centered in $\rho = 0$. The path integral $\oint \vec{p} \cdot \vec{dl} = p_\theta \rho d\theta$ gives the angular momentum L_p of the pairs forming the superfluid phase. Its value is $L_p = hl$. For $l = 1$, $L_p = h = 6.67\,10^{-27}\,g\,cm^2\,s$. The conclusion is that an averaged superfluid motion can mimic the macroscopic rotation of the fluid. In the present case (rigid rotation) superfluid vortices must be parallel with the rotation axis and have an uniform space distribution.

S must be an harmonic function ($\triangle S = 0$) [11].

Consider now two contiguous vortices rotating in the same sense, the velocity of the fluid in a point of the segment joining the centers of the vortices is smaller then the one computed at the symmetric point "outside" the two vortices. Because the Bernoulli theorem, the pressure of the fluid inside the vortices is larger then the one outside the vortices. The net result is a repulsive forces F_v between the vortexes. This force is proportional to the square of the velocity, consequently $F_v \propto 1/\rho^2$ is a long range force. Vortices will form a crystal in order to minimize their potential energy. [j]

The topology implies that a vortex either forms a ring or starts and ends where a discontinuity in the fluid is present. Closed loops does not contribute to the total angular momentum, consequently open vortices starting and ending on the crust are the source of the angular momentum. Moreover the vortexes are pinned on the crust. The physics of the pinning is badly understood.

It is easy to compute the vortex line density in a rotating NS of mass M. Let be Ω its angular velocity, its total angular momentum is $L = 2\pi I\Omega$. The NS contains $N_n \sim 1.2\,10^{57}\,M/M_\odot$ nuclei a pair of which has $L = h$ (the case $m = 1$ is the most energetically favorable). It follows that $N_v = 2I\Omega/(hN_n) \sim 1.5\,10^{15}\,(\Omega/Hz)(M/M_\odot)$ are necessary to have the required angular momentum. The average distance λ_v between two vortices is $\lambda_v = \sqrt{\pi R^2/N_v}, \sim 2.6\,10^{-8}\,\sqrt{(M/M_\odot)(\Omega/Hz)}\,cm$. Note that the distance between two vortices is orders of magnitude larger than the distant between two nuclei $\sim 10^{-14}\,cm$. In spite of this fact, the interaction energy between two vortices is important, because, as already said, of the long range hydrodynamical forces. Finally, the vortex lattice has its own oscillations, called Tkachenko waves, that can have some importance in the GW radiation emission of rotating NS

Magnetized Neutron stars

In presence of magnetic field, the hydrostatic equations Eq. (29) must be modified by tacking into account the magnetic pressure:

$$grad\,U + grad\,\epsilon = -\frac{1}{n}\,(curl\,B) \wedge B \tag{35}$$

The integrability conditions require that the extra term (Lorentz force) on the R.H.S of the Eq.(35) to be a gradient of a potential M. This condition can be easily fulfilled in the case of an axi-symmetric star with a poloydal magnetic

[j] A lattice of vortices was observed in rotating superfluid H_e^4 A picture of this lattice can be found in the already quoted Tilley bock [11].

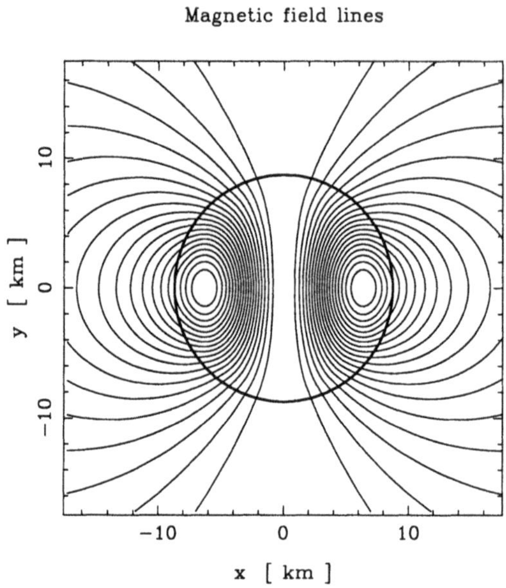

Figure 6. Magnetic field configuration in the case of an high conductivity non superconductor core [13]: the function F is a constant. Note that this macroscopic magnetic configuration could be the result of a averaged ensemble of a type 2 supraconductor fluxoides

field. In this case, we can choose a vector potential $\vec{A}$ of spherical components $A_\rho = A_\theta = 0$, $A_\phi(\rho, \theta)$ to describe the magnetic field $\vec{B} = curl\,\vec{A}$. It is easy to show that if A_ϕ satisfies to the Grad-Shafranov equation

$$\Delta A_\phi - 2A_\phi/r^2 = n\,F(A_\phi) \tag{36}$$

the integrability conditions are fulfilled. Here F is an arbitrary function depending on the evolutionary track of the magnetic field. Fig. (6) and Fig. (7) show two examples of the magnetic field configurations [12].

The problem of stability of the above magnetic configurations is not yet solved, it is highly plausible that stable magnetic configurations does not exist in a perfect fluid. The observed magnetic field at the surface of the pulsars shows that a stabilization mechanism must exist. The crust can stabilize the magnetic field because the magnetic field is frozen in a solid surface.

As it was already said, nuclear matter behaves as a superconductor: if the temperature of the nuclear matter is less than a critical temperature $T_c \sim 10^{10}\,K$, the nuclear matter expels the magnetic field. In nuclear matter,

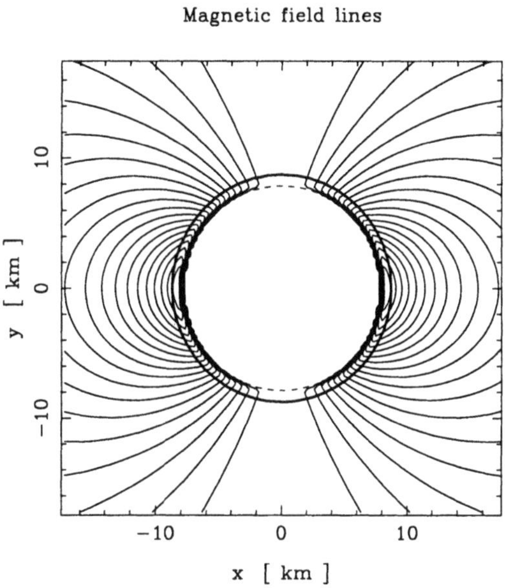

Figure 7. Magnetic field configuration in the case of type 1 superconductor core: the magnetic field is frozen in the outer normal crust of the star [13].

the origin of the superconductivity is due to the pairing of protons. A pair of protons behaves as a boson and a macroscopic quantum state is formed by boson condensation, consequently protons can be superconductor and superfluid at the same times. Actually, from laboratory experiments we know that there exists two kind of superconductor : the type 1 and type 2. Type 1 superconductor is characterized by the fact that magnetic field is completely expelled, during the transition toward the superconducting phase, from the bulk of the sampling. In the superconductor of type 2, the magnetic field is concentrated in small regions called fluxoides, characterized by quantized magnetic flux $\Phi_{mic} = lch/(2e)$ where l is an integer number(the factor of two at the denominator is due to the pairing of two fermions). [k]

The magnetic field is concentrated in a small cylinder whose radius is

[k]The only difference between the neutronic superfluidity and the protonic one, is the charge of the pairs. Therefore we must replace the conjugate momentum $i\hbar\nabla$ by $i\hbar\nabla + 2e\vec{A}/c$ If S is the phases of the wave function the current density is $\vec{J} = e/m_p(\hbar\nabla S + 2e\vec{A})\psi\psi^*$ where $\psi\psi^*$ is the pair density n_p. By taking the curl of the above equation and by bearing in

the penetration length λ_p. It is easy to compute the number of fluxoides N_F required to generate an averaged flux $B\pi R^2$ across an equatorial surface of the NS. We have $N_f = \pi BR^2/\Phi_{mic} = 1.4\,10^{31}B_{12}(\frac{R}{10\,km})^2$, from which they density $N_F/(\pi R^2) = 4.6\,10^{18}B_{12}(\frac{R}{10\,km})^{-2}\,cm^{-2}$ and they average distance $d_F = 4.6\,10^{-10}B_{12}^{1/2}(\frac{R}{10\,km})\,cm$. Therefore the distance between two fluxoides is much greater than the region where the magnetic field is different from zero and their magnetic interaction can be neglected. From the energetic point of view, the increasing of energy due to the deformation of the magnetic field is compensated by the binding energy of the proton-proton paring. From what is said, it is easy to understand that there exists a critical magnetic field B_c above of which the transition toward the superconductor state is forbidden.

Rheology

If the problem of the equilibrium configuration is well settled and turns to be quite simple to handle, the problem of the flow inside of a NS is much harder. In fact in a rotating and magnetized supercondutive type 2 superfluid NS the situation looks like nuclear matter in which a bunch of fluxoides and a bunch of vortices cross each other and interact when nuclear matter flows. Moreover the vortices and fluxoides are pinned in the solid part of the star. The final goal would be to find some macroscopic equation of motion, able to describe in a reliable way, the flow of a so messy nuclear matter. Rheology is the branch of the physics devoted to this kind of problems. In more detail, consider the Euler equations

$$\partial_t \vec{v} + \vec{v}\cdot\nabla\vec{v} + \frac{1}{\mathrm{n}}grad\,P + grad\,U = 0 \tag{38}$$

These equations, that describe the conservation of the momentum, can be easily generalized to the case of non isotropic fluid. The dissipative terms are the main problem. The basic idea, is to modify the pressure tensor by adding a stress tensor σ_{ij} depending on the velocity of the fluid. The total stress tensor P_{ij} is $P_{ij} = P\delta_{ij} + \sigma_{ij}$ where P is the ordinary pressure and δ_{ij} id the Kronecker symbol. Newtonian fluid are the one for which the stress tensor

---

mind that $curl\,\vec{H} = 4\pi\vec{J}/c$, ($\vec{B} = \vec{H}$) we obtain the London equation

$$\triangle\vec{B} = -8\pi e^2 n_p/(mc^2)\vec{B} \tag{37}$$

The requirement that the wave function is single-valued implies that the magnetic flux is quantized. A solution of the London equation shows that it exists a penetration length $\lambda_p = c\sqrt{m/(8\pi e^2 n_p)} = 2\,10^{-12}(\frac{n_p}{10^{14}\,gcm^{-3}})^{-2}$,i.e. the magnetic field can penetrate at this depth from the outside of the sample before vanishing.

σ_{ij} is an isotropic tensor depending on the first derivatives of the velocity:

$$\sigma_{ij} = \mu(\nabla_i v_j + \nabla_j v_i - \frac{2}{3}\nabla_l v^l \,\delta_{ij}) + \zeta\delta_{ij}\nabla_l v^l \tag{39}$$

μ and ζ are positive functions depending on the parameters of the fluid. μ is called the shear viscosity and ζ the bulk viscosity. For quite general fluids ζ vanishes. Nuclear matter has no vanishing bulk viscosity: at temperature larger then $T = 10^{10}\,K$ the bulk viscosity dominates. [l]

Non-Newtonian fluid are more complicated, and the skill of the physicist is to find smart approximations. Honey is an example: it seems viscous but it behaves in different way from oil which is a Newtonian fluid: one drop of honey starts to strain from the spoon, and when it breaks the part straining from the spoon goes up. A such a behavior can be modeled with the visco-elastic approximation. An other example is the oil painting, a major technological improvement in the art. Oil painting has the interesting properties that its viscosity is very low when the stress is high (therefore it is easily spread out) and its viscosity is high when the stress is low, therefore the fresh painting does not strain. Opposite behavior: maïzena. This behavior can be described by introducing non linearities in the viscous term.
How to treat nuclear matter of a rotating suprerluid rotating star is an open problem. Until now, people use a perfect fluid approximation with a bulk and shear viscosity. This approximation is not satisfactory, and results obtained on the instabilities of the modes in NS. must be taken cautiously.

5 Sources of gravitational waves

As it was already said in the introduction,in order to avoid uninteresting repetitions, we summarize the different type of source. The reader will find more details in the already quoted paper ([1]). Only recent progress obtained either by observations or by modeling will be described in some detail.

5.1 Collapsing objects

Type 2 supernonvae (SN) were the first candidates as sources of GW. There exists a consensus that massive stars ($8\,M_\odot \leq M \leq 25\,M_\odot$) end their life as

[l]The origin of the bulk viscosity is the finite time needed to the nuclear reactions to restore the chemical equilibrium. When density n varies, the response of the pressure can be retarded. Let be τ this time delay. At the first order, a series expansion gives: $P(\mathrm{n}(t-\tau)) = P(\mathrm{n}(t) - d\,P/d\,\mathrm{n}D\,\mathrm{n}/dt\,\tau$. By taking in to account the continuity equation $D\,\mathrm{n}/dt + \mathrm{n}div\,\vec{v} = 0$, we obtain $P(\mathrm{n},(t-\tau)) = P(\mathrm{n}(t)) + \mathrm{n}\tau\,d\,P/d\,\mathrm{n}\,div\,\vec{v}$. It follows that the bulk viscosity coefficient ζ is $\zeta = \mathrm{n}\tau\,d\,P/d\mathrm{n}$

NS. Stars more massive then 25 $M_\odot$ form black holes. Recent observations of very bright γ rays bursts support strongly the hypothesis that B.H. can be formed at the end of the collapse of a very massive core (hypernovae).
Optimistic order of magnitude estimations of energy radiated ($10^{-2} M/M_\odot c^2$), during the core collapse have suggested the possibility to detect SN at Virgo cluster distances ($20\,Mpc$). Unfortunately numerical simulations of rotating collapsing cores have shown that GW energy radiated is 3 orders of magnitudes lower than the previous estimations (cf ([1]) and literature quoted there) [14]. There are few reasons that explain this discrepancy: consider at first the supernovae that form NS: the EOS of the pre-collapse core is soft, consequently the compactness of core is low and the deformation due to the rotation is important where the density is low. When the core reaches the nuclear density, the nuclear matter is very hot ($T \sim 10^{11}\,K$), consequently, the bouncing or oscillating core has low compactness and the energy losses via GW are small with respect to neutrino losses. Remember that a hot NS have a radius of $100\,km$ and its compactness factor $\mathcal{C} \sim 0.03$ (cf. Eq.(6)). Finally, as it was explained in the already quoted article [1] ,there exists a subtle compensation between the variation of the inertia momentum and the deformation parameter.
All these factors go in the wrong direction, and the GW emission is highly reduced.
What it was said above still holds for massive supernovae that form a BH. The main difference being the phase of BH formation that could be in principle an efficient GW emitter. Unfortunately, this is not the case: numerical simulations show that the the efficiency is very low indeed. ([15]) The reason is that the the radiation is swallowed up by the BH.

5.2 Rotating Neutron Stars

In the recent years a lot of work was performed on rotating NS. As it was already said,the breaking of the axi-symmetry of a NS is required in order to have a GW emitter. There are two type of mechanisms able to break the symmetry of a NS , the first one is based on the anisotropy of the matter forming the NS. This anisotropy can be due to the magnetic field, or to the solid part of the NS. (For example the existence of mountains) The magnetic field is the more promising mechanism. In fact its existence is proved, its intensity is known for a few NS, and its dipole momentum is well estimated for many pulsars. Among the known pulsars, the Crab is the most promising source. The deformation of the Crab, and consequently the amplitude of the GW, depends on the structure of the magnetic field. With the structure showed in

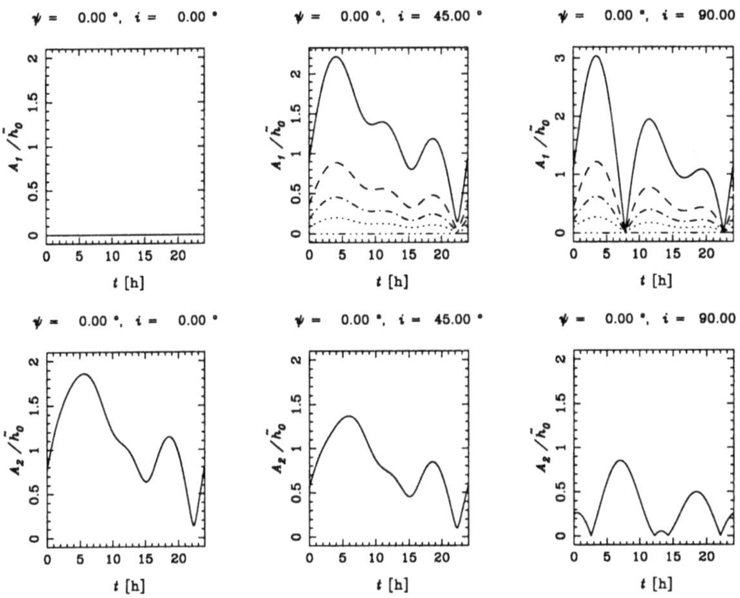

Figure 8.
Relative amplitude of the signal at the frequency Ω (A_1, top) and 2Ω (A_2, bottom) as a function of the local UT time t (assuming a vanishing local sidereal time at 0 h UT), for the specific case of the VIRGO detector and the Crab pulsar. The polarization angle is $\psi = 0^o$. Each plot corresponds to a given value of the inclination angle i. Various lines in the same plot correspond to various values for the angle α between the magnetic axis and the rotation axis: solid line: $\alpha = 15.00^o$, dashed line: $\alpha = 33.75^o$, dot-dash-dot-dash line: $\alpha = 52.50^o$, dotted line: $\alpha = 71.25^o$, dash-dot-dot-dot line: $\alpha = 90.00^o$ [13].

Fig. (5) and for a magnetic dipole moment equal to the one estimated for the Crab Pulsar, 4 years of observation should be enough to detect the continuous GW signal of the Crab pulsar.

The recent discovery of the X-ray binary source SAX J 1808.4-3658 showing a pulsating emission at the frequency of $400\,Hz$ shows that accreeting NS stars can be also a promising source of GW. All the deformation mechanisms working for isolated pulsar are worth for accreeting NS in binary systems. Moreover plasma straining across the magnetic along the surface can accumulates close to the magnetic poles and the corresponding GW emission could be detected by the first generation of GW detectors ([16]).

5.3 *Spontaneous Breaking of Symmetry of axi-symmetric NS*

Spontaneous breaking symmetry (SBS) is an important mechanism in generating GW by a rotating NS.
SBS is typical of systems in some steady state configurations having some symmetry, and bifurcating toward on other steady-state configuration having a different symmetry when some internal parameter changes. A typical example is a cylindrical stick, that bends when a compression force acts axially on its extremities. More generally, a dynamical system is said to break its symmetry if there exists solutions with lower symmetries then those of the Lagrangian from which the equations of motion have been derived.
We are interested in two types of SBS: the first one can happen in insulated stars, the second one in stars coupled with radiation.
The first one is often called *viscosity driven* instability mechanism, the second type is known as *Chandrasekhar -Friedman-Schutz* (CFS) instability. A wide discussion of the first mechanism can be found in the Appendix 2).
In studying the equilibrium configuration of rigidly self rotating incompressible fluid stars, Jacobi found that for an angular velocity higher than a critical value, tri-axial rigidly rotating steady state configurations exists. The fundamental parameter is the ratio T/W between the rotation kinetic energy T and the gravitational energy W. If $T/W \geq 0.3175$ then the 3-axis configuration (ellipsoids) exists, in the opposite case, only axi-symmetric one. The breaking mechanism is easy to understand: the kinetic energy T of a rotating system can be written as $T = L/I$ where L and I are respectively the angular momentum and the inertial momentum with respect to the axis of rotation of the body. The total energy E of the system is $E = T - W = L/I - W$. If the system is isolated L stays constant during an arbitrary evolution of the system. In order to reduce the total energy more interesting for the system to evolve in a such a way that I increases. It can happen that a 3-axis configuration minimizes the total energy. Viscosity plays a capital role: Consider a 2-axis configuration $\mathbf{C_2}$ and a system evolving adiabatically from $\mathbf{C_2}$ to a 3-axis configuration $\mathbf{C_3}$. The kinetic energy T decreases because the inertia momentum increases, and the resulting surplus of mechanical energy is transformed in thermal energy and radiated. When the system reaches the $\mathbf{C_3}$ configuration its energy is lower than the one at $\mathbf{C_2}$ configuration. In other words, viscosity helps the 3-axial instability.
Numerical simulations shows that only EOS stiff enough allow to obtain 3-axis stars, General relativistic effects tend to stabilize the star. Among the different EOS, only the stiffest one allows to obtain a 3-axis star in the General Relativity frame. On the contrary 3-axis strange stars, could exist.

CFS instability mechanism

The origin of this kind of instabilities is the possibility for a rotating star to loose kinetic energy and angular momentum via gravitational waves. A simple example illustrates the mechanism of this type of instabilities. Consider the wind blowing on the surface of water. A small wave propagating along the direction of the wind will be amplified if its propagation velocity is less then the velocity of the wind. In fact, in this case the wind will accomplish a positive work on the wave. Gravity waves, in deep water propagate with a velocity that is proportional to the inverse of the square root of its wave length. Consequently, for any velocity of the wind there exists a gravity wave growing with the wind. (The stronger the wind the greater the wave length). In other words, the possibility to exchange linear momentum between water and wind opens the possibility of instabilities.

A rotating star can loose angular momentum via gravitational waves (gravitational waves play the role of the wind). That means that there exists an exchange of angular momentum between a perturbation traveling along azimuthal direction and the vacuum, therefore instabilities are possible. The necessary conditions allowing for the growth of the perturbations are:
-The star must rotate and the perturbation must counter-rotates with a velocity lower than the rotation of the star. In other words an observer in a rest frame must see the perturbation dragged by the star.
-The perturbation must be coupled with the vacuum via some kind of radiation (gravitational or electro-magnetic one)
- The rising time τ_r and dissipation time τ_d must satisfy the inequality $1/\tau_r - 1/\tau_d \geq 0$
From what was said above it appears that any kind of perturbations satisfying the three above conditions,can break the axial symmetry and radiates. A pioneering work by Chandrasekhar,showed that the f modes (oscillations for which the restoring force is the gravity) are instable if the star rotates at an angular velocity grater then a critical velocity depending on the quantum number of the wave. More recently, Anderssonn [17] discovered that the called inertial modes (r modes) for which the restoring force is the Coriolis force are also instable for any angular velocity of the star. The coupling with the gravitation is weaker than the one of the f modes, consequently the conditions on the dissipative terms are more severe (see a review paper by Frieedman and Lockitch [18]). In superfluid stars, we have seen other modes can exist (Tkachenko modes) and their instability must be studied. In magnetized NS magnetic field tends to rigidify the star and tends to kill the

above instability. Fortunately in presence of a magnetic field other modes are possible. The magnetized star is an unexplored realm.

Conclusions
Deformed rotating NS are promising candidates as continuous GW emitters. The main problem is the knowledge of the dissipative mechanisms. As far we know, now only "naive" shear viscosity and bulk viscosity of non-superfluid non-superconducting nuclear matter were used in studying this kind of instabilities. A lot of work must be done in order to explore this wide field of the physics of NS.

6 Coalescing Objects

Coalescing NS or coalescing BH were considered as the most powerful sources of GW. VIRGO and Improved LIGO should detect coalescing NS within a distance of $20\,Mpc$ with a signal to noise ratio (S/N) of 7. (a $S/N \geq 7$ is required to have an acceptable low false alarm probability). Recent estimations by Yungelson and all,([19]) gives a NS coalescing rate of about $1\,y^{-1}\,(D/200Mps)^{-3}$ within a sphere of radius D. This rate is about 10 times greater then the detection rate of γ- rays bursts. The discrepancy, under the hypothesis that a fraction of γ ray bursts is generated by coalescing NS, is explained by invoking a beaming of the electro-magnetic emission. Since the observations of a few violent γ- ray bursts, we know that different mechanism than the NS coalescence, are required to have the observed fluency of the most violent events, in other words, not all γ rays bursts are generated by coalescence of NS. The main conclusion is that coalescence rate of NS could be lower than the one quoted above, therefore they detection requires more sensitive detectors.
Coalescence of binary BH of masses M_1, M_2 is a more promising sources, in fact for these sources $S/N \propto \mathcal{M}^{5/6}$ where $\mathcal{M} = (M_1 + M_2)^{-1/5})(M_1 M_2)^{3/5}$ is the chirp mass, therefore a coalescence within a distance of $1\,Gpb$ would be detected if $M_1 \sim M_2 = 50\,M_{\odot}$. The problem is to know if a such a system exist and how many. The main problem is to know how they form. Following Yungelson et al. (already quoted article), stellar evolution can not produce binary massive BK systems, Postnov et al. [20]) obtain the opposite result. The existence of primordial binary BH was proposed by Ioka et al [21]. Finally, possibility of formation of binary massive BH in high star density regions, as star cluster or at the center of galaxies should be examined.
From what we said, a certain confusion exists on this subject. The main conclusion and the most important message is that the first generation of GW

detectors will be sensitive enough to clarify this domain of the astrophysics.

In recent years important progresses were accomplished in simulating coalescence of NS. The pre-merging phase is now well understood and modeled with high accuracy in the frame of GR.[22] The coalescence phase and the consequent BH formation had also been computed numerically [23].

It is possible to predict with high accuracy at which orbital period two NS of a given mass and given EOS start to merge. Remember that when the two stars start to merge the GW radiation is abruptly reduced (loss of coherence in the hydro-flow !) therefore, it turns to be possible, in principle, to have informations on EOS of the nuclear matter at very high density. Unfortunately, very little progress were accomplished in modeling coalescence of BH. We hope that what we have learned in treating numerically the problem of coalescing NS can be applied to the more urgent problem of coalescing BH.

7 Conclusions

At this point, a natural question of a surviving reader is "is it all ?" "Are you sure that other kind of sources cannot exist ?" Making predictions in astrophysics is a hazardous job: Nature is more imaginative then men. Consider for example the history of Radioastronomy. Radio-waves where detected by Jansky coming from our galaxy. Nevertheless the sun was not considered a possible source of radio-waves. In fact the radio emission of a black body at the temperature of the sun is negligible. Astrophysics have forgotten that plasma *coherent* or collective modes can exist and can radiate in a very efficient way. The radio-emission of the sun was discovered during the last war by the radar operators. On the other size of the spectrum, on X-rays range, there was the same consensus: the X-rays emission of a star cannot be detected: compute the black body radiation of a star having a surface temperature of few thousand of K and etc. etc.. Thanks to influence of B. Rossi a rocket with a X-ray detector aboard was launched and the source Sco-X1 was discovered: X-rays astronomy was born. Theoretician have forgotten that *compact* stars can exist and matter can fall down and to be hot enough to radiate in the X-rays range.
In studying the possible sources of GW we have used all the ingredients that were forgotten in the electromagnetic astronomy i.e: *compactness* , *coherence* of the motion and of course massive objects. Moreover, all the computed GW energy emissions are close to the maximum allowed energy Mc^2. Finally, NS and BH are the most compact objects we can have, at least in the frame of the present physics. If we have to bet against Nature, we can see that

Nature has a little number of degrees of freedom in finding something of exceptionally new.
The main message is the following, NS and BH are the only candidates as GW sources. It is natural for the experimentalist to ask to the theoretician "What is the coalescing rate of massive BH ?" or "Are rotating NS deformed enough to be detectable with the first generation of GW detectors ?" We think that the correct answer is "it cannot excluded: experiments will answer to this question" and we want to add *The first generation of detectors* will do it.

Appendix 1 : The Magneto Hydrodynamics equations

The Magneto-hydrodynamics approximation consists in neglecting the displacement current in Maxwell equations and in considering fluids with vanishing resistivity. These equations are the hydrodynamical equations coupled with the Maxwell equations. The hydro-equations contain the extra term $(curl\,\vec{B}) \wedge \vec{B}$ and the equations of magnetic field are:

$$\frac{1}{c}\partial_t \vec{B} + curl\,\vec{E} = 0 \tag{40}$$

Under the above hypothesis, $\vec{E}$ vanishes in a frame at rest with the matter. A Lorentz transform allows us to write the MHD Eq. in the general case. By neglecting terms $O((v/c)^2)$ the Eq. (40) becomes

$$\partial_t \vec{B} - curl\,(\vec{v} \wedge \vec{B}) = 0 \tag{41}$$

An important consequence of these equations is that the magnetic flux across a loop co-moving with the fluid is conserved. Proof.
Lemma: If C is a closed line co-moving with the fluid the following identity holds:

$$\frac{d}{dt}\oint_C A_i dl^i \equiv \oint (\frac{D}{dt}A_j + A_i \nabla_j v^i) dl^j \tag{42}$$

$$\equiv \oint_C (\partial_t A_j + v^i(\nabla_i A_j - \nabla_j A_i) + \nabla_j (A_i v^i)) dl^j \tag{43}$$

$$= \oint_C (\partial_t \vec{A} - \vec{v} \wedge curl\,\vec{A}) \cdot \vec{dl} \tag{44}$$

Let be $\vec{A}$ the vector potential defined by $\vec{B} = curl\,\vec{A}$ The Eq.(41) can be rewritten

$$\partial_t \vec{A} - \vec{v} \wedge \vec{B} = grad\,Q \tag{45}$$

where Q is an arbitrary function. Taking into account the identities (42) we have

$$\oint_C (\partial_t \vec{A} - \vec{v} \wedge curl\, \vec{A}) \cdot \vec{dl} = \frac{d}{dt} \int_S \vec{B} \cdot \vec{n} dS = 0$$

where S is an arbitrary surface defined by the line C, and $\vec{n}$ is the unit vector normal to S. The magnetic lines are "frozen" in the matter.

Appendix 2 : Spontaneous symmetry breaking

Review of classical results about MacLaurin/Jacobi ellipsoids

Let us recall some classical results from the theory of rotating Newtonian homogeneous bodies. It is well known that a self-gravitating incompressible fluid rotating rigidly at some moderate velocity takes the shape of an axisymmetric ellipsoid: the so-called *MacLaurin spheroid.* At the critical point $T/|W| = 0.1375$ in the MacLaurin sequence, two families of triaxial ellipsoids branch off: the *Jacobi ellipsoids* and the *Dedekind ellipsoids.* The former are triaxial ellipsoids rotating rigidly about their smallest axis in an inertial frame, whereas the latter have a fixed triaxial figure in an inertial frame, with some internal fluid circulation at constant vorticity (see Tassoul [24] for a review of these classical results). The MacLaurin spheroids are dynamically unstable for $T/|W| \geq 0.2738$. Thus the Jacobi/Dedekind bifurcation point $T/|W| = 0.1375$ is dynamically stable. However, in presence of some dissipative mechanism such as viscosity or gravitational radiation (CFS instability) that breaks the circulation or angular momentum conservation, the bifurcation point becomes secularly unstable against the $l = 2, m = 2$ "bar" mode. Note also that a non-dissipative mechanism such as the one due to a magnetic field with a component parallel to the rotation axis breaks the circulation conservation end may generate a spontaneous symmetry breaking. If one takes into account only the viscosity, the growth of the bar mode leads to the deformation of the MacLaurin spheroid along a sequence of figures close to some Riemann S ellipsoids[m] and whose final state is a Jacobi ellipsoid [26]. On the opposite, if the gravitational radiation reaction is taken into account but not the viscosity, the MacLaurin spheroid evolves close to another Riemann S sequence towards a Dedekind ellipsoid [27].

---

[m]The *Riemann S* family is formed by homogeneous bodies whose fluid motion can be decomposed into a rigid rotation about a principal axis and a uniform circulation whose vorticity is parallel to the rotation vector. MacLaurin, Jacobi and Dedekind ellipsoids are all special cases of Riemann S ellipsoids (for more details, cf. Chap. 7 of ref. [25] .

As we shall see in the next section, this symmetry breaking is a particular case of a more general phenomenon.

Spontaneous breaking of symmetry: a general phenomenon

A toy model

Let us start with a simple example. Consider the toy model of Fig. 9 : two heavy points of mass m, rigidly connected and sliding on a circular guide of radius R, of negligible mass and rotating with the angular velocity Ω. It is straightforward to write down the expression for the potential energy of this system and to find the equilibrium configuration in a frame rotating with the guide. With the notations of Fig. 9, the potential (potential energy per unit mass) writes:

$$U(\theta) = -(R\Omega)^2(\sin^2\alpha + \cos 2\alpha \sin^2\theta) - 2gR\cos\alpha\cos\theta \ , \qquad (46)$$

g being the gravity acceleration. In spite of the invariance of the potential U with respect the transformation $\theta \mapsto -\theta$, two equilibrium configurations exist if the angular velocity is larger than the value $\Omega_{\text{crit}} = [\, g\cos\alpha/(R\cos 2\alpha)\,]^{1/2}$: $\theta = 0$ and $\cos\theta = g\cos\alpha/(R\Omega^2\cos 2\alpha)$.

This behavior is typical of the phenomenon of spontaneous symmetry breaking. More generally, a dynamical system is said to break spontaneously its symmetry if there exist solutions with lower symmetries than those of the Lagrangian from which the equations of motion have been derived. Other examples of spontaneous symmetry breaking are the phase transitions (ferromagnetism at the Curie point, melting of solids), the famous walking stick in Charlie Chapling movies (the stick is axially symmetric but when the pressure of the hand on its top is larger than a critical value, the stick bends), the buckling collapse of metallic structures and so on.

Coming back to the toy model, it is easy to see that, when two equilibrium configurations exist, the one is unstable and the other one is stable. For what follows it is more convenient to take the angular momentum as a free parameter instead of the angular velocity. Let us suppose that some viscous force acts on the two mass points sliding on the circular guide. If at time $t = 0$, the system is in the equilibrium configuration $\theta = 0$ and its angular momentum L is larger than the critical value L_{crit} corresponding to Ω_{crit}, then the system will move to the second equilibrium configuration. In this example, we have all the main ingredients of the spontaneous symmetry breaking mechanism: a parameter describing the equilibrium configuration of the system, a dynamical quantity that is not conserved (the kinetic energy) and a total

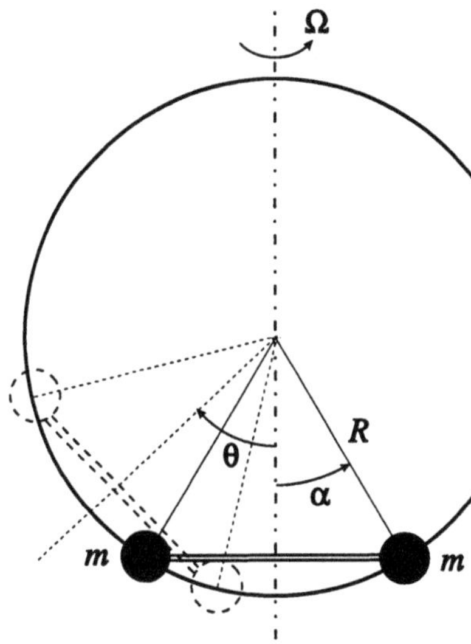

Figure 9. Toy model illustrating the phenomenon of spontaneous symmetry breaking: the configuration depicted with dashed lines violates the symmetry with respect to the rotation axis and is a stable equilibrium configuration if $\Omega > \Omega_{\rm crit}$ [12].

energy that must be at its minimum. In order to understand the existence of two equilibrium configurations, it is worth writing the total energy (per unit mass) with respect to the inertial frame in terms of the angular momentum L: $E = 1/2\, L^2/I + U_{\rm grav}$, where $I(\theta)$ is the moment of inertia of the system with respect to the rotation axis and $U_{\rm grav}$ is the gravitational potential: $U_{\rm grav} = -2gR\cos\alpha\cos\theta$. It is clear that, at fixed L, L^2/I decreases when I increases, i.e. when θ increases. On the contrary, $U_{\rm grav}$ is an increasing function of θ. When L is large enough, the first term prevails and E decreases when θ increases.

In order to see the link existing between the symmetry breaking mechanism of the above toy model and the second order phase transition theory, let us briefly recall the Landau theory of these phenomena (see Chap. 14 of [28]). In a second order phase transition, a thermodynamical system makes a continuous transition from a state with a given symmetry to a state with a lower symmetry. The leading idea in the Landau theory is the expansion of the thermodynamical potential $\Phi(P,T)$ in a series of a parameter η (the *order parameter*) which measures how far the thermodynamical system is from the state of higher symmetry:

$$\Phi(P,T,\eta) = \Phi_0(P,T) + \alpha\eta + A\eta^2 + B\eta^3 + C\eta^4 + \cdots \tag{47}$$

From invariance considerations, α must be identically equal to zero, B must also be equal to zero in order to obtain a stable system. On the contrary, C must be positive. A must be larger than zero in the symmetric phase because the system is stable. On the contrary, in the less symmetric phase, values of η different from zero correspond to a stable state. That is possible only if $A < 0$, therefore A must vanish at the critical temperature: $A = a(T - T_c)$. In conclusion we can write:

$$\Phi(P,T,\eta) = \Phi_0(P,T) + a(T - T_c)\eta^2 + C\eta^4 \ . \tag{48}$$

The value of η in the vicinity of T_c is:

$$\eta^2 = a(T_c - T)/(2C) \ . \tag{49}$$

From the expression (48) for the thermodynamical potential all the interesting thermodynamical quantities and their jumps at the critical temperature can be derived.

Coming back to the toy model, consider the total potential (eq. (46)) and expand it as function of θ around the value $\theta = 0$. The result is

$$\begin{aligned} U = &-(R\Omega \sin\alpha)^2 + R(g\cos\alpha - R\Omega^2 \cos 2\alpha)\ \theta^2 \\ &+R[R\Omega^2 \cos 2\alpha - (g\cos\alpha)/4]\ \theta^4/3 + O(\theta^6) \ . \end{aligned} \tag{50}$$

It is easy to see the analogy between expressions (48) and (50). In fact θ is the order parameter of the toy model: when $\theta = 0$ the system is axisymmetric and it breaks this symmetry for $\theta > 0$. The potential U is the analogous of the thermodynamical potential Φ. No extra terms with odd power of θ appear, there is a critical value of Ω for which the quadratic terms vanish and, last but not least, the quartic term is positive in the vicinity of Ω_{crit}. It is straightforward to define a "temperature" of the system. Let us put $T = 1/(R\Omega)^2$; then $T_c = \cos 2\alpha\,/(gR\cos\alpha)$, $a = 2(gR\cos\alpha)^3/\cos^2 2\alpha$ and $C = (gR\cos\alpha)^2/\cos 2\alpha$. The reader can push the analogy further by defining the equivalent "thermodynamical quantities" and compute their discontinuities at the transition point.

An equivalent treatment (but slightly more complicated) can be employed by replacing Ω by the angular momentum L. In this case α must be different from zero, in order to have a finite angular momentum when $\theta = 0$.

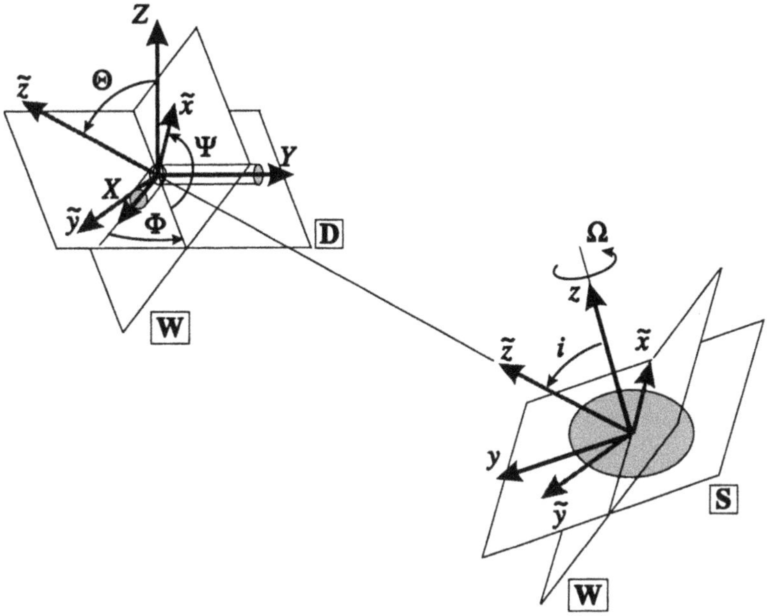

Figure 10. Relative orientation of various orthonormal frames introduced in the text: $(\mathbf{e}_x, \mathbf{e}_y, \mathbf{e}_z)$ is the neutron star inertial (non-rotating) frame, associated with the ACMC coordinates (x, y, z); $(\mathbf{e}_x, \mathbf{e}_y)$ define the plane **S** perpendicular to the rotation axis. $(\mathbf{e}_{\tilde{x}}, \mathbf{e}_{\tilde{y}}, \mathbf{e}_{\tilde{z}})$ is the frame associated with the gravitational wave received on Earth; $(\mathbf{e}_{\tilde{x}}, \mathbf{e}_{\tilde{y}})$ define the wave plane **W**. Note that $\mathbf{e}_{\tilde{x}} = \mathbf{e}_x$. $(\mathbf{e}_X, \mathbf{e}_Y, \mathbf{e}_Z)$ is the interferometric detector frame; $(\mathbf{e}_X, \mathbf{e}_Y)$ define the detector plane **D** [13].

Appendix 3 : Signal received by an interferometric detector

Due to the weakness of the expected gravitational signal from pulsars, long integration times, typically of the order of the year, are required to extract the signal out of the noise. For such a long observing time, the motion of the Earth must be taken into account for it modifies the position of the pulsar with respect to the antenna pattern of the interferometric detector. It results in a modulation of both the amplitude and the frequency of the signal (Jotania et al [29]) Note that if one is searching for a known pulsar, the frequency modulation (Doppler shift) can be obtained readily from the radio observations.

In this section, we examine the daily modulation of the signal amplitude due to the Earth's rotation following[13]. The special case of an interferometric detector situated on the equator with arms symmetrically placed about the North-South direction and a gravitational wave coming from the Northern celestial pole has been treated by Jotania et al. [29].

Beam-pattern factors

Let $(\mathbf{e}_{\tilde{x}}, \mathbf{e}_{\tilde{y}}, \mathbf{e}_{\tilde{z}})$ be the orthonormal frame associated with the gravitational wave: $\mathbf{e}_{\tilde{z}}$ is perpendicular to the wave plane and parallel to the "line of sight", being orientated from the neutron star center to the Earth center; $\mathbf{e}_{\tilde{x}}$ is the unit vector of the wave plane which is perpendicular to the star's rotation axis. In the $(\mathbf{e}_{\tilde{x}}, \mathbf{e}_{\tilde{y}}, \mathbf{e}_{\tilde{z}})$ frame the transverse traceless gravitational wave is expressible from its two polarization modes h_+ and $h_\times$

$$h_+ = h_0 \sin\alpha[\frac{1}{2}\cos\alpha \sin i \cos i \cos\Omega(t - t_0) - \frac{\sin\alpha}{2}(1 + \cos^2 i)\cos 2\Omega(T - t_0)] \tag{51}$$

$$h_- = h_0 \sin\alpha[\frac{1}{2}\cos\alpha \sin i \cos i \sin\Omega(t - t_0) - \sin\alpha \cos i \sin 2\Omega(T - t_0)] \tag{52}$$

$$\mathbf{h} = h_+ (\mathbf{e}_{\tilde{x}} \otimes \mathbf{e}_{\tilde{x}} - \mathbf{e}_{\tilde{y}} \otimes \mathbf{e}_{\tilde{y}}) + h_\times (\mathbf{e}_{\tilde{x}} \otimes \mathbf{e}_{\tilde{y}} + \mathbf{e}_{\tilde{y}} \otimes \mathbf{e}_{\tilde{x}}) \,. \tag{53}$$

where i, α and Ω are indicated in Fig. (11).

The response of an interferometric detector of the VIRGO/LIGO type to the above gravitational wave depends upon the relative orientation of wave frame with respect to the detector's arms. Let $(\mathbf{e}_X, \mathbf{e}_Y, \mathbf{e}_Z)$ be the orthonormal frame such that $\mathbf{e}_Z$ is perpendicular to the detector plane, pointing toward the zenith and $\mathbf{e}_X$ and $\mathbf{e}_Y$ are unit vectors along the two detector's arms. Let (Θ, Φ, Ψ) be the three Euler angles which specify the position of the wave frame $(\mathbf{e}_{\tilde{x}}, \mathbf{e}_{\tilde{y}}, \mathbf{e}_{\tilde{z}})$ with respect to the detector frame $(\mathbf{e}_X, \mathbf{e}_Y, \mathbf{e}_Z)$ (cf. Fig. 10). The signal measured by the detector is

$$h(t) = F_+(\Theta, \Phi, \Psi)\, h_+(t) + F_\times(\Theta, \Phi, \Psi)\, h_\times(t) \,, \tag{54}$$

with the following beam-pattern factors cf. Eq. (7a) of Dhurandhar & Tinto (1988) [30] :

$$F_+(\Theta, \Phi, \Psi) = \frac{1}{2}(1 + \cos^2\Theta)\cos 2\Phi \cos 2\Psi - \cos\Theta \sin 2\Phi \sin 2\Psi \tag{55}$$

$$F_\times(\Theta, \Phi, \Psi) = \frac{1}{2}(1 + \cos^2\Theta)\cos 2\Phi \sin 2\Psi + \cos\Theta \sin 2\Phi \cos 2\Psi \,. \tag{56}$$

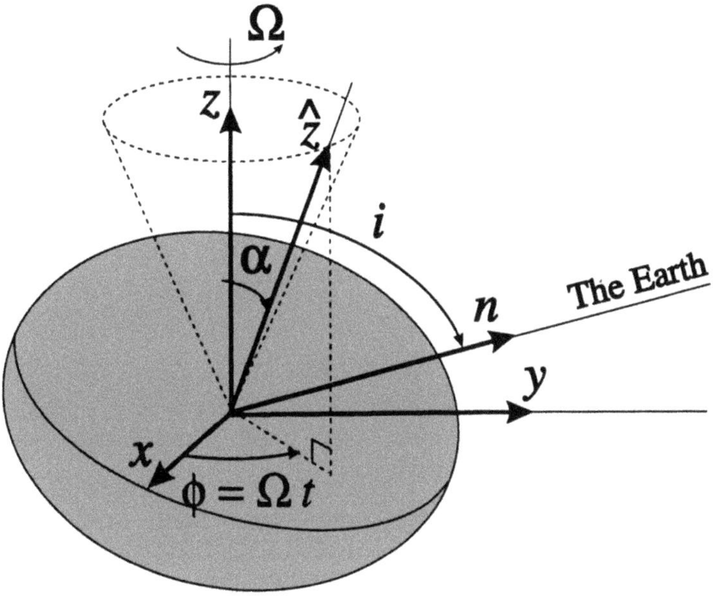

Figure 11. Definition of the angle α and i (See Eq. (51) and (52))

The Euler angles (Θ, Φ, Ψ) are not constant because of the motion of the detector with respect to the source induced by the Earth's diurnal rotation and revolution around the Sun. To compute their variations let us introduce the "celestial sphere frame" $(\mathbf{e}_{X'}, \mathbf{e}_{Y'}, \mathbf{e}_{Z'})$ such that $\mathbf{e}_{Z'}$ is along the Earth's rotation axis, pointing toward the North pole, $\mathbf{e}_{X'}$ and $\mathbf{e}_{Y'}$ are in the Earth's equatorial plane, $\mathbf{e}_{X'}$ pointing toward the vernal point (i.e. $\mathbf{e}_{X'}$ is along the intersection of the Earth's equatorial plane with the Earth's orbital plane). For time scales of the order of the year, $(\mathbf{e}_{X'}, \mathbf{e}_{Y'}, \mathbf{e}_{Z'})$ can be considered as fixed with respect to the (approximatively) inertial frame containing the solar system barycentre and the neutron star center. The wave frame $(\mathbf{e}_{\tilde{x}}, \mathbf{e}_{\tilde{y}}, \mathbf{e}_{\tilde{z}})$ is also fixed with respect to this inertial frame. Then let (Θ', Φ', ψ) be the Euler angles which specify the position of the wave frame with respect to the celestial sphere frame (cf. Fig. 12). Θ' and Φ' are simply related to the equatorial coordinates of the neutron star on the celestial sphere (the right ascension[n] $\bar{\alpha}$ and the declination $\bar{\delta}$) by (cf. Fig. 12):

$$\Theta' = \bar{\delta} + \frac{\pi}{2} \qquad \text{and} \qquad \Phi' = \bar{\alpha} - \frac{\pi}{2} \,. \tag{57}$$

The triad $(\mathbf{e}_{\bar{x}}, \mathbf{e}_{\bar{y}}, \mathbf{e}_{\bar{z}})$ is related to the triad $(\mathbf{e}_{X'}, \mathbf{e}_{Y'}, \mathbf{e}_{Z'})$ by

$$\mathbf{e}_{\bar{i}} = A_{\bar{i}J'} \, \mathbf{e}_{J'} \,, \tag{58}$$

with the orthogonal matrix [cf. e.g. Eq. (4-46) of Goldstein (1980) with the substitution (57)]

$$A_{\bar{i}J'} = \begin{pmatrix} \sin\bar{\alpha}\cos\psi - \cos\bar{\alpha}\sin\bar{\delta}\sin\psi & -\cos\bar{\alpha}\cos\psi - \sin\bar{\alpha}\sin\bar{\delta}\sin\psi & \cos\bar{\delta}\sin\psi \\ -\sin\bar{\alpha}\sin\psi - \cos\bar{\alpha}\sin\bar{\delta}\cos\psi & \cos\bar{\alpha}\sin\psi - \sin\bar{\alpha}\sin\bar{\delta}\cos\psi & \cos\bar{\delta}\cos\psi \\ -\cos\bar{\alpha}\cos\bar{\delta} & -\sin\bar{\alpha}\cos\bar{\delta} & -\sin\bar{\delta} \end{pmatrix} . \tag{59}$$

Let $(\mathbf{e}_{X''}, \mathbf{e}_{Y''}, \mathbf{e}_{Z''})$ be the orthonormal frame linked to the geographical location of the detector site, such that $\mathbf{e}_{Z''} = \mathbf{e}_Z$ is the local vertical, $\mathbf{e}_{X''}$ is in the North-South direction and $\mathbf{e}_{Y''}$ is in the West-East direction (cf. Fig. 13). The position of the wave frame $(\mathbf{e}_{\bar{x}}, \mathbf{e}_{\bar{y}}, \mathbf{e}_{\bar{z}})$ with respect to the cardinal frame $(\mathbf{e}_{X''}, \mathbf{e}_{Y''}, \mathbf{e}_{Z''})$ is determined by the three Euler angles (Θ, Φ'', Ψ) where Θ and Ψ are the same angles as those relative to the detector frame and entering Eqs. (55)-(56). The angle Φ'' is related to Φ by

$$\Phi = \Phi'' + \lambda \,, \tag{60}$$

where λ is the azimuth of the X-arm of the detector, i.e. the angle between the South direction and the X-arm, as measured in the retrograde way. Φ'' is directly related to the azimuth a of the source by

$$\Phi'' = -a - \frac{\pi}{2} \,. \tag{61}$$

The minus sign which occurs in this relation comes from the fact that the azimuth is measured *westwards* from the South, hence in the inverse trigonometric way. Putting together relations (60) and (61) we obtain

$$\Phi = -a + \lambda - \frac{\pi}{2} \,. \tag{62}$$

Let $(\theta'', \varphi'', \psi'')$ be the Euler angles which specify the position of the cardinal frame $(\mathbf{e}_{X''}, \mathbf{e}_{Y''}, \mathbf{e}_{Z''})$ with respect to the celestial sphere frame

[n] a bar is put on α to distinguish it from the angle α between the magnetic and rotation axis introduced earlier in the text.

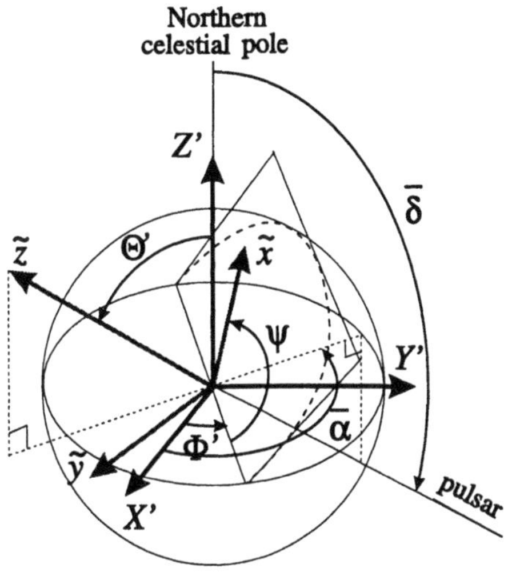

Figure 12. Relative orientation of the gravitational wave frame $(\mathbf{e}_{\tilde{x}}, \mathbf{e}_{\tilde{y}}, \mathbf{e}_{\tilde{z}})$ and the celestial sphere frame $(\mathbf{e}_{X'}, \mathbf{e}_{Y'}, \mathbf{e}_{Z'})$.

$(\mathbf{e}_{X'}, \mathbf{e}_{Y'}, \mathbf{e}_{Z'})$. One has immediately $\psi'' = -\pi/2$ because $\mathbf{e}_{X''}$ is orientated in the North-South direction. θ'' is simply related to the latitude l of the detector site by

$$\theta'' = \frac{\pi}{2} - l \, . \tag{63}$$

φ'' is linked to the local sidereal time T (i.e. the angle between the local meridian and the vernal point) by

$$\varphi'' = T + \frac{\pi}{2} \, . \tag{64}$$

T can be expressed in terms of the sidereal time at Greenwich at 0h UT, $T_{\rm Green}(0)$, and the local UT time t by

$$T(t) = \sigma t + T_{\rm Green}(0) - L \, , \tag{65}$$

where $\sigma = 1.00273790935 \times 15^{\rm o}/{\rm h}$ (Meeus 1991) is a conversion factor from mean solar time to sidereal time (hence σ accounts for the revolution of the

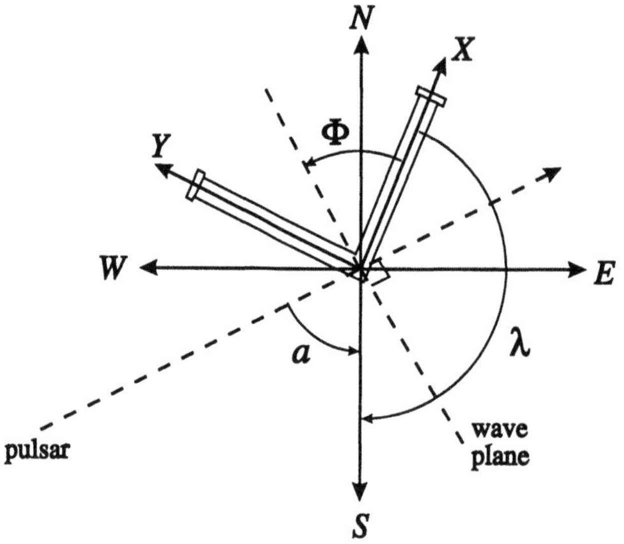

Figure 13. Orientation of the detector arms with respect to the local North-South and West-East directions. The dashed line is the projection of the gravitational wave propagation direction on the horizontal plane. The depicted configuration corresponds to that of the VIRGO detector ($\lambda = -161^\circ\,33'$).

Earth around the Sun) and L is the longitude of the detector site (the minus sign in front of L comes from the fact that the geographical longitudes are measured positively *westwards*). The triad $(\mathbf{e}_{X''}, \mathbf{e}_{Y''}, \mathbf{e}_{Z''})$ is related to the triad $(\mathbf{e}_{X'}, \mathbf{e}_{Y'}, \mathbf{e}_{Z'})$ by

$$\mathbf{e}_{I''} = B_{I''J'}\ \mathbf{e}_{J'}\ , \tag{66}$$

with the orthogonal matrix

$$B_{I''J'} = \begin{pmatrix} \sin l \cos T & \sin l \sin T & -\cos l \\ -\sin T & \cos T & 0 \\ \cos l \cos T & \cos l \sin T & \sin l \end{pmatrix} . \tag{67}$$

Now the wave frame $(\mathbf{e}_{\tilde{x}}, \mathbf{e}_{\tilde{y}}, \mathbf{e}_{\tilde{z}})$ is related to the triad $(\mathbf{e}_{X''}, \mathbf{e}_{Y''}, \mathbf{e}_{Z''})$ by

$$\mathbf{e}_{\tilde{i}} = C_{\tilde{i}J''}\ \mathbf{e}_{J''}\ , \tag{68}$$

with the orthogonal matrix

$$C_{\bar{i}J''} = \begin{pmatrix} \cos\Psi\cos\Phi'' - \cos\Theta\sin\Phi''\sin\Psi & \cos\Psi\sin\Phi'' + \cos\Theta\cos\Phi''\sin\Psi & \sin\Psi\sin\Theta \\ -\sin\Psi\cos\Phi'' - \cos\Theta\sin\Phi''\cos\Psi & -\sin\Psi\sin\Phi'' + \cos\Theta\cos\Phi''\cos\Psi & \cos\Psi\sin\Theta \\ \sin\Theta\sin\Phi'' & -\sin\Theta\cos\Phi'' & \cos\Theta \end{pmatrix} . \tag{69}$$

From equations (58), (66) and (68), the matrix C is given by the product

$$C = A \cdot {}^tB . \tag{70}$$

Performing the right-hand-side product and identifying each matrix element by those of expression (69) leads to the following trigonometrical relations

$$\cos\Theta = -\cos\bar{\delta}\cos l\cos H(t) - \sin\bar{\delta}\sin l \tag{71}$$

$$\sin\Theta\cos a = \cos\bar{\delta}\sin l\cos H(t) - \sin\bar{\delta}\cos l \tag{72}$$

$$\sin\Theta\sin a = \cos\bar{\delta}\sin H(t) \tag{73}$$

$$\begin{aligned}\sin\Theta\sin\Psi = & -\cos\psi\cos l\sin H(t) \\ & -\sin\psi\sin\bar{\delta}\cos l\cos H(t) + \sin\psi\cos\bar{\delta}\sin l\end{aligned} \tag{74}$$

$$\begin{aligned}\sin\Theta\cos\Psi = & \cos\psi\cos\bar{\delta}\sin l \\ & +\sin\psi\cos l\sin H(t) - \cos\psi\sin\bar{\delta}\cos l\cos H(t),\end{aligned} \tag{75}$$

where we have used the relation (61) between Φ'' and a and have introduced the local hour angle of the source:

$$H(t) = T(t) - \bar{\alpha} = \sigma t + T_{\text{Green}}(0) - L - \bar{\alpha} . \tag{76}$$

From the above relations the beam-pattern factors $F_+(\Theta, \Phi, \Psi)$ and $F_\times(\Theta, \Phi, \Psi)$ can be computed for any instant t, given the position (l, L) of the detector on the Earth, its orientation λ, and the position $(\bar{\alpha}, \bar{\delta})$ of the source on the sky, as well as the angle ψ that the vector $\mathbf{e}_{\bar{x}}$ forms with the intersection line of the wave plane and the Earth's equatorial plane. First one computes the local hour angle $H(t)$ by Eq. (76), and the angle Θ by Eq. (71). The angle Φ is computed by means of Eqs. (62), (72) and (73). Finally, the local polarization angle Ψ is deduced from Eqs. (74) and (75).

References

1. Bonazzola S. Marck J.A. *Ann. Rev. Part. Sci.* **45**, 635 (1994)
2. Finzi B.,*Atti Accad. Naz. Lincei Cl. Sc. Fis. Mat. Nat.* **6**, 18 (1949)
3. Pirani F.A.E. ,*Acta Phys. Pol.* **15**, 389 (1956)
4. Hulse R.A., Taylor J. H,*Astrophys. Jour.* **195**, L51 (1975)
5. Damour T., Deruelle N. ,*Phys. Lett.* **87A**, 81 (1981)
6. J,A. Marck and Lasota J.P. *Relativistic gravitation and gravitational radiation*, (Cambridge University Press, Cambridge, 1997).
7. Blanchet L. in: *Relativistic gravitation and gravitational radiation*, (Cambridge University Press, Cambridge, 1997).
8. Tourrenc Ph. (*This volume.*
9. Zel'dovich Ya.B. and I.D. Novikov *Relativistic Astrophysics: Stars and Relativity*, Vol. !, Midway reprint (1978)
10. Haensel P. in: *Relativistic gravitation and gravitational radiation*, (Cambridge University Press, Cambridge, 1997).
11. Tilley D.R and Tilley J. *Superfluidity and Superconductivity* Douglas F.Brewer Ed. (1990)
12. Bocquet M., Bonazzola S. and Gourgoulhon E. in *Relativistic gravitation and gravitational radiation*, (Cambridge University Press, Cambridge, 1997).
13. Bonazzola S. and Gourgoulhon E. *Astron. and Astrophys.* **312**, 675 (1986)
14. Müller E. in *Relativistic gravitation and gravitational radiation*, (Cambridge University Press, Cambridge,1997)
15. Stark R.F., Piran T., *Phys. Rev. Lett.* **55**, 891 (1985).
16. see S.Bonazzola and E.Gourgoulhon in *Gravitational waves - Sources and Detectors*, Ciufolini Fidecaro eds., Word Scintfig pubs.(1997),
17. Andersson N, and Kokkotas K., Astrophys. Journ. **516**, 307 (1999).,
18. Friedman JU.L. and Lockitch K.H. Preprint-gr- qc/9908083 (1999)
19. Yungelson L., Portegies Zwart S.F., in *Proceedings of the Second Workshop on Gravitational Wave Data Analysis* (Orsay, 13-15 November 1997), Eds. M. Davier & P. Hello, (Editions Frontières, Gif-sur-Yvette,1998).
20. Postnov K.A., Prokhorov M.E., in *Proc. XXXIV Rencontres de Moriond*, to be published (preprint: astro-ph/9903193)
21. Ioka K., Tanaka T., Nakamura T., *Phys. Rev.* D **60**, 083512 (1999)
22. Bonazzola S. Gourgoulhon E. Marck J.A. Phys. Rev. Lett. **82**, 1892 (1999).
23. Shibata M, Uryu K. Preprint-gr-qc/9911058 (1999).

24. Tassoul J.L. *Theory of rotating stars* Princeton University Press: Princeton (1978)
25. Chandrasekhar S. *Ellipsoidal figures of equilibrium* Yale University Press. New Haven, (1969)
26. Press W.H., Teukolsky S.A. *Astrophys. J.* **181**, 513 (1973)
27. Miller B.D. *Astrophys. J.* **287**, 809 (1974)
28. Landau and Lifchitz E. *Physique statistique*, Ed. Mir (Moscow) (1984)
29. Jotania K. Balluri S.R. Dhurandhar S.V Astron and Astrophys. **306**, 317 (1996)
30. Dhurandhar S.V and Timto M. MRAS. **234**, 663 (1996)

True color image of a 36"x 36" field centered on SN 1987A. It is the combination of blue, green and red images obtained with the Wide Field Camera on the Hubble Space Telescope and processed by Dr. S. Scuderi. (courtesy of N. Panagia, ESA-STScI, Baltimore).

SUPERNOVAE

N. PANAGIA
ESA-STScI, 3700 San Martin Drive, Baltimore,
MD 21218, USA
E-mail: panagia@stsci.edu

The properties of supernovae (SNe) are reviewed. It is shown that the observed characteristics of the morphological classes of SNe (types Ia, Ib/c, II) can be explained in terms of two basic explosion mechanisms, *i.e.* core collapse of massive stars and thermonuclear explosion of low/moderate mass stars. The study of SNe can provide valuable insight in the late phases of the evolution of their progenitors and, therefore, can constrain the theory of stellar evolution quite tightly. Also, bright SNe of type Ia can be used to probe the Universe up to high redshifts, allowing us to measure cosmological constants and to gain valuable information on the formation and evolution of galaxies. The case of SN 1987A is discussed in some detail: it appears that in this event a number of predictions were astonishingly well verified but still many aspects were completely at variance with "common wisdom" expectations, indicating that the SN phenomenon is still not fully understood.

1 Introduction

As their name indicates, *supernovae* (SNe) are discovered in the sky as "new stars" (*-novae*) of exceptionally high brightness (*super-*). The fact that SNe are formidable explosions completely different from, and vastly more energetic than classical *novae* [a], was first recognized by Baade and Zwicky (1934). They noticed that novae during explosion become no brighter than about 1 million times (*i.e.* 15 mag) they are in a quiescent phase. Therefore, any historical event in our Galaxy that had reached a magnitude as bright as 0 or brighter but is not detectable at present, had to belong to a separate class of intrinsically brighter objects. And indeed, the distribution of observed magnitudes of explosive events detected in galaxies of the Local Group indicated the presence of two peaks, one at the expected brightness of classical novae and another at more than thousand times brighter luminosities.

Supernovae represent the explosive death of both low mass stars (type Ia) and moderate and high mass stars (types Ib/c and II). They are extremely bright, (roughly 10^9 $L_\odot$ [b] rivalling, for a few days, the combined light of the entire host galaxy. In all cases, a SN explosion injects highly metal-enriched

[a] Novae are produced by sudden nuclear ignition of a very thin layer of hydrogen near the surface of a degenerate star accreting matter from a binary companion.
[b] In Astronomy the symbol $\odot$ is used to denote the Sun. Thus, $L_\odot = 3.8 \times 10^{33}$ $erg\ s^{-1}$ is the solar luminosity and $M_\odot = 2.0 \times 10^{33}$ g is the solar mass.

material (at least 1 $M_\odot$) and a conspicuous amount of kinetic energy (about 10^{51} ergs) into the surrounding medium (see Section 2.1). In addition, the blast waves from SN explosions produce powerful sources of radio and X-ray emission – supernova remnants – that can be seen and studied many thousands of years after the event (see Section 2.2). Therefore, it is clear that SN explosions are crucial events that determine most of the aspects of the evolution of galaxies, *i.e.* most of the visible Universe.

Some SNe in our Milky Way galaxy have been close enough to be visible to the naked eye, and records of their occurrence can be found in ancient annals. In particular, during the past 2000 years 9 such events have been recorded. A few of these events were very bright. The supernova of 1006 AD, for example, was about 1/10 as bright as the full moon! The last supernova to be seen in our Galaxy was discovered in 1604 by the famous astronomer Kepler. On the basis of these historical records one may infer that the average rate of SN explosions in the Galaxy be of the order of 5 per millennium. However, one has to allow for the fact that most SNe are either too far or are too obscured by dark dust clouds of the galactic disk to be visible. Actually, one can estimate that only about 10% have been close enough and bright enough to be detectable by naked eye. Therefore, a more realistic SN explosion rate for our Galaxy is about one every twenty years (see also Section 2.3).

Being so bright, SNe are ideal probes of the distant Universe. And indeed studies of SNIa up to redshifts ~ 1.2 have allowed us to explicitely measure both the local expansion rate of the Universe and other cosmological parameters (see Section 2.4). The brightest supernova discovered in the last three centuries is supernova 1987A in the Large Magellanic Cloud, a small satellite galaxy to the Milky Way. Section 3 is devoted to it.

2 Properties of Supernovae

2.1 Supernova Types

Morphologically, Supernovae are distinguished into two main classes, Type I and Type II according to the main criterion of whether their spectra (thus, their ejecta) contain Hydrogen (Type II) or no Hydrogen (Type I).

Type II SNe are produced by the core collapse of massive stars, say, more massive than 8 $M_\odot$ and at least as massive as 20 $M_\odot$ (SN 1987A) or even 30 or more $M_\odot$ (SN 1986J). Thus, the lifetime of a SNII progenitor is shorter than about 100 million years (and can be as short as a FEW million years). Therefore, SNII can be found only in galaxies that are either just formed or that have efficient, ongoing star formation, such as spiral and irregular

Table 1. GENERAL PROPERTIES OF SUPERNOVAE

Type	Ia	Ib/c	II
Hydrogen	NO	NO	YES
Optical Spectrum	Metal lines deep 6150	Metal lines no 6150	P Cyg lines Balmer series
Absolute Luminosity at max light	$\sim 4 \times 10^9 L_\odot$ small dispersion standard candles	$\sim 10^9 L_\odot$ small disp.?	$\sim 10^9 L_\odot$ large disp.
Optical Light Curve	homogeneous	rather homogeneous	heterogeneous
UV spectrum	very weak	weak	strong
Radio Emission	no detection	strong fast decay	strong slow decay
Location	all galaxies	spirals	spirals & irregulars
Stellar Population	old	young	young
Progenitors	white dwarfs in binary systems	moderately massive stars	massive stars

galaxies.

The class of Type I supernovae has been recognized (*e.g.*, Panagia 1985) to consist of two subclasses, Type Ia and Type Ib/c that, although sharing the common absence of Hydrogen, are widely apart in other properties and, especially, in their origins. The spectroscopic criterion to discern the two subclasses from each other is the presence (Ia) or absence (Ib/c)[c] of a strong Si^+ 6150 absorption feature which is prominent in their early epoch spectra. The astrophysical difference between Type Ia and Ib/c SNe is that the former are found in all type of galaxies, from ellipticals through spirals to irregulars, whereas the latter are found exclusively in spiral galaxies, mostly associated

[c]They are classified Ib if strong He lines are present in their spectra, and Ic otherwise.

with spiral arms and frequently in the vicinities of large ionized nebulae (giant HII regions). These characteristics indicate that SNIb/c are the end result of a relatively young population of stars (ages less than 100 million years) while SNIa progenitors must be stellar systems that have considerably longer lifetimes, of the order of 10^9 years or more.

The progenitors of SNIa are believed to be stars that would not produce a SN explosion if there were single stars but that end up exploding because, after reaching the white dwarf stage, they accrete enough mass from a binary companion to exceed the Chandrasekhar mass, and ignite explosive nucleosynthesis in their cores. This process of "nuclear bomb" is expected to disrupt the entire star while synthetizing about 0.6 $M_\odot$ (Ia) of radioactive ^{56}Ni, which will power the SN optical light curves. SNIa are very luminous objects and form a quite homogeneous class of SNe, both in their maximum brightness and their time evolution. Thus, SNIa constitute ideal "standard candles" for distance determinations on cosmological scales (see Sect. 2.4).

Type Ib/c, on the other hand, must be significantly more massive because they are only found in spiral galaxies, and often associated with their spiral arms: this suggests progenitor masses in excess of $5M_\odot$. Therefore, either they represent the upper end of the SNIa class or they are a subclass of core collapse supernovae, possibly massive stars that occur in binary systems and are able to shed most of their outer H-rich layers before undergoing the explosion.

2.2 Radio Properties

A series of papers published over the past 18 years on radio supernovae (RSNe) has established the radio detection and/or radio evolution for 25 objects: 2 Type Ib supernovae, 5 Type Ic supernovae, and 18 Type II supernovae. A much larger list of almost 80 more SNe have low radio upper limits (*e.g.*, Weiler *et al.* 1986, 1998). A summary of the radio information can be found at: *http://rsd-www.nrl.navy.mil/7214/weiler/sne-home.html.*

All known RSNe appear to share common properties of: 1) non-thermal synchrotron emission with high brightness temperature; 2) a decrease in absorption with time, resulting in a smooth, rapid turn-on first at shorter wavelengths and later at longer wavelengths; 3) a power-law decline of the flux density with time at each wavelength after maximum flux density (optical depth ≈ 1) is reached at that wavelength; and 4) a final, asymptotic approach of spectral index α to an optically thin, non-thermal, constant negative value.

The current model for radio supernovae includes acceleration of relativistic electrons and compression of the magnetic field, necessary for synchrotron emission. These processes occur at the SN shock interface with a relatively high-density circumstellar medium (CSM) which has been ionized and heated by the initial UV/X-ray flash Chevalier (1982a,b). This CSM, which is also the source of the initial absorption, is presumed to have been established by a constant mass-loss ($\dot{M}$) rate, constant velocity (w) wind (i.e., $\rho \propto r^{-2}$) from a red supergiant (RSG) progenitor or a binary companion.

In our extensive study of the radio emission from SNe, several effects have been noted: 1) Type Ia are not radio emitters to the detection limit of the VLA[d]; 2) Type Ib/c are radio luminous with steeper spectral indices and a fast turn-on/turn-off, usually peaking at 6 cm near or before optical maximum; and 3) Type II show a range of radio luminosities with flatter spectral indices and a relatively slow turn-on/turn-off. These results lead to the conclusion that most SNII progenitors were RSGs, SNIb/c result from the explosion of more compact stars, members of relatively massive binary systems, and SNIa progenitors had little or no appreciable mass loss before exploding, excluding scenarios that involve binary systems with red giant companions. In some individual cases, it has also been possible to detect thermal hydrogen along the line of sight (Montes, Weiler & Panagia 1997, Chu *et al.* 1999), to demonstrate binary properties of the stellar system, and to show clumpiness of the circumstellar material (*e.g.*, Weiler, Sramek & Panagia 1990). More speculatively, it may be possible to provide distance estimates to radio supernovae (Weiler *et al.* 1998).

As an illustration we show that case of SN 1979C that exploded in April 1979 in the spiral galaxy NGC 4321=M100. This supernova was first detected in the radio in early 1980 (Weiler *et al.* 1981) and is still bright enough to be accurately measured at different frequencies, thus offering a unique opportunity to do a very thorough study of its radio properties, the nature of the radio emission mechanisms and the late evolution of the SN progenitor. Figure 1 displays the time evolution of SN 1979C radio flux at two frequencies (1.47 and 4.88 GHz). One can recognize the "canonical" properties (non-thermal spectral index, flux peaking at later times for lower frequencies, asymptotic power law decline) that allows one to estimate the circumstellar material distribution, corresponding to a constant velocity pre-SN wind with a mass loss rate of $\sim 2 \times 10^{-4} M_\odot$ /year and a probable $20 M_\odot$ progenitor. In addition, the almost sinosoidal modulation of the light curves reveals the presence of a $5 M_\odot$ binary companion in a slightly elliptical orbit (Weiler *et al.* 1992). And the marked jump up of the flux about ten years after the explosion (Montes *et*

[d]The VLA is operated by the NRAO of the AUI under a cooperative agreement with the NSF.

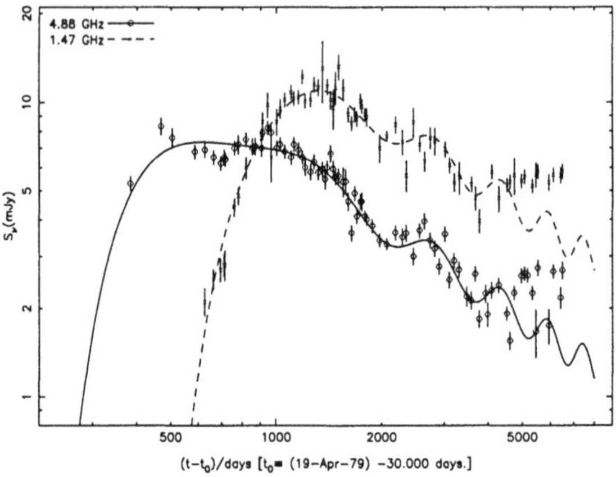

Figure 1. The 1.47 and 4.88 GHz radio emission of SN 1979C as a function of time

*al.*2000) suggests that the progenitor had a rather sudden change in its mass loss rate about 10,000 years before exploding, possibly due to pulsational instability (Bono & Panagia 1999, in preparation).

2.3 Supernova Rates

Determining the rates of SN explosions in galaxies requires knowing how many SNe have exploded in a large number of galaxies over the period of time during which they were monitored. Although it sounds easy, this process is rather tricky because data collected from literature usually do not report the control times over which the searches were conducted. On the other hand, more systematic searches that record all needed information have been started rather recently and the number of events thus recorded is rather limited, so that the statistics is still rather uncertain. In a recent study, Cappellaro *et al.*(1999) have thoroughly discussed this problem and, from the analysis of all combined data set available, have derived the most reliable SN rates for different types of galaxies. We have taken their rates and, for each galaxy class, we have renormalized them to the appropriate H-band ($\sim 1.65\mu m$) luminosity rather than the B-band ($\sim 0.45\mu m$) luminosity as done by Cappellaro *et al.* (1999). These new rates, displayed in Table 2, are essentially rates per unit galaxy mass because the H-band luminosity of a galaxy is roughly proportional to its mass. We see that SN rates closely reflect the star formation activity

Table 2. SUPERNOVA RATES *(in units of SN per century & per 10^{10} $L^H_\odot$)*

Galaxy Type	SNIa	SNIb/c	SNII	All SNe
E-S0	0.05±0.02	< 0.01	< 0.02	0.05±0.02
S0a-Sb	0.10±0.04	0.06±0.03	0.24±0.111	0.40±0.12
Sbc-Sd	0.21±0.08	0.14±0.07	0.86±0.35	1.21±0.37
Sm, Irr	0.59±0.24	0.33±0.24	0.97±0.60	1.87±0.67

of the various classes, not only for type II and Ib/c SNe but also for SNIa. In particular, the rates for SNII-Ib/c are 3-4 times higher in late type spirals (Sbc-d) and irregulars than they are in early type spirals (S0-Sb): this is clear evidence that star formation is considerably more active in the former than it is in the latter group. Also, we notice that late type galaxies (*i.e.* the ones with most active star formation, Sbc through Irr) have SNIa rates which are 4-10 times higher that the earliest type galaxies (*i.e.* E-S0). This is a new result (Panagia 1999, in preparation) and implies that SNIa progenitors are intermediate mass stars (say, $8 > M/M_\odot > 3$) and that early type galaxies are likely to capture and accrete star forming galaxies on a time scale of one to few billion years to replenish their reservoir of SNIa progenitors.

Recent estimates of the global history of star formation in the Universe were used by Madau, Della Valle & Panagia (1998) to compute the theoretical Type Ia and Type II SN rates as a function of cosmic time from the present epoch to high redshifts. They show that accurate measurements of the frequency of SN events already in the range $0 < z < 1$, and even more so at higher redshifts, will be valuable probes of the nature of Type Ia progenitors and the evolution of the stellar birthrate in the Universe.

2.4 Cosmological Applications

As mentioned before, SNIa are virtually ideal standard candles (*e.g.*, Hamuy *et al.* 1996) to measure distances of truly distant galaxies, currently up to redshift around 1 and, considerably more in the foreseeable future (for a review, see Macchetto and Panagia 1999). In particular, Hubble Space Telescope observations of Cepheids in parent galaxies of SNe Ia (an international project lead by Allan Sandage) have lead to very accurate determinations of their distances and the absolute magnitudes of SNIa at maximum light, *i.e.* $M_B = -19.50 \pm 0.06$ and $M_V = -19.49 \pm 0.06$ (*e.g.*, Sandage *et al.* 1996, Saha *et al.* 1999). Using these calibrations it is possible to determine the distances of much more distant SNe Ia. A direct comparison with the Hubble diagram (*i.e.* a plot of the observed magnitudes of SNIa versus their cosmological velocities) of distant SNe Ia ($30,000\ km\ s^{-1} > v > 3,000\ km\ s^{-1}$) gives a Hubble constant (*i.e.* the expansion rate of the local Universe) of $H_0 = 60 \pm 6\ km\ s^{-1}\ Mpc^{-1}$ (Saha *et al.* 1999). Studying more distant SNIa (*i.e.* $z > 0.1$) it has benn possible to extend our knowledge to other cosmological parameters. The preliminary results of two competing teams (Riess *et al.* 1998, Perlmutter *et al.* 1999) agree in indicating a non-empty inflationary Universe with parameters lying along the line $0.8\Omega_M - 0.6\Omega_\Lambda = -0.2 \pm 0.1$. Correspondingly, the age of the Universe can be bracketed within the interval 12.3–15.3 Gyrs to a 99.7% confidence level (Perlmutter *et al.* 1999).

3 Supernova 1987A in the Large Magellanic Cloud

3.1 The Early Story

Supernova 1987A was discovered on February 24, 1987 in the nearby, irregular galaxy, the Large Magellanic Cloud, which is located in the southern sky. SN 1987A is the first supernova to reach naked eye visibility after the one studied by Kepler in 1604 AD and is undoubtedly the supernova event best studied ever by the astronomers. Actually, despite the fact that SN 1987A has been more than hundred times fainter than its illustrious predecessors in the last millennium, it has been observed in such a detail and with such an accuracy that we can define this event as a *first* under many aspects (e.g. neutrino flux, identification of its progenitor, gamma ray flux) and in any case as the *best* of all. Reviews of both early and more recent observations and their implications can be found in Arnett *et al.* (1989) and Gilmozzi and Panagia (1999), respectively.

SN 1987A early evolution has been highly unusual and completely at variance with the *wisest* expectations. It brightened much faster than any other known supernova: in about one day it jumped from 12th up to 5th magnitude at optical wavelengths, corresponding to an increase of about a factor of thousand in luminosity. However, equally soon its rise leveled off and took a much slower pace indicating that this supernova would have never reached those peaks in luminosity as the astronomers were expecting. Similarly, in the ultraviolet, the flux initially was very high, even higher than in the optical. But since the very first observation, made with the International Ultraviolet Explorer (IUE in short) satellite less than fourteen hours after the discovery, the ultraviolet flux declined very quickly, by almost a factor of ten per day for several days. It looked as if it was going to be a quite disappointing event and, for sure, quite peculiar, thus not suited to provide any useful information about the other more common types of supernova explosions. But, fortunately, this proved not to be true and soon it became apparent that SN 1987A is the most valuable mean to test our ideas and theories about the explosion of supernovae.

And even particle emission was directly measured from Earth: on February 23, around 7:36 Greenwich time, the neutrino telescope ("Kamiokande II", a big cylindrical "tub" of water, 16 m in diameter and 17 m in height, containing about 3300 m^3 of water, located in the Kamioka mine in Japan, about 1000 m underground) recorded the arrival of 9 neutrinos within an interval of 2 seconds and 3 more 9 to 13 seconds after the first one. Simultaneously, the same event was revealed by the IMB detector (located in the Morton-Thiokol salt mine near Faiport, Ohio) and by the "Baksan" neutrino telescope (located in the North Caucasus Mountains, under Mount Andyrchi) which recorded 8 and 5 neutrinos, respectively, within few seconds from each other. This makes a total of 25 neutrinos from an explosion that allegedly produces 10 billions of billions of billions of billions of billions of billions of them! But a little more than two dozens neutrinos was more than enough to verify and confirm the theoretical predictions made for the core collapse of a massive star (*e.g.*, Arnett *et al.* 1989 and references therein). This process was believed to be the cause of the explosion of massive stars at the end of their lives, and SN 1987A provided the experimental proof that the theoretical model was sound and correct, promoting it from a nice theory to the description of the truth.

3.2 SN 1987A Progenitor Star

From both the presence of hydrogen in the ejected matter and the conspicuous flux of neutrinos, it was clear that the star which had exploded was quite massive, about twenty times more than our Sun. And all of the disappointing peculiarities were due to the fact that just before the explosion the supernova progenitor was a blue supergiant star instead of being a red supergiant as common wisdom was predicting. There is no doubt about this explanation because SN 1987A is exactly at the same position as that of a well known blue supergiant, Sk $-69°$ 202. And the IUE indicated that such a star was not shining any more after the explosion: the blue supergiant had gone BANG (Gilmozzi *et al.* 1987, Kirshner *et al.* 1987).

On the other hand, common wisdom cannot be wrong and it was *not* quite wrong, after all. At later times, in late May 1987, the IUE revealed the presence of emission lines of nitrogen, oxygen, carbon and helium in the ultraviolet spectrum. They kept increasing in intensity with time and proved to be quite narrow, indicating that the emitting matter was moving at much lower speeds (less than a factor of hundred slower) than the supernova ejecta. The chemical abundances and the slow motion were clear sign that that was matter ejected by a *red* supergiant in the form of a gentle wind. But there was no such a star in sight just before the explosion. Therefore, the same star that exploded, had also been a red supergiant, less than hundred thousand years before the explosion itself: a short time in the history of the star but quite enough to make all the difference.

3.3 Explosive Nucleosynthesis

The optical flux reached a maximum around mid-May, 1987, and declined at a quick pace until the end of June, 1987, when rather abruptly it slowed down, setting at a much more gentle decline of about 1% a day (Pun *et al.* 1995). Such a decay has been followed since then quite regularly: a perfect constant decay with a characteristic time of 114 days, just the same as that of the radioactive isotope of cobalt, ^{56}Co, while transforming into iron. This is the best evidence for the occurrence of nucleosynthesis during the very explosion: ^{56}Co is in fact the result of ^{56}Ni and this latter can be formed at the high temperatures which occur after the core collapse of a massive star. So now, not only are we sure that such a process is operating in a supernova explosion, just as theorists predicted, but we can also determine the amount of nickel produced in the explosion, slightly less than 8/100 of a solar mass or, approximately, 1% of the mass of the stellar core before the explosion. And the hard X-ray emission detected since July 1987 and the subsequent detection of gamma-ray emission confirm the reality of this process and provide more details about its exact occurrence (*e.g.*, Arnett *et al.* 1989 and references therein).

3.4 HST Observations

The Hubble Space Telescope was not in operation when the supernova exploded, but it did not miss its opportunity in due time and its first images, taken with the ESA-FOC in August 23 and 24, 1990, revealed the inner circumstellar ring in all its "glory" and detail (*cf.* Jakobsen *et al.* 1991), showing that, despite spherical aberration, HST was not a complete disaster, after all.

Since those early times, Hubble has kept an attentive eye on SN 1987A, obtaining both imaging and spectrographic observations (*e.g.*, Fig. 2) at least

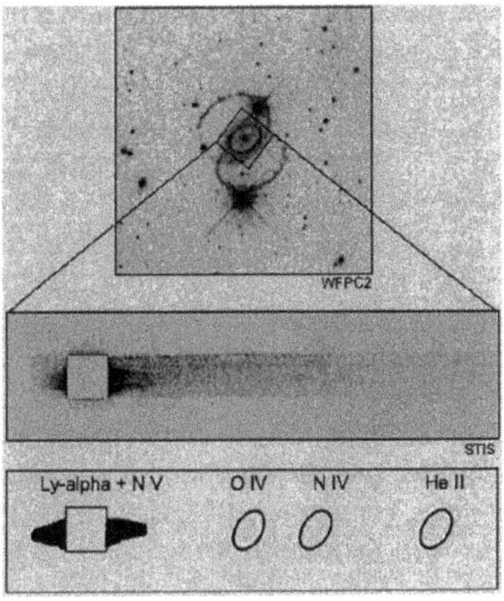

Figure 2. The field of view centered on SN 1987A, as viewed by the HST-WFC2, was observed with HST-STIS and the resulting ultraviolet spectrum is shown in the bottom panels.

once a year, accumulating valuable data and revealing quite a number of interesting results (see Gilmozzi & Panagia 1999), such as:
- The sequence of images obtained over more than 8 years has allowed us to measure the expansion of the supernova material directly: this the first time it has ever been possible and has permitted to identify the correct models to understand the explosion phenomenon (Pun *et al.* 1999, in preparation).
- The origin and the nature of the beautiful circumstellar rings are still partly a mistery. They have been measured to expand rather slowly, about 10-20 $km\ s^{-1}$, *i.e.* 100-2000 times slower than the SN ejecta, and to be highly N rich: both these aspects indicate that the rings were expelled from the progenitor star when it was a red supergiant, about 20,000 years before the explosion (Panagia *et al.* 1996). However, one would have expected such a star to eject material in a more regular fashion, just pushing away material gently in all directions rather than puffing rings like a pipe smoker. Another puzzle is that the star was observed to be a "blue" supergiant in the years before the explosion, and not a red supergiant anymore. This forces one to admit that the star had a rather fast evolution, which was not predicted by

"standard" stellar evolution theory, and still is hard to understand fully.
- The highest velocity material expelled in SN 1987A explosion has been detected for the first time by the Space Telescope Imaging Spectrograph (STIS) (*e.g.,* Sonneborn *et al.* 1998). The spectrograph has found the first direct evidence for material from SN 1987A colliding with its inner circumstellar ring. The fastest debris, moving at 15,000 $km\ s^{-1}$ are now colliding with the slower moving gas of the inner circumstellar ring (Fig. 2) .

In less than a decade the full force of the supernova fast material will hit the inner ring, heating and exciting its gas and producing a new series of cosmic fireworks that will offer a spectacular view for several years. This is going the "beginning of the end" because in about another century most, if not all, the material in the rings will be swept away and disappear, loosing their identities and merging into the interstellar medium of the Large Magellanic Cloud. This is not a complete loss, however, because by studying this destructive process, we will be able to probe the ring material with a detail and an accuracy which are not possible with current observations.

References

1. W.D. Arnett *et al.*, *ARA&A* **27**, 629 (1989).
2. W. Baade and F. Zwicky, *Proc. Nat. Acad. Sci. U.S.* **20**, 254 (1934).
3. E. Cappellaro, R. Evans and M. Turatto, *Astron. & Astrophys.* **351**, 459 (1999).
4. R.A. Chevalier, *Astrophys. J.* **259**, 302 (1982a).
5. R.A. Chevalier, *Astrophys. J. Lett.* **259**, L85 (1982b).
6. Y.H. Chu *et al.*, *Astrophys. J. Lett.* **512**, L51 (1999).
7. R. Gilmozzi *et al.*, *Nature* **328**, 318 (1987).
8. R. Gilmozzi and N. Panagia, *Mem. Soc. Astron. It.* **70**, 583 (1999).
9. M. Hamuy *et al.*, *Astron. J.* **112**, 2391 (1994).
10. R.P. Kirshner *et al.*, *Astrophys. J.* **320**, 602 (1987).
11. P. Jakobsen *et al.*, *Astrophys. J.* **369**, L63 (1991).
12. F.D. Macchetto and N. Panagia, in *Post-Hipparcos Cosmic Candles*, eds. A. Heck & F. Caputo, p. 225-245 (Kluwer-Holland, 1999).
13. P. Madau, M. Della Valle and N. Panagia, *Mon. Not. Roy. Astron. Soc* **297**, L17 (1998).
14. M.J. Montes, K.W. Weiler, & N. Panagia, *Astrophys. J.* **488**, 792 (1997).
15. M.J. Montes *et al.*, *Astrophys. J.*, in press (2000).
16. N. Panagia, 1985, in *Supernovae As Distance Indicators*, Lect. Notes Phys. Vol. 224, p. 226-240 (Springer-Verlag-Berlin, 1985).
17. N. Panagia *et al.*, *Astrophys. J.* **380**, L23 (1991).

18. N. Panagia *et al.*, *Astrophys. J.* **457**, 604 (1996).
19. S. Perlmutter *et al.*, *Astrophys. J.* **517**, 565 (1999).
20. C.S.J. Pun *et al.*, *Astrophys. J. Suppl.* **99**, 223 (1995).
21. A.G. Riess *et al.*, *Astrophys. J.* **504**, 935 (1998).
22. A. Saha *et al.*, *Astrophys. J.* **522**, 802 (1999).
23. A. Sandage *et al.*, *Astrophys. J.* **460**, L15 (1996).
24. G. Sonneborn *et al.*, *Astrophys. J.* **492**, 139 (1998).
25. K.W. Weiler *et al.*, *Astrophys. J.* **243**, L151 (1991).
26. K.W. Weiler *et al.*, *Astrophys. J.* **301**, 790 (1986).
27. K.W. Weiler, N. Panagia, & R.A. Sramek, *Astrophys. J.* **364**, 611 (1990).
28. K.W. Weiler *et al.*, *Astrophys. J.* **500**, 51 (1998).

WHAT HAVE WE LEARNED ABOUT GAMMA RAY BURSTS FROM THEIR AFTERGLOWS?

MARIO VIETRI

Dip. Fisica E. Amaldi, via della Vasca Navale 84, 00146 Roma, Italy
E-mail: vietri@corelli.fis.uniroma3.it

The discovery of GRBs' afterglows has allowed us to establish several facts: their distance and energy scales, the fact that they are due to explosions, that the explosions are relativistic, and that the afterglow emission mechanism is synchrotron radiation. On the other hand, recent data have shown that the fireball model is wrong when it comes to the emission mechanism of the true burst (which is unlikely to be synchrotron again) and that shocks are not external. Besides these relatively tame points, I will also discuss the less well established physics of the energy deposition mechanism, as well as the possible burst progenitors.

1 Introduction

Gamma ray bursts (GRBs) were discovered in 1969 (Klebesadel, Strong and Olson 1973) by American satellites of the *Vela* class aimed at verifying Russian compliance with the nuclear atmospheric test ban treaty. Though the discovery was made in 1969, the paper appeared only four years later because the authors had lingering doubts about the reality of the effects they had discovered. Since then, several thousands of bursts have been observed by a more than a dozen different satellites, but it is remarkable that the basic burst features outlined in the abstract of the 1969 paper (photons in the range $0.2-1.5$ MeV, durations of $0.1-30$ s, fluences in the range $10^{-5}-2\times10^{-4}$ $ergs\ cm^{-2}$) have remained substantially unchanged.

Current evidence (Fishman and Meegan 1995) has highlighted a wide ($0.01-100$ s) duration distribution, with hints of a bimodality which is claimed to correlate (at the 2.5σ level) with spectral properties. All bursts' spectra observed so far are strictly non–thermal, and there has never been any confirmation by BATSE of a supposed thermal component (nor of cyclotron lines or precursors, for this matter) claimed in previous reports. A remarkable feature reported by BATSE is the bewildering diversity of light curves, ranging from impulsive ones (a spike followed by a slower decay, nicknamed FREDs for Fast Rise-Exponential Decay), to smooth ones, to long ones with amazingly sharp fluctuations, including even some with a strongly periodic appearance (two such examples are the 'hand' and the 'comb', so nicknamed from the number of high–Q, regularly repeating sharp spikes).

The most exceptional result from BATSE, though, was the sky distri-

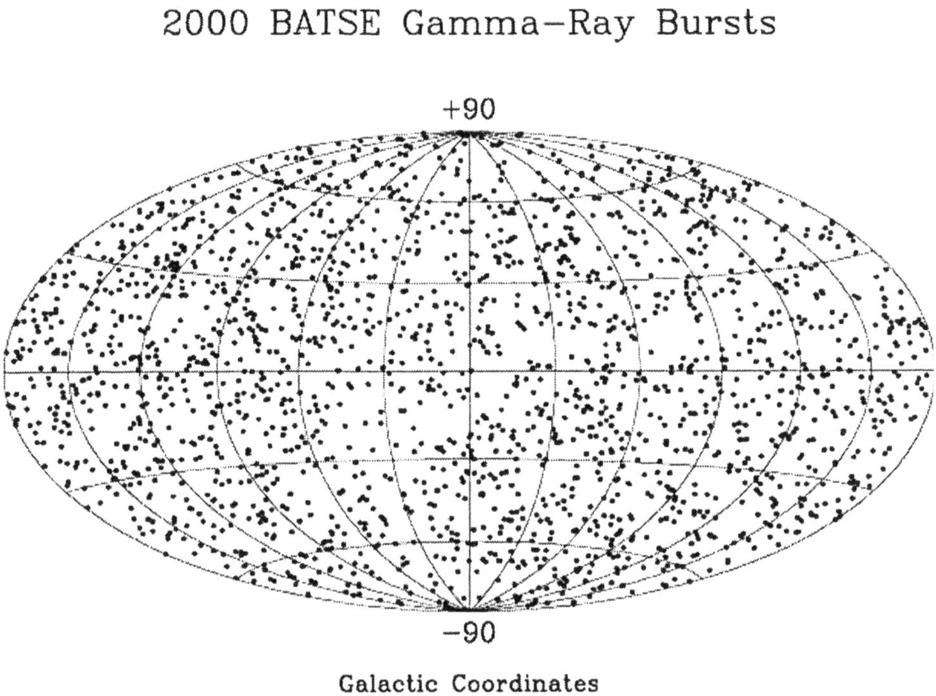

Figure 1. Burst distribution on the plane of the sky.

bution of the bursts (Fig.1). It was obvious from it that the bursts *had* to be extragalactic, as already discussed by theorists (Usov and Chibisov 1975, Paczyński 1986).

2 Afterglows

The next major step was triggered by BeppoSAX: in the summer of 1996, L. Piro and his coworkers located in archival data of the satellite the soft X-ray counterpart of a GRB (GRB 960720). They immediately conceived the idea of implementing a procedure to follow the next burst in real time, by re-orienting the whole satellite, after the initial detection by the Wide Field Cameras, so that the more sensitive Narrow Field Instruments could pinpoint the burst location to within 45 arcsecs, a feat never achieved in such short times, and by a single satellite. After an initial snafu (GRB 970111), the gigantic effort paid off with the discovery of the X-ray afterglow of GRB 970228 (Costa *et al.*,

GRB	z	E_{iso}
970228	0.695	5×10^{51} erg
970508	0.835	2×10^{51} erg
971214	3.4	3×10^{53} erg
980703	0.93	3×10^{53} erg
990123	1.7	4×10^{54} erg
990510	1.6	2×10^{53} erg
990712	0.43	

1997), immediately followed by the discovery of its fading optical counterpart (van Paradijs *et al.*, 1997), obtained through a search inside the WFC error box, in perfect agreement with theoretical predictions (Vietri 1997a, Mèszàros and Rees 1997).

After the detection of the optical counterpart, the door was open to find the bursts' redshifts: Table I summarizes the status of our current knowledge (September 1999); bursts' luminosities are for isotropic sources. Two comments are in order. First, the bursts have *prima facie* a redshift distribution not unlike that of AGNs and of the Star Formation Rate (SFR). The initial hope that they might trace an even more distant and elusive Pop III, triggered by the fact that the second redshift detected was also the largest so far (GRB 971214, $z = 3.4$), has now vanished. Second, in order to place the energy release of GRB 990123 in context, one should notice that 4×10^{54} $ergs$ is the energy obtained by converting the rest–mass of two solar masses, or, alternatively, the energy emitted by the whole Universe out to $z \approx 1$ within the burst duration. So, a single (perhaps double) star outshines the whole Universe.

Besides the distance and energy scales, the major impact of the discovery of afterglows has been the establishment of some key features of the fireball model (Rees and Mèszàros 1992):

1. bursts are due to explosions, as evidenced by their power–law behaviour;

2. the explosions are relativistic, as proved by the disappearence of radio flares;

3. the burst emission is due to synchrotron emission, as shown by the afterglow spectrum, and its optical polarization.

I will illustrate these points in the following, but, lest we become too proud, we should also remember that the fireball model has met some failures. The original version of the model (Mèszàros and Rees 1993) advocated the dissipation of the explosion energy at external shocks (*i.e.*, those with the interstellar medium). Sari and Piran (1997), following a point originally made by Ruderman (1975) showed that these shocks smooth out millisecond timescale variability, which can only be maintained by the internal shocks proposed by Paczyǹski and Xu (1994). Also, the fireball model originally ascribed even the emission from the burst proper (as opposed to the afterglow) to optically thin synchrotron processes; I will discuss in the section *Embarrassments* why this is exceedingly unlikely. Furthermore, even the last tenet of mid–90s common wisdom, *i.e.*, that bursts are due to neutron binary mergers, does not look too promising at the moment (since some bursts seem to be located inside star forming regions, incompatible with the long spiral–in time), though of course it is by no means ruled out yet.

2.1 The fireball model

Here, one may assume that an unknown agent deposits $10^{51} - 10^{54}$ *ergs* inside a small volume of linear dimension $\approx 10^6 - 10^7$ *cm*. The resulting typical energy density corresponds to a temperature of a few MeVs, so that electrons and positrons cannot be bound by any known gravitational field. In these conditions, optical depths for all known processes exceed 10^{10}. The fluid expands because of its purely thermal pressure, converting internal into bulk kinetic energy. Parametrizing the baryon component mass as $M_b \equiv E/\eta c^2$, it can be shown that, for $1 \leq \eta \leq 3 \times 10^5$ (Mèszàros, Laguna, Rees 1993) the fluid achieves quickly (the fluid Lorenz factor increases as $\gamma \propto r$) a coasting Lorenz factor of $\gamma \approx \eta$.

The requisite asymptotic Lorenz factor is dictated by observations: photons up to $\epsilon_{ex} \approx 18\ GeV$ have been observed by EGRET from bursts (Fishman and Meegan 1995). For these photons to evade collisions with other photons, and thus electron/positron pair production, it is necessary that, in the reference frame in which a typical burst photon (with $\epsilon \approx 1\ MeV$) and the exceptional photon are emitted, they appear as below pair production threshold: thus we must have $\epsilon' \epsilon'_{ex} \leq 2 m_e c^2$. Since $\epsilon' \approx \epsilon/\gamma$, and similarly for the other photons, we find (Baring 1993)

$$\gamma \approx 300 \left(\frac{\epsilon}{1\ MeV} \frac{\epsilon_{ex}}{10\ GeV} \right)^{1/2} . \tag{1}$$

From what we said above, we thus require a maximum baryon contamination, in an explosion of energy E, of $M_b < 10^{-6} M_{\odot} (E/10^{51}\ erg)(300/\eta)$.

The energy release is now assumed to be in the form of an inhomogeneous wind, with parts having a Lorenz factor larger than parts emitted previously.

This leads to shell collisions (the internal shock model) at radii r_{sh} which allow a time–scale variablity $\delta t \approx r_{sh}/2\gamma^2 c$; for $\delta t = 1\ ms$, $r_{sh} \approx 10^{13}\ cm$, which fixes the internal shock radii. Particle acceleration at these internal shocks and ensuing non–thermal emission is thought to lead to the formation of the burst proper. At larger radii, a shock with the surrounding ISM forms, and shell deceleration begins at a radius $r_{ag} = (3E/4\pi n m_p c^2 \gamma^2)^{1/3} \approx 10^{17}\ cm$ for a $n = 1\ cm^{-3}$ particle density typical of galactic disks. It is thought that the afterglow begins when the shell begins the slowdown, as this drives a marginally relativistic shock into the ejecta, thusly extracting a further fraction of their bulk kinetic energy.

2.2 Why explosions

The success of the fireball model lies in this, that it decouples the problem of the energy injection mechanism from the following evolution, which is, furthermore, an essentially hydrodynamical problem. It can be shown, in fact (Waxman 1997) that the evolution of the external shock is adiabatic, that the shock Lorenz factor decreases as $\gamma \propto r^{-3/2}$ because of the inertia of the swept–up matter, and thus r scales with observer's time as $t = r/\gamma^2 c \rightarrow \gamma \propto t^{-3/8}$ (for a radiative solution $\gamma \propto r^{-3/7}$, Vietri 1997b). If afterglow emission is due to optically thin synchrotron in a magnetic field in near–equipartition with post–shock energy density, it can be shown that $B \propto \gamma$, that the typical synchrotron frequency at the spectral peak $\nu_m \propto \gamma B \gamma_e^2 \propto \gamma^4$ (where $\gamma_e \propto \gamma$ is the lowest post–shock electron Lorenz factor), and that $F(\nu_m) \propto t^{-3\beta/2}$, where β is the afterglow spectral slope. As it can be seen, these expectations are based exclusively upon the hydrodynamical evolution (and the synchrotron spectrum), and are thus reasonably robust.

We thus expect power–law time decays, a characteristic of strong explosions (see the Sedov analogy!), with time– and spectral–indices closely related. This is what is observed everywhere, from the X–ray through the optical to the radio, (see Piro and Fruchter, this volume), the few exceptions being discussed later on. In fact, the equality of the time–decay index of the X–ray and optical data in afterglows of individual sources has been taken as the key element to show that emission in the different bands is due to the same source. Time indices in the X–ray are in the range 0.7 – 2.2 (Frontera *et al.*, in preparation).

2.3 Why synchrotron spectrum in the afterglow

After having established that bursts are due to explosions, we happily learn that afterglows emit through synchrotron processes. In fig. 2 (Galama *et al.*,

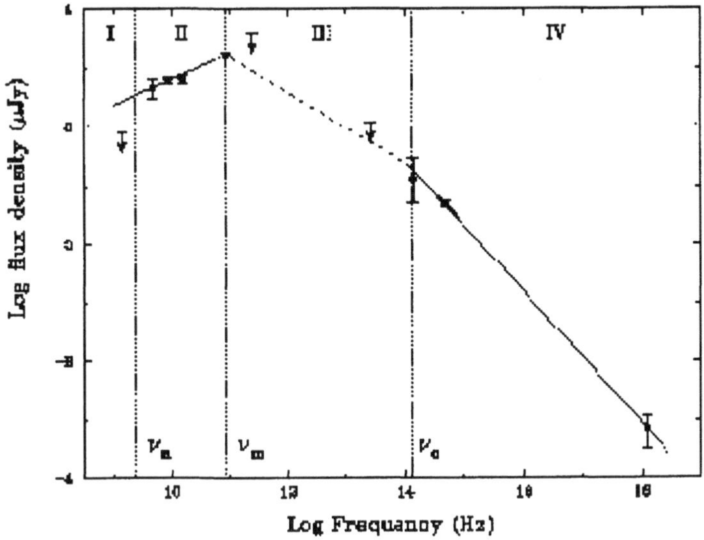

Figure 2. Simultaneous spectrum of the afterglow of GRB 970508, from Galama *et al.*, 1998.

1998), we show the superposition of theoretical expectations for an optically thin synchrotron spectrum (including the cooling break at $\nu \approx 10^{14}\ Hz$) with observations for GRB 970508. The remarkable agreement is even more exciting as we remark that observations are not truly simultaneous, but are scaled back to the same time by means of the theoretically expected laws for time–decay, thus simultaneoulsy testing the correctness of our hydro. Another piece of evidence comes from the discovery of polarization in the optical afterglow of GRB 990510 (Fig. 3, Covino *et al.*, 1999, Wijers *et al.*, 1999). This polarization may appear small ($\approx 2\%$), but it is surely not due to Galactic effects: stars in the same field show a comparable degree of polarization, but along an axis different by about 50°. Also, polarization in the source galaxy is unlikely, because of a very stringent upper limit on the reddening due to this galaxy (Covino *et al.*, 1999). The only remaining question mark is emission from an anisotropic source, but this would require a disk of $10^{18}\ cm$ to survive the intense γ ray (and X, and UV) flash: though not excluded, it does not look likely.

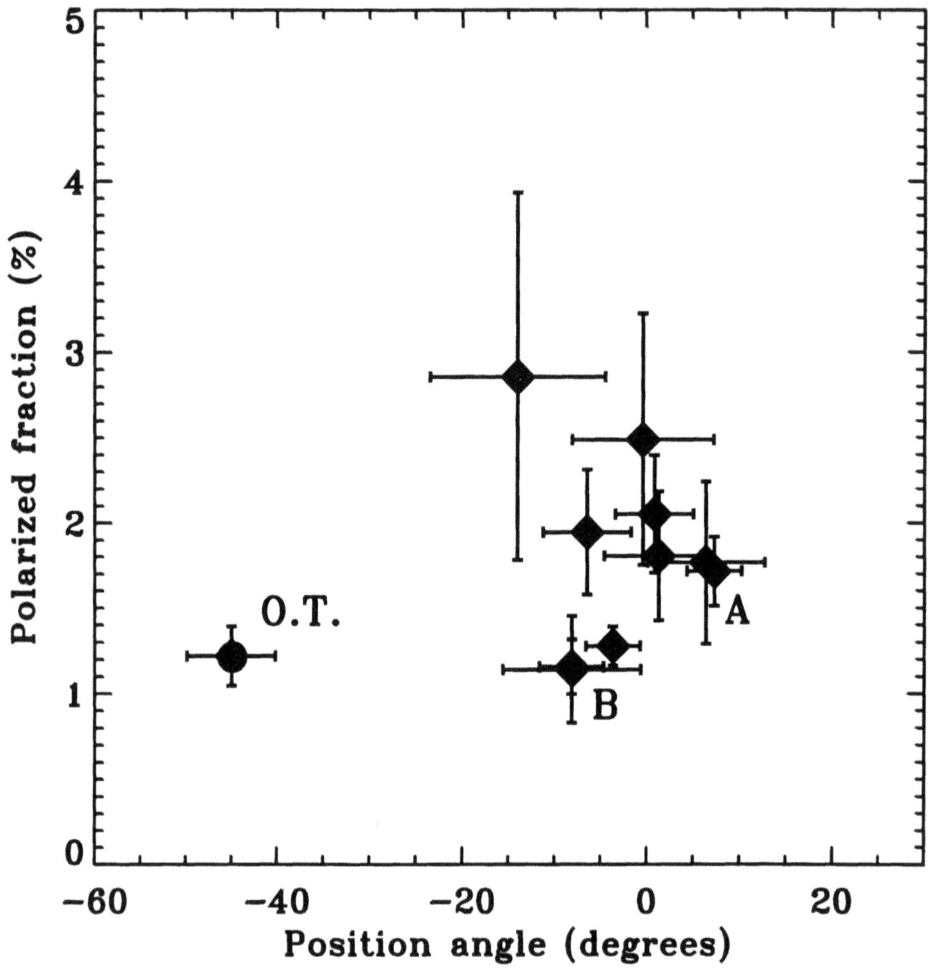

Figure 3. Polarization amplitude and position angle for optical afterglow of GRB 990510, from Covino *et al.*, 1999

2.4 Why relativistic expansion

Radio observations of the first burst observed so far (GRB 970508, Frail *et al.*, 1997) showed puzzling fluctuations by about a factor of 2 in the flux, over a time–scale of days, disappearing after about 30 days from the burst (Fig. 4).

This extreme, and unique behaviour, was explained by Goodman (1997), who showed that it is due to interference of rays travelling along different paths through the ISM, and randomly deflected by the spatially varying refractive index of the turbulent ISM. The wonderful upshot of this otherwise marginal phenomenon, is that these effects cease whenever the source expands beyond a radius

$$R = 10^{17}\ cm \frac{\nu_{10}^{6/5}}{d_{sc,kpc} h_{75}} \left(\frac{SM}{10^{-2.5} m^{-20/3}\ kpc} \right)^{-3/5}, \tag{2}$$

where ν_{10} is the radio observing frequency in units of $10^{10}\ Hz$, $d_{sc,kpc}$ is the distance of the ISM from us (assumed to be a uniform scattering screen), and SM is the Galactic scattering measure, scaled to a typical Galactic value. The existence of interference effects is made more convincing by the amplitude of the average increase (a factor of 2, as observed), the correctness in the prediction of the time–interval between different peaks, and of the decorrelation bandwidth. Since flares disappear after about 30 days, it means that the average speed of the radio source is $R/30days = 3 \times 10^{10}\ cm\ s^{-1}$. So we see directly that GRB 970508 expanded at an average speed of c over a whole month, giving us a direct observational proof that the source is highly relativistic. This proof is completely equivalent to superluminal motions in blazars, and is the strongest evidence in favor of the fireball model.

2.5 GRB 970508: our best case

The afterglow of GRB 970508 is our best case so far: it is in fact a burst for which not only do we know the redshift, but also a radio source that has been monitored for more than 400 days after the explosion (Frail, Waxman and Kulkarni 2000). Through these observations we can see the transition to a sub–relativistic regime at $t \approx 100\ d$, measure the total energetics of the following Sedov phase (unencumbered by relativistic effects!) $E_{New} = 5 \times 10^{50}\ ergs$, determine two elusive parameters, $\epsilon_{eq} = 0.5$ and $\epsilon_B = 0.5$ (the efficiencies with which energy is transfered to post–shock electrons by protons, and with which an equipartition field is built up), and the density of the surrounding medium $n \approx 0.4\ cm^{-3}$. All of these values look reasonable (perhaps ϵ_{eq} and ϵ_B exceed our expectations by a factor of 10, a fact that could be remedied by introducing a slight density gradient which would keep the shock more efficient), so that our confidence in the external–shock–in–the–ISM model is boosted.

Another precious consequence of these late–time observations is that they yield information on beaming and energetics. In fact, GRB 970508 appeared

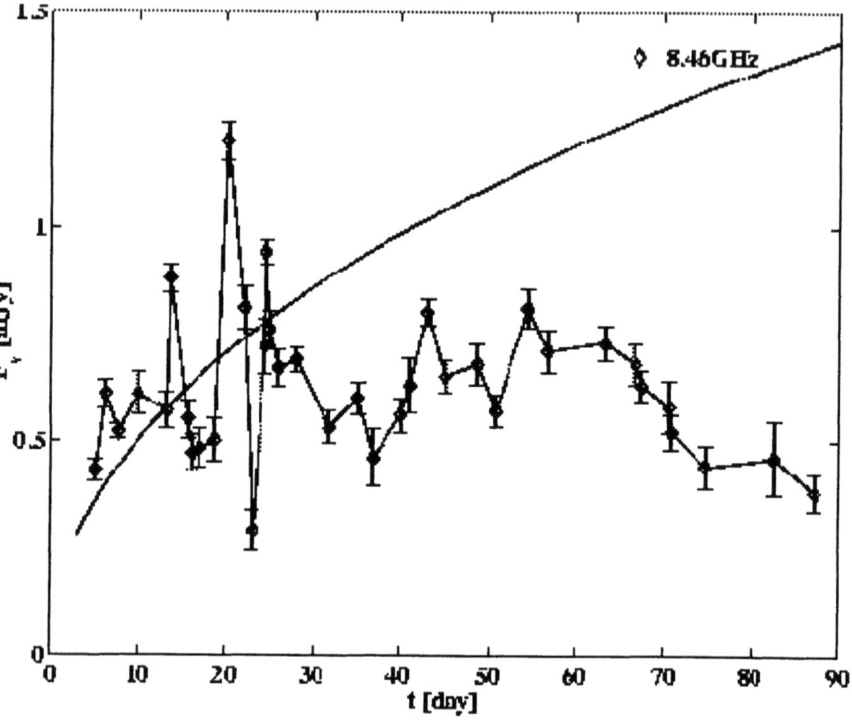

Figure 4. VLA observations at 8.46 GHz of the afterglow of GRB 970508, from Waxman,Frail and Kulkarni 1998.

to have a kinetic energy of $E_{rel} = 5 \times 10^{51}$ erg when in the relativistic phase, a measurement which can be reconciled with E_{New} (remember that the expansion is adiabatic, so that we must have $E_{New} = E_{rel}$!) only if the unknown beaming angle, assumed $= 4\pi$ in deriving E_{rel}, is smaller than 4π by the factor E_{New}/E_{rel}; we thus have the only measurement of $\delta\Omega/4\pi = 0.1$, so far. This already rules out all classes of models requiring unplausibly large amounts of beaming, 10^{-8} or even beyond. Hopefully, more such measurements will come in the future, since this observationally heavy method is subject to many fewer uncertainties than the competing method of trying to locate breaks in the time–decay of afterglows. Also, the radiative efficiency of the burst can be estimated: correcting the observed burst energy release $E_{GRB} = 2 \times 10^{51}$ erg for the same beaming factor, the radiative efficiency is $E_{GRB}\delta\Omega/4\pi/(E_{New} + E_{rel}\delta\Omega/4\pi) = 0.3$, again a unique determination. Notice however that this figure is subject to a systematic uncertainty: we do not know whether the beaming fraction is the same for the burst proper and for the afterglow.

3 Embarrassments

Something is rotten in the fireball kingdom as well, namely, departures from pure power–law behaviours, and the spectra of the bursts proper.

3.1 Unpowerlawness

Departures from power–laws are expected when one considers the extremely idealized character of the solutions discussed so far: perfect spherical symmetry, uniform surrounding medium, smooth wind from the explosion, ϵ_{eq} and ϵ_B constant in space and time. The tricky point here is to disentangle these distinct factors. In GRB 970508 and GRB 970828 (Piro *et al.*, 1999, Yoshida *et al.*, 1999) a major departure was observed in the X–ray emission, within a couple of days from the burst; they constitute the single, largest violations observed so far, in terms of number of photons. It is remarkable that spectral variations were simultaneously observed, and that both bursts showed traces (at the 2.7σ significance level) of an iron emission line. The similarity of the bursts' behaviour argues in favor of the reality of these spectral features, which have been interpreted as thermal emission from a surrounding stellar–size leftover, pre–expelled by the burst's progenitor (Lazzati *et al.*, 1999, Vietri *et al.*, 1999). Clearly, these departures hold major pieces of information on the bursts' surroundings, and the nature of bursts' progenitors.

It has been argued (Rhoads 1997) that, whenever the afterglow shell decelerates to below $\gamma \approx 1/\theta$, where θ is the beam semi–opening angle, emission should decrease because of the lack of emitting surface, compared to an isotropic source. But, in view of the existence of clear environmental effects (GRB 970508 and GRB 970828), it appears premature to put much stock in the interpretation of time–power–law breaks as due to beaming effects. And equally, it appears to this reviewer that the same comment applies to the interpretation of a resurgence of flux as due to the appearance of a SN remnant behind the shell. The major uncertainty here is the non–uniqueness of the interpretation: Waxman and Draine (2000) have shown that effects due to dust can mimic the same phenomenon.

3.2 Bursts' spectra

A clear prediction of the emission of optically thin synchrotron is that the low–photon–energy spectra should scale like $dN_\nu/d\nu \propto \nu^\alpha$, with $\alpha = -3/2$, since the emission is in the fast cooling regime. Within thin synchrotron, there is no way to obtain $\alpha > -3/2$. This early–recognized requirement (Katz 1994) is so inescapable that it has been dubbed the 'line of death'. Observations are notoriously discordant with this prediction. Preece *et al.*(1999) have shown that, for more than 1000 bursts, α is distributed like a bell between -2 and 0, with mean $\bar{\alpha} \approx -1$. The tail of this distribution also contains a few tens of objects with $\alpha \approx +1$. An example of these can be found in Frontera *et al.*, 1999 (GRB 970111), which is instructive since BeppoSAX has better coverage of the critical, low–photon–energy region. In particular, BATSE seems to loose sensitivity below $\approx 30\ keV$, but this is still not enough to explain away the discrepancy with the theory. Also, Preece *et al.*, 1999, showed that the time–integrated spectral energy distribution has a peak at a photon energy $\epsilon_{pk} \approx 200\ keV$, and that ϵ_{pk} has a very small variance from burst to burst. Again, this does not seem dependent upon BATSE's lack of sensitivity above $700\ keV$, and again this has no explanation within the classic fireball model.

Many people agree that the neglect of Inverse Compton processes may be the root of the disagreement: the trick here is to devise a fireball model that smoothly incorporates them. One should remember that the details of the fireball evolution are *generic*, *i.e.*, they do not depend upon any detailed property of the source, so that things like the radius at which the fireball becomes optically thin (to pairs or baryonic electrons), the radius at which acceleration ends, the equipartition magnetic field, and so on, are all reliably and inescapably fixed by the outflow's global properties. A step toward the solution has been made by Ghisellini and Celotti (1999) who remarked that at least some bursts have compactness parameters $l = 10(L/10^{53}\ erg\ s^{-1})(300/\gamma)^5 \gg 1$. Under these conditions, a pair plasma will form, nearly thermalized at $kT \approx m_e c^2$, and with Thomson optical depth $\tau_T \approx 10$. The modifications which this plasma will bring to the burst's spectrum are currently unknown, but it may be remarked that this configuration will be optically thick to both high–energy synchrotron photons due to non–thermal electrons accelerated at the internal shocks, and to low–energy cyclotron photons emitted by the thermal plasma, but it will be optically thin in the intermediate region reached by cyclotron photons up-scattered via IC processes off non–thermal electrons. A model along this line (*i.e.*, upscattering of cyclotron photons by highly relativistic electrons) is in preparation (Vietri 2000a), but it remains to be seen whether it (like any other model, of course) can simultaneously explain the spectral shape and the narrow range of the spectral distribution peak energy ϵ_{pk}.

4 On the central engine

As remarked several times already, the fireball evolution is independent of the source nature. The only existing constraint is the maximum amount of baryon contamination, which is

$$M_b = \frac{E}{\eta c^2} = 10^{-6} M_\odot \frac{E}{10^{51}\ erg} \frac{300}{\gamma}\ . \tag{3}$$

This is a remarkably small value: since the inferred luminosities exceed the Eddington luminosity by 13 orders of magnitude, they clearly have all it takes to disrupt a whole star, no matter how compact. Yet, the energy deposition must somehow occur outside the main mass, lest the explosion be slowed down to less relativistic, or even possibly Newtonian speeds. In order to satisfy this constraint, it has emerged that the most favorable configuration has a stellar–mass black hole ($M_{BH} \approx 3 - 10 M_\odot$) surrounded by a thick torus of matter ($M_t \approx 0.01 - 1 M_\odot$, with $\rho \approx 10^{10}\ g\ cm^{-3}$). The presence of a black hole is *not* required by observations in any way: models involving neutron stars are still admissible, the advantage of having a black hole being only the deeper potential well: you may get more energy out per unit accreted mass. The configuration thusly envisaged has a cone surrounding the symmetry axis devoid of baryons, since all models leading to this configuration have large amounts of specific angular momentum, and thus baryons close to the rotation axis either are not there, or have accreted onto the black hole due to their lack of centrifugal support.

4.1 Energy release mechanism

There are two major mechanisms for energy release discussed in the literature, the first one to be proposed (Berezinsky and Prilutskii 1986) being the reaction $\nu + \bar{\nu} \rightarrow e^- + e^+$. Neutrinos have non–negligible mean free paths in the tori envisaged here, so that this annihilation reaction will take place not inside tori themselves, where they are preferentially generated because densities are highest, but in a larger volume surrounding the source. This is both a blessing and a disgrace: by occupying a larger volume, the probability that every neutrino finds its antiparticle to annihilate decreases, but then the energy is released in baryon–cleaner environments. The problem, though complex, is eminently suitable for numerical simulations, showing (Janka *et al.*, 1999, and references therein) that about 10^{50} $ergs$ can be released this way, above the poles of a black hole where less than $10^{-5}M_{\odot}$ are found.

Highly energetic bursts cannot be reproduced by this mechanism, due to its low efficiency: the second mechanism proposed involves the conversion of Poynting flux into a magnetized wind. The basic physical mechanisms are well–known (Usov 1992) since they have been studied in the context of pulsar emission: electrons are accelerated by a motional electric field $\vec{E} = \vec{v} \wedge \vec{B}/c$ due to the rotation of a sufficiently strong magnetic dipole, attached either to a black hole, or to the torus. Photons are then produced by synchrotron or curvature radiation, and photon/photon collisions produce pairs, to close the circle and allow looping. In order to carry away 10^{51} $erg\ s^{-1}$, a magnetic field of $\approx 10^{15}$ G is required. This is not excessive, since it is about three orders of magnitude below equipartition with torus matter, and because such fields already exist in nature, see SGR 1806-20 and SGR 1900+140: the key point is to understand whether some kind of dynamo effect can lead to these high values within the short allotted time.

Depending upon whether the open magnetic field lines extending to infinity are connected to the black hole or to the torus, the source of the energy of the outflow will be the rotational energy of the black hole (the so–called Blandford–Znajek effect) or of the torus. The first case is traditionally discussed in the context of AGNs (Rees, Blandford, Begelman and Phinney 1984), but it is harshly disputed whether the energy outflow may be actually dominated by the black hole rather than by the disk (Ghosh and Abramowicz, 1997, Livio, Ogilvie and Pringle 1998). On the other hand, the torus looks ideal as the source of a dynamo: its large shear rate, the presence of the Balbus-Hawley instability to convert polidal into toroidal flux, and the possible presence of the anti–floating mechanism inhibiting ballooning of the magnetic field (Kluzniak and Ruderman 1998), all seem to favor the existence of a fast dynamo. It should also be remarked that the configuration of the magnetic field in this problem is known: in fact, the configuration discussed in Thorne *et al.*, 1986 for black holes, only uses the assumptions of steady–state and axial symmetry, and is thus immediately extended to magnetic fields anchored to the torus. What is really required here is a first order study, of the sort published by Tout and Pringle (1992) on angular momentum removal from young, pre–main–sequence stars via magnetic stresses, and on the associated $\alpha - \omega$ dynamo. Until such studies are made, it will be premature to claim that neutrino annihilations are responsible for the powering of GRBs.

4.2 Progenitors

There is no lack of proposed progenitors, but I will discuss only binary neutron mergers (Narayan, Paczyński and Piran 1992), collapsars (Woosley 1993, Paczyński 1998) and SupraNovae (Vietri and Stella 1998, 1999).

Clearly, NS/NS mergers is the best model on paper: it involves objects which have been detected already, orbital decay induced by gravitational wave emission is shown by observations to work as per the theory, and numerical simulations by Janka's group show that a neutrino–powered outflow in baryon–poor matter can be initiated. The major theoretical uncertainties here concern bursts' durations and energetics: all numerical models produce short bursts ($\approx 0.1\ s$) with modest energetics, $E < 10^{51}\ erg$. This is a direct consequence of the mechanism for powering the burst: large, super–Eddington luminosities are carried away by neutrinos, leading to a large mass influx, but only a small fraction, $1-3\%$, can be harnessed for the production of the burst. Furthermore, we cannot invoke large beaming factors in this case: the outflow is only marginally collimated, in agreement with expectations that an accretion disk with inner and outer radii $R_{out}/R_{in} \approx$ a few (for the case at hand) can only produce a beam semi–opening angle of R_{in}/R_{out}. So, perhaps, this model may account for the short bursts, but it should be remembered that nothing of what was discussed above pertains to this subclass: BeppoSAX (and thus all BeppoSAX–triggered observations) can only detect long bursts.

On the other hand, future space missions, whether or not able to locate short bursts, can provide a decisive test of this model, provided they can follow with sufficient sensitivity a given burst for several hours. This model, in fact, is the only one proposed so far according to which some explosions should take place outside galaxies: according to Bloom, Sigurdsson and Pols (1999), about 50% of all bursts will be located more than 8 kpc from a galaxy, and 15% in the IGM. This characteristic is testable without recourse to optical observations. In fact, the afterglow begins with a delay (as seen by an outside observer) of $t_d = (r_{ag} - r_{sh})/\gamma^2 c \approx r_{ag}/\gamma^2 c$, which varies greatly depending upon the environment in which the burst takes place:

$$t_d = \begin{cases} 15\ s & \text{ISM, n} = 1\ cm^{-3} \\ 5\ min & \text{galactic halo, n} = 10^{-4}\ cm^{-3} \\ 4\ h & \text{IGM, n} = 10^{-8}\ cm^{-3} \end{cases} \tag{4}$$

Between the burst proper and the beginning of the power–law–like afterglow, thus a silence of recognizable duration is expected (Vietri 2000b).

Collapsars are currently in great vogue as a possible source of GRBs: the large amount of energy available as the core of a supermassive star collapses directly to a black hole is in fact very attractive, even though (again!) the limited efficiency of the reaction $\nu + \bar{\nu} \rightarrow e^- + e^+$ makes most of this energy unavailable. Here too there is some evidence that these objects must exist (Paczyński 1998), and numerical simulations again showing energy preferentially deposited along the hole rotation axis are also available (McFayden and Woosley 1999). Here however, what is truly puzzling is how the outflow can pierce the star's outer layers without loading itself with baryons: we should remember that at most $10^{-6} M_\odot$ can be added to 10^{51} *erg*: more baryons imply a proportionately slower outflow. The argument is that the dynamical timescale of the outer layers of a massive stars is of order of a few hours, so that, even if the core collapses and pressure support is removed, nothing will happen during the energy release phase: the outflow must pierce its way through. Two processes seem especially dangerous: Rayleigh–Taylor instability of the fluid heated–up by neutrino annihilations as it is weighed upon by the colder, denser outer layers, and Kelvin–Helmholtz instability after the hot fluid has pierced the outer layers and is passing through the hole. It is well–known that the non–linear development of these instabilities leads to mass entrainment, and that the time–scale for the development of these instabilities is very fast. Furthermore, the baryon–free outflow may be 'poisoned' by baryons to a deadly extent, even if numerical simulations, with their finite resolution, were to detect nothing of the kind.

The third class of models, SupraNovae, concerns supramassive neutron stars which are stabilized against self–gravity by fast rotation, to such an extent that they cannot be spun down to $\omega = 0$ because they implode to a black hole. As the star's residual magnetic dipole sheds angular momentum, this is exactly the fate to be expected for the whole star, except for a small equatorial belt , whose later accretion will power the burst. It is easy to show that this implosion must take place in a very baryon–clean environment. The major uncertainties here concern the channels of formation and the existence of this equatorial belt. Two channels of formation have been proposed: direct collapse to a supramassive configuration (Vietri and Stella 1998) and slow mass accretion in a low–mass X–ray binary (Vietri and Stella 1999). Both are possible, though none yet is supported by observations. The existence of the left–over belt has recently been questioned by Shibata, Baumgarte and Shapiro (1999), who however simulated the collapse of neutron stars with intermediate equations of state, which are entirely (or nearly exactly so) contained inside the marginally stable orbit even before collapse: clearly, these must be swallowed whole by the resulting black hole. Soft equations of state are free of this objection, and are thus much more likely to leave behind an equatorial belt. The soft EoSs are especially favored since the neutron stars must survive the r–mode instability, and thus soft EoSs (Weber 1999) would be in any case required. So one might say that the existence of these stars hinges on one uncertainty only, the EoS of nuclear matter. Besides the baryon–clean environments, SupraNovae have another advantage over rival models: only the lowest density regions would be left behind, precisely those with the smallest neutrino losses. The powering of the burst can thus occur through accretion caused by removal of angular momentum by magnetic stresses, without the parallel, unproductive, neutrino generation.

5 Conclusions

It is difficult to end on an upbeat note: we cannot expect in the near future a rate of progress similar to the one we witnessed in the past three years. In particular, it may be expected that the next flurry of excitement will come with the beginning of the SWIFT mission, which promises to collect relevant data (redshifts, galaxy types, location within or without galaxies, absorption or emission features in the optical and in the X–ray) for a few hundred bursts. This data will nail the major characteristics of the environment (at large) in which bursts take place, and we may be able to rule out a few models. On the other hand, the energy release process, shrouded as it is in optical depths $> 10^{10}$, will remain mysterious, our only hope in this direction being gravitational waves.

Judging by the analogy with radio pulsars, this will correspond to the flattening of the learning curve. Aside from this, we may hope to locate the equivalent of the binary radio pulsar, but, differently from Jo Taylor, we have to be awfully quick in grabbing it.

Acknowledgements

Thanks are due to Gabriele Ghisellini, who wisely steered me away from synchrotron emission, and toward the true light of Inverse Compton.

References

1. Baring, M., 1993, Ap.J., 418, 391.
2. Berezinsky, V.S.,, Prilutskii, O.F., 1986, Astr.Ap., 175, 309.
3. Bloom, J.S., Sigurdsson, S., Pols, O.R., 1999, MNRAS, 305, 763.
4. Costa, E., *et al.*, 1997, Nature, 387, 783.
5. Covino, S., *et al.*, 1999, Astr.Ap., 348, L1.
6. Fishman, G.J., Meegan, C.A., 1995, ARAA, 33, 415.
7. Frail, D., *et al.*, 1997, Nature, 389, 261.

8. Frail, D., Waxman, E., Kulkarni, S., 2000, Ap.J., submitted, astro–ph 9910319.
9. Frontera, F., *et al.*, 1999, Ap.J.S., in press, astro–ph 9911228.
10. Galama, T., *et al.*, 1998, Ap.J.L., 500, L97.
11. Ghisellini, G., Celotti, A., 1998, Ap.J.L., 511, L93.
12. Ghosh, P., Abramowicz, M.A., 1997, MNRAS, 292, 887.
13. Goodman, J.J., 1997, New Astr., 2, 449.
14. Janka, T., Eberl, T., Ruffert, M., Fryer, C.L., 1999, Ap.J.L., 527, L39.
15. Katz, J., 1994, Ap.J., 422, 248.
16. Klebesadel, R.W., Strong, I.B., Olson, R.A., 1973, Ap.J.L., 182, L85.
17. Kluzniak, W., Ruderman, M., 1998, Ap.J.L., 505, L113.
18. Livio, M., Ogilvie, G.I., Pringle, J.E., 1999, Ap.J., 512, 100.
19. Lazzati, D., Campana, S., Ghisellini, G., MNRAS, 1999, 304, L31.
20. Mc Fayden, A., Woosley, S.E., 1999, Ap.J., 524, 262.
21. Mészáros, P. Laguna, P., Rees, M.J., 1993, Ap.J., 415, 181.
22. Mészáros, Rees, M.J., 1993, Ap.J., 405, 278.
23. Mészáros, Rees, M.J., 1997, Ap.J., 476, 232.
24. Narayan, R., Paczynski, B., Piran, T., 1992, Ap.J.L., 395, L83.
25. Paczyǹski, B., 1986, Ap.J.L., 308, L43.
26. Paczyǹski, B., 1998, Ap.J.L., 494, L45.
27. Paczyǹski, B., Xu, G., 1994, Ap.J., 427, 708.
28. Piro, L., *et al.*, 1999, Ap.J.L., 514, L73.
29. Preece, R.D., *et al.*, 1999, Ap.J.S., in press, astro–ph. 9908119.
30. Rees, M.J., Begelman, M.C., Blandford, R.D., Phinney E.S., 1984, Nature, 295, 17.
31. Rees, M.J., Mèszàros, P., 1992, MNRAS, 258, P41.
32. Rhoads, J., 1997, Ap.J.L., 487, L1.
33. Ruderman, R., 1975, Ann. NY. Acad. Sci., 262, 164.
34. Sari, R., Piran, T., 1997, Ap.J., 485, 270.
35. Shibata, M., Baumgarte, T.W., Shapiro, S.L., 1999, Phys.Rev.D, submitted, astro–ph. 9911308.
36. Thorne, K.S., Price, R.H., MacDonald, D.A., 1986, *Black holes: the membrane paradigm*, Yale:New Haven, Yale Univ. Press.
37. Tout, C.A., Pringle, J.E., 1992, MNRAS, 256, 269.
38. Usov, V.V., 1992, Nature, 357, 472.
39. Usov, V.V., Chibisov, G.V. Soviet Astr., 19, 115.
40. van Paradijs, J., *et al.*, 1997, Nature, 386, 686.
41. Vietri, M., 1997a, Ap.J.L., 478, L9.
42. Vietri, M., 1997b, Ap.J.L., 488, L105.
43. Vietri, M., 2000a, in preparation.

44. Vietri, M., 2000b, Ap.J.L., submitted.
45. Vietri, M., Perola, G.C., Piro, L., Stella, L., 1999, MNRAS, 308, P29.
46. Vietri, M., Stella, L., 1998, Ap.J.L., 507, L45.
47. Vietri, M., Stella, L., 1999, Ap.J.L., 527, L43.
48. Waxman, E., 1997, Ap.J.L., 489, L33.
49. Waxman, E., Draine, B.T., 2000, Ap.J., in press, astro-ph. 9909020.
50. Waxman, E., Frail, D., Kulkarni, D., 1998, Ap.J., 497, 288.
51. Weber, F., 1999, in *Pulsars as astrophysical laboratories for nuclear and particle physics*, Bristol, U.K.; Institute of Physics.
52. Wijers, R.A.M.J., *et al.*, 1999, Ap.J.L., 523, L33.
53. Woosley, S., 1993, Ap.J., 405, 273.
54. Yoshida, A., *et al.*, 1999, Astr.Ap.S., 138, 433.

THE MYSTERY OF ULTRA-HIGH ENERGY COSMIC RAYS

A. V. OLINTO

Department of Astronomy & Astrophysics,
& Enrico Fermi Institute,
The University of Chicago, Chicago, IL 60637, USA
E-mail: olinto@oddjob.uchicago.edu

The origin of cosmic rays with energies higher than 10^{20} eV remains a mystery. Accelerating particles up to these energies is a challenge even for the most energetic astrophysical objects known. While the isotropy in arrival directions argues for an extra-galactic origin, the photon-pion production off the cosmic background radiation limits the sources of such particles to systems less than 50 Mpc away from us. The combination of large gyroradii, efficient energy losses, and isotropic arrival directions defies most of the proposed astrophysical accelerators as well as the more exotic alternatives. I briefly review theoretical models for the acceleration and propagation of ultra-high-energy cosmic-rays and discuss the potential for future observatories to resolve this cosmic mystery.

1 Introduction

The origin of cosmic rays with energies above 10^{20} eV is an intriguing mystery. At present, about 20 events above 10^{20} eV have been reported worldwide by experiments such as the High Resolution Fly's Eye, AGASA, Fly's Eye, Haverah Park, Yakutsk, and Volcano Ranch. (For recent reviews of these observations see, e.g., [1]). The unexpected flux above $\sim 7 \times 10^{19}$ eV [2] shows no sign of the Greisen-Zatsepin-Kuzmin (GZK) cutoff.[3] A cutoff should be present if these ultra-high energy particles are protons, nuclei, or photons from extragalactic sources. Cosmic ray protons of energies above a few 10^{19} eV lose energy to photopion production off the cosmic microwave background (CMB) and cannot originate further than about 50 Mpc away from Earth. Nuclei are photodisintegrated on shorter distances due to the infrared background while the radio background constrains photons to originate from even closer systems.

In addition to the presence of events past the GZK cutoff, the arrival directions of the highest energy events show no clear angular correlation with any of the plausible optical counterparts such as sources in the Galactic plane, the Local Group, or the Local Supercluster. If these events are protons, their arrival direction should point back to their sources, but unlike luminous structures in a 50 Mpc radius around us, the distribution of the highest energy events is isotropic.

At these high energies the Galactic and extragalactic magnetic fields should not affect the orbits significantly. Protons at 10^{20} eV propagate mainly in straight lines as they traverse the Galaxy since their gyroradii are ~ 100 kpc in μG fields which is typical in the Galactic disk so they should point back to their sources within a few degrees. Extragalactic fields are expected to be $\ll \mu$G, and induce at most $\sim 1^o$ deviation from the source. Even if the Local Supercluster has relatively strong fields, the highest energy events may deviate at most $\sim 10^o$.[4]

If astrophysical sources cannot explain these observations, the exciting alternative involves physics beyond the standard model of particle physics. Not only the origin of these particles may be due to physics beyond the standard model, but their existence can be used to constrain extensions of the standard model such as violations of Lorentz invariance.

The absence of a GZK cutoff and the isotropy of arrival directions are some of the challenges that models for the origin of UHECRs face. This mystery has generated a number of proposals but no model can claim victory at this point. The exact shape of the spectrum at the highest energies as well as a clear composition determination awaits future observatories such as the Pierre Auger Project and the proposed satellites OWL and Airwatch.

In this talk, I briefly review the models that attempt to solve this mystery. For more extensive reviews, see [5].

2 Astrophysical Zevatrons

These challenging observations have generated two different proposals to reaching a solution: A bottom-up approach involves looking for *Zevatrons*, possible acceleration sites in known astrophysical objects that can reach ZeV energies, while a top-down approach involves the decay of very high mass relics from the early universe and physics beyond the standard model of particle physics.

Cosmic rays can be accelerated in astrophysical plasmas when large-scale macroscopic motions, such as shocks and turbulent flows, are transferred to individual particles. The maximum energy of accelerated particles, $E_{\rm max}$, can be estimated by requiring that the gyroradius of the particle be contained in the acceleration region: $E_{\rm max} = Ze\,B\,L$, where Ze is the charge of the particle, B is the strength and L the coherence length of the magnetic field embedded in the plasma. For $E_{max} \gtrsim 10^{20}$ eV and $Z \sim 1$, the only known astrophysical sources with reasonable BL products are neutron stars ($B \sim 10^{13}$ G, $L \sim 10$ km), active galactic nuclei (AGNs) ($B \sim 10^4$ G, $L \sim 10$ AU), radio lobes of AGNs ($B \sim 0.1\mu$G, $L \sim 10$ kpc), and clusters of galaxies ($B \sim \mu$G, $L \sim 100$

kpc).

Clusters of Galaxies: Cluster shocks are reasonable sites to consider for ultra-high energy cosmic ray (UHECR) acceleration, since particles with energy up to E_{max} can be contained by cluster fields. However, efficient losses due to photopion production off the CMB during the propagation inside the cluster limit UHECRs in cluster shocks to reach at most $\sim$ 10 EeV.[5]

AGN Radio Lobes: Next on the list of plausible Zevatrons are extremely powerful radio galaxies [6]. Jets from the central black-hole of an active galaxy end at a termination shock where the interaction of the jet with the intergalactic medium forms radio lobes and 'hot spots'. Of special interest are the most powerful AGNs where shocks can accelerate particles to energies well above an EeV via the first-order Fermi mechanism. These sources may be responsible for the flux of UHECRs up to the GZK cutoff.[7]

A nearby specially powerful source may be able to reach energies past the cutoff. However, extremely powerful AGNs with radio lobes and hot spots are rare and far apart. The closest known object is M87 in the Virgo cluster ($\sim$ 18 Mpc away) and could be a main source of UHECRs. Although a single nearby source can fit the spectrum for a given strength and structure of the intergalactic magnetic field [8], it is unlikely to match the observed arrival direction distribution. After M87, the next known nearby source is NGC315 which is already too far at a distance of $\sim$ 80 Mpc.

A recent proposal tries to get around this challenge by invoking a Galactic wind with a strongly magnetized azimuthal component [9]. Such a wind can significantly alter the paths of UHECRs such that the observed arrival directions of events above 10^{20} eV would trace back to the Virgo cluster close to M87. If our Galaxy has a such a wind is yet to be determined. The proposed wind seems hard to support physically and would focus most events into the northern Galactic pole and render point source identification fruitless.[10] Future observations of UHECRs from the Southern Hemisphere by the Southern Auger Site will provide data on previously unobserved parts of the sky and help distinguish plausible proposals for the effect of local magnetic fields on arrival directions. Full sky coverage is a key discriminator of such proposals.

AGN - Central Regions: The powerful engines that give rise to the observed jets and radio lobes are located in the central regions of active galaxies and are powered by the accretion of matter onto supermassive black holes. It is reasonable to consider the central engines themselves as the likely accelerators.[11,?] In principle, the nuclei of generic active galaxies (not only the ones with hot spots) can accelerate particles via a unipolar inductor not unlike the one operating in pulsars. In the case of AGNs, the magnetic field is provided by the infalling matter and the spinning black hole horizon provides the imperfect conductor for the unipolar induction.

The problem with AGNs as UHECR sources is two-fold: first, UHE particles face debilitating losses in the acceleration region due to the intense radiation field present in AGNs, and second, the spatial distribution of objects should give rise to a GZK cutoff of the observed spectrum. In the central regions of AGNs, loss processes are expected to downgrade particle energies well below the maximum achievable energy. This limitation has led to the proposal that quasar remnants, supermassive black holes in centers of inactive galaxies, are more effective UHECR accelerators.[12] In this case, losses are not as significant but the distribution of sources should still lead to a clear GZK cutoff unless the spectrum is fairly hard.

Neutron Stars Another astrophysical system capable of accelerating UHECRs is a neutron star.[?,13,14] Acceleration processes inside the neutron star light cylinder are bound to fail much like the AGN central region case: ambient magnetic and radiation fields induce significant losses. However, the plasma that expands beyond the light cylinder is freer from the main loss processes and may be accelerated to ultra high energies.

One possible source of UHECR past the GZK cutoff is the early evolution of neutron stars. In particular, newly formed, rapidly rotating neutron stars may accelerate iron nuclei to UHEs through relativistic MHD winds beyond their light cylinders.[14] In this case, UHECRs originate mostly in the Galaxy and the arrival directions require that the primaries be heavier nuclei. Depending on the structure of Galactic magnetic fields, the trajectories of iron nuclei from Galactic neutron stars may be consistent with the observed arrival directions of the highest energy events.[15] Moreover, if cosmic rays of a few times 10^{18} eV are protons of Galactic origin, the isotropic distribution observed at these energies is indicative of the diffusive effect of the Galactic magnetic fields on iron at $\sim 10^{20}$ eV. This proposal awaits a clear composition determination.

Gamma-Ray Bursts Transient high energy phenomena such as gamma-ray bursts may accelerate protons to ultra-high energies .[16] Aside from both having unknown origins, GRBs and UHECRs have some similarities that argue for a common origin. Like UHECRs, GRBs are distributed isotropically in the sky, and the average rate of γ-ray energy emitted by GRBs is comparable to the energy generation rate of UHECRs of energy $> 10^{19}$ eV in a redshift independent cosmological distribution of sources, both have $\approx 10^{44}$erg /Mpc^3/yr.

However, the distribution of UHECR arrival directions and arrival times argues against the GRB–UHECR common origin. Events past the GZK cutoff require that only GRBs from $\lesssim$50 Mpc contribute. Since less than about *one* burst is expected to have occurred within this region over a period of 100 yr, the source would appear as a concentration of UHECR events. Therefore, a very large dispersion of $\gtrsim$ 100 yr in the arrival time of protons produced in a single burst is necessary. The deflection by random magnetic fields combined with the energy spread of the particles is usually invoked to reach the required dispersion.[16] If the dispersion in time and space is achieved, the energy spectrum for the nearby source(s) becomes very narrowly peaked $\Delta E/E \sim 1$. Finally, if the observed small scale clustering of arrival directions is confirmed by future experiments with clusters having lower energy events precede higher energy ones, bursts would be invalidated.[17]

3 Hybrid Models

The UHECR puzzle has inspired proposals that use Zevatrons to generate UHE particles other than protons, nuclei, and photons. These use physics beyond the standard model in a bottom-up approach, thus, named hybrid models.

The most economical among such proposals involves a familiar extension of the standard model, namely, neutrino masses. If some flavor of neutrinos have masses $\sim$ 1 eV, the relic neutrino background will cluster in halos of galaxies and clusters of galaxies. High energy neutrinos ($\sim 10^{21}$ eV) accelerated in Zevatrons can annihilate on the neutrino background and form UHECRs through the hadronic Z-boson decay.[18]

This proposal is aimed at generating UHECRs nearby (in the Galactic halo and Local Group halos) while using Zevatrons that can be much further than the GZK limited volume, since neutrinos do not suffer the GZK losses.The weak link in this proposal is the nature of a Zevatron powerful enough to accelerate protons above ZeVs that can produce ZeV neutrinos as secondaries. This Zevatron is quite spectacular, requiring an energy generation in excess of presently known highest energy sources.

Another suggestion is that the UHECR primary is a new particle. The mass of a hypothetical hadronic primary can be limited by the shower development of the Fly's Eye highest energy event to be below $\lesssim$50 GeV.[19] Both a long lived new particle and the neutrino Z-pole proposals involve neutral particles which are usually harder to accelerate (they are created as secondaries of even higher energy charged primariess) but can traverse large distances without being affected by the cosmic magnetic fields. Thus, a signature of such hybrid models for future experiments is a clear correlation between the position of powerful Zevatrons in the sky such as distant compact radio quasars and the arrival direction of UHE events.[20]

Another exotic primary that can use a Zevatron to reach ultra high energies is the vorton. Vortons are small loops of superconducting cosmic string stabilized by the angular momentum of charge carriers.[21] Vortons can be a component of the dark matter in galactic halos and be accelerated in astrophysical Zevatrons [22]. Although not yet clearly demonstrated, the shower development profile is also the likely liability of this model.

4 Top-Down Models

It is possible that none of the astrophysical scenarios are able to meet the challenge posed by the UHECR data as more observations are accumulated. In that case, one alternative is to consider top-down models. This proposal dates back to the work on monopolonia of Hill and Schramm.[23] Other top-down proposals involve the decay of ordinary and superconducting cosmic strings, cosmic necklaces, vortons, and superheavy long-lived relic particles. The idea behind these models is that relics of the very early universe, topological defects (TDs) or superheavy relic (SHR) particles, produced after or at the end of inflation, can decay today and generate UHECRs. Defects, such as cosmic strings, domain walls, and magnetic monopoles, can be generated through the Kibble mechanism as symmetries are broken with the expansion and cooling of the universe.[?] Topologically stable defects can survive to the present and decompose into their constituent fields as they collapse, annihilate, or reach critical current in the case of superconducting cosmic strings. The decay products, superheavy gauge and higgs bosons, decay into jets of hadrons, mostly pions. Pions in the jets subsequently decay into γ-rays, electrons, and neutrinos. Only a few percent of the hadrons are expected to be nucleons. Typical features of these scenarios are a predominant release of γ-rays and neutrinos and a QCD fragmentation spectrum which is considerably harder than the case of shock acceleration.

ZeV energies are not a challenge for top-down models since symmetry breaking scales at the end of inflation typically are $\gg 10^{21}$ eV (typical X-particle masses vary between $\sim 10^{22} - 10^{25}$ eV) . Fitting the observed flux of UHECRs is the real challenge since the typical distances between TDs is the Horizon scale, $H_0^{-1} \simeq 3h^{-1}$ Gpc. The low flux hurts proposals based on ordinary and superconducting cosmic strings. Monopoles usually suffer the opposite problem, they would in general be too numerous. Inflation succeeds in diluting the number density of monopoles usually making them too rare for UHECR production. To reach the observed UHECR flux, monopole models usually involve some degree of fine tuning. If enough monopoles and antimonopoles survive from the early universe, they may form a bound state, named monopolonium, that can decay generating UHECRs. The lifetime of monopolonia may be too short for this scenario to succeed unless they are connected by strings.[24]

Once two symmetry breaking scales are invoked, a combination of horizon scales gives room to reasonable number densities. This can be arranged for cosmic strings that end in monopoles making a monopole string network or even more clearly for cosmic necklaces.[25] Cosmic necklaces are hybrid defects where each monopole is connected to two strings resembling beads on a cosmic string necklace. Necklace networks may evolve to configurations that can fit the UHECR flux which is ultimately generated by the annihilation of monopoles with antimonopoles trapped in the string.[25,26] In these scenarios, protons dominate the flux in the lower energy side of the GZK cutoff while photons tend to dominate at higher energies depending on the radio background. If future data can settle the composition of UHECRs from 0.01 to 1 ZeV, these models can be well constrained. In addition to fitting the UHECR flux, topological defect models are constrained by limits on the flux of high energy photons, from 10 MeV to 100 GeV, observed by EGRET.

Another interesting possibility is the recent proposal that UHECRs are produced by the decay of unstable superheavy relics that live much longer than the age of the universe.[27] SHRs may be produced at the end of inflation by non-thermal effects such as a varying gravitational field, parametric resonances during preheating, instant preheating, or the decay of topological defects. These models need to invoke special symmetries to insure unusually long lifetimes for SHRs and that a sufficiently small percentage decays today producing UHECRs.[27,28] As in the topological defects case, the decay of these relics also generate jets of hadrons. These particles behave like cold dark matter and could constitute a fair fraction of the halo of our Galaxy. Therefore, their halo decay products would not be limited by the GZK cutoff allowing for a large flux at UHEs.

Future experiments should be able to probe these hypotheses. For instance, in the case of SHR and monopolonium decays, the arrival direction distribution should be close to isotropic but show an asymmetry due to the position of the Earth in the Galactic Halo.[26] Studying plausible halo models and the expected asymmetry will help constrain halo distributions especially when larger data sets are available from future experiments. High energy gamma ray experiments such as GLAST will also help constrain the SHR models due to the products of the electromagnetic cascade.[29]

5 Conclusion

Next generation experiments such as the High Resolution Fly's Eye which recently started operating, the Pierre Auger Project which is now under construction, the proposed Telescope Array, and the OWL and Airwatch satellites will significantly improve the data at the extremely-high end of the cosmic ray spectrum.[1] With these observatories a clear determination of the spectrum and spatial distribution of UHECR sources is within reach.

The lack of a GZK cutoff should become clear with HiRes and Auger and most extragalactic Zevatrons may be ruled out. The observed spectrum will distinguish Zevatrons from top-down models by testing power laws versus QCD fragmentation fits. The cosmography of sources should also become clear and able to discriminate between plausible populations for UHECR sources. The correlation of arrival directions for events with energies above 10^{20} eV with some known structure such as the Galaxy, the Galactic halo, the Local Group or the Local Supercluster would be key in differentiating between different models. For instance, a correlation with the Galactic center and disk should become apparent at extremely high energies for the case of young neutron star winds, while a correlation with the large scale galaxy distribution should become clear for the case of quasar remnants. If SHRs or monopolonia are responsible for UHECR production, the arrival directions should correlate with the dark matter distribution and show the halo asymmetry. For these signatures to be tested, full sky coverage is essential. Finally, an excellent discriminator would be an unambiguous composition determination of the primaries. In general, Galactic disk models invoke iron nuclei to be consistent with the isotropic distribution, extragalactic Zevatrons tend to favor proton primaries, while photon primaries are more common for early universe relics. The hybrid detector of the Auger Project should help determine the composition by measuring simultaneously the depth of shower maximum and the muon content of the same shower. The prospect of testing extremely high energy physics as well as solving the UHECR mystery awaits improved observations that should be coming in the next decade with experiments under construction or in the planning stages.[1]

This work was supported by NSF through grant AST 94-20759 and DOE grant DE-FG0291 ER40606.

References

1. J. W. Cronin, *Nucl. Phys.* B X. Bertou, M. Boratav, and A. Letessier-Selvon, astro-ph/001516, *Int. J. of Modern Physics A* (2000).

2. M. Takeda et al., *Phys. Rev. Lett.*
3. K. Greisen, *Phys. Rev. Lett.* **16** 748 (1966); G. T. Zatsepin and V. A. Kuzmin, *Sov. Phys. JETP Lett.* **4** 78 (1966).
4. D. Ryu, H. Kang and P. L. Bierman, *Astron. Astrophys.* **335** (1998) 19; G. Sigl, M. Lemoine, and P. Biermann, *Astropart. Phys.* **10** (1999) 141. S. Berezinsky, et al., *Astrophysics of Cosmic Rays*, (Amsterdam: North Holland, 1990); R. D. Blandford, *Particle Physics and the Universe*, eds. Bergstrom, Carlson and Fransson (World Scientific, 1999); V. S. Berezinsky, *Nucl. Phys.* B **70** (1999) 419; P. Bhattacharjee and G. Sigl, *Phys. Reps.* in press. (2000); A.V. Olinto, *Phys. Reps.* in press (2000).
5. H. Kang, D. Ryu, T.W. Jones, *Astropart. Phys. MNRAS*, **286** (1997) 257.
6. P.L. Biermann and P. Strittmatter, *Astropart. Phys.* **322** (1987) 643; P. L. Biermann, *J. Phys. G: Nucl. Part. Phys.* **23** (1997) 1.
7. J. P. Rachen and P. L. Biermann, *Astron. Astrophys.* **272** (1993) 161.
8. P. Blasi and A. V. Olinto, *Phys. Rev.* D **59**, 023001 (1999).
9. E. J. Ahn, et al., astro-ph/9911123
10. P. Billoir and A Letessier-Selvon, astroph/000142 (2000).
11. K.S. Thorne, R.M. Price, and D. MacDonals, *Black Holes: The Membrane Paradigm* (New Haven: Yale Press) (1986).
12. E. Boldt and P. Ghosh, *Mon. Not. R. Astron. Soc.*, in press (1999).
13. A. R. Bell, *Mon. Not. R. Astron. Soc.* **257**, 493 (1992).
14. A. V. Olinto, R. I. Epstein, and P. Blasi, *Proceedings of 26th ICRC*, Salt Lake City, **4**, 361 (1999); P. Blasi, R. I. Epstein, and A. V. Olinto, astro-ph/9912240.
15. V. N. Zirakashvili, et al., *Astron. Lett.* **24**, 139 (1998).
16. E. Waxman, *Phys. Rev. Lett.* **75**, 386 (1995); ibid., Astrophys. J. **452**, L1 (1995); M. Vietri, *Astrophys. J.* **453**, 883 (1995).
17. G. Sigl, M. Lemoine, and A. V. Olinto, *Phys. Rev.* D
18. D. Fargion, B. Mele, and A. Salis, astro-ph/9710029; T. Weiler, *Astropar. Phys.* **11**, 303 (1999).
19. I. F. M. Albuquerque, G. R. Farrar, and E. W. Kolb, *Phys. Rev.* D **59**, 015021 (1999).
20. G. R. Farrar and P. L. Biermann, *Phys. Rev. Lett.* **81**, 3579 (1998).
21. R. L. Davis and E. P. S. Shellard, *Nucl. Phys.* B
22. S. Bonazzola and P. Peter, *Astropart. Phys.* **7**, 161 (1997).
23. C. T. Hill, *Nucl. Phys.* B **224**, 469 (1983); D. N. Schramm and C.T. Hill, Proc. 18th ICRC (Bangalore) **2**, 393 (1983); C. T. Hill and D. N. Schramm, *Phys. Lett.* B **131**, 247 (1983).
24. J. J. Blanco-Pillado and K. D. Olum, astro-ph/9904315.

25. V. Berezinsky and A. Vilenkin, *Phys. Rev. Lett.* **79**, 5202 (1997).
26. V. Berezinsky, P. Blasi, and A. Vilenkin, *Phys. Rev.* D **58**, 103515-1 (1998).
27. V. Berezinsky, M. Kachelrieß and A. Vilenkin, *Phys. Rev. Lett.* **79**, 4302 (1997); V. Kuzmin and V. Rubakov, Yad. Fisika **61**, 1122 (1998).
28. D. J. H. Chung, E. W. Kolb, and A. Riotto, *Phys. Rev.* D **59**, 023501 (1999); *Phys. Rev. Lett.* **81**, 4048 (1998); V. Kuzmin and I. Tkachev, *Phys. Rev.* D **59**, 123006 (1999); JETP. Lett.
29. P. Blasi, *Phys. Rev.* D **60**, 023514 (1999).

Optical Modeling of Gravitational Wave Interferometers

Jean-Yves VINET
I.L.G.A. , Observatoire de la Côte d'Azur
06304 NICE (France)
vinet@obs-nice.fr

We present the principles of the modeling works that have been carried out for a better understanding of large GW interferometers, in the domain of interferometry itself, then in the field of optics

1 Introduction

The aim of this course is to give an idea of the different theoretical and numerical studies that have been carried out for defining the features of present long base line interferometers like Virgo. We first discuss the principles of ideal GW interferometers, we then see that the quality of the Fabry-Perot cavities plays a central role, which justifies a numerical investigation on real optical systems and the corresponding techniques are discussed in the second part.

2 Theory of GW Interferometers

2.1 Conventions used thoughout this section

We assume a monochromatic light source, and we describe in the predent section the (ideal) light beam circulating inside the interferometer as a plane wave, and moreover, we consider a given component of the electric field, so that the optical field at any place x of an optical system is of the scalar form

$$A(t,x) \;=\; A(x)\;\mathrm{e}^{-i\omega t} \tag{1}$$

A simple propagation step along a path of length L in a vacuum is therefore represented by a phase factor, and the relation between amplitudes will be

$$A(x+L) \;=\; \mathrm{e}^{ikL}\;A(x) \tag{2}$$

with $k \;=\; \omega/c$, c being the velocity of light. When a light ray encounters a mirror, it is partially reflected, transmitted and absorbed. We adopt the following convention : A_{in} being the incoming amplitude, A_{ref} the reflected, A_{trans} the transmitted, we have :

$$A_{ref} \;=\; ir\;A_{in} \quad , \quad A_{trans} \;=\; t\;A_{in} \tag{3}$$

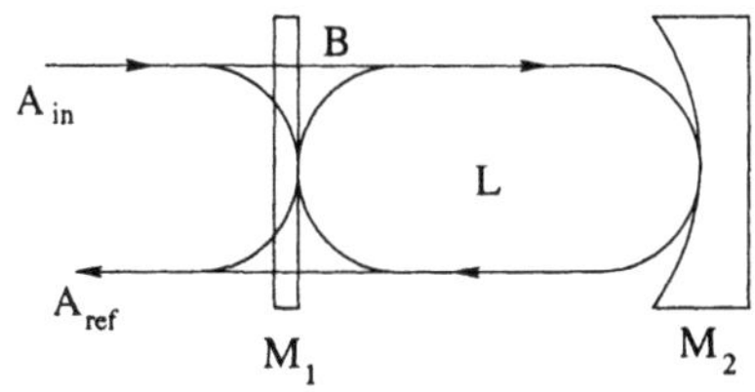

Figure 1: Fabry-Perot cavity

r, t being respectively the reflection and transmission coefficients of the mirror (real numbers). The $\pi/2$ phase lag between the two amplitudes is necessary for satisfying conservation laws. We have the optical balance :

$$r^2 + t^2 = 1 - p \tag{4}$$

where p is the loss coefficient, accounting for absorption in the coating or scattering into a different mode due to mirror geometrical imperfections (p can be as low as a few ppm (10^{-6}) for supermirrors as Virgo's).

2.2 The Fabry-Perot resonant cavity

A Fabry-Perot cavity is made of two parallel mirrors. When light enters the cavity through mirror 1, it is partially reflected and partially transmitted. The transmitted wave is reflected by mirror 2, the returns to mirror 1 where it is recombined with the incoming wave and partially transmitted to the exterior. On Fig.2.2, we have spatially separated the left and right propagating waves for the sake of clarity. If the phase after a round trip in the cavity allows it, the interference of the incoming wave and the returning wave is constructive and a strong intracavity wave builds up. light can be stored. We call r_i, t_i ,p_i ($i = 1, 2$), the parameters of the mirrors, and A_{in} the incoming wave. The length of the cavity is L. We can write the interference at M_1 for the intracavity wave as :

$$B = t_1 \, A_{\text{in}} - r_1 r_2 \, \mathrm{e}^{2ikL} B$$

so that

$$B = \frac{t_1}{1 + r_1 r_2 \mathrm{e}^{2ikL}} A_{\text{in}}$$

Clearly a resonance occurs when $\mathrm{e}^{2ikL} = -1$. We first discuss the case when the length of the cavity is fixed, and the frequency of light variable. The inverse case will be presented later. For a given L, we have a series of resonant frequencies

$$\nu_n = \left(n + \frac{1}{2}\right) \frac{c}{2L}$$

The spacing between two successive resonances is called *Free spectral Range* (FSR), and noted $\Delta\nu_{FSR}$.

$$\Delta\nu_{FSR} = \frac{c}{2L}$$

For a 3 kilometers cavity (as in VIRGO), the FSR is close to 50 kHz, whereas the optical frequency (at $\lambda = 1.06\,\mu$m) is about 3×10^{14} Hz, so that the integer n is close to 6×10^9. The ratio $S = B/A_{\text{in}}$ is called *surtension factor* . Its maximum value is

$$S_{\max} = \frac{t_1}{1 - r_1 r_2}$$

The width of the resonance line may be evaluated as follows. We assume that the frequency is close a resonance, so that

$$\nu = \nu_n + \delta\nu$$

with $\delta\nu \ll \Delta\nu_{FSR}$. We have

$$2kL = (2n+1)\pi + 2\pi \frac{\delta\nu}{\Delta\nu_{FSR}}$$

The surtension coefficient takes on the form

$$S = \frac{t_1}{1 - r_1 r_2 \exp\left(2i\pi \frac{\delta\nu}{\Delta\nu_{FSR}}\right)}$$

Its square modulus gives the ratio between the intensities :

$$|S|^2 = \frac{t_1^2}{(1 - r_1 r_2)^2 + 4 r_1 r_2 \sin\left(\pi \frac{\delta\nu}{\Delta\nu_{FSR}}\right)}$$

This is

$$|S|^2 = S_{\max}^2 \times \frac{1}{1 + \left[\frac{2\sqrt{r_1 r_2}}{1 - r_1 r_2} \sin\left(\pi \frac{\delta\nu}{\Delta\nu_{FSR}}\right)\right]^2}$$

Proximity of the resonance allows to replace the sine by its argument, so that

$$|S|^2 = S_{\max}^2 \times \frac{1}{1 + \left[2\mathcal{F} \frac{\delta\nu}{\Delta\nu_{FSR}}\right]^2}$$

with the following definition

$$\mathcal{F} = \frac{\pi\sqrt{r_1 r_2}}{1 - r_1 r_2} \tag{5}$$

for the *finesse* of the cavity. The values of $\delta\nu$ such that the surtension is half its maximum are :

$$\delta\nu = \pm\frac{\Delta\nu_{FSR}}{2\mathcal{F}}$$

and the *Full Width at Half Maximum* (FWHM) of the resonance is finally :

$$\Delta\nu_{FWHM} = \frac{\Delta\nu_{FSR}}{\mathcal{F}}$$

One can note that we have described the cavity by an extra set of parameters $\mathcal{F}$ and $\Delta\nu_{FSR}$ equivalent to $r_1 r_2$ and L. $\mathcal{F}$ contains only a photometric information about mirrors, whereas $\Delta\nu_{FSR}$ contains a geometrical information about the cavity. The exact expression for the resonance can be written under the form

$$|S|^2/S^2_{\text{max}} = \frac{1}{1+\left[\frac{2\mathcal{F}}{\pi}\sin\left(\pi\,\frac{\delta\nu}{\Delta\nu_{FSR}}\right)\right]^2}$$

The wave reflected off the cavity can be computed by

$$A_{\text{ref}} = ir_1 A_{\text{in}} + ir_2 t_1 \text{e}^{2ikL} B$$

by substituting the value of B, we get

$$A_{\text{ref}} = i\,\mathcal{R}\,A_{\text{in}}$$

Where $\mathcal{R}$ is the reflectance of the cavity, defined as

$$\mathcal{R} = \frac{r_1 + (1-p_1) r_2\,\text{e}^{2ikL}}{1 + r_1 r_2\,\text{e}^{2ikL}} \tag{6}$$

For a cavity operated in the reflection mode, having a finite reflectivity of the input mirror (M_1), a high reflectivity end mirror (M_2) and reasonable losses (p_1, p_2), it can be seen that the global reflectance is about unity, with a small peak of absorption at resonance. The phase of the reflected wave undergoes a rapid transition of 2π when crossing the resonance (see Fig.3 and Fig.4). This is classical in all oscillators, and can be better understood in a simplified model. Note that $\delta\nu = \pm\,0.5 \times \delta\nu_{FWHM}$ correspond to half the maximum absorption and to a dephasing of $\pm\pi/2$ with respect to resonance.

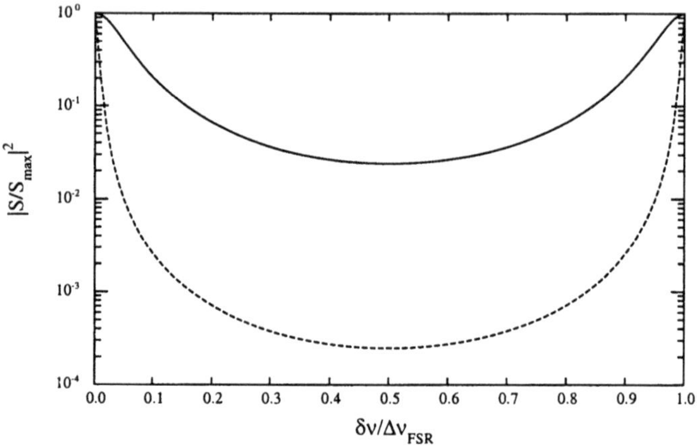

Figure 2: Resonance line shape for a finesse of $\mathcal{F}$ = 10 (solid line), and $\mathcal{F}$ = 100 (dotted line). Frequency unit is $\Delta\nu_{FSR}$

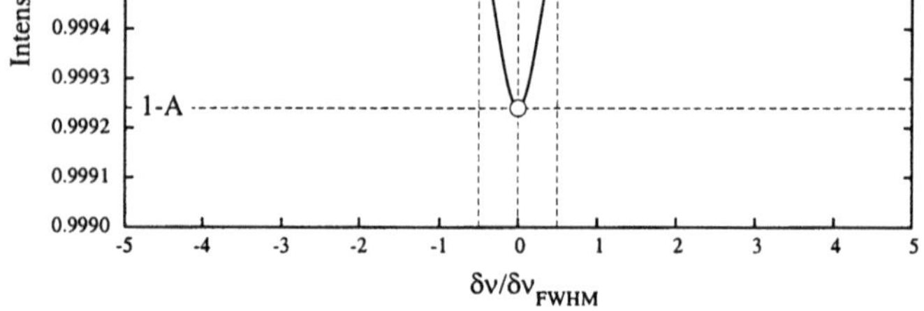

Figure 3: Absorption line of a cavity for r_1=0.85 and r_2=0.99998.(Finesse $\mathcal{F} \simeq 19.3$ A is the maximum of absorption.

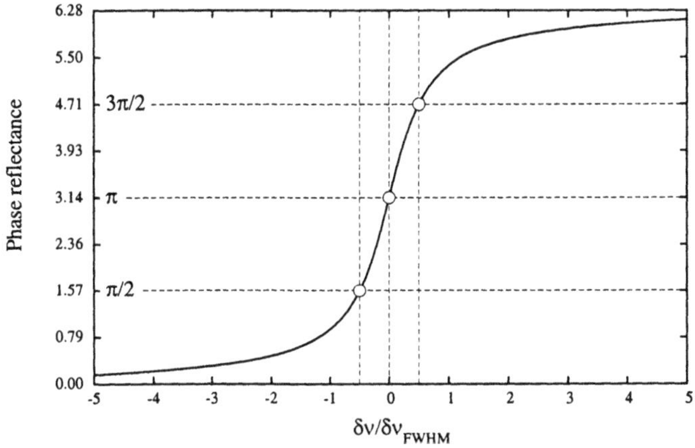

Figure 4: Phase reflectance of a cavity for r_1=0.85 and r_2=0.99998

If now, the frequency of the light source is fixed and the length of the cavity variable, which is ideally the case in a GW interferometer, instead of resonant frequencies, we have resonant lengths given by

$$L_n = \left(n + \frac{1}{2}\right) \frac{\lambda}{2}$$

showing that the displacement separating two successive resonances is :

$$\Delta L_{FSR} = \frac{\lambda}{2}$$

It is easy to show that the width of the resonance, in terms of displacement, is

$$\delta L_{FWHM} = \frac{\lambda}{2\mathcal{F}}$$

We develop now an approximate model of a cavity relying on the fact that the finesse is large compared to unity. It will prove useful for our further discussions of more complex systems involving cavities. A key parameter is indeed the finesse, defined by eq.(5) and depending only on the parameter $r_1 r_2$. Conversely, it is possible to compute $r_1 r_2$ from $\mathcal{F}$:

$$r_1 r_2 = 1 - \frac{\pi}{\mathcal{F}}\sqrt{1 + \frac{\pi^2}{4\mathcal{F}^2}} - \frac{\pi^2}{2\mathcal{F}^2}$$

If $\mathcal{F}$ is much larger than 1, we can limit the expression at the first order in $1/\mathcal{F}$, and take

$$r_1 r_2 \; = \; 1 - \frac{\pi}{\mathcal{F}}$$

Consider now the reflectance of the cavity Eq.(6), and the phase factor $2kL$. We assume a frequency ν that is slightly detuned with respect ot resonance by an amount $\delta\nu$ so that :

$$2kL \; = \; 2k_0 L + 2\pi \frac{\delta\nu}{\Delta\nu_{FSR}} \; = \; 2\pi \frac{1}{\mathcal{F}} f$$

where the *reduced frequency f* is the ratio of the offset with respect to resonance to linewidth :

$$f \equiv \frac{\delta\nu}{\delta\nu_{FWHM}}$$

with $2k_0 L \equiv \pi \mathrm{mod}[2\pi]$, we have :

$$r_2 \mathcal{R} = \frac{r_1 r_2 - (1-p_1) r_2^2 \, \mathrm{e}^{2i\pi f/\mathcal{F}}}{1 - r_1 r_2 \, \mathrm{e}^{2i\pi f/\mathcal{F}}} \tag{7}$$

We set $(1-p_1) r_2^2 = (1-p)$, where p accounts for all losses in the cavity. By expanding $r_2 \mathcal{R}$ at first order in $1/\mathcal{F}$ we get :

$$r_2 \mathcal{R} \; = \; - \frac{1 - p\mathcal{F}/\pi + 2if}{1 - 2if}$$

The quantity $\sigma = p\mathcal{F}/\pi$ is called *coupling rate* and it is easily seen that $0 < \sigma < 2$. We see that most of the properties of the FP cavity can be known by only estimating its coupling rate. The reflectance of the cavity can thus be written (by putting $r_2 \simeq 1$ at this point) :

$$\mathcal{R} \; = \; - \frac{1 - \sigma + 2if}{1 - 2if}$$

We see that the reflectance at resonance is

$$\mathcal{R}(0) \; = \; -(1 - \sigma)$$

so that $\sigma = 1$ corresponds to total absorption of light, or *optimal coupling*. For σ running from 0 to 1 the cavity is overcoupled, this means that, at resonance, the incoming field is increasingly absorbed by the cavity until total absorption. then past 1, the field is decreasingly absorbed until total reflection. The intensity reflection coefficient is :

$$|\mathcal{R}|^2 \; = \; 1 - \frac{\sigma(2-\sigma)}{1 + 4f^2}$$

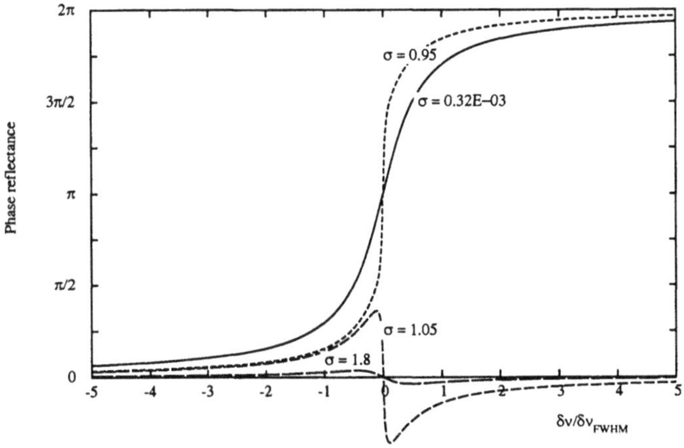

Figure 5: Transition from over to undercoupling. (NB : For σ=1, there is no reflected wave)

The reflected phase is :

$$\mathrm{Arg}\left[\frac{\mathcal{R}}{|\mathcal{R}|}\right] = \pi + \tan^{-1}\left(\frac{2f}{1-\sigma}\right) + \tan^{-1}(2f) \tag{8}$$

Increasing values of σ progressively decouples the cavity from the incoming field, the reflectivity becomes near unity (because the input mirror becomes the only part of the FP visible from the exterior) , and the reflected phase becomes more and more unsensitive to frequency detuning (see Fig.5). Note that the coupling is strong when the coupling constant is weak, and vice-versa. We should better therefore, call σ the *undercoupling constant*, but we keep the present definition for the sake of brevity. The surtension coefficient is defined by

$$\mathcal{S} = \frac{t_1^2}{1 + r_1 r_2 e^{2ikL}}$$

In the coupling rate notations, it becomes

$$\mathcal{S} = \frac{\sigma(2-\sigma)}{p\,(1+4f^2)}$$

showing that the maximum of intracavity power is reached at optimal coupling ($\sigma = 1$). The phase reflectance (see eq.8) is

$$\Phi(f) = \pi + \tan^{-1}\left(\frac{2f}{1-\sigma}\right) + \tan^{-1}(2f)$$

For f very small, this is at first order :

$$\Phi(f) = \frac{2(2-\sigma)}{1-\sigma} f = \frac{2(2-\sigma)}{1-\sigma} \mathcal{F} \frac{\delta\nu}{\Delta\nu_{FSR}}$$

In terms of the coupling rate, and absolute frequency detuning, the slope of $\Phi(\delta\nu)$ is

$$\frac{d\Phi}{d\delta\nu} = \frac{2\sigma(2-\sigma)}{1-\sigma} \frac{\pi}{p\,\Delta\nu_{FSR}}$$

In terms of the coupling rate and absolute displacement of the mirrors, we have the slope

$$\frac{d\Phi}{d\delta L} = \frac{2\sigma(2-\sigma)}{1-\sigma} \frac{2\pi}{p\lambda}$$

It could seem that the optimum detectivity is near the optimal coupling, where the slope is a maximum. The infinite slope at optimal coupling is very appealing, but unfortunately corresponds to total absorption, so that there is no reflected wave... Thsi will be discussed later, in the Michelson interferometer section, when we shall study the conversion of a phase change into an amplitude change, detectable by a diode. If the coupling rate is small (i.e. the losses small and the finesse moderate), which is the current case in GW interferometers, the slope is simply :

$$\frac{d\Phi}{d\delta L} = \frac{8\mathcal{F}}{\lambda}$$

In the undercoupling regime ($1 < \sigma < 2$), the phase reflectance has two extrema, for

$$f = \pm\frac{1}{2}\sqrt{\sigma - 1}$$

these two extrema being

$$\Phi = \pm \tan^{-1}\left(\frac{2-\sigma}{\sigma\sqrt{\sigma-1}}\right)$$

showing that the phase reflectance becomes flat as $\sigma \to 2$.

2.3 Optics in a wave Space-Time

Retarded time in a GW

When studying gravitational waves (GW), some arbitrary in the choice of the coordinates allows simplifications by partially removing this arbitrariness. In the so-called TT-gauge, the Space-Time metrics is of the form

$$g_{\mu\nu} = \eta_{\mu\nu} + h_{\mu\nu}$$

where $\eta_{\mu\nu} = \mathrm{diag}(1, -1, -1, -1)$ is the Minkowski tensor of Special Relativity, and $h_{\mu\nu} \ll 1$ the gravitational perturbation propagating as a wave. This tensor reduces to two independent components, called h_+ and $h_\times$. Assume that the GW is propagating along the z direction, then we have :

$$h_{\mu\nu} = \begin{pmatrix} 0 & 0 & 0 & 0 \\ 0 & h_+ & h_\times & 0 \\ 0 & h_\times & -h_+ & 0 \\ 0 & 0 & 0 & 0 \end{pmatrix}$$

We shall assume in what follows that the z direction is orthogonal to the plane of the optics laboratory (or of the antenna), and consider the propagation of a light ray along the x, y directions. We know that in a vacuum, light follows a null geodesic, i.e. if dx^μ represents the space-time elementary vector separating two events encountered by the light ray, we can write :

$$g_{\mu\nu}\, dx^\mu dx^\nu = 0$$

or in detail, taking $dx^\mu = (c\,dt,\ dx,\ dy,\ dz)$,

$$0 = c^2dt^2 - dx^2 - dy^2 - dz^2 + 2h_\times dx\, dy + h_+ \left(dx^2 - dy^2\right)$$

where $h_{+,\times}$ are functions of t, z only. It can be shown that there is no change of direction of the light ray during its interaction with the GW as long as the GW frequency is negligible compared to the EM frequency, which is safely verified for known GW sources. In this case, the only effect is a phase change during propagation. Let us see this in detail : For a path lying along the x direction, we have simply :

$$0 = c^2dt^2 - dx^2 + h_+ dx^2$$

or as well, h_+ being so small,

$$dx = \pm c dt \left[1 + \frac{1}{2} h_+(t)\right] \tag{9}$$

where the sign depends obviously on the propagation direction. Now, consider the round trip experiment, in which a light ray is firstly emitted from abscissa 0 at time t_0, then received at abscissa $L > 0$ at time t_1 : we have using (9) with the + sign :

$$L = c(t_1 - t_0) + \frac{1}{2}\, c \int_{t_0}^{t_1} h_+(u)\, du$$

then the light ray is reflected back and returns to the origin at time t_2, we have then, using again (9) but with the - sign :

$$-L = c(-t_2 + t_1) - \frac{1}{2} c \int_{t_1}^{t_2} h_+(u)\, du$$

by subtracting the last equation to the preceding, we get

$$2L = t_2 - t_0 + \frac{1}{2} c \int_{t_0}^{t_2} h_+(u)\, du$$

In the sequel we shall omit the index + in the GW amplitude and write simply $h(t)$ instead of $h_+(t)$. Assume now that $t_2 = t$ is the detection time, and $t_0 = t_r$ the unknown time at which the light ray was emitted (retarded time). We have the relation :

$$t_r = t - \frac{2L}{c} + \frac{1}{2} \int_{t_r}^{t} h(u)\, du$$

This is an implicit equation in t_r but very easy to solve at first order in h :

$$t_r = t - \frac{2L}{c} + \frac{1}{2} \int_{t-2L/c}^{t} h(u)\, du$$

If we consider a monochromatic wave of frequency $\nu_g = \Omega/2\pi$, such that $h(t) = h\cos(\Omega t)$, the result is

$$t_r = t - \frac{2L}{c} + h\frac{L}{c}\mathrm{sinc}(\Omega L/c)\cos\left(\Omega(t - L/c)\right)$$

Clearly the result is identical for a round trip along the y axis, except that the first order term must be changed of sign. We have generally :

$$t_r = t - \frac{2L}{c} + \epsilon h\frac{L}{c}\mathrm{sinc}(\Omega L/c)\cos\left(\Omega(t - L/c)\right)$$

where $\epsilon = 1$ along the x axis and $\epsilon = -1$ along the y axis. One way of detecting gravitational waves could be to measure the excess time delay between emission and back detection of light. Some experiments have been proposed using this principle, for instance by analyzing solar system radar ranging data (in the microwave domain, but the principle is the same).

The A133 Algebra

Let us now turn to wave optics. Our light ray is in fact a monochromatic plane wave of frequency $\nu = \omega/2\pi$. Call $B(t)$ the (complex) amplitude at the end of the round trip, and $A(t)$ its value at the beginning. We have

$$B(t) = A(t_r)$$

If we note

$$A(t) = A\mathrm{e}^{-i\omega t}$$

we get

$$B(t) = A\mathrm{e}^{-i\omega t_r} = A\mathrm{e}^{-i\omega(t-2L/c)} \exp\left[ih\frac{\omega L}{c}\mathrm{sinc}(\Omega L/c)\cos\left(\Omega(t-L/c)\right)\right]$$

Since we are always at first order in h, we write

$$B(t) = A\mathrm{e}^{-i\omega t}\mathrm{e}^{2i\omega L/c} +$$

$$\tfrac{i}{2}hA\,\tfrac{\omega L}{c}\mathrm{sinc}(\Omega L/c)\;\mathrm{e}^{2i\omega L/c}\mathrm{e}^{i\Omega L/c}\mathrm{e}^{-i(\omega+\Omega)t} +$$

$$\tfrac{i}{2}hA\,\tfrac{\omega L}{c}\mathrm{sinc}(\Omega L/c)\;\mathrm{e}^{2i\omega L/c}\mathrm{e}^{-i\Omega L/c}\mathrm{e}^{-i(\omega-\Omega)t}$$

It clearly appears that the action of the GW was to create two sidebands of very low amplitude, of frequencies $\nu \pm \nu_g$ from one single frequency ν. Now let us see what happens if the incoming optical wave is already modulated and exhibits two sidebands. This is necessary because in interferometers, light undergoes several times the action of the GW in order to enhance the signal production. Let the incoming amplitude be of the form

$$A(t) = \left(A_0 + \frac{1}{2}h\,A_1\mathrm{e}^{-i\Omega t} + \frac{1}{2}h\,A_2\mathrm{e}^{i\Omega t}\right)\mathrm{e}^{-i\omega t}$$

The scaling factor is h because we assume the GW to be the only cause of generation of sidebands in the whole (unknown) optical system. We have then

$$B(t) = A(t_r) = \left[A_0 + \frac{1}{2}hA_1\mathrm{e}^{-i\Omega t}\mathrm{e}^{2i\eta} + \frac{1}{2}hA_2\mathrm{e}^{i\Omega t}\mathrm{e}^{-2i\eta}\right] \times$$

$$\times\ \mathrm{e}^{-i\omega t}\mathrm{e}^{2i\xi}\mathrm{e}^{-i\epsilon h\xi\mathrm{sinc}(\eta)\cos(\Omega t-\eta)}$$

For shortening the formula, we have used the abbreviations: $\xi \equiv \omega L/c$ and $\eta \equiv \Omega L/c$. After a 1st order expansion of the exponential, we get

$$B(t) = \left(B_0 + \frac{1}{2}h\,B_1\mathrm{e}^{-i\Omega t} + \frac{1}{2}h\,B_2\mathrm{e}^{i\Omega t}\right)\mathrm{e}^{-i\omega t}$$

with the following notation :

$$B_0 = e^{2i\xi} A_0$$

$$\begin{aligned} B_1 &= e^{2i(\xi+\eta)} A_1 - i\epsilon\xi \mathrm{sinc}(\eta) e^{i(2\xi+\eta)} A_0 \\ B_2 &= e^{2i(\xi-\eta)} A_1 - i\epsilon\xi \mathrm{sinc}(\eta) e^{i(2\xi-\eta)} A_0 \end{aligned}$$

We see that if we define "generalized amplitudes" as rank 3 vectors having the carrier amplitude, the upper sideband and the lower sideband respectively as coordinates, by setting

$$\mathcal{A} = (A_0,\ A_1,\ A_2)$$

and

$$\mathcal{B} = (B_0,\ B_1,\ B_2)$$

the amplitude after a round trip that we have precedently computed may be written in the form :

$$\mathcal{B} = \mathbf{X}\mathcal{A}$$

where $\mathbf{X}$ is the linear round trip operator defined as

$$\mathbf{X} = \begin{pmatrix} e^{2i\xi} & 0 & 0 \\ -i\epsilon\xi \mathrm{sinc}(\eta) e^{i(2\xi+\eta)} & e^{2i(\xi+\eta)} & 0 \\ -i\epsilon\xi \mathrm{sinc}(\eta) e^{i(2\xi-\eta)} & 0 & e^{2i(\xi-\eta)} \end{pmatrix} \tag{10}$$

It is easy to check that the set of all operators having the form

$$\mathbf{O} = \begin{pmatrix} O_{00} & 0 & 0 \\ O_{10} & O_{11} & 0 \\ O_{20} & 0 & O_{22} \end{pmatrix}$$

is stable for any algebraic operation, and even may be given a structure of non-commutative algebra isomorphous to the algebra of first order expansions. We call it "A133" for brevity. The basic algebraic operations are defined by

- The sum :

$$(\mathbf{A}+\mathbf{B})_{ij} = \mathbf{A}_{ij} + \mathbf{B}_{ij}$$

- The product :

$$(\mathbf{A}\ \mathbf{B})_{ii} = \mathbf{A}_{ii}\mathbf{B}_{ii}$$

$$(\mathbf{A}\ \mathbf{B})_{i0} = \mathbf{A}_{i0}\mathbf{B}_{00} + \mathbf{A}_{ii}\mathbf{B}_{i0}$$

- The inverse :

$$(\mathbf{A}^{-1})_{ii} = \frac{1}{\mathbf{A}_{ii}}$$

$$(\mathbf{A}^{-1})_{i0} = -\frac{\mathbf{A}_{i0}}{\mathbf{A}_{00}\ \mathbf{A}_{ii}}$$

An A133 operator may be associated to any optical element of a complex optical system. The diagonal elements $\mathbf{O}_{ii}$ represent action of that element on the carrier and the sidebands. Often (mirrors, lenses) there is no frequency dependence because the gravitational perturbation causes a negligible frequency shift, well inside the tolerances of the mirror coatings, and in this case, the corresponding operator is simply scalar. In fact the only non-diagonal operators are those corresponding to propagation of light in a vacuum over long distances. The result is that, after some (A133) algebra, the whole opical system has an associated A133 operator describing its behaviour.

2.4 Signal to Noise Ratio

We can start with a pure monochromatic wave

$$\mathcal{A}_{\text{in}} = (A,\ 0\ ,0)$$

$\mathbf{S}$ being the A133 system operator, we know that the output wave is given by :

$$\mathbf{A}_{\text{out}} = A\left[S_{00} + \frac{h}{2}S_{10}\,\mathrm{e}^{-i\Omega t} + \frac{h}{2}S_{20}\,\mathrm{e}^{i\Omega t}\right]\mathrm{e}^{-i\omega t}$$

The corresponding detectable power is, up to a normalization factor, and calling P_{in} the incoming power :

$$P(t) = \mathbf{A}_{\text{out}}\,\overline{\mathbf{A}_{\text{out}}} =$$

$$= P_{\text{in}}\left[|S_{00}|^2 + \frac{h}{2}\ (S_{10}\overline{S_{00}} + \overline{S_{20}}S_{00})\,\mathrm{e}^{-i\Omega t} + \frac{h}{2}\ (S_{20}\overline{S_{00}} + \overline{S_{10}}S_{00})\,\mathrm{e}^{i\Omega t}\right]$$

The signal amplitude at frequency ν_g is thus

$$S(\nu_g) = |S_{10}\overline{S_{00}} + \overline{S_{20}}S_{00}|$$

The DC component of the output is proportional to $|S_{00}|^2$, so that our main concern, the SNR is proportional to :

$$\mathrm{SNR}(\nu_g) \propto |S_{10}\,\mathrm{e}^{-i\varphi_{00}} + \overline{S_{20}}\,\mathrm{e}^{i\varphi_{00}}|$$

where φ_{ij} is the argument of S_{ij}. We have as well, with the correct normalisation :

$$\mathrm{SNR}(\nu_g) = \sqrt{\frac{P_{\mathrm{in}}}{h\nu}}\, ||S_{10}| + |\overline{S_{20}}|\, e^{i(\varphi_{10}+\varphi_{20}-2\varphi_{00})}|\, h(\nu_g) \tag{11}$$

Inversely, the spectral density $h_{\mathrm{SN}}(\nu_g)$ equivalent to the quantum noise is obtained by taking a unitary SNR :

$$h_{\mathrm{SN}}(\nu_g) = \sqrt{\frac{h\nu}{P_{\mathrm{in}}}}\, \frac{|S_{00}|}{|S_{10}\overline{S_{00}} + \overline{S_{20}}S_{00}|}$$

We see that evaluation of the SNR of any optical GW detector eventually reduces to calculation of the S_{i0} of the whole system.

2.5 Resonant cavities in a GW

The first element we need, before addressing more complex structures, is the A133 operator associated to a Fabry-Perot cavity. We take the same notations as in Fig.2. The intracavity (vector) amplitude $\mathcal{B}$ obeys ;

$$\mathcal{B} = t_1 \mathcal{A}_{\mathrm{in}} - r_1 r_2 \mathbf{X}\, \mathcal{B}$$

where $\mathbf{X}$ is the round trip operator just defined above (Eq.10. We have thus

$$\mathcal{B} = [1 + r_1 r_2 \mathbf{X}]^{-1}\, t_1\, \mathcal{A}_{\mathrm{in}}$$

The reflected amplitude is :

$$\mathcal{A}_{\mathrm{ref}} = i\, r_1\, \mathcal{A}_{\mathrm{in}} + i\, t_1 r_2\, \mathbf{X}\, \mathcal{B}$$

$$= i\, [r_1 + (1-p_1) r_2\, \mathbf{X}]\, [1 + r_1 r_2 \mathbf{X}]^{-1}\, \mathcal{A}_{\mathrm{in}}$$

so that the reflectance of the cavity is the operator

$$\mathbf{F} = [r_1 + (1-p_1) r_2\, \mathbf{X}]\, [1 + r_1 r_2 \mathbf{X}]^{-1} \tag{12}$$

It is possible to compute the components of $\mathbf{F}$:

$$\mathbf{F} = \begin{pmatrix} F & 0 & 0 \\ G_+ & F_+ & 0 \\ G_- & 0 & F_- \end{pmatrix}$$

F is the ordinary reflectance of the FP for the carrier, $F_\pm$ the ordinary reflectance of the FP for the upper and lower sidebands respectively. For the sake of simplicity, we use again the notation :

$$\begin{aligned} \xi &= kL \\ \eta &= \Omega L/c \end{aligned}$$

(recall that $\Omega/2\pi$ is the GW frequency). We have then, after direct evaluation of $\mathbf{F}$ according to Eq.12 :

$$F = \frac{r_1 + (1-p_1)r_2 e^{2i\xi}}{1 + r_1 r_2 e^{2i\xi}}$$

$$F_\pm = \frac{r_1 + (1-p_1)r_2 e^{2i(\xi\pm\eta)}}{1 + r_1 r_2 e^{2i(\xi\pm\eta)}}$$

$$G_\pm = -i\epsilon \frac{t_1^2 r_2 \xi \mathrm{sinc}(\eta) e^{i(2\xi\pm\eta)}}{\left(1 + r_1 r_2 e^{2i\xi}\right)\left(1 + r_1 r_2 e^{2i(\xi\pm\eta)}\right)}$$

In the coupling rate (σ) formalism, this can be approximated by

$$F = -\frac{1 - \sigma + 2i\Delta f}{1 - 2i\Delta f}$$

$$F_\pm = -\frac{1 - \sigma + 2i(\Delta f \pm f_g)}{1 - 2i(\Delta f \pm f_g)}$$

$$G_\pm = i\epsilon \frac{2\mathcal{F}L}{\lambda} \frac{2 - \sigma}{(1 - 2i\Delta f)\left[1 - 2i(\Delta f \pm f_g)\right]}$$

where $\Delta f = \delta\nu/\delta\nu_{\mathrm{FWHM}}$ is the reduced detuning of the light source from resonance, and $f_g = \nu_g/\delta\nu_{\mathrm{FWHM}}$ the reduced gravitational frequency. When we vary the detuning, we see that the modulus of G_+ has a resonance for $\Delta f = 0$ (resonance of the carrier) and a second resonance when $\Delta f = -f_g$, the upper sideband becoming resonant. The modulus of G_- has also a resonance for $\Delta f = 0$ and for $\Delta f = f_g$, the lower sideband becoming resonant (see Fig.6). A symmetrical figure can be obtained with $|G_+|$.

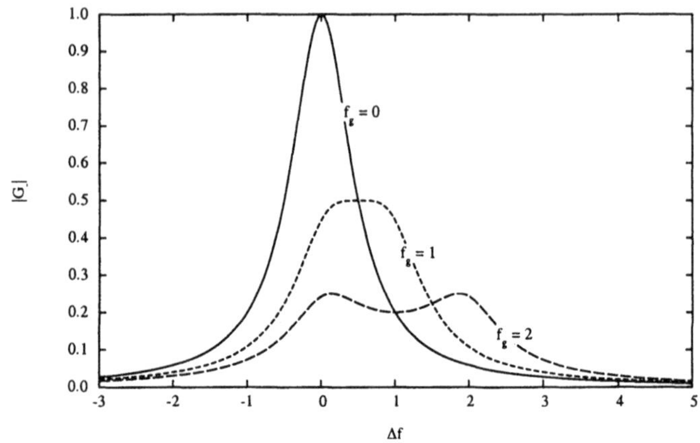

Figure 6: Efficiency of lower sideband generation vs detuning of the source for three reduced GW frequencies. Solid line :$f_g = 0$, short dashed line : $f_g = 1$, long dashed line : $f_g = 2$

2.6 Michelson Interferometer involving FP cavities

We take the classical Michelson geometry, but replace the end mirrors by two identical Fabry-Perot cavities, F_1 and F_2. Note that even when optically identical, the effect of a GW on them will be different, and consequently we must denote the corresponding operators by different notations (see Fig.7). The transmitted amplitude is

$$\mathcal{A}_{\text{trans}} = -r_s t_s \left(\mathrm{e}^{2ika}\mathbf{F}_1 + \mathrm{e}^{2ikb}\mathbf{F}_2\right) \mathcal{A}_{\text{in}}$$

whereas the reflected amplitude is

$$\mathcal{A}_{\text{ref}} = i \left(t_s^2 \mathrm{e}^{2ika}\mathbf{F}_1 - r_s^2 \mathrm{e}^{2ikb}\mathbf{F}_2\right) \mathcal{A}_{\text{in}}$$

Note that we neglect phases of the order of $2\pi\nu_g a/c$. The expressions of $\mathbf{F_1}$ and $\mathbf{F_2}$ for perfectly identical but orthogonal cavities lying respectively along the x and y directions, are :

$$\mathbf{F}_1 = \begin{pmatrix} F & 0 & 0 \\ G_+ & F_+ & 0 \\ G_- & 0 & F_- \end{pmatrix} \quad , \quad \mathbf{F}_2 = \begin{pmatrix} F & 0 & 0 \\ -G_+ & F_+ & 0 \\ -G_- & 0 & F_- \end{pmatrix}$$

We can define a transmittance and a reflectance A133 operator an obvious way, by

$$\mathcal{A}_{\text{trans}} = \mathbf{T}_{\text{Mic}} \mathcal{A}_{\text{in}}$$

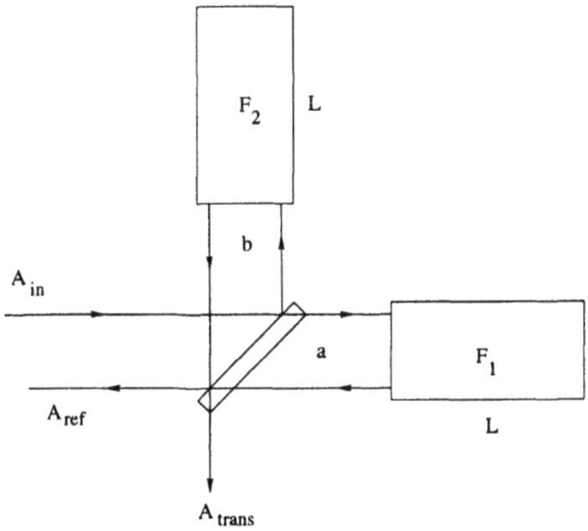

Figure 7: Geometry of a Michelson with FP cavities

$$\mathcal{A}_{\rm ref} \; = i\,\mathbf{R}_{\rm Mic}\,\mathcal{A}_{\rm in}$$

The elements of these operators are as follows, assuming a perfectly symmetrical splitter ($r_s = t_s = \sqrt{1-p_s}/2$) for the transmittance:

$$\begin{aligned}
T_{\rm Mic,00} &= -(1-p_s)\mathrm{e}^{ik(a+b)}\cos[k(a-b)]\,F \\
T_{\rm Mic,11} &= -(1-p_s)\mathrm{e}^{ik(a+b)}\cos[k(a-b)]\,F_+ \\
T_{\rm Mic,22} &= -(1-p_s)\mathrm{e}^{ik(a+b)}\cos[k(a-b)]\,F_- \\
T_{\rm Mic10} &= -i(1-p_s)\mathrm{e}^{ik(a+b)}\sin[k(a-b)]\,G_+ \\
T_{\rm Mic20} &= -i(1-p_s)\mathrm{e}^{ik(a+b)}\sin[k(a-b)]\,G_-
\end{aligned}$$

and for the reflectance :

$$\begin{aligned}
R_{\rm Mic,00} &= i(1-p_s)\mathrm{e}^{ik(a+b)}\sin[k(a-b)]\,F \\
R_{\rm Mic,11} &= i(1-p_s)\mathrm{e}^{ik(a+b)}\sin[k(a-b)]\,F_+ \\
R_{\rm Mic,22} &= i(1-p_s)\mathrm{e}^{ik(a+b)}\sin[k(a-b)]\,F_- \\
R_{\rm Mic,10} &= (1-p_s)\mathrm{e}^{ik(a+b)}\cos[k(a-b)]\,G_+ \\
R_{\rm Mic,20} &= (1-p_s)\mathrm{e}^{ik(a+b)}\cos[k(a-b)]\,G_-
\end{aligned}$$

It is evident that when the interferometer is tuned at a dark fringe for the carrier, the sidebands are transmitted, and conversely. The SNR takes the form :

$$\mathrm{SNR}(\nu_g) \;\propto\; (1-p_s)\sin[k(a-b)\left|G_+\frac{\overline{F}}{|F|}-\overline{G_-}\frac{F}{|F|}\right| \tag{13}$$

If we assume the carrier at a dark fringe, we get

$$\mathbf{T}_{\mathrm{Mic}} \;=\; (1-p_s)\mathrm{e}^{ik(a+b)}\begin{pmatrix} 0 & 0\,0 \\ -iG_+ & 0\,0 \\ -iG_- & 0\,0 \end{pmatrix} \quad , \; \mathbf{R}_{\mathrm{Mic}} \;=\; (1-p_s)\mathrm{e}^{ik(a+b)}\begin{pmatrix} iF & 0 & 0 \\ 0 & iF_+ & 0 \\ 0 & 0 & iF_- \end{pmatrix}$$

This allows to study the SNR of a simple Michelson having FP cavities as arms. We have in the coupling rate formalism, neglecting p_s at this level :

$$\mathrm{SNR}(f_g) \;=\; \frac{4\mathcal{F}L}{\lambda}\frac{2-\sigma}{\sqrt{1+4\Delta f^2}}\frac{1}{2}\left|\frac{\mathrm{e}^{i\Psi_+}}{\sqrt{1+4(\Delta f+f_g)^2}}+\frac{\mathrm{e}^{-i\Psi_-}}{\sqrt{1+4(\Delta f-f_g)^2}}\right|$$

where

$$\Psi_+ \;=\; \tan^{-1}\left(2(\Delta f+f_g)\right)-\tan^{-1}\left(\frac{2\Delta f}{1-\sigma}\right)$$

$$\Psi_- \;=\; \tan^{-1}\left(2(\Delta f-f_g)\right)-\tan^{-1}\left(\frac{2\Delta f}{1-\sigma}\right)$$

After some algebra, we find the following result :

$$\mathrm{SNR}(f_g) \;=\; \frac{8(1-\sigma/2)\mathcal{F}L}{\lambda}\times$$

$$\left[\frac{(1-\sigma+4\Delta f)^2+4(1-\sigma)^2 f_g^2}{(1+4\Delta f)^2\left((1-\sigma)^2+4\Delta f)^2\right)\left(1+8(\Delta f^2+f_g^2)+16(\Delta f^2-f_g^2)^2\right)}\right]^{1/2} \tag{14}$$

if the cavities are at resonance ($\Delta f = 0$), we have simply

$$\mathrm{SNR}(f_g) \;=\; \frac{8\mathcal{F}L}{\lambda}\frac{1-\sigma/2}{\sqrt{1+4f_g^2}}$$

we plot hereafter the spectral density of equivalent h for various values of $\mathcal{F}$ for a 20W light source at $\lambda = 1.064\,\mu$m. (see Fig.8). The sensitivity at low frequency is a function of $\mathcal{F}$. The optimum value of $\mathcal{F}$ occurs theoretically for $\sigma = 1$, i.e. for the optimal coupling of the cavities. This corresponds to

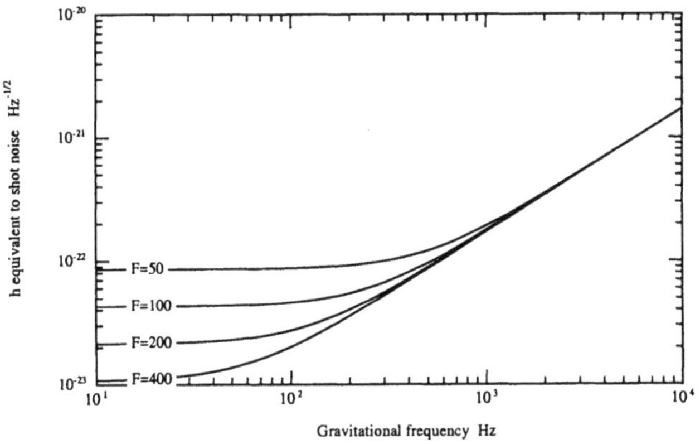

Figure 8: Simple Michelson with FP cavities : Spectral density of h equivalent to shot noise

$\mathcal{F} = \pi/p$. For $p = 3\,10^{-5}$, this corresponds to a finesse of 10^5. On the other hand, when $\sigma = 1$, the surtension coefficient is $\mathcal{S} = 1/p$, and this means here a surtension of $\simeq 3\,10^4$. For a 10 W laser source, this is is 0.3 MW stored light power. Let us keep however in mind that the improvement due to increasing the finesse occurs only at low frequency. But at low frequency, the limitation of the sensitivity is due to thermal noise, and it is worthless to try higher finesses as long as a means of reducing thermal noise has'nt been found . Better idea is to incrase the laser power, because all the curve is globally lowered. But 20W (as assumed in Fig.8) is the maximum presently reasonable for 1 CW monomode, stabilized laser. For gaining 1 order of magnitude, we would have to lock in phase an array of 3 such lasers. This is quite feasible, but the result can be achieved with a much more elegant and convenient solution, as explained hereafter. Let us remark that for given ν_g, the SNR is of the form

$$SNR = \frac{8\pi L}{\lambda}\frac{1}{p}\frac{\sigma(1-\sigma/2)}{\sqrt{1+q^2\sigma^2}}$$

with $q \equiv 2\pi\nu_g/p\Delta\nu_{\rm FSR}$ and consequently is a maximum for a finite value of σ. The parameter q is very high even for $\nu_g = 10\,\text{Hz}$, and a good approximation of the optimal coupling rate is :

$$\sigma_{opt} = \left(\frac{2}{q^2}\right)^{1/3} = \left(\frac{p\Delta\nu_{\rm FSR}}{\sqrt{2}\,\pi\nu_g}\right)^{2/3}$$

The optimal finesse is therefore :

$$\mathcal{F}_{opt}(\nu_g) = \left(\frac{\pi}{p}\right)^{1/3} \left(\frac{\Delta\nu_{\mathrm{FSR}}}{\nu_g}\right)^{2/3}$$

For instance, with $p = 3\,10^{-5}$, $\Delta\nu_{\mathrm{FSR}} = 50\,\mathrm{kHz}$, this gives

$$\mathcal{F}_{opt}(\nu_g) = 13782 \times \left(\frac{10\mathrm{Hz}}{\nu_g}\right)^{2/3}$$

But the maximum is very flat, and it is not necessary to require the true optimum. A value of σ such that $q\sigma = 2$ is quite sufficient, the SNR differing from its true optimum by only 10%. this corresponds to

$$\mathcal{F}_{opt}(\nu_g) = \frac{\Delta\nu_{\mathrm{FSR}}}{\nu_g}$$

The pseudo-optimal finesse for $\nu_g = 1$ kHz is for instance $\mathcal{F} = 50$. The pseudo-optimal finesse depends of a reference frequency $\nu_g^{(0)}$ which is an equivalent parameter, the length of the cavities being fixed. In terms of this reference frequency, we have :

$$\mathrm{SNR}(\nu_g) = \frac{4\,\frac{\nu_{\mathrm{opt}}}{\nu_g^{(0)}}}{\sqrt{1 + \left(2\,\frac{\nu_g}{\nu_g^{(0)}}\right)^2}}$$

where ν_{opt} is the optical frequency. This formula is valid except for too small values of $\nu_g^{(0)}$. For the interval [10 Hz, 10kHz], it is valid. We see the huge scale factor provided by the cavities. When the two cavities have a common detuning, the SNR is reduced, as can be red directy on Eq.(14). But a resonance occurs when the upper sideband created by the GW becomes resonant (for $f_g = \Delta f$). At this frequency, the loss due to the frequency offset of the carrier is somewhat compensated by the resonance (see Fig2.6) One important point is that, working out of resonance, the reflectances of the cavities are much higher than in the tuned case. This regime of operation, of no benefit in the simple Michelson configuration, becomes interesting when recycling is applied, as will be shown later.

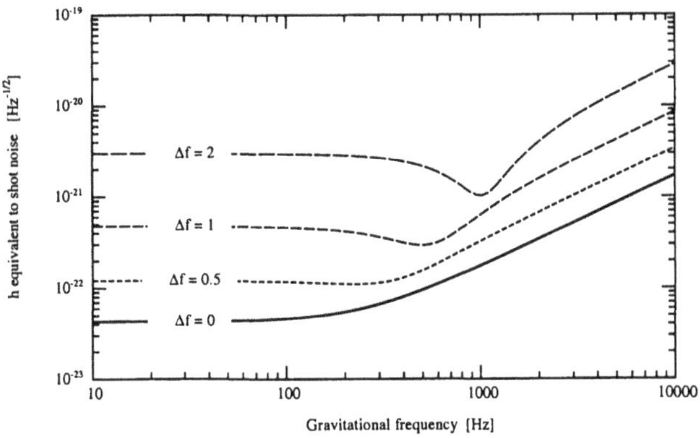

Figure 9: Michelson with detuned cavities (F=100)

2.7 Recycling

standard power recycling

In the so called *standard recycling* sheme, we assume the FP cavities at resonance ($\Delta f = 0$). The SNR takes on the simple form

$$\text{SNR} = \frac{4\frac{\mathcal{F}L}{\lambda}(2-\sigma)}{\sqrt{1+4f_g^2}} \frac{t_r(1-p_s)}{1-r_r(1-p_s)|1-\sigma|}$$

Where we see directly how increasing the coupling factor increases the Michelson SNR, but decreases the recycling factor. Anyway, we are free to choose the best recycling reflectance r_r, i.e. that maximizing the recycling surtension factor. This happens when

$$r_{r\ opt} = (1-p_r)(1-p_s)|1-\sigma|$$

giving

$$S_{r\ opt} = (1-p_s)\sqrt{\frac{1-p_r}{1-(1-p_r)(1-p_s)^2(1-\sigma)^2}}\ \frac{4\frac{\mathcal{F}L}{\lambda}(2-\sigma)}{\sqrt{1+4f_g^2}}$$

The mirror losses will be taken very small (of the order of 10 ppm), and we have seen that the coupling rate in a simple Michelson must be relatively small. It will be even smaller here, because the recycling factor would be destroyed by a large cavity absorption. It is therefore not unrealistic to consider that the total losses are dominated by the cavity resonant absorption, and however, small ($p_r + 2p_s \ll 2\sigma \ll 1$). The optimal SNR is then

$$\mathrm{SNR}(\nu_g) = \frac{4\pi L}{\sqrt{2}\lambda}\frac{1}{p}\frac{\sigma^{1/2}(2-\sigma)}{\sqrt{1+\left(2\pi\frac{\nu_g}{p\,\Delta\nu_{\mathrm{FSR}}}\sigma\right)^2}}$$

When searching for the optimal value of σ, we get the following equation, with $q = 2\pi\nu_g/p\Delta\nu_{\mathrm{FSR}}$:

$$\frac{1}{2}q^2\sigma^3 + q^2\sigma^2 + \frac{3}{2}\sigma^2 - 1 = 0$$

for avoiding an exact but useless and cumbersome resolution of this equation, we rather solve it in q :

$$q^2 = \frac{1-3\sigma^2/2}{\sigma^2(1+\sigma/2)}$$

Now we remark that, even for low GW frequencies (10 Hz), q^2 is very large : Consequently, σ must be very small, and we can take the approximation

$$\sigma_{opt} = \frac{1}{q}$$

or, in terms of finesse,

$$\mathcal{F}_{opt} = \frac{\Delta\nu_{\mathrm{FSR}}}{2\nu_g^{(0)}}$$

Where $\nu_g^{(0)}$ is the GW frequency for which the SNR is optimized. But here, the maximum is sharp (see Fig.10). Remark that this value is half the pseudo-optimum for the simple Michelson. This sharp maximum makes the SNR very sensitive to the GW frequency at which the SNR is optimized. With physically significant parameters (frequencies in the detection range [10 Hz,10 kHz], and small losses), the SNR can be approximated by a simple formula. Call p_{ITF} the losses encountered in the recycling mirror and the splitter, i.e. the losses external to FP's : we have

$$1 - p_{\mathrm{ITF}} = (1-p_r)(1-p_s)^2 \Rightarrow p_{\mathrm{ITF}} \simeq p_r + 2p_s$$

The losses internal to FP's are still $p = 1-(1-p_1)r_2^2$. Neglecting non essential small terms leads to :

$$\mathrm{SNR}(\nu_g) = \frac{1}{\sqrt{p_{\mathrm{ITF}} + \frac{p\Delta\nu_{\mathrm{FSR}}}{\pi\nu_g^{(0)}}}}\frac{2\frac{\nu_{\mathrm{opt}}}{\nu_g^{(0)}}}{\sqrt{1+\left(\frac{\nu_g}{\nu_g^{(0)}}\right)^2}} \tag{15}$$

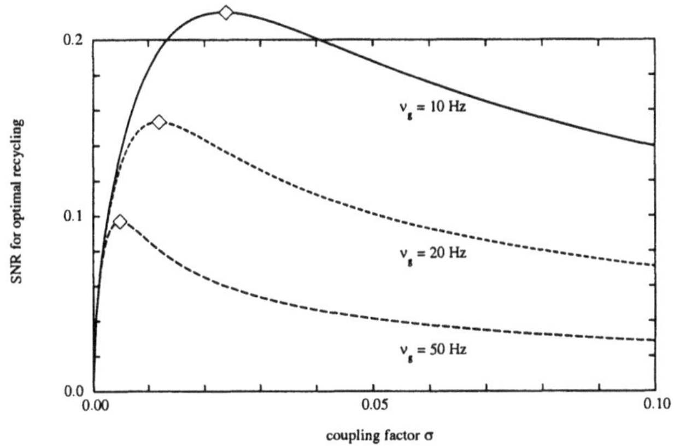

Figure 10: SNR vs σ for three GW frequencies. The small diamonds show the approximate optima theoretically derived

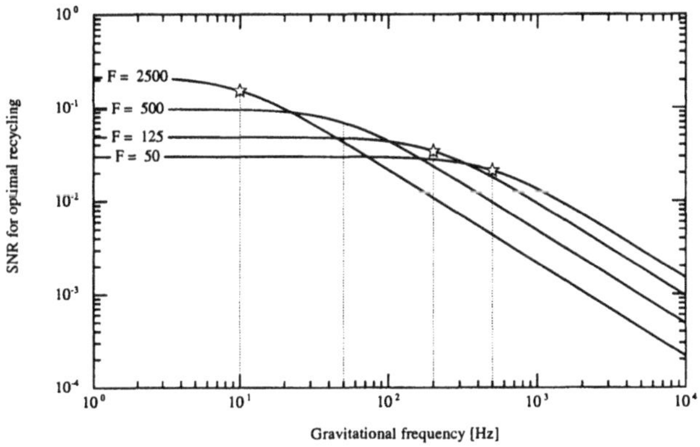

Figure 11: SNR vs σ for four finesses. The small stars point the GW frequency at which the SNR was optimized.

the parameter $p\Delta\nu_{\rm FSR}/2\pi$ has the dimension of a frequency, and is of order 1 Hz. The first term represents the gain due to optimal recycling, the second is the SNR of a simple Michelson. We can conclude that a power recycled Michelson, having an optimal recycling rate, and an optimal finesse for a given GW frequency is not significantly better that a simple Michelson when that frequency is very low. In this subsection and in the next one, we see how the reflectivity of the Fabry-Perot cavities play a central role. The efficiency of recycling crucially depends on the qulity of the reflectivity. This is the reason why at low frequency, a high finesse being needed, the coupling rate increases, the reflectivity decreases, and the effect of recycling becomes negligible. This strong requirement of very reflecting cavities was the cause of a number of numerical optics studies that in turn, motivated section 3.

detuned power recycling

We consider the case of a power recycled Michelson with detuned cavities. The basic idea is to exploit at the same time the resonance (frequency ν_0) of a cavity for one sideband (such that $\nu_L \pm \nu_g = \nu_0$) and the fact that the carrier being out of resonance, the reflectivity of the cavities is enhanced, and consequently the recycling efficiency also. We restrict our attention to two special cases giving the same result for the SNR : The *symmetrical* detuning, in which the two cavities have the same detuning Δf, and the *antisymmetrical* detuning, in which one cavity is detuned by Δf, and the other by $-\Delta f$. In the first case, the upper sideband is resonant in the two arms, and the lower sideband never, in the second case, the upper sideband is resonant in the first arm, and the lower sideband in the second arm, so that finally, the effect is identical. We develop the symmetrical case. Owing to the general Eq.14, the SNR for a detuned, power recycled Michelson is :

$$\mathrm{SNR}(f_g) = \frac{8\mathcal{F}L(1-\sigma/2)}{\lambda}\,\frac{t_r}{1-r_r(1-p_s)\rho}\sqrt{S}$$

where $\rho(\Delta f)$ is the FP's modulus reflectance, and

$$S = \frac{(1-\sigma+4\Delta f^2)^2+4(1-\sigma)^2 f_g^2}{(1+4\Delta f^2)((1-\sigma)^2+4\Delta f^2)(1+8(\Delta f^2+f_g^2)+16(\Delta f^2+f_g^2)^2)}$$

recall that

$$\rho(\Delta f) = \sqrt{1-\frac{\sigma(2-\sigma)}{1+4\Delta f^2}}$$

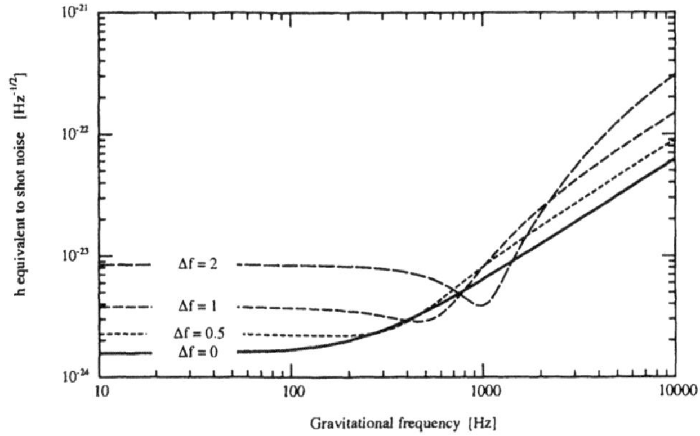

Figure 12: Detuned recycled Michelson (F=100)

The optimal recycling is obtained when

$$r_r = (1 - p_r)(1 - p_s)^2 \rho^2$$

The efficiency of recycling essentially depends on the reflectivity of the cavities. When the detuning is not zero, it simultaneously happens, for $f_g = \Delta f$ that one of the sidebands is resonant, and the reflectivity of the cavities, higher than when the carrier is resonant. This is the reason why it is possible to have a better SNR for f_g in the neighborhood of Δf. (see Fig.12).

The maximum SNR is

$$\mathrm{SNR_{max}} = \frac{8\mathcal{F}L(1-\sigma/2)}{\lambda} \frac{1}{\sqrt{1-(1-p_r)(1-p_s)\rho^2}} \times$$

$$\times \sqrt{\frac{(1-\sigma)^2 + 4(1-\sigma)(3-\sigma)\Delta f^2 + 16\Delta f^4}{(1+4\Delta f^2)((1-\sigma)^2 + 4\Delta f^2)(1+16\Delta f^2)}}$$

Synchronous Recycling

The title of the present section could have been "how to make a narrow band optical detector by 6 orders of magnitude better than bar detectors". The basic idea of synchronous recycling is to have two identical cavities, and a coupling. In such a system a system of *supermodes* exists, corresponding to combinations of the individual eigenmodes of one cavity. For instance, to a given TEM_{00} mode of frequency ν_0, corresponds two supermodes, a symmetrical (S) and an

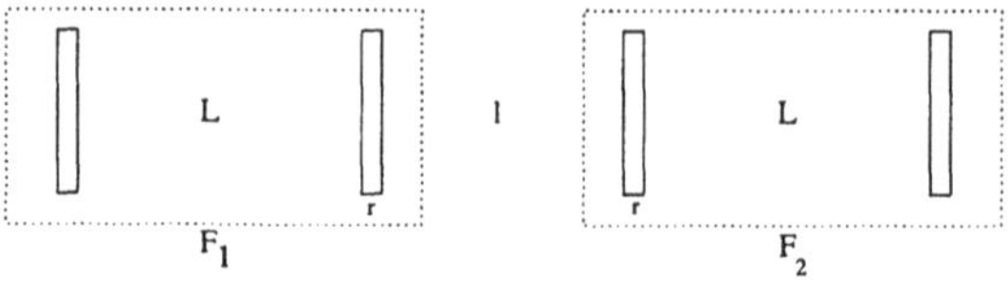

Figure 13: System of coupled cavities

antisymmetrical (A). The eigenfrequencies ν_S, ν_A differ from ν_0 by an amount depending of the coupling. When the coupling tends to zero, the frequencies ν_S, ν_A tend to the same limit ν_0, and to degeneracy. If the coupling is very weak, the difference $\nu_S - \nu_A$ may fall in the audio range, and a gravitational perturbation is able to pump energy from one mode in the other. The principle of operation is thus to tune the coupling at a minimum, the light source on the A mode, and waiting the signal on the S frequency (or vice-versa). Another way of understanding what happens in coupled cavities is to consider the beat note between these A and S modes. The result is that the stored energy is periodically exchanged between the two cavities, at a frequency which is the gap $\nu_S - \nu_A$ (Think to coupled pendulums). We feel that if the GW frequency is exactly this beat note, the light will accumulate positive phase shifts during the first half GW period, then will be transferred to the second cavity at the moment when the phase becomes negative in the first, and positive in the second, so that, roughly speaking, it sees always a long arm, (or a short one) and we can expect the phase modulation to increase indefinitely. It has been first proposed by Ph. Bernard and E. Picasso [4] to use this effect in high Q superconducting microwave cavities.

It is worth studying the effect on a simplified model involving only two coupled optical cavities (see Fig.13). The two cavities (of length L) are facing each other. The light can be transmitted through the central region of length l. In fact, this region is itself a cavity and we call it the central cavity. Without changing the two FP's, it is possible to tune the central cavity by changing the distance l. When the central cavity is at resonance, its tranmittance is a maximum, and the coupling is strong. When the central cavity is at antiresonance, its transmittance is a minimum, and the coupling is weak. We assume in the following simple model no losses, a reflectivity of 1 for the two end mirrors, and of r for the two inner mirrors. Let us consider the resonance condition for a wave to remain stored in the system. If we call F the reflectances of the (identical) cavities, we have for a round trip in the central cavity :

$$(iF\,\mathrm{e}^{ikl})^2 = 1$$

Two series of solutions can be obtained by taking

$$iF\,\mathrm{e}^{ikl} = 1 \quad \text{symmetrical mode}$$

$$iF\,\mathrm{e}^{ikl} = -1 \quad \text{antisymmetrical mode}$$

in case of zero losses, the reflectance of one cavity is of modulus 1 :

$$F = \frac{r + \mathrm{e}^{2ikL}}{1 + r\,\mathrm{e}^{2ikL}} = \mathrm{e}^{2ikL}\,\frac{1 + r\,\mathrm{e}^{-2ikL}}{1 + r\,\mathrm{e}^{2ikL}}$$

If we take the resonance as a reference frequency, we can write

$$2kL = \frac{4\pi\nu_0 L}{c} + \frac{4\pi L\delta\nu}{c}$$

where ν_0 is the resonance frequency of the (isolated) cavity, and $\delta\nu$ the unknown detuning giving a resonance in the coupled system. We have thus $4\pi\nu_0/\lambda \equiv \pi\,[\text{mod}2\pi]$, and we can work with the reduced detuning already used above, $\Delta f = \delta\nu/\delta\nu_{\mathrm{FWHM}}$ which is simply the ratio of the detuning to the linewidth of the cavity. The round trip phase becomes simply

$$2kL = \pi + \frac{2\pi}{\mathcal{F}}\,\Delta f$$

so that the reflectance reduces to the pure phase factor

$$\mathrm{Arg}(F) = \pi + \frac{2\pi}{\mathcal{F}}\,\Delta f + 2\,\tan^{-1}\left[\frac{r\,\cos(2\pi\Delta f/\mathcal{F})}{1 - r\,\sin(2\pi\Delta f/\mathcal{F})}\right]$$

For the phase factor corresponding to the central cavity, we have

$$kl = \frac{2\pi\nu_0 l}{c} + \frac{\pi l}{\mathcal{F}L}\,\Delta f$$

The constant phase $\varphi = 2\pi\nu_0 l/c$ can be considered as the tuning of the central cavity. The resonance conditions become

$$2\,\tan^{-1}\left[\frac{r\,\cos(2\pi\Delta f/\mathcal{F})}{1 - r\,\sin(2\pi\Delta f/\mathcal{F})}\right] = (2n+1)\pi - \frac{\pi}{2} - \varphi - \frac{2\pi}{\mathcal{F}}\Delta f - \frac{\pi l}{\mathcal{F}L}\,\Delta f$$

leading to the S-modes equation :

$$\frac{r\,\cos(2\pi\Delta f/\mathcal{F})}{1 - r\,\sin(2\pi\Delta f/\mathcal{F})} = \tan\left[\frac{\varphi}{2} + \frac{\pi}{4} + \left(1 + \frac{l}{2L}\right)\Delta f\right]^{-1} \qquad (16)$$

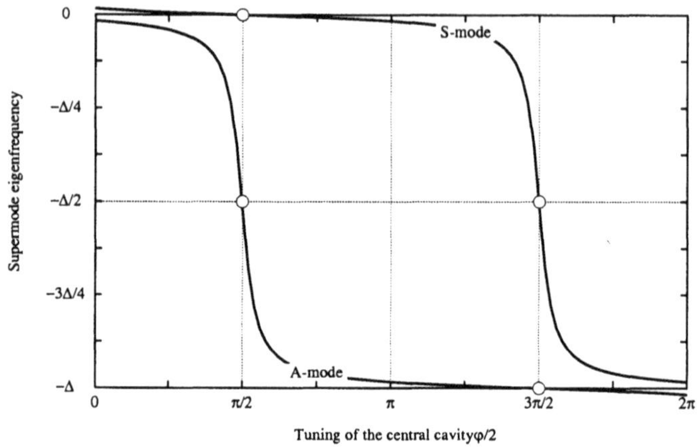

Figure 14: Relative detuning of the A and S supermodes vs tuning of the central cavity

The A-modes equation can be obtained a similar way :

$$\frac{r\,\cos(2\pi\Delta f/\mathcal{F})}{1-r\,\sin(2\pi\Delta f/\mathcal{F})} \;=\; -\tan\left[\frac{\varphi}{2}+\frac{\pi}{4}+\left(1+\frac{l}{2L}\right)\Delta f\right] \qquad (17)$$

These are implicit equations in the unknown detuning Δf. The numerical solutions are plotted on Fig.14. The round trip phase in the central cavity is equal to 2φ. The value $\varphi = 0$ corresponds thus to *antiresonance*, then to a minimum of coupling, and a weak splitting of the resonance lines. The tuning has period π, so that we retrieve a similar situation at $\varphi = \pi$ where the S-frequency is near the preceding A-frequency. The value $\varphi = \pi/2$ corresponds to *resonance* of the central cavity, thus to a maximum of coupling, and a maximum of line splitting This maximum is half the FSR (the interval between the two white spots on the figure). In order to study the minimum of coupling, and the frequency gap between the A and S modes at this tuning, we turn to our simplified model, which will be of some use anyway in the sequel. For zero losses, the parameter σ is zero, and we have for the phases :

$$\frac{\pi}{2}+2\,\tan^{-1}(2\Delta f)+\frac{2\pi l}{c}\,(\nu_0+\delta\nu)\;\equiv\;0\;\;(\mathrm{S-modes})$$

$$\frac{\pi}{2}+2\,\tan^{-1}(2\Delta f)+\frac{2\pi l}{c}\,(\nu_0+\delta\nu)\;\equiv\;\pi\;\;(\mathrm{A-modes})$$

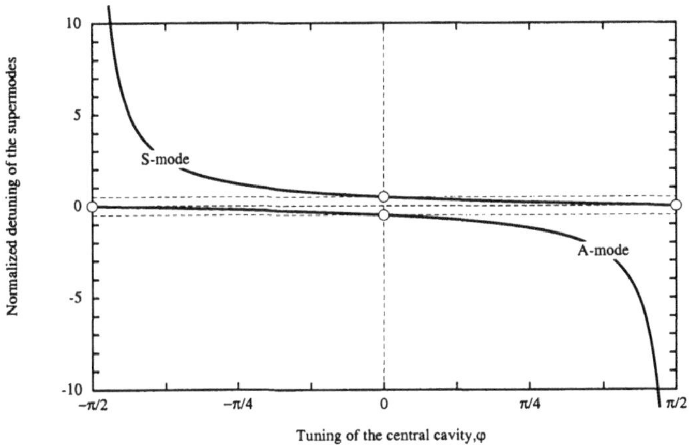

Figure 15: Approximate model of degeneracy removing by coupling

We can write as well

$$\Delta f_S = \frac{1}{2}\tan\left[\frac{\varphi}{2}+\frac{\pi}{4}+\frac{\pi l}{2\mathcal{F}L}\Delta f_S\right]^{-1}$$

The term $\pi l/2\mathcal{F}L$ is very small for kilometric cavities of finesse $\simeq 100$ and a metric central cavity. If we neglect it, we have the very simple results :

$$\Delta f_S = \frac{1}{2}\tan\left[\frac{\varphi}{2}+\frac{\pi}{4}\right]^{-1}$$

$$\Delta f_A = \frac{1}{2}\tan\left[\frac{\varphi}{2}+\frac{\pi}{4}\right]$$

The following plot (Fig.15) is to be compared with the preceding. The approximation used is valid only for detunings much smaller than the FSR. For $\varphi = \pi/2$, we have seen that the detuning of the A-mode is half the FSR, the model consequently fails, this is the reason of the divergence of the A-mode at this point. The same reason causes the divergence of the S-mode at $-\pi/2$. If we restrict our attention to the neighbouring of $\varphi = 0$, i.e. the validity range of the present model, we can see a good agreement with the exact calculation. It is in particular easy to compute the minimum line splitting :

$$[\Delta f_S - \Delta f_A]_{\min} = \frac{1}{2}\left[\frac{1}{\tan(\pi/4)}+\tan(\pi/4)\right] = 1$$

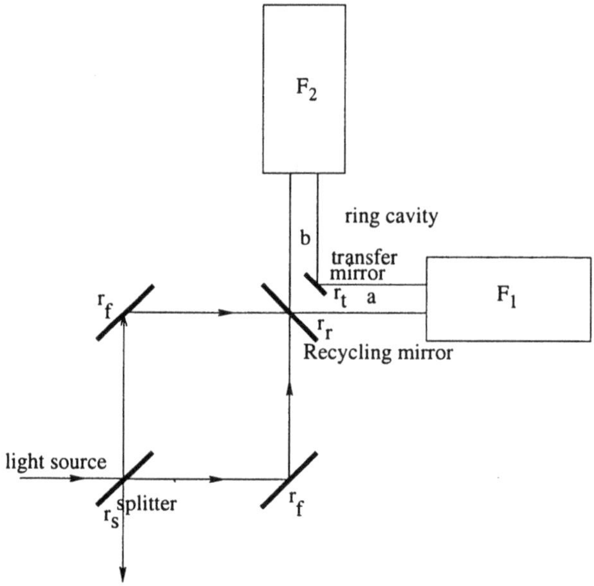

Figure 16: Sketch of the synchronous recycling configuration

corresponding, in terms of frequency, to

$$[\delta\nu_S - \delta\nu_A]_{\min} = \delta\nu_{\mathrm{FWHM}} = \frac{c}{2\mathcal{F}L}$$

In other words, the minimum splitting is nothing but the linewidth of the cavity. If we intend to use this device to detect GW by coupling the A and S modes with the gravitational perturbation, we see that we have to use high finesse and long cavities. For the current situation (L=3 km and $\mathcal{F}$=100), the frequency gap is ν_g=500 Hz. Higher values can be obtained by a different tuning of the central cavity : The general result is

$$\nu_g = \delta\nu_S - \delta\nu_A = \frac{1}{\cos\varphi}\frac{c}{2\mathcal{F}L}$$

We have now to study the response of a real system involving a light source and a detector. The scheme of Fig.16 was suggested years ago by R. Drever [3] after a very different approach than Ph. Bernard & E. Picasso. The coupled cavities are in what we call *ring cavity* on the figure. The optical path has been split for clarity, and it could seem strange to separate between

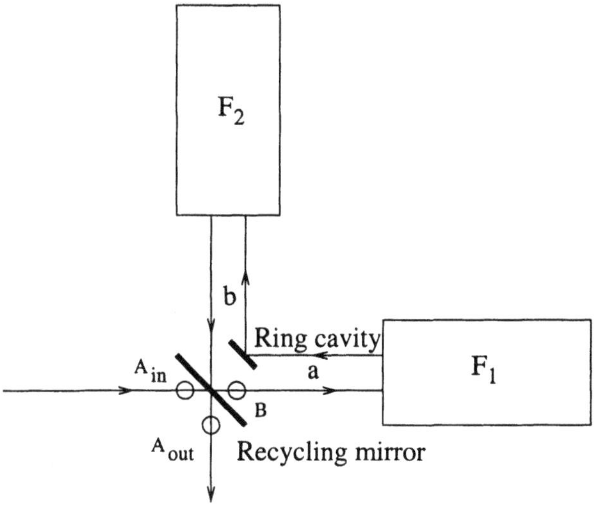

Figure 17: Ring cavity and counterclockwise optical path

the incident and the reflected wave off a cavity. It is however possible by using polarization rotators and polarization sensitive reflectors, so that the situation is almost that of the figure. If (as likely) these switching elements induce losses, these losses can be localized in the mirror r_t. The splitter and the square path allow to launch two rotating waves in the ring cavity, one clockwise and one counterclockwise, these waves are recombined on the splitter. We first consider the counterclockwise wave (see Fig.17) and evaluate the A133 reflection operator. We have firstly for the intracavity wave :

$$B = t_r A_{in} + r_r r_t e^{2ikl} \mathbf{F_2 F_1} \, B$$

or :

$$B = t_r \left[1 - r_r r_t e^{2ikl} \mathbf{F_2 F_1}\right]^{-1}$$

then

$$A_{out} = i\, r_r \, A_{in} - i\, t_r r_t e^{2ikl} \mathbf{F_2 F_1} \, B$$

so that the reflectance is :

$$\mathbf{R}_+ = r_r - t_r^2 r_t e^{2ikl} \mathbf{F_2 F_1} \left[1 - r_r r_t e^{2ikl} \mathbf{F_2 F_1}\right]^{-1}$$

$$= \left[r_r - (1 - p_r) r_t e^{2ikl} \mathbf{F_2 F_1}\right] \left[1 - r_r r_t e^{2ikl} \mathbf{F_2 F_1}\right]^{-1}$$

Obviously, for the clockwise optical path, we have the reflectance :

$$\mathbf{R}_- \;=\; \left[r_r - (1-p_r)r_t e^{2ikl}\mathbf{F_1F_2}\right]\left[1 - r_r r_t e^{2ikl}\mathbf{F_1F_2}\right]^{-1}$$

Now, if we return to the splitter, we can compute the transmittance of the whole system :

$$\mathbf{T} \;=\; -i\,t_s^2 r_f^2\, e^{i\phi}\mathbf{R}_- + i r_s^2 r_f^2\, e^{i\phi}\mathbf{R}_+$$

Where the two transfer mirrors of the square cavity have been assumed identical. ϕ is the optical path inside the square cavity. If further we assume a perfectly symmetrical splitter, we can write

$$\mathbf{T} \;=\; -i\,\frac{1}{2}(1-p_s)r_f^2 e^{i\phi}\,(\mathbf{R}_- - \mathbf{R}_+)$$

A direct caclulation gives

$$[R_- - R_+]_{10} \;=\; -t_r^2 r_t\, e^{2ikl}\,\frac{2G_+(F_+ - F)}{DD_+}$$

Where the definitions of $G_\pm$, $F_\pm$ are the same as in section 7 , and

$$D_a = 1 - r_r r_t\, e^{2ikl} F_a^2 \quad (a = -1, 0, 1)$$

The [10] component of $\mathbf{T}$ is thus :

$$T_{10} = \frac{\mathcal{F}L}{\lambda}\,\frac{4i(2-\sigma)^2(1-p_s)r_f^2 r_t e^{i\phi} t_r^2 f_g}{(1-2i\Delta f)^2[1-2i(\Delta f + f_g)]^2\left[1 - r_r r_t e^{2ikl}F^2\right]\left[1 - r_r r_t e^{2ikl}F_+^2\right]}$$

One would obtain a similar xpression for T_{20} by changing the sign of f_g. The preceding expression exhibits a sharp resonance peak when the resonance condition

$$2kl + 2\mathrm{Arg}(F) \equiv 0 \;\; [\mathrm{mod} 2\pi]$$

is met The difference of π with respect to the preceding subsection (two isolated cavities) is due to the fact that we have now two extra mirrors, for recycling (M_r) and transfer (M_t) each adding a phase of $\pi/2$. This should be kept in mind in any comparison. In particular, the resonance condition for the central cavity is now $2kl \equiv 0$, and the antiresonance is $2kl \equiv \pi$. We assume the laser frequency given, so that kl is a constant, representing the tuning of the central cavity. The long cavities are detuned by a microscopic change in length making their new resonance shifted by an amount Δf. For Δf corresponding to a resonance of the ring cavity, we have thus to solve

$$\tan^{-1}\left(\frac{2\Delta f}{1-\sigma}\right) + \tan^{-1}(2\Delta f) \;=\; -kl$$

or,

$$\frac{2(2-\sigma)\Delta f}{1-\sigma-4\Delta f^2} = -\tan(kl)$$

this gives two solutions :

$$\Delta f_S = \frac{1}{2}\left[\frac{1-\sigma/2}{\tan(kl)} + \sqrt{\left(\frac{1-\sigma/2}{\tan(kl)}\right)^2 + 1-\sigma}\right]$$

and

$$\Delta f_A = \frac{1}{2}\left[\frac{1-\sigma/2}{\tan(kl)} - \sqrt{\left(\frac{1-\sigma/2}{\tan(kl)}\right)^2 + 1-\sigma}\right]$$

Note that

$$\Delta f_A \times \Delta f_S = -\frac{1-\sigma}{4} \tag{18}$$

and

$$\Delta f_S - \Delta f_A = \sqrt{\left(\frac{1-\sigma/2}{\tan(kl)}\right)^2 + 1-\sigma} \tag{19}$$

We remark that the minimum frequency gap, is

$$[\Delta f_S - \Delta f_A]_{\min} = \sqrt{1-\sigma}$$

For having a large SNR at the normalized GW frequency $f_g^{(0)} > \sqrt{1-\sigma}$, we follow the following scheme :

- Tune the central cavity in such a way that $\Delta f_S - \Delta f_A = f_g^{(0)}$, which happens for

$$kl = \tan^{-1}\left[\frac{1-\sigma/2}{\sqrt{f_g^{(0)2} - (1-\sigma)}}\right]$$

- Δf_A and Δf_S are now determined. Tune the cavities in such a way that $\Delta f = \Delta f_A$.

Obviously, a similar process can be carried out starting from the upper frequency Δf_S and going to the lower level Δf_A : this is described by the T_{20} component of the A133 transmittance (see Fig.18). Unfortunately, only one of the two components T_{10} and T_{20} can be made resonant. If we work on T_{10}, we

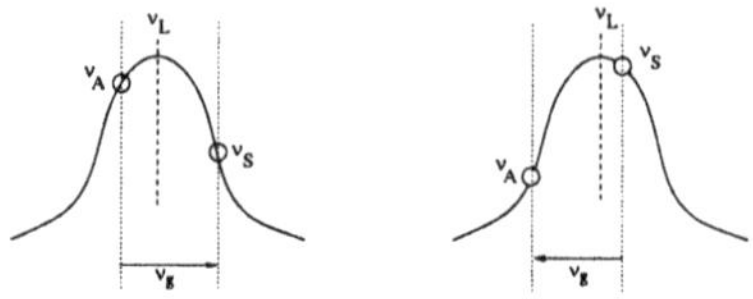

Figure 18: Gravito-optical pumping

can therefore neglect T_{20}, and vice-versa. At the end of this process, we see that the resonance condition is met simultaneously, for

$$1 - r_r r_t e^{2ikl} F^2 = 1 - r_r r_t \rho_A^2$$

because the tuning satisfies the modes A equation, and for

$$1 - r_r r_t e^{2ikl} F_+^2 = 1 - r_r r_t \rho_S^2$$

because the tuning satisfies the modes S equation when $f_g = f_g^{(0)}$. ρ_A and ρ_S are the modulus reflectances :

$$\rho_{A,S}^2 = 1 - \frac{\sigma(2-\sigma)}{1 + 4\Delta f_{A,S}^2}$$

We have then, at resonance, when $f_g = f_g^{(0)}$,

$$|T_{10}| = \frac{\mathcal{F}L}{\lambda} \frac{16(1-\sigma/2)^2 t_r^2 f_g^{(0)}}{(1+4\Delta f_A^2)(1+4\Delta f_S^2)(1-r_r r_t \rho_A^2)(1-r_r r_t \rho_S^2)}$$

We can still fix the recycling rate by opimizing T_{10} with respect to r_r. In the general case, when the gravitational frequency is larger than the minimum gap, it is cumbersome to derive the optimal r_r. We have :

$$[r_r]_{\text{opt}} = \frac{1 + (1-p_r) r_t^2 \rho_A^2 \rho_S^2}{r_t(\rho_A^2 + \rho_S^2)} - \sqrt{\left(\frac{1 + (1-p_r) r_t^2 \rho_A^2 \rho_S^2}{r_t(\rho_A^2 + \rho_S^2)}\right)^2 - (1-p_r)}$$

If we use the definitions of ρ_A and ρ_S, it is possible to show that

$$\rho_A \times \rho_S = 1 - \sigma$$

which is particularly remarkable, being independent on the tuning of the cental cavity. After that, the sum $\rho_A^2 + \rho_S^2$ will obviously depend on the tuning :

$$\rho_A^2 + \rho_S^2 = 1 + (1-\sigma)^2 - \frac{\sigma^2(2-\sigma)^2}{\sigma^2 + 4 f_g^{(0)2}}$$

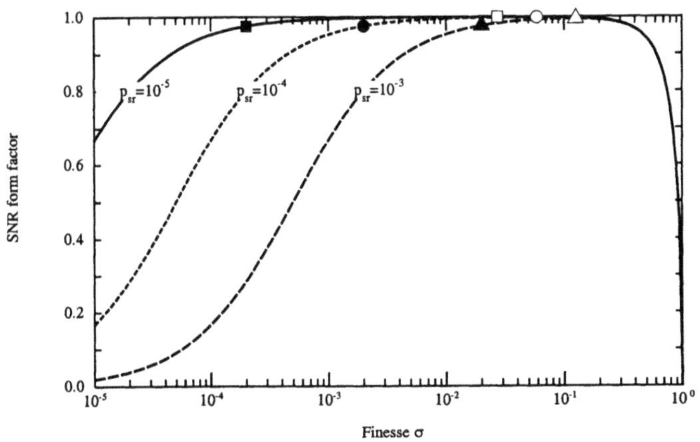

Figure 19: Form factor K for the SNR vs finesse $\sigma = p\mathcal{F}/\pi$. The white dots are the strict optima, the black dots mark the pseudo-optima

It is however reasonable to optimize the SNR for the lowest possible GW frequency, i.e when $f_g^{(0)} = f_m = \sqrt{1-\sigma}$, situation in which we have $\rho_A = \rho_S = \sqrt{1-\sigma}$. The optimum value of the recycling mirror reflection coefficient is simply

$$[r_r]_{\text{opt}} = (1-p_r)r_t(1-\sigma)$$

And the optimal peak value of $|T_{10}|$ is

$$|T_{10}|_{\text{peak}} = \frac{2\pi L}{\lambda p} K(\sigma)$$

with the form factor

$$K(\sigma) = \frac{2\sigma\sqrt{1-\sigma}}{1-(1-p_{RC})(1-\sigma)^2}$$

and $1 - p_{RC} = (1-p_r)r_t^2$. The form factor K has the maximum value 1, obtained for the approximate value $\sigma = (2p_{RC})^{1/3}$. But the shape of the curve is so flat, that this value is misleading, a value of K very close to 1 is obtained already for the pseudo- optinum $\sigma \simeq 20p_{sr}$ (see fig.19). The peak SNR at resonance is therefore

$$\text{SNR}_{\text{peak,Max}} = \frac{2\pi L}{\lambda p}$$

for a wide range of reference GW frequencies. For $f^{(0)}$ too small, however, the SNR falls to zero. Remark that this peak value scales as $1/p$, whereas the zero frequency limit of the (wideband) power recycling scales as $1/\sqrt{p}$. The spectral density of h equivalent to shot noise is :

$$h(f_0) \;=\; \sqrt{\frac{h\nu}{P_L}}\,\frac{1}{\mathrm{SNR}_{\mathrm{peak,Max}}}$$

For cavity losses $p = 3\,10^{-5}$, l=3 km, $\lambda = 1.064\,10^{-6}$ m, P_L=20W, we have

$$h(f_0) \;\simeq\; 1.6\;10^{-25}\mathrm{Hz}^{-1/2}$$

It is now necessary to study the width of the resonance. For this purpose, we assume the laser being locked on the antisymmetric resonance, and the gravitational frequency in the neighbourhood of the gap $f_g^{(0)}$, i.e. $f_g = f_g^{(0)} + \delta f$, so that $\Delta f_A + f_g \;=\; \Delta f_S + \delta f$. Consider the SNR :

$$\mathrm{SNR}(\delta f) = \frac{16\mathcal{F}L}{\lambda}\,\frac{(1-\sigma/2)^2 t_r^2 f_g}{(1+4\Delta f_A^2)(1+4(\Delta f_A+f_g)^2)(1-r_r r_t \rho_A^2)\left(1-r_r r_t \rho_+^2 \mathrm{e}^{2i\Phi}\right)}$$

In this expression, the varying terms are :

- The fast varying phase

$$\Phi \;=\; \tan^{-1}\left[\frac{2(\Delta f_S+\delta f)}{1-\sigma}\right] + \tan^{-1}\left[2(\Delta f_S+\delta f)\right] - \tan^{-1}\left[\frac{2\Delta f_S}{1-\sigma}\right] - \tan^{-1}\left[2\Delta f_S\right]$$

  expanded at first order in δf this gives

$$\Phi \;=\; A\delta f \quad \text{with} \quad A \;=\; \frac{4(2-\sigma)(1-\sigma+4\Delta f_S^2)}{[(1-\sigma)^2+4\Delta f_S^2]\,[1+4\Delta f_S^2]}$$

- The reflectivity of the cavities for the upper sideband :

$$\rho_+^2 \;=\; 1 - \frac{\sigma(2-\sigma)}{1+4(\Delta f_S+\delta f)^2}$$

  This differs from unity by a small amount, whose variation is like second order. More specifically, the second order expansion gives

$$\rho_+^2 \;=\; \rho_S^2 + \frac{8\sigma(2-\sigma)\Delta f_S}{(1+4\Delta f_S^2)}\,\delta f + \frac{4\sigma(2-\sigma)(1-12\Delta f_S^2)}{(1+4\Delta f_S^2)^3}$$

  We have already seen that the best value of σ is very small, in order to have a good reflectivity of the cavities. It can be thus understood, and numerically checked that the variations of ρ_+^2 around the S resonance can be neglected.

- The term,

$$\frac{f_g^{(0)} + \delta f}{1 + 4(\Delta f_S + \delta f)^2}$$

which varies very little.

The study of the shape of the resonance line can thus be carried out on the only term :

$$\begin{aligned}|1 - r_r r_t \rho_+^2 e^{2i\Phi}| &\simeq \left[(1 - r_r r_t \rho_S^2)^2 + 4 r_r r_t \sin^2(A\delta f)\right]^{1/2} \\ &= (1 - r_r r_t \rho_S^2)\left[1 + \left(\frac{2\sqrt{r_r r_t \rho_S^2}\sin(A\delta f)}{1 - r_r r_t \rho_S^2}\right)^2\right]^{1/2}\end{aligned}$$

expression very similar to a cavity resonance, with the *superfinesse*

$$\mathcal{SF} = \frac{\pi\sqrt{r_r r_t \rho_S^2}}{1 - r_r r_t \rho_S^2}$$

The linewidth (FWHM) od the SNR is thus :

$$\Delta f_g = \frac{\sqrt{3}(1 - r_r r_t \rho_s^2)}{A\sqrt{r_r r_t \rho_S^2}}$$

A simplified model will help to get simple estimates of the peak value and the linewidth of the SNR. We have seen that small values of σ are pseudo-optimal. We can then try a first order approximation in σ and a fortiori in the various losses. The SNR (with optimized recycling rate) becomes :

$$\mathrm{SNR}(f_g) = \frac{16\mathcal{F}L}{\lambda}$$

$$\frac{(1-\sigma) f_g (p_{RC} + 2\sigma)}{(1 + 4\Delta f_A^2)(1 + 4\Delta f_S^2)(1 - (1 - p_{RC})(1 - \sigma)\rho_A^2)|1 - (1 - p_{RC})(1 - \sigma)\rho_S^2\, e^{2i\Phi}|}$$

It is easy to see that the 1st order expressions for the S and A detunings are respectively :

$$\Delta f_S = \frac{1 - \sigma/2}{2\tan(z/4)}$$

$$\Delta f_A = -\frac{(1 - \sigma/2)\tan(z/4)}{2}$$

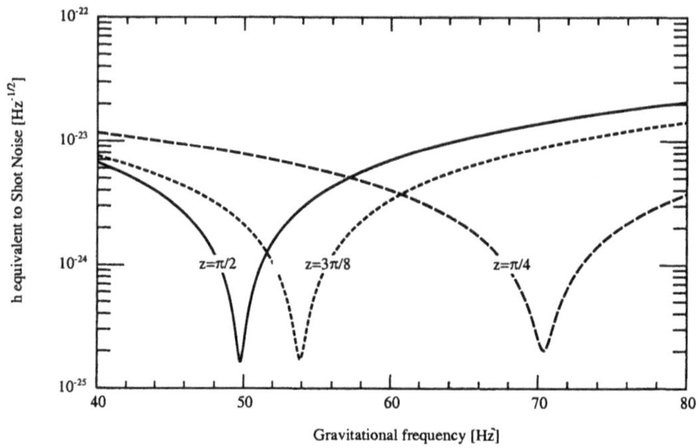

Figure 20: Effect of the detuning of the central cavity on the response of synchronous recycling interferometer

where $z = 2kl$ is the tuning of the central cavity. we find then

$$1 + 4\Delta f_S^2 = \frac{1 - \sigma\cos^2(z/4)}{\sin^2(z/4)}$$

$$1 + 4\Delta f_A^2 = \frac{1 - \sigma\sin^2(z/4)}{\cos^2(z/4)}$$

whence

$$(1 + 4\Delta f_S^2)(1 + 4\Delta f_A^2) = \frac{4(1-\sigma)}{\sin^2(z/2)}$$

and also

$$\rho_S^2 = 1 - 2\sigma\sin^2(z/4)$$
$$\rho_A^2 = 1 - 2\sigma\cos^2(z/4)$$

The gravitational resonance frequency is

$$f_g^{(0)} = \Delta f_S - \Delta f_A = \frac{1 - \sigma/2}{\sin(z/2)}$$

If we can neglect the ring cavity losses p_{RC} (a few 10^{-5}) with respect to σ (up to 1%), we have simply

$$1 - (1 - p_{RC})(1 - \sigma)\rho_S^2 \simeq \sigma[1 + 2\sin^2(z/4)]$$

$$1-(1-p_{RC})(1-\sigma)\rho_A^2 \simeq \sigma[1+2\cos^2(z/4)]$$

For the varying phase factor, we have

$$\Phi = 4(1-\sigma/2)\sin^2(z/4)\,\delta f$$

The SNR is :

$$\text{SNR} = \frac{8\pi L}{\lambda p}\frac{(1-\sigma/2)f_g^{(0)}\sin(z/2)}{(1-\sigma)[1+2\cos^2(z/4)][1+2\sin^2(z/4)]}$$

$$\times\left[1+\left(\frac{8\sin^2(z/4)}{\sigma[1+2\sin^2(z/4)]}\,\delta f\right)^2\right]^{-1/2}$$

and finally :

$$\text{SNR} = \frac{2\pi L}{\lambda}\frac{1}{p}(1-\sigma/2)\frac{4\sin(z/2)}{3+\sin^2(z/2)}$$

$$\times\left[1+\left(\frac{8\sin^2(z/4)}{\sigma[1+2\sin^2(z/4)]}\,\delta f\right)^2\right]^{-1/2}$$

from where we conclude that the overall peak value, corresponding to $z=\pi/2$ is

$$\text{SNR}_{peak,max} = \frac{2\pi L}{\lambda}\frac{1}{p} \tag{20}$$

This peak corresponds to the resonance frequency

$$f_g^{(0)} = 1-\sigma/2$$

in other words, the minimum resonance frequency is given by the linewidth of the cavity, and the minimum gravitational linewidth (FWHM) :

$$\delta f_{FWHM,min} = \sigma\sqrt{3}$$

In terms of gravitational frequencies, we find the relation with the cavity linewidth :

$$\Delta\nu_{g,FWHM,min} = \sqrt{3}\times\delta\nu_{FWHM}$$

Let us summarize the results for small σ and z not far $\pi/2$:

- By varying the tuning of the central cavity, it is possible to adjust the resonance for a GW frequency equal to or larger than the linewidth of the cavities. The general formula is :

$$\nu_g^{(0)} = \frac{1-\sigma/2}{\sin(z/2)}\times\delta\nu_{FWHM} \tag{21}$$

- The best response of the interferometer is obtained for the lowest GW frequency, when the central cavity is exactly antiresonant ($z = \pi/2$), the value of the SNR resonance peak is:

$$\mathrm{SNR}_{peak} = \mathrm{SNR}_{peak,max} \times P(z) \tag{22}$$

where the maximum peak value has been expressed above (Eq. 20), and $P(z)$ is a form factor, taking the value 1 for $z = \pi/2$:

$$P(z) = \frac{4\sin(z/2)}{3 + \sin^2(z/2)} \tag{23}$$

- When the central cavity is progressively detuned from antiresonance, the GW resonance frequency increases, the sensitivity decreases, and the GW linewidth increases. The general formula for the GW linewidth is :

$$\delta\nu_{g,FWHM} = \sigma\sqrt{3} \times \frac{1 + 2\sin^2(z/4)}{4\sin^2(z/4)} \times \delta\nu_{FWHM} \tag{24}$$

These approximations remain true as long as σ does'nt exceed a few %. For very low gravitational frequencies, the linewidth has to be very thin, and the finesse very high, σ cannot more be kept small and the approximation fails. In fact we already know from the preceding study that the SNR tends to zero when the resonance peak tends to zero. The ratio $\nu_g/\delta\nu_{g,FWHM}$ gives an idea of the equivalent Q of the resonator. For the optimal operation point ($z = \pi/2$), we have

$$Q \simeq \frac{1}{\sigma\sqrt{3}} = \frac{\pi}{\sqrt{3}p\mathcal{F}}$$

Signal recycling

Dual or Signal recycling was proposed some years ago by B. Meers. The idea is to add one more mirror after the output port of the interferometer in order to store the sidebands generated by the GW. The out port plus the dual recycling mirror form a resonant cavity, and the sidebands that are near resonance are enhanced. This allows to modify the sensitivity curve and have a gain factor at a given frequency range of special interest. We have seen other methods giving a comparable result. Here, one more benefit is to enhance the constrast of the interferometer by the spatial filtering effect of the extra Fabry-Perot installed at the output (But this is out of the scope of the present document). The sketch of the setup and the notation are shown on Fig.21. The lengths

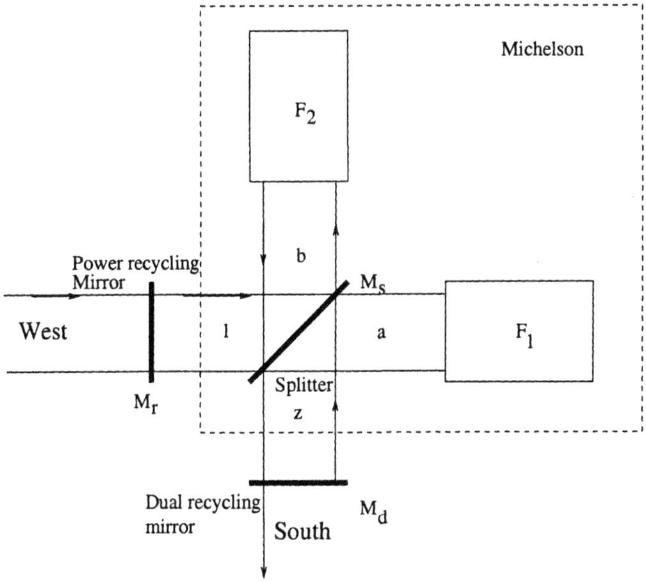

Figure 21: sketch of the Dual recycling configuration

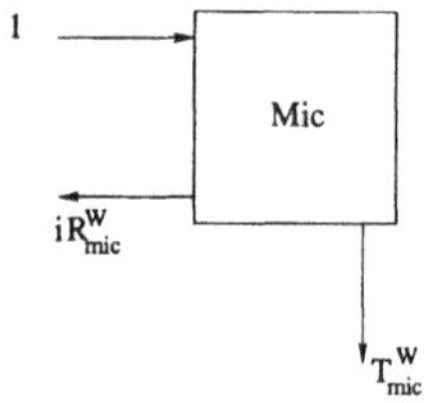

Figure 22: West input

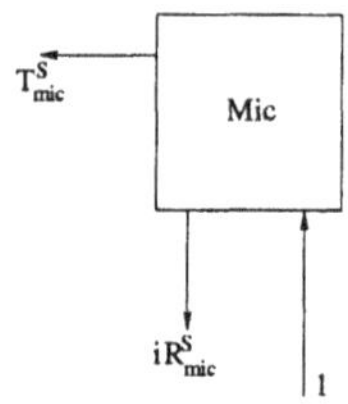

Figure 23: South input

of the short arms are a and b, the length of the power recycling cavity is l, and the length of the dual recycling cavity is z. The parameters of the mirrors are labeled by r,s,d. The A133 operator corresponding to the whole setup may be constructed by successive shells. We first consider the Michelson (mic) as a black box having two inputs, West (as in the preceding sections) and South (because the dual recycling reinjects from the South). It has therefore an A133 reflectance R^W_{mic} and a transmittance $T^W_m ic$ (see Fig.22). It has also a reflectance R^S_{mic} and a transmittance T^S_{mic} for the South input port : The corresponding operators are easy to compute :

$$\begin{aligned}
R^W_{mic} &= t_s^2 e^{2ika} F_1 - r_s^2 e^{2ikb} F_2 \\
T^W_{mic} &= -r_s t_s \left(e^{2ika} F_1 + e^{2ikb} F_2\right) \\
R^S_{mic} &= -r_s^2 e^{2ika} F_1 + t_s^2 e^{2ikb} F_2 \\
T^S_{mic} &= -r_s t_s \left(e^{2ika} F_1 + e^{2ikb} F_2\right)
\end{aligned}$$

If we assume the splitter to be strictly symmetrical, and the Michelson tuned at a black fringe, we have simply, after setting $m = (a+b)/2$,

$$R^W_{mic} = -R^S_{mic} = \frac{i}{2}(1-p_s)e^{2ikm}(F_1 + F_2) = R_{mic}$$

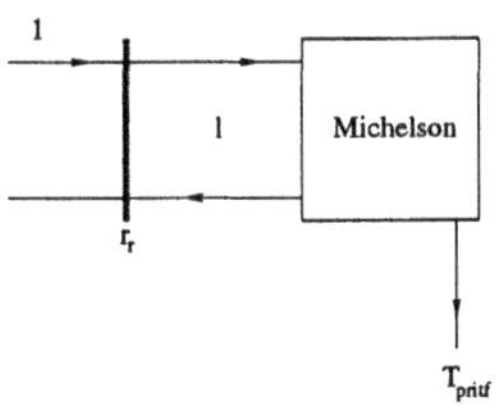

Figure 24: West input on a power recycled Michelson

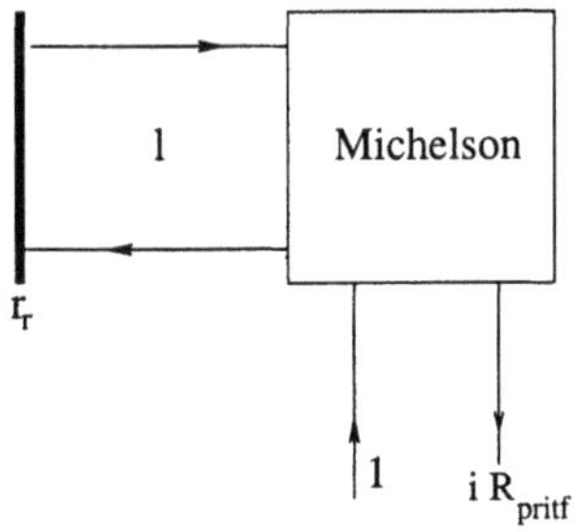

Figure 25: South input on a power recycled Michelson

$$T^W_{mic} \;=\; T^S_{mic} \;=\; -\frac{i}{2}(1-p_s)e^{2ikm}(F_1 - F_2) \;=\; T_{mic}$$

with

$$F_1 \;=\; \begin{pmatrix} F & 0 & 0 \\ G_+ & F_+ & 0 \\ G_- & 0 & F_- \end{pmatrix} \;,\; F_2 \;=\; \begin{pmatrix} F & 0 & 0 \\ -G_+ & F_+ & 0 \\ -G_- & 0 & F_- \end{pmatrix}$$

F, $F_\pm$ and $G_\pm$ having the definitions set in section 8. Now the power recycled interferometer (pritf) (see Fig.24 and Fig.25) has a transmittance T_{pritf} for a west input, and a reflectance R_{pritf} for a south input.

$$T_D \;=\; t_d e^{ikz}\left[1 + r_d e^{2ikz} R_{pritf}\right]^{-1} T_{pritf}$$

After some elementary algebra, we find $9m \equiv (a+b)/2)$:

$$T_{D,10} \;=\; -i\,\frac{t_r t_D (1-p_s) e^{ik(l+2m+z)} G_+}{\left[1 - i(1-p_s) r_D e^{2ik(m+z)} F_+\right]\left[1 + i(1-p_s) r_r e^{2ik(l+m)} F\right]} \tag{25}$$

$$T_{D,20} \;=\; -i\,\frac{t_r t_D (1-p_s) e^{ik(l+2m+z)} G_-}{\left[1 - i(1-p_s) r_D e^{2ik(m+z)} F_-\right]\left[1 + i(1-p_s) r_r e^{2ik(l+m)} F\right]} \tag{26}$$

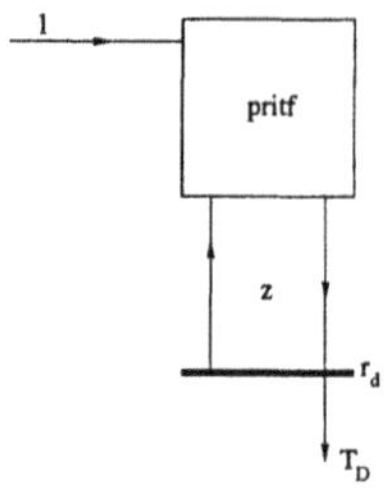

Figure 26: West input on a power and dual recycled Michelson

It is easy ro recognize in these formulas the SNR for a power recycled Michelson, as already derived in sect. 9, multiplied by an extra surtension factor :

$$S_D = \frac{t_D}{1 - i(1-p_s)r_D e^{2ik(m+z)} F_+}$$

if we choose to make the 20 component resonant. It is possible

- to tune the long Fabry-Perot's at resonance, so that $Arg(F) = \pi$,
- to tune the power recycling cavity so as to obtain resonance, by taking

$$2k(l+m) + \frac{\pi}{2} \equiv \pi$$

the surtension factor may be written as

$$S_D = \frac{t_D}{1 - e^{i(\pi/2 + 2k(m+z) + Arg(F_+)}\rho_+}$$

where, f being the gravitational reduced frequency,

$$\rho_+ = 1 - \frac{\sigma(2-\sigma)}{1+4f^2}$$

$$Arg(F_+) = \tan^{-1}\left[\frac{2f}{1-\sigma}\right] + \tan^{-1}[2f]$$

clearly, it is always possible to tune the dual recycling cavity to meet resonance, with the condition :

$$\frac{\pi}{2} + 2k(m+z) + \tan^{-1}\left[\frac{2f}{1-\sigma}\right] + \tan^{-1}[2f] \equiv 0$$

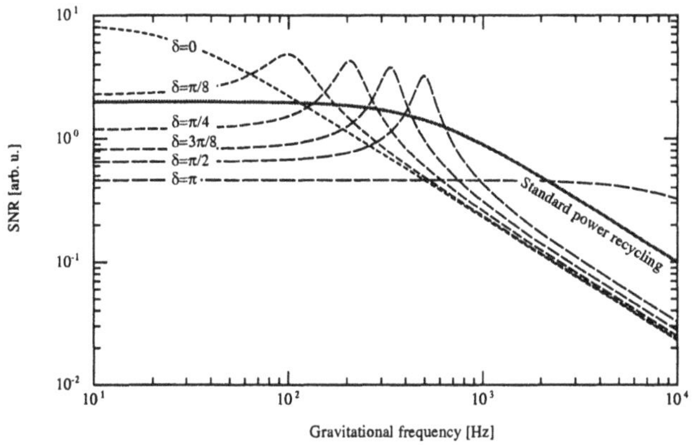

Figure 27: SNR of dual recycling configuration for various detuning of the signal recycling cavity $\delta \equiv 4\pi(m+z)/\lambda \ [\mathrm{mod}\ 2\pi]$

It is possible also to adjust the value of t_D for having a more or less sharp resonance. We remark that for $\delta = \pi$ the cavity built by the corner cavity mirror and the dual recycling mirror is at resonance. This results in a broadband response, larger than the simple Fabry-Perot. This special regime has been named "resonant signal extraction" [5].

3 Numerical Optics and GW Interferometers

3.1 introduction

In interferometers, we need to store light in long cavities in which light propagates back and forth. Wehave seen that a good reflectivity of these cavities is a key condition for efficiency in recycling. This reflectivity, combinerd with the ability to achieve a dark fringe have been an actual worry at the beginning of interferometer projects, and have triggered a lot of optical simulations of FP cavities and interferometers. It was essential to have theoretical models for light propagation. The theory used up to now for this purpose is the Scalar Diffraction Theory (SDT) (this seemed sufficient, owing to the very weak departure of the optical elements from an ideal shape). The basis of the SDT is the Kirchhoff equation, it seemed therefore useful to recall it and its derivation, in order to see clearly what means the paraxial approximation which is in fact more widely used.

The Helmholtz equation

A component of the the real Electric field, say $\mathcal{E}(x,y,z)$ in a homogeneous dielectric medium of refractive index n, obeys the wave equation ($c = 2.997925 \times 10^8$km.s^{-1} being the speed of light in a vacuum)

$$\left[\Delta - \frac{n^2}{c^2}\frac{\partial^2}{\partial t^2}\right] \mathcal{E}(t,x,y,z) = 0 \tag{27}$$

The light coming from a laser can be viewed, in a naive representation as a pure monochromatic wave. In fact, real lasers have a finite linewidth, and a finite coherence range. So far as the dimensions of an optical system are small compared to the coherence range, the monochromatic approximation remains valid. The lasers used in gravitational wave interferometers are highly stabilised in frequency and have huge coherence ranges. Even kilometric optical systems may be treated assuming purely monochromatic waves. For a monochromatic wave of frequency $\nu = \omega/2\pi$, we can set

$$\mathcal{E}(x,y,z) = \frac{1}{2}\left(E(x,y,z)\,\mathrm{e}^{-i\omega t} + \overline{E}(x,y,z)\,\mathrm{e}^{i\omega t}\right) \tag{28}$$

and for the amplitude E of the electric field, we obtain the Helmholtz equation

$$\left[\Delta + k^2\right] E(x,y,z) = 0 \tag{29}$$

where $k \equiv n\omega/c$.

The Kirchhoff integral

This is usually the most delicate part in optics books, and often skipped by stressed readers. The role and the status of the Kirchhoff theory is therefore seldom known : Is it "exact" or "approximate", then with respect to what ? We try to address these issues and give answers at the end of the present section. Recall that for two arbitrary fields A and B, we have a relation between a volume integral and an integral on the surface bounding the same volume, known as Green's theorem :

$$\int_{\mathcal{V}} (A\Delta B - B\Delta A)\,d\vec{r} = -\oint_{\mathcal{S}} \left[A\,\frac{\partial B}{\partial n} - B\,\frac{\partial A}{\partial n}\right] ds \tag{30}$$

$\vec{n}$ represents the inward normal to the surface $\mathcal{S}$ surrounding the volume $\mathcal{V}$ (see Fig.28). The following notation has been used :

$$\frac{\partial}{\partial n} \equiv \vec{n}\cdot\vec{\nabla}$$

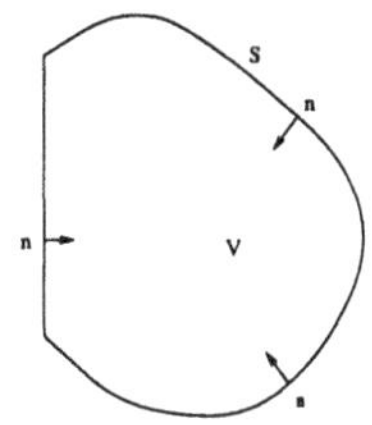

Figure 28:

Consider now a solution $E(x,y,z) = E(\vec{r})$ of the Helmholtz equation (29) , and $G(\vec{r})$ a Green function, i.e. satisfying

$$\left[\Delta + k^2\right] G(\vec{r}) = -\delta(\vec{r}) \tag{31}$$

We have as well, Δ' involving 2d order derivatives,

$$\left[\Delta' + k^2\right] G(\vec{r} - \vec{r'}) = -\delta(\vec{r} - \vec{r'}) \tag{32}$$

by multiplying both sides by $E(\vec{r'})$, we can write

$$E(\vec{r'}) \left[\Delta' + k^2\right] G(\vec{r} - \vec{r'}) = -\delta(\vec{r} - \vec{r'})\, E(\vec{r'})$$

and obviously, with (29) :

$$G(\vec{r} - \vec{r'}) \left[\Delta' + k^2\right] E(\vec{r'}) = 0$$

by subtracting these two equations, we get :

$$E(\vec{r'})\, \Delta'\, G(\vec{r} - \vec{r'}) - G(\vec{r} - \vec{r'})\, \Delta'\, E(\vec{r'}) = -\delta(\vec{r} - \vec{r'})\, E(r')$$

Consider now a volume $\mathcal{V}$ in which Eq.(29) holds, bounded by a closed surface $\mathcal{S}$. By integrating the preceding equation over the volume, we get :

$$E(\vec{r}) = -\int_{\mathcal{V}} \left[E(\vec{r'})\, \Delta'\, G(\vec{r} - \vec{r'}) - G(\vec{r} - \vec{r'})\, \Delta'\, E(\vec{r'})\right] d\vec{r'}$$

by using Green's theorem, this becomes :

$$E(\vec{r}) = \oint_{\mathcal{S}} \left[E(\vec{r'})\, \vec{n'} \cdot \vec{\nabla}' G(\vec{r} - \vec{r'}) - G(\vec{r} - \vec{r'}) \vec{n'} \cdot \vec{\nabla}' E(\vec{r'})\right] ds'$$

provided that the point $\vec{r}$ is inside the closed surface $\mathcal{S}$. If the point is outside, the integral vanishes, and this property will be exploited below. Assume now that the surface $\mathcal{S}$ extends to infinity. There are thus two half spaces that we refer to as the left half space, and the right half space respectively. On the surface at infinity, it can be shown that the surface integral vanishes, due to the radiation condition on the field (outgoing waves, no source at infinity). The result is that at any point $\vec{r}$ in the right half space we have :

$$E(\vec{r}) = \oint_S \left[E(\vec{r'})\, \vec{n'} \cdot \vec{\nabla}' G(\vec{r} - \vec{r'}) - G(\vec{r} - \vec{r'})\vec{n'} \cdot \vec{\nabla}' E(\vec{r'})\right] ds' \tag{33}$$

whereas for $\vec{r}$ in the left half space, we have :

$$\oint_S \left[E(\vec{r'})\, \vec{n'} \cdot \vec{\nabla}' G(\vec{r} - \vec{r'}) - G(\vec{r} - \vec{r'})\vec{n'} \cdot \vec{\nabla}' E(\vec{r'})\right] ds' = 0 \tag{34}$$

Now it is well known that a solution of Eq.(31) is :

$$G(\vec{r}) = \frac{\mathrm{e}^{ikr}}{4\pi r} \tag{35}$$

so that by taking

$$G_1(\vec{r} - \vec{r'}) = \frac{\mathrm{e}^{ik\rho'}}{4\pi\rho'}$$

with $\rho' = |\vec{r} - \vec{r'}|$, we have a Green function. Let us define $\vec{r'} = [x', y', z']$. Now, if we consider

$$G_2(\vec{r} - \vec{r''}) = \frac{\mathrm{e}^{ik\rho''}}{4\pi\rho''} \tag{36}$$

with $\rho'' = |\vec{r} - \vec{r''}|$ and $\vec{r''} = [x', y', -z']$, we note that it is the symmetrical of the preceding with respect to the plane $z = 0$. If the surface $\mathcal{S}$ is this plane, any point in the right half space, will give a non-zero contribution by the G_1 integral, and a zero contribution by the G_2 integral, the equations (33,34) being exchanged. We may thus add any multiple of G_2 to G_1 without changing the result :

$$E(\vec{r}) = \iint_{z=0} \left[E(\vec{r'})\, \vec{n'} \cdot \vec{\nabla}' G(\vec{r}, \vec{r'}) - G(\vec{r}, \vec{r''})\vec{n'} \cdot \vec{\nabla}' E(\vec{r'})\right] dx'\, dy' \tag{37}$$

where $G = G_1 + \beta G_2$ with β arbitrary. The special choice

$$G(\vec{r},\vec{r'}) = G_1(\vec{r} - \vec{r'}) - G_2(\vec{r} - \vec{r''}) \tag{38}$$

is especially interesting, because it gives a Green function that is zero on the surface $z = 0$, which greatly simplifies the equation. We obtain simply

$$E(\vec{r}) = \iint_{z=0} E(\vec{r'})\, \vec{n'} \cdot \vec{\nabla} G(\vec{r},\vec{r'})\, dx'\, dy' \tag{39}$$

this is the Kirchhoff equation

Application of the Kirchhoff equation

The preceding equation, establishing a relation between the field inside a volume and the field at the boundary is exact, but taken in the strict sense, of almost no practical interest : It could seem that in order to compute E, we need first to know E. It can however be widely exploited, with an approximation, in the following situation. Assume that the surface $z = 0$ contains a hole, and that a primary electromagnetic wave is coming from the left (see Fig.29). We can assume that at the immediate right of the surface $z = 0$, the field is simply the field at the left, transmitted through the hole. This means that on the right side of the plane, the field is zero outside the hole, and identical to the field coming from the left, within the hole. We can change the sense of the Kirchhoff equation (39) by introducing two fields, one is the coming one $E_1(\vec{r})$, which is assumed to be given throughout the aperture D, and the second one $E_2(\vec{r})$, which is to be computed from the preceding one by using a Kirchhoff-like formula :

$$E_2(\vec{r}) = \iint_D E_1(\vec{r'})\, \vec{n'} \cdot \vec{\nabla} G(\vec{r},\vec{r'})\, dx'\, dy' \tag{40}$$

The function $K(\vec{r},\vec{r'}) \equiv \vec{n'} \cdot \vec{\nabla} G(\vec{r},\vec{r'})$ is called the Diffraction Kernel. It can be explicitly computed :

$$K(\vec{r},\vec{r'}) = -\frac{i}{\lambda}\, \frac{e^{ik\rho}}{\rho} \left(1 + \frac{i}{k\rho}\right) \frac{z}{\rho}$$

where $\rho \equiv \sqrt{(x-x')^2 + (y-y')^2 + z^2}$.
With this explicit formula, we obtain:

$$E_2(\vec{r}) = -\frac{i}{\lambda} \iint_D E_1(\vec{r'})\, \frac{e^{ik\rho}}{\rho} \left(1 + \frac{i}{k\rho}\right) \frac{z}{\rho}\, dx'\, dy' \tag{41}$$

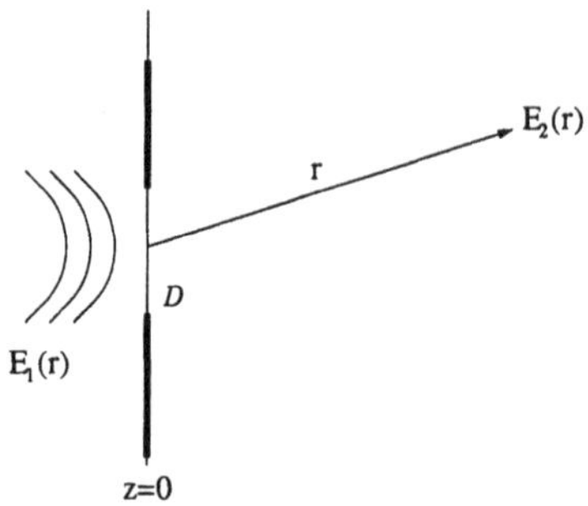

Figure 29:

For a numerical implementation of Eq.(41), it is convenient to extract the rapidly oscillating term in $\exp(ik\rho)$ by writing, with $\mu^2 \equiv (x-x')^2+(y-y')^2$,

$$\rho = z + \sqrt{\mu^2+z^2} - z = z + \frac{\mu^2}{\sqrt{z^2+\mu^2} + z}$$

so that

$$\mathrm{e}^{ik\rho} = \mathrm{e}^{ikz} \times \exp\left[\frac{ik\mu^2}{z+\sqrt{z^2+\mu^2}}\right] \tag{42}$$

As a first example, we compute the diffraction pattern of a rectangular aperture illuminated by a constant amplitude. The aperture has its length [-b=-1cm,b=1cm] directed along y, and its width [-a=-0.5cm,a=0.5cm] along x. The distance of the observation plane is z=1 km, the wavelength is $\lambda = 1$ μm. The computational window containing the source was [-1 cm, 1 cm] × [-1 cm, 1 cm],and the discretization grid was 200 × 200 points. The well known far field theory (see [6]) gives a central spot of rectangular shape, with its longer dimension along x, and its shorter along y. The dark lines correspond to solutions of

$$\sin\left[\frac{kax}{z}\right] = 0$$

i.e.

$$x_n = \frac{n\lambda z}{2a} = n \times 10.6 \text{ cm}$$

and

$$\sin\left[\frac{kby}{z}\right] = 0$$

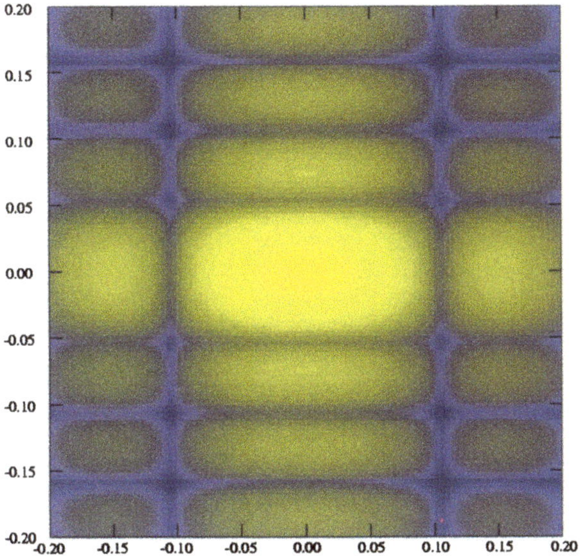

Figure 30: Diffraction pattern of a uniform rectangular aperture a =1 cm, b=0.5cm, at z=1 km for the Nd:YAG wavelength : Distribution of log(I)

i.e

$$y_m = \frac{m\lambda z}{2b} = m \times 5.3 \text{ cm}$$

The plot is logarithmic with respect of the light intensity (Fig.30)

As a second example we compute using Eq.(41) the diffraction pattern of a circular aperture illuminated by a constant amplitude. The radius of the aperture is $a = 1$ cm, the distance of the observator is $z = 1$ km, the wavelength is $\lambda = 1\ \mu$m. The window containing the source was [-1 cm, 1 cm] × [-1 cm, 1 cm], and the discretization grid used for numerical integration was 200 × 200 points. The far field theory foresees a central spot surrounded by rings, the dark rings correspond to solutions r_{dark} of

$$J_1\left(\frac{2\pi a}{\lambda z}\, r\right) = 0$$

the first zeros of the Bessel functions J_1 are

$$\zeta_1 = 3.83171 \leadsto r_{\text{dark},1} \approx 6.5 \text{ cm}$$

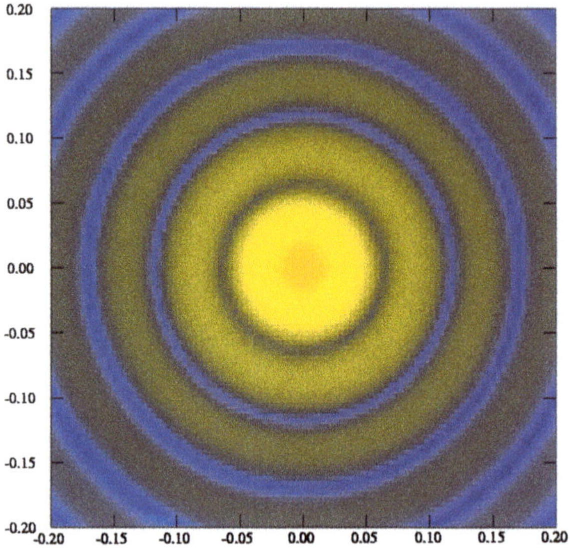

Figure 31: Diffraction pattern of a uniform circular aperture a =1 cm at z=1 km for the Nd:YAG wavelength : Distribution of log(I)

$$\begin{aligned}\zeta_2 &= 7.01559 \rightsquigarrow r_{\text{dark},2} \approx 11.9 \text{ cm}\\ \zeta_3 &= 10.17347 \rightsquigarrow r_{\text{dark},2} \approx 17.2 \text{ cm}\end{aligned}$$

The plot (Fig.31) is logarithmic with respect to the light intensity. it can be seen that the dark rings coming out of the numerical calculation are in agreement with the far field theory. The near field theory of this case is analytically difficult. Numerical exploitation of the Kirchhoff formula, give access to the near field. See for instance a longitudinal (i.e. along the propagation direction z) cut of the intensity distribution (Fig.32). In this case, the circular aperture had 0.1mm radius. The transmitted field is computed starting from z = 0.01 mm. Finally let we consider the gaussian beam (this will be studied later in detail), of amplitude

$$E(x, y, 0) \propto \exp\left[-\frac{x^2+y^2}{w_0^2}\right]$$

in the plane $z = 0$. For w_0=2 cm, this is the amplitude of the light taken at the input mirrors of the Virgo cavities. It seems to diffract without any lobe, as can be seen on Fig.33. In fact, the amplitude extends to infinity, even if it

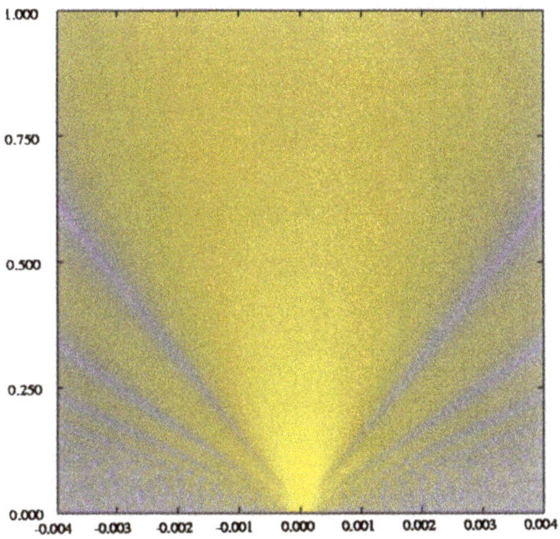

Figure 32: Diffraction pattern of a uniform circular aperture $a = 0.1$ mm for the Nd:YAG wavelength : Distribution of log(I)

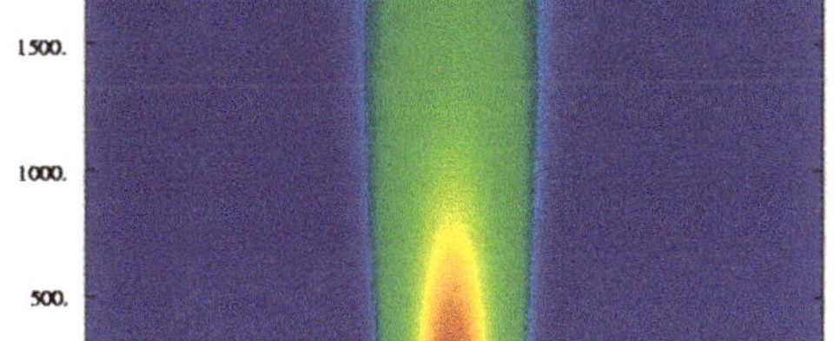

Figure 33: Diffraction of a gaussian wave from 100 m to 3 km

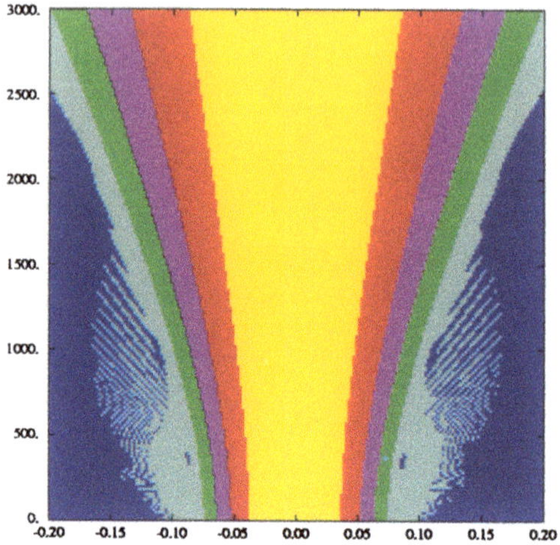

Figure 34: Diffraction of a gaussian wave from 100 m to 3 km,logarithmic plot

becomes negligible for radial distances larger than w_0, whereas the computing window is finite. It is necessary to take a window much larger than the gaussian radius of the beam. Too narrow windows are understood like a diaphragm, and spurious rings are generated. In the preceding case, even with a computing window as wide as 30 cm, faint lobes can be observed on a logarithmic plot (see Fig.34).

Consistency of the Kirchhoff equation

It is not obvious by only looking at the Kirchhoff formula (41), that the secondary field E_2 will reduce to the input field E_1 when $z \rightarrow 0$, i.e. that the diffraction kernel tends to a delta function for $z \rightarrow 0$. Because we have changed the meaning of the surface integral, the question of the agreement of the approximation done in the preceding subsection with reality could be raisen : Let us try to discuss this issue. The Kirchhoff equation can be written under the form :

$$E_2(x,y,z) = \int_{-\infty}^{\infty} dx' \int_{-\infty}^{\infty} dy' \left[\partial_{z'} \left(\frac{e^{ik\rho'}}{4\pi\rho'} - \frac{e^{ik\rho''}}{4\pi\rho''} \right) \right]_{z'=0} E_1(x',y',0) \tag{43}$$

with

$$\begin{aligned}\rho' &= \sqrt{(x-x')^2+(y-y')^2+(z-z')^2}\\ \rho'' &= \sqrt{(x-x')^2+(y-y')^2+(z+z')^2}\end{aligned}$$

It is clear that the integral may be viewed as a 2D convolution product, and therefore may be transformed into a simple algebraic product by a 2D Fourier tranform in the variables (x', y'). Recall that the 2D Fourier Transform of any function $f(x, y)$ of integrable square modulus is defined by :

$$\tilde{f}(p,q) = \int_{\mathbf{R}} dx \int_{\mathbf{R}} dy \; \mathrm{e}^{ipx}\mathrm{e}^{iqy} \; f(x,y) \tag{44}$$

and the reciprocal transform by

$$f(x,y) = \frac{1}{4\pi^2} \int_{\mathbf{R}} dp \int_{\mathbf{R}} dq \; \mathrm{e}^{-ipx}\mathrm{e}^{-iqy} \; \tilde{f}(p,q) \tag{45}$$

A useful result can be found in [8]. The function

$$g(x,y,z) = \frac{\mathrm{e}^{ik\sqrt{x^2+y^2+z^2}}}{2\pi\sqrt{x^2+y^2+z^2}}$$

has the following FT :

$$\tilde{g}(p,q,z) = i\,\frac{\mathrm{e}^{iz\sqrt{k^2-p^2-q^2}}}{\sqrt{k^2-p^2-q^2}}$$

After a Fourier transform, the Kirchhoff equation becomes, using this result :

$$\widetilde{E_2}(p,q,z) = \left[\partial_{z'}\left(i\,\frac{\mathrm{e}^{i(z-z')\sqrt{k^2-p^2-q^2}}}{2\sqrt{k^2-p^2-q^2}} - i\,\frac{\mathrm{e}^{i(z+z')\sqrt{k^2-p^2-q^2}}}{2\sqrt{k^2-p^2-q^2}}\right)\right]_{z'=0} \times$$
$$\times\ \widetilde{E_1}(p,q,0)$$

and eventually reduces to

$$\widetilde{E_2}(p,q,z) = \mathrm{e}^{iz\sqrt{k^2-p^2-q^2}} \times \widetilde{E_1}(p,q,0) \tag{46}$$

which shows that the propagator, defined as the Fourier Transform of the diffraction kernel has the very simple form

$$\tilde{G}(p,q,z) \; = \; \mathrm{e}^{iz\sqrt{k^2-p^2-q^2}} \tag{47}$$

We see that this is perfectly consistent with the Helmholtz equation, which becomes, after a FT :

$$(\partial_z^2 \, + \, k^2 - \, p^2 \, - q^2 \,) \; \tilde{E}(p,q,z) \; = \; 0$$

Anyway, for $z \to 0$, the propagator reduces to 1, showing that the diffraction kernel reduces to $\delta(\vec{r} \, - \, \vec{r'})$, and the diffraction to

$$\widetilde{E_2}(p,q,0) \; = \; \widetilde{E_1}(p,q,0) \; \Rightarrow E_2(x,y,0) \; = \; E_1(x,y,0)$$

as could be expected. We can conclude that, despite a serious change of meaning with respect to the Green theorem, the Kirchhoff formula is strictly equivalent to the wave equation, at least in the case where initial data are given on a plane screen. It follows that if convenient, it is possible to split space into successive slices along the propagation direction, the final data of slice #n being the initial data for slice #(n+1), provided that reflections at each cut do not exist or are ignored. This scheme can be used in compound systems with interfaces, and as will be seen later on, in resonant cavities. We can add the following remark : If we interpret the 2D Fourier transform in the transverse plane as a continuous expansion on plane waves of various directions, by identifying

$$p \; = \; k \sin\theta \, \cos\phi \;\; , \;\; q \; = \; k \sin\theta \, \sin\phi$$

where (θ,ϕ) denote that direction, we see that the propagator is nothing but the phase change along the z axis of this special plane wave :

$$\tilde{G}(p,q,z) \; = \; \tilde{G}(\theta,\phi,z) \; = \; \mathrm{e}^{ikz\cos\theta}$$

The Fresnel approximation and the paraxial diffraction equation (PDE)

As soon as the distance z separating the input aperture from the observation plane is much larger than the wavelength, the $1/k\rho$ term in Eq.(41) becomes negligible, and we can write :

$$E_2(\vec{r}) \; = - \, \frac{i}{\lambda} \oint_D E_1(\vec{r'}) \, \frac{\mathrm{e}^{ik\rho}}{\rho} \, \cos(\theta) \, ds' \tag{48}$$

where θ is the angle under which the element of aperture centered at (x',y') is seen from the observation point (x,y). Eq.48 is often referred to as the "Huyghens-Fresnel" equation. It can be (and was) derived heuristically by considering all points of the aperture as elementary sources of spherical waves : At any point of the right hand side half space, the amplitude is the sum of all these wavelets, and we can for instance say that the elementary amplitude created at $\vec{r}$ by the small elementary source

$$ds(x', y') = E_1(x', y')dx'\, dy'$$

is :

$$dE_2(x, y) = \kappa \int\!\!\int_D \frac{e^{ik\rho}}{\rho} E_1(x', y')\, dx'\, dy'$$

where κ is some coefficient to be determined. For this purpose, we can require that the propagation of an indefinite plane wave is the same plane wave, up to a phase factor. This means that

$$e^{ikz} = \kappa \int\!\!\int_{R^2} \frac{e^{ik\rho}}{\rho}\, dx'\, dy'$$

The integral is easy to compute, being the value at $p = q = 0$ of the Fourier transform of $e^{ik\rho}/\rho$ that is known, as said above, we have thus

$$e^{ikz} = \kappa \left[2i\pi \frac{e^{ik\sqrt{k^2-p^2-q^2}}}{\sqrt{k^2 - p^2 - q^2}} \right]_{p=q=0} = \frac{2i\pi e^{ikz}}{k}$$

It is therefore necessary that $\kappa = -i/\lambda$. This was known long before Kirchhoff's theory, which is the mathematical justification to the Huyghens principle and to the Fresnel formula. If θ in Eq.48 is small, we are in the paraxial regime. If the observation point is near the optical axis, and the distance Z long enough, we can neglect the quantity

$$\sqrt{(x - x')^2 + (y - y')^2}$$

with respect to z, except in the phase factor. This leads to the paraxial diffraction integral :

$$E_2(x, y, z) = -\frac{i}{\lambda z} \exp(ikz) \times \tag{49}$$

$$\times \int\!\!\int_D E_1(x', y', 0) \exp\left[ik\frac{(x - x')^2 + (y - y')^2}{2z}\right] dx'\, dy'$$

All consequences of this formula are said having been obtained within the Fresnel approximation. Remark that this equation is the convolution product of the field $E(z = 0)$ with the simplified (paraxial) diffraction kernel

$$K_{\mathrm{P}}(x,y,z) \;=\; -\,\frac{i}{\lambda z}\,\exp\left[ik\frac{x^2+y^2}{2z}\right]$$

Use of the Fourier transform is especially convenient here, because the Fourier transform of K is easy to compute. For a function of the form

$$G(x,y) \;=\; \mathrm{e}^{-Z(x^2+y^2)}$$

where Z is any complex number of positive real part, it can be shown that

$$\widetilde{G}(p,q) \;=\; \frac{\pi}{Z}\,\mathrm{e}^{-\frac{p^2+q^2}{4Z}} \tag{50}$$

in particular the propagator is

$$\widetilde{K_{\mathrm{P}}}(p,q,z) \;=\; \exp\left[-i\frac{z(p^2+q^2)}{2k}\right] \tag{51}$$

Obviously, we could have deduced it from the "exact" propagator

$$\tilde{G}_{\mathrm{exact}}(p,q,z) \;=\; \mathrm{e}^{iz\sqrt{k^2-p^2-q^2}}$$

by assuming that the values of p,q are restricted to small values due to the behavior of the function to be propagated (small means $p,q \ll k$). This is one more version of the paraxial approximation, the diffraction is "adiabatic" along z , so that the angles of the rays are small. $p, q, \sqrt{k^2-p^2-q^2}$ may be thought of as the coordinates of the wave vector of an elementary plane wave. Then , θ being the direction of that elementary wave, we have $\theta \simeq \sin\theta = \sqrt{p^2+q^2}/\sqrt{k^2-p^2-q^2}$. If θ is small, we can thus write :

$$\tilde{G}(p,q,z) \;=\; \mathrm{e}^{ikz} \times \mathrm{e}^{-i\frac{z(p^2+q^2)}{2k}}$$

An alternative way of computing E_2 is therefore :

$$E_2(x,y,z) \;=\; \widetilde{\widetilde{K_{\mathrm{P}}}(p,q,z) \times \widetilde{E_1}(p,q,0)} \tag{52}$$

This is a very convenient way, as will be shown later. The paraxial diffraction integral is related to the paraxial diffraction equation. Consider the Helmholtz equation :

$$\left[\Delta + k^2\right]\mathcal{E} \;=\; 0 \tag{53}$$

If the field is expected to propagate mainly in the z direction, with a slow expansion in the transverse plane, we can use the slowly varying envelope approximation scheme, i.e.

$$\mathcal{E}(x,y,z) \;=\; \mathrm{e}^{ikz} \times \; E(x,y,z)$$

in which the envelope $E(x,y,z)$ is assumed to depend slowly on z, the rapidly oscillating factor having been extracted. More specifically, we intend to use the approximation

$$\frac{\partial E}{\partial z} \;\ll\; k\, E$$

for neglecting second order derivatives of E, so that the Helmholtz equation becomes :

$$[2ik\, \partial_z \;+\; \Delta_{\mathrm{T}}]\; E \;=\; 0 \tag{54}$$

where

$$\Delta_{\mathrm{T}} \;\equiv\; \partial_x^2 + \partial_y^2$$

is the transverse Laplace operator. By taking the Fourier transform of Eq.(54) with respect to x, y, we obtain :

$$\left[2ik\, \partial_z \;-\; (p^2+q^2)\right] \tilde{E}(p,q,z) \;=\; 0$$

the solution of which is of the form

$$\tilde{E}(p,q,z+\Delta z) \;=\; \tilde{E}(p,q,z) \times \exp\left[-\, i\frac{(p^2+q^2)\Delta z}{2k}\right]$$

in which we recover the propagator (51).

The Fraunhofer approximation

The ultimate approximation for a diffracted wave holds when the very far field is considered. The Fresnel-Huyghens integral can be written as :

$$E(x,y,z) \;=\; -\,\frac{i}{\lambda z}\; \exp\left[i\pi\frac{x^2+y^2}{\lambda z}\right] \;\times$$

$$\int_0^\infty \; \exp\left[i\pi\frac{x'^2+y'^2}{\lambda z}\right] \exp\left[-i\pi\frac{xx'}{\lambda z}\right] \exp\left[-i\pi\frac{yy'}{\lambda z}\right] \; E(x',y',0)\; dx'\, dy'$$

If we assume the transverse extension of the initial amplitude bounded by a radius a, the order of magnitude of the argument of the quadratic term in the complex exponential is

$$\delta < \pi \frac{a^2}{\lambda z} = \pi \times N_F$$

N_F is called *Fresnel number.* If the observation distance is so large that N_F my be neglected, we can write simply

$$E(x, y, z) = -\frac{i}{\lambda z} \exp\left[i\pi \frac{x^2 + y^2}{\lambda z}\right] \times$$

$$\int_0^\infty \exp\left[-2i\pi \frac{xx'}{\lambda z}\right] \exp\left[-2i\pi \frac{yy'}{\lambda z}\right] E(x', y', 0)\, dx'\, dy'$$

which is nothing but the Fourier transform of the incoming amplitude :

$$E(x, y, z) = -\frac{i}{\lambda z} \exp\left[i\pi \frac{x^2 + y^2}{\lambda z}\right] \tilde{E}\left(\frac{2\pi x}{\lambda z}, \frac{2\pi y}{\lambda z}, 0\right)$$

This is the Fraunhoffer approximation, and allows to compute quickly the properties of the diffracted field for z very large. For instance, for the rectangular aperture $[-a, a] \times [-b, b]$, one finds immediately

$$|E(x, y, z)|^2 = \frac{16a^2b^2}{\lambda^2 z^2} \left[\text{sinc}\left(\frac{2\pi x}{\lambda z}\right) \text{sinc}\left(\frac{2\pi y}{\lambda z}\right)\right]^2$$

explaining the pattern of Fig.30. The Fraunhofer approximation holds also at the focus of a mirror or of a lens.

Representation of optical elements

The action of thin optical elements, like thin lenses or nearly flat mirrors on the optical amplitudes can be modelled without using a diffraction integral. Consider for instance the reflection off a curved mirror of curvature radius R_c and diameter D. Assume the mirror to close the aperture in the plane $z = 0$ (see Fig.35). Strictly speaking, the field arriving on the mirror's surface should be computed fromn the field in the plane by Kirchhoff's equation. It is more convenient to discuss in the Fourier space. Calling $E_1(x, y, 0)$ the field in the plane $z = 0$, and $E_2(x, y, z)$ the field on the mirror's surface, we have as seen in (46) :

$$\widetilde{E_2}(p, q, z) = e^{iz\sqrt{k^2 - p^2 - q^2}} \times \widetilde{E_1}(p, q, 0)$$

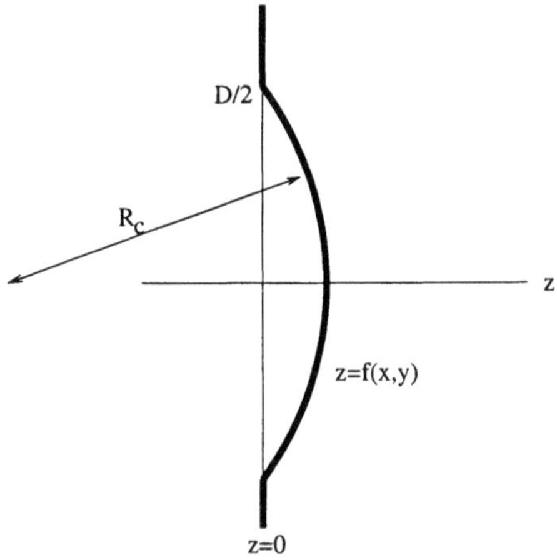

Figure 35: reflection off a curved mirror

The argument of the imaginary exponential can be written as :

$$z\sqrt{k^2 - p^2 - q^2} \;=\; kz \;+\; z\left(\sqrt{k^2 - p^2 - q^2} - k\right)$$

or

$$z\sqrt{k^2 - p^2 - q^2} \;=\; kz \;-\; \frac{z(p^2 + q^2)}{2k}\,\frac{2}{1 + \sqrt{1 - \frac{p^2+q^2}{k^2}}}$$

It is necessary to estimate the different orders of magnitude of these terms.

- The quantity $p^2 + q^2$ is determined by the spatial behavior of the input wave. If the spatial frequencies are of the order of magnitude of w_0, for instance for a TEM mode of Fourier transform

  $$\tilde{\phi}(p, q) \;=\; \exp\left(-\;\frac{w_0^2(p^2 + q^2)}{4}\right)$$

  we have

  $$\frac{p^2 + q^2}{k^2} \;\sim\; \frac{4}{k^2 w_0^2} \;\sim\; \left(\frac{\lambda}{\pi w_0}\right)^2 \;\sim\; \theta_g^2$$

  θ_g being nothing but the divergence angle of the beam. For long baseline interferometers, this aperture is of the order of tens of microradians, for Virgo, $\theta_g^2 \;=\; 1.7 \;\times\; 10^{-5}$Rd. It is clear that we can neglect this term.

- The argument of the imaginary exponential therefore reduces to

$$z\sqrt{k^2 - p^2 - q^2} = kz - \frac{z(p^2 + q^2)}{2k}$$

but

$$\frac{z(p^2 + q^2)}{2k} \sim \frac{z\lambda}{\pi w_0^2} \sim \frac{z}{z_R}$$

where z_R is the Rayleigh parameter of the beam, about 1km for GW interferometers, whereas z is of the order of tens of micrometers. More precisely, we have for a parabolic mirror $z_{\text{max}} = D^2/8R_c$, on the other hand, the Rayleigh parameter is related to the curvature radius (in a flat/parabolic cavity of length L) by

$$z_R = \sqrt{L(R_c - L)} = \frac{1}{\alpha} \times R_c$$

where

$$\alpha = \frac{R/L}{\sqrt{R/L - 1}}$$

the factor α is of the order of the unity (For Virgo, $\alpha \simeq 2.97$). We have thus :

$$\frac{z(p^2 + q^2)}{2k} \sim \frac{\alpha}{8}\left(\frac{D}{R_c}\right)^2$$

Taking again Virgo figures (D=35 cm, R_c=3.45 km) this is :

$$\frac{z(p^2 + q^2)}{2k} \sim 3.8 \times 10^{-9}$$

The conclusion is that we can write with a good accuracy

$$\widetilde{E_2}(p, q, z) = e^{ikz} \times \widetilde{E_1}(p, q, 0)$$

which by inverse Fourier transform gives simply

$$E_2(x, y, z) = e^{ikz} \times E_1(x, y, 0)$$

So that if we consider the amplitude on a surface of equation $z = f(x,y)$, we can write simply

$$E_2(x,y) \;=\; \mathrm{e}^{ikf(x,y)} \;\times\; E_1(x,y)$$

For the reflected wave $E_3(x,y,z)$, we reverse the point of view. If we would compute E_2 knowing E_3 we would find (the propagation direction being reverse)

$$E_2(x,y) \;=\; \mathrm{e}^{-ikf(x,y)} \;\times\; E_3(x,y)$$

Therefore

$$E_3(x,y) \;=\; \mathrm{e}^{2ikf(x,y)} \;\times\; E_1(x,y)$$

and the reflection operator is simply the phase factor ;

$$R \;=\; \mathrm{e}^{2ikf(x,y)} \tag{55}$$

In particular, for a parabolic mirror, well adapted to gaussian beams in the paraxial approximation, we have

$$R \;=\; \exp\left[i\;\frac{2\pi(x^2+y^2)}{\lambda R_c}\right] \tag{56}$$

discrete bases for free space propagation

The set of all complex functions $f(x,y)$ of integrable square modulus may be given the structure of a Hilbert vector space, by introducing the scalar product :

$$< f\;,\;g >= \int_{\mathbf{R}^2} dx\;dy\;f(x,y)\;\overline{g(x,y)} \tag{57}$$

If we think to these functions in terms of optical amplitudes at a given point of the path of a light beam having the preferred propagation direction z, we see that

$$\|f\|^2 \;=< f\;,\;f >= \int_{\mathbf{R}^2} dx\;dy\;\|f(x,y)\|^2 \tag{58}$$

is nothing but the light power of the beam crossing the transverse plane at z. Obviously, a number of bases can be constructed for this Hilbert space. It is possible to find several discrete bases, and the corresponding vectors are called Transverse Electromagnetic Modes, and are labelled by two indices :

$\text{TEM}_{(m,n)}$. The more often employed bases for studying cavities and laser beams, are the Hermite-Gauss modes $\text{HG}_{(m,n)}$ when rectangular coordinates are convenient, and the Laguerre-Gauss modes $\text{LG}_{(m,n)}$ when polar coordinates are convenient.

The fundamental mode $\text{TEM}_{(0,0)}(x,y;z) = \text{HG}_{(0,0)}(x,y;z) = \text{LG}_{(0,0)}(x,y;z)$ is, with $r^2 \equiv x^2+y^2$,

$$\text{TEM}_{(0,0)}(x,y;z) = \sqrt{\frac{2}{\pi w(z)^2}}\ \mathrm{e}^{ikz}\ \mathrm{e}^{-i\arctan(z/b)}\mathrm{e}^{-r^2/w(z)^2}\ \mathrm{e}^{ikr^2/2R(z)} \tag{59}$$

where $w(z)$ gives the radius of the beam, and $R(z)$ the curvature radius of the phase surface. b is the Rayleigh range :

$$b = \frac{\pi w_0^2}{\lambda}$$

where the z coordinate was chosen such that $w_0 = w(0)$. Then

$$w(z) = w_0\sqrt{1+\frac{z^2}{b^2}}$$

$$R(z) = z\sqrt{1+\frac{b^2}{z^2}}$$

The Hermite-Gauss and Laguerre-Gauss functions

It can be shown that the PDE has Hermite-Gauss solutions of the form

$$\text{HG}_{(m,n)}(x,y;z) = c_{m,n}\ \mathrm{e}^{ikz}\ H_m\left(\sqrt{2}\frac{x}{w(z)}\right)\ H_n\left(\sqrt{2}\frac{y}{w(z)}\right)\ \times$$

$$\mathrm{e}^{-i(m+n+1)\arctan(z/b)}\ \mathrm{e}^{-r^2/w(z)^2}\ \mathrm{e}^{ikr^2/2R(z)} \tag{60}$$

where the functions $H_n(X)$ are the Hermite polynomials and $c_{m,n}$ a normalization constant to be defined later.

Several properties of these functions are very convenient, and we recall them herafter without proof.

- The Hermite polynomials are defined by :

$$H_n(x) = \mathrm{e}^{x^2}\left(\frac{d}{dx}\right)^n \mathrm{e}^{-x^2}$$

- They obey a recursion relation :

$$H_{n+1}(x) \;=\; 2x\, H_n(x) \;-\; 2n\, H_{n-1}(x)$$

- They obey an orthogonality relation

$$\int_{-\infty}^{\infty} H_m(x)\, H_n(x)\, \mathrm{e}^{-x^2} dx \;=\; \sqrt{\pi} 2^m m!\, \delta_{mn}$$

The normalization constants for the HG modes are therefore :

$$c_{m,n} \;=\; \left[\frac{2}{\pi w^2}\; \frac{1}{2^{m+n} m!\; n!}\right]^{1/2}$$

- There is a translation formula :

$$H_n(x+\Delta/2) \;=\; \sum_{k=0}^{n} C_n^k\; H_{n-k}(x)\; \Delta^k \tag{61}$$

(it can be shown using the recursion formula)

- It is possible to give the general expression of the Fourier Transform of any mode ; write the modal amplitude under the general form

$$\mathrm{HG}_{(m,n)}(x,y) \;=\; H_m\left(\sqrt{2}\frac{x}{w}\right)\; H_n\left(\sqrt{2}\frac{y}{w}\right)\; \exp\left(-Z\; \frac{x^2+y^2}{w^2}\right)$$

where Z is any complex number of positive real part. The Fourier Transform is :

$$\widetilde{\mathrm{HG}_{(m,n)}}(p,q) \;=\; \frac{\pi w^2}{Z}\; \left(\frac{i}{Z}\right)^{m+n}\; \left(2Z-Z^2\right)^{(m+n)/2}\; \times$$

$$H_m\left(\frac{pw}{\sqrt{2}\sqrt{2Z-Z^2}}\right)\; H_n\left(\frac{qw}{\sqrt{2}\sqrt{2Z-Z^2}}\right)\; \exp\left[-\frac{w^2(p^2+q^2)}{4Z}\right] \tag{62}$$

In a certain sense, we see that the HG modes are eigenvectors of the Fourier transform.

- The first Hermite polynomials are :

$$H_0(x) \;=\; 1$$

$$
\begin{aligned}
H_1(x) &= 2x \\
H_2(x) &= 4x^2 - 2 \\
H_3(x) &= 8x^3 - 12x \\
H_4(x) &= 16x^4 - 48x^2 + 12
\end{aligned}
$$

etc...

Using polar coordinates (r, θ) in the (x, y) plane, a new class of solutions to the PDE can be found, of the form

$$
\begin{aligned}
\mathrm{LG}_{\mathrm{m,n}}(r, \theta; z) &= c_{m,n}\, \mathrm{e}^{ikz} \left(\sqrt{2}\frac{r}{w(z)}\right)^n L_m^{(n)}(2r^2/w(z)^2)\times \\
&\mathrm{e}^{-i(2m+n+1)\arctan(z/b)}\, \mathrm{e}^{-r^2/w(z)^2}\, \mathrm{e}^{ikr^2/2R(z)} \cos(n\theta) \qquad (63)
\end{aligned}
$$

The functions $L_m^{(n)}(X)$ are the generalized Laguerre polynomials. They are defined by

$$
L_m^{(n)}(x) = \frac{\mathrm{e}^x}{m!\, x^n} \left(\frac{d}{dx}\right)^m \left(x^{n+m}\mathrm{e}^{-x}\right)
$$

They obey the recurrence relation :

$$
(m+1)L_{m+1}^{(n)}(x) = (2m+n+1-x)L_m^{(n)}(x) - (m+n)L_{m-1}^{(n)}(x)
$$

The first ones are as follows :

$$
\begin{aligned}
L_0^{(0)}(x) &= 1 \\
L_1^{(0)}(x) &= 1 - x \\
L_2^{(0)}(x) &= 1 - 2x + \frac{x^2}{2} \\
L_1^{(1)}(x) &= 2 - x \\
L_2^{(1)}(x) &= 3 - 3x + \frac{x^2}{2} \\
L_1^{(2)}(x) &= 3 - x \\
L_2^{(2)}(x) &= 6 - 4x + \frac{x^2}{2}
\end{aligned}
$$

The normalization relation for the Laguerre polynomials are given by [10] :

$$\int_0^\infty L_m^n(x)^2\, x^n\, e^{-x}\, dx = \frac{(m+n)!}{m!}$$

so that the normalization constants c_{mn} are :

$$c_{mn} = \frac{2}{w}\sqrt{\frac{m!}{\pi\,(1+\delta_{n0})\,(m+n)!}}$$

Properties of the Fabry-Perot resonator in the paraxial approximation

It has been seen that the free space propagation operator has, in a certain sense eigenmodes called $\mathrm{TEM_{m,n}}$ modes. They have the significant two following properties :

- They are of finite transverse extension, there is already a storage in the transverse plane
- They have a parabolic equiphase surface

The second property allows to make "matched" mirrors, of shape adapted to the equiphase surface, reflecting the mode on itself.

Consider for instance a plane-spherical cavity with a plane input mirror M_1, and a spherical mirror M_2 of curvature radius R_c, at a distance L. in order to be matched to M_1, the stored wave must be at its waist at $z = 0$ on the input plane. Then, the stored wave must have a phase curvature radius of $R-c$ at $z = L$, so that we can write

$$R_c = L\left(1 + \frac{b^2}{L^2}\right)$$

b (Rayleigh parameter) having the definition previously encountered. This gives

$$b = \sqrt{L(R_c - L)}$$

Clearly, this is possible only if $R_c > L$. This is a stability condition for that type of cavity. If this condition is fulfilled, the cavity is able to store any $\mathrm{TEM_{m,n}}$ mode, provided it is near resonance. The size of the waist is

$$w_0 = \sqrt{\lambda b/\pi}$$

The resonance condition, assuming a $\pi/2$ dephasing at each reflection, is :

$$\pi \;-\; 2(m+n+1)\tan^{-1}\left(\frac{L}{b}\right) \;\equiv\; 2p\pi$$

the eigenmodes of the cavity are thus labeled by 3 integers. The frequency spacing between modes is a very important feature in a cavity. If two modes have by chance close eigenfrequencies, a class of perturbations of the mirrors having the right symmetry will pump power from one mode to the other due to the finite linewidths (see below). In particular, if the cavity is operated on its fundamental mode, it is better to choose the geometrical parameters in such a way that the nearest transverse modes $(m, n) \neq (0,0)$ are well separated from the reference mode. The $TEM_{0,1}$ and $TEM_{1,0}$ modes are especially well coupled with the $TEM_{0,0}$ in case of misalignment of mirrors. Let us discuss this issue now. We call $\Phi_{m,n,p}$ the total dephasing of the m, n, p mode over a round trip in the cavity. We have :

$$\Phi_{m,n,p} \;=\; \frac{4\pi\nu_{m,n,p}L}{c} \;-\; 2(m+n+1)\tan^{-1}\left(\frac{L}{b}\right) \;+\; \pi \;=\; 2p\pi$$

We see that the frequency gap between two successive longitudinal resonances, or Free Spectral Range (FSR), $(\Delta p = \pm 1)$, is $\Delta\nu_{FSR} \;=\; c/2L$. we see that the frequencies of the modes are given by

$$\nu_{m,n,p} \;=\; \Delta\nu_{FSR}\left(p - \frac{1}{2} + (m+n+1)\alpha\right)$$

with $\alpha \;=\; \tan^{-1}(L/b)/\pi$. Assume the operation mode has frequency $\nu_{0,0,p_0}$, the distances of the other modes are :

$$\delta\nu_{m,n,p} \;=\; \nu_{m,n,p} - \nu_{0,0,p_0} \;=\; (p - p_0 + (m+n)\alpha)\,\Delta\nu_{FSR} \qquad (64)$$

The distribution of resonances being periodic, it is sufficient to study it over a FSR. Given the length L of the cavity, the curvature radius R_c can be chosen under the following constraints :

- it must be larger than L
- it must not cause a too large magnification factor between the input mirror and the end mirror
- it must give a value of α such that equation 64 has no zero solutions for (m, n) small.

In the case of Virgo, the length of the arms is $L = 3$km and the curvature radius of the end mirror $R_c = 3.45$ km, so that $\alpha \simeq 0.38238$. The frequency offsets of the 15 nearest tranverse modes are given in the following table.

Mode order (m+n)	*Frequency offset (Hz)*
8	2950.62
3	7352.16
11	10302.78
6	14704.32
14	17654.94
1	19105.86
9	22056.48
4	26458.02
12	29408.64
7	33810.18
15	36760.79
2	38211.72
10	41162.34
5	45563.88
13	48514.49

Remark that the (0,1) and (1,0) modes are well separated from the (0,0), and that there is no coincidence for orders lower than 15. The nearest are the family (m+n=8) which are not easy to couple to (0,0) by a simple perturbation.

3.2 Numerical propagation

An optical instrument is generally composed of optical elements like lenses, mirrors, and of space between them. For studying any property of the instrument, we have to represent the action of each of these elements on a light beam. Everyone understands that space produces a diffraction of the beam, thin lenses or almost flat mirrors pure refraction or reflection, and thick lenses both. There are a lot of ways of representing numerically a light beam, for, instance :

- by sampling its complex amplitude on a rectangular grid in the x, y plane
- by mapping it on a polar mesh in the (r, θ) plane
- by expanding it on a discrete basis of modes

to each of these decisions correspond a special way of computing the field diffracted at a distance z. The method to choose depends obviously of the type of effects we want to analyze. In the case of axial symmetry, the polar representation will be convenient. In case of very small misalignments of mirrors in a resonant cavity, the modal expansion will do the job. Anyway, each of the corresponding algorithm belong to the class of "spectral methods" expanding the optical amplitudes on a basis of simple functions for which the diffraction/refraction problem is already solved. These simple functions will be

- plane waves in the case of Fourier (or Hankel) transform based methods
- $\mathrm{TEM_{m,n}}$ modes (HG or LG) in the modal methods

FFT based numerical propagation

When the complex amplitude is sampled on a rectangular grid with equally spaced sampling points, it is possible to use a discrete 2D Fourier Transform to propagate the field.

The DFT

The Discrete Fourier Transform (DFT) is obtained by considering a function $\Phi(t)$ which is zero outside the interval $[0,T]$. The Fourier Transform then reduces to :

$$\tilde{\Phi}(f) = \int_0^T e^{2i\pi ft}\,\Phi(t)\,dt$$

For a numerical integration, we can cut the interval in N slices of width $\Delta t = T/n$, and write approximately :

$$\tilde{\Phi}(f) = \frac{T}{N}\sum_{j=0}^{N-1} e^{2i\pi fjT/N}\,\Phi(jT/N)$$

Now it is possible to sample the function in the frequency domain too. The smallest frequency interval we can consider is obviously $\Delta f = 1/T$, because the longest time interval on which the function Φ can be studied is T. The sampling will thus be :

$$f_m = m \times \frac{1}{T}$$

and the samples of the Fourier Transform are :

$$\widetilde{\Phi}_m \equiv \widetilde{\Phi}(m/T) = \sum_{j=0}^{N-1} e^{2i\pi mj/N}\,\Phi_j \tag{65}$$

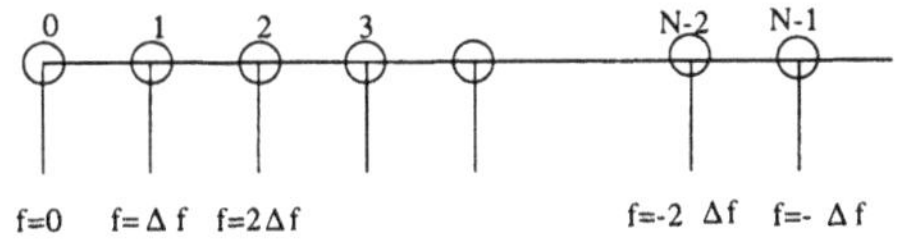

Figure 36: Assign frequencies to the DFT samples

with the notation $\Phi_j = \Phi(jT/N)$. Eq.(65) expresses the DFT.
If we consider the vector $\widetilde{\Phi}_m$, several remarks arise

- It si easily seen that

$$\widetilde{\Phi}_{m+N} = \widetilde{\Phi}_m$$

  showing that the DFT has period N with respect to m, it is therefore sufficient to compute $\{\widetilde{\Phi}_m \ ; \ m = 0, \ldots, N-1\}$.

- Clearly $m = 0$ corresponds to the mean of the function $\Phi(t)$, and thus to the value at $f = 0$ of its Fourier transform. Now, it is easily seen that

$$\widetilde{\Phi}_{N-m} = \widetilde{\Phi}_{-m}$$

  and as a result, the second half of the vector $\{\widetilde{\Phi}_m\}$ contains the negative frequencies (see Fig.(36))

- The maximum frequency is thus

$$f_{\max} = \left[\frac{N}{2}\right]\frac{1}{T}$$

- Consider a Fourier transform followed by the reciprocal :

$$\widetilde{\Phi}_m = \frac{1}{N}\sum_{j=0}^{N-1} \mathrm{e}^{2i\pi mj/N}\,\Phi_j$$

  (as already seen, the time element is T/N, and the frequency element is $1/T$, so that the time $\times$ frequency element is $1/N$) then

$$\widetilde{\widetilde{\Phi}}_n = \frac{1}{N}\sum_{m=0}^{N-1}\sum_{j=0}^{N-1} \mathrm{e}^{2i\pi m(j-n)/N}\,\Phi_j$$

but $-N+1 \leq j-n \leq N-1$, so that

$$\sum_{m=0}^{N-1} \mathrm{e}^{2i\pi m(j-n)/N} = \frac{\mathrm{e}^{2i\pi(j-n)} - 1}{\mathrm{e}^{2i\pi(j-n)/N} - 1} = \begin{cases} 0 \text{ if } j \neq n \\ N \text{ if } j = n \end{cases}$$

then

$$\tilde{\tilde{\Phi}}_n = \Phi_n \tag{66}$$

which shows that the "approximate" of the inverse FT is the exact inverse of the "approximate" FT. Practically, when implementing any DFT algorithm, this is the first property to check.

- Let us denote here by

$$\widetilde{\Phi_{N,m}}$$

the N-points DFT of Φ. Assume that $N = 2N'$ even, and $m = 2m'$ too. We can write

$$\widetilde{\Phi_{N,m}} = \frac{T}{N} \sum_{j=0}^{N-1} \mathrm{e}^{2i\pi mj/N} \Phi_j = \frac{T}{2N'} \sum_{j=0}^{2N'-1} \mathrm{e}^{2i\pi m'j/N'} \Phi_j$$

By splitting the sum into two segments we get

$$\widetilde{\Phi_{N,m}} = \frac{1}{2}\frac{T}{N'} \sum_{j=0}^{N'-1} \mathrm{e}^{2i\pi m'j/N'} \Phi_j + \frac{1}{2}\frac{T}{N'} \sum_{j=N'}^{2N'-1} \mathrm{e}^{2i\pi m'j/N'} \Phi_j$$

and by renaming $j = j' + N'$ in the second sum,

$$\widetilde{\Phi_{N,m}} = \frac{1}{2}\frac{T}{N'} \sum_{j=0}^{N'-1} \mathrm{e}^{2i\pi m'j/N'} \Phi_j + \frac{1}{2}\frac{T}{N'} \sum_{j'=0}^{N'-1} \mathrm{e}^{2i\pi m'j'/N'} \Phi_{j'+m}$$

if we note $\Phi^{(1)}$ and $\Phi^{(2)}$ the two halves (of lengths $N' = N/2$) of the input vector Φ (of length N), we have the following property

$$\widetilde{\Phi_{2N',2m'}} = \frac{1}{2}\left(\widetilde{\Phi^{(1)}_{N',m'}} + \widetilde{\Phi^{(2)}_{N',m'}}\right) \tag{67}$$

In the case where $m = 2m' + 1$ is odd, we have :

$$\widetilde{\Phi_{N,m}} = \frac{1}{2}\frac{T}{N'}\sum_{j=0}^{N'-1} e^{2i\pi m'j/N'} e^{i\pi j/N'} \Phi_j - \frac{1}{2}\sum_{j'=0}^{N'-1} e^{2i\pi m'j'/N'} e^{i\pi j/N'} \Phi_{j'+m}$$

If we introduce the new function $\Psi_j \equiv e^{2i\pi j/N} \times \Phi_j$, we see that the preceding equation reads :

$$\widetilde{\Phi_{2N',2m'}} = \frac{1}{2}\left(\widetilde{\Psi^{(1)}_{N',m'}} + \widetilde{\Psi^{(2)}_{N',m'}}\right) \tag{68}$$

and the conclusion is that the N-ranks Discrete Fourier Transform essentially reduces to two $N/2$-rank partial transformations of the two halves of the input vector.

It must be clear that the DFT is not a sampling of the continuous transformation. The finite step integration which was at the starting point of the algorithm can only converge towards the true FT as N increases. The linear algebra involving vectors of size N and rank N DFT's is perfectly closed due to eq.(66), but it represents a world different from reality. We give for instance the result of a basic experiment. We consider the function

$$F(t) = \exp\left(-\frac{t^2}{2\tau^2}\right)$$

in our experiment, we take the time constant τ=1s. Its Fourier transform is :

$$\tilde{F}(f) = \sqrt{2\pi}\,\tau \exp\left(-\frac{1}{2}(2\pi f\tau)^2\right)$$

and we compare a N-sample of the continous FT of F to the DFT of a N-sample of F (see table (3.2)). The interval over which the function is sampled is called window. The size of the window must be chosen such that the function takes vanishing values near the ends of the window. Moreover, there is an optimal density of samples, which implies that when the size of the sample is changed, the size of the window giving the optimal agreement changes too.

Sample size	*Optimal window*	*rms error*
16	10s	$6.9\ 10^{-7}$
32	14s	$1.8\ 10^{-12}$
64	17s	$7.2\ 10^{-17}$

In this case, it can be seen that increasing the size of the sample is useless, since at N=64, the ultimate precision of the computer is reached.

We have seen that the DFT has the dichotomic property, reducing in principle a N-DFT to two $N/2$-DFT's. The basic of the FFT is to recursively compute any N-DFT from a series of initial 2-DFT's. In this elementary scheme (due to J. Cooley and J. Tukey [9]), N has obviously to be an integer power of 2. Moreover, it is clear that the number of recursions is $\log_2(N)$, so that the number of operations, grows as $N \times \log_2(N)$, which is a tremendous improvement with respect to the naive DFT scheme. For instance, in a 2D Fourier Transform, if a 1024×1024 2D sample is needed, FFT provides a factor of roughly 10^6 gain in CPU time. All other properties are exactly those described for the DFT.

The Fourier Transform of the paraxial diffraction kernel

$$K(x,y,\Delta z) \;=\; -\;\frac{i}{\lambda z}\;\exp\left(-i\frac{k(x^2+y^2)}{2\Delta z}\right)$$

is :

$$\widetilde{K}(p,q,\Delta z) \;=\exp\left(-i\frac{\Delta z(p^2+q^2)}{2k}\right)$$

Now, if we intend to use a DFT for computing the diffraction integral according to the scheme

$$E_2(x,y,z+\Delta z) \;=\; \widetilde{\widetilde{K_{\mathrm{P}}}(p,q,\Delta z) \;\times\; \widetilde{E_1}(p,q,z)} \tag{69}$$

we need the discretization of $\widetilde{K}(p,q,\Delta z)$ with respect to p,q. We remember that the frequency increment in the DFT is $\delta f \;=\; 1/T$ where T is the time window. In terms of spatial circular frequencies, the increment will be $\delta p \;=\; 2\pi/F_x$, where F_x is the x side of the 2D spatial window. We have also $\delta q \;=\; 2\pi/F_y$. The discretization is therefore of the form :

$$\widetilde{K}(m,n,\Delta z) \;=\; \exp\left[-\;i\;\pi\lambda\Delta z\left(\frac{m^2}{F_x^2}\;+\;\frac{n^2}{F_y^2}\right)\right]$$

If we remember that the DFT has a specific way of sorting the frequencies, we could think that we have to correctly arrange the FFT of the input field, before multiplying by $\widetilde{K}$. Clearly, it is more efficient not to correct the FFT's, but rather write the propagator according to the same convention. For a square computation window of size `window` the FORTRAN sequence calculating $\widetilde{K}(i,j,z)$, could be :

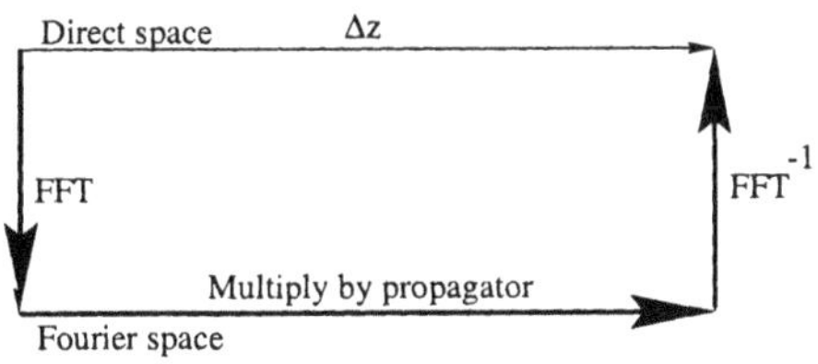

Figure 37: propagation step

```
      mil=n/2+1
      do i=1,n
         if (i.le.mil) then
            ind1=i-1
         else
            ind1=i-1-n
         endif
         do j=1,n
            if (j.le.mil) then
               ind2=j-1
            else
               ind2=j-1-n
            endif
            square=ind1**2+ind2**2
            phase=-pi*lambda*z*square/window**2
            ktilde(i,j)=dcmplx(dcos(phase),dsin(phase))
         enddo
      enddo
```

FFT propagation algorithms

A step Δz of propagation will be carried out following the scheme showed on Fig.37, and propagation steps can be linked into series corresponding to the various interfaces of an optical system. The optical amplitudes are sampled on a rectangular grid (it is not necessary to use a square grid. If we have spatially squeezed beams, one direction can be larger and more sampled than the other). Call `a(i,j)` the complex array (of size $n \times n$) representing the amplitude at $z = 0$. The propagation step is for instance

```
c-----------------------------------------------------------
c     the subroutine cfft2d(m,n,ar,iflag) carries out the FT
c     m,n : size of the array to be transformed
```

```
c     ar : the array to be transformed, and on return, the
c          transformed array
c     iflag = 1 : direct transform
c     iflag = -1: inverse transform
c------------------------------------------------------------
      call cfft2d(n,n,a,1)
      do i=1,n
        do j=1,n
           a(i,j)=a(i,j)*ktilde(i,j)
        enddo
      enddo
      call cfft2d(n,n,a,-1)
c------------------------------------------------------------
```

after what a(i,j) represents the propagated amplitude. Mirrors are sampled on the same grid, and reprsented by complex arrays :

```
c------------------------------------------------------------
      subroutine defmirr(n,window,reflect,radius,curvature,mir)
c
c     returns an array of samples of the mirror equivalent
c     phase for given parameters
c
c       n         : rank of the arrays
c       window    : computation window
c       reflect   : photometric amplitude reflectivity
c       radius    : radius (half size) of the mirror
c       curvature : inverse of curvature radius
c       mir       : array representing the mirror
c
      implicit none
c
      integer n,i,j
      real*8 reflect,radius,curvrad,pi,x,y,rp2
      real*8 window,lambda,dx,phase
      complex*16 ci,mir(n,n)
c
      data pi/3.141592653589793d0/
      data ci/(0.d0,1.d0)/
      data lambda/1.064d-6/
c
      dx=window/n
```

```
      do i=1,n
         x=(i-1)*dx-window/2
         do j=1,n
            y=(j-1)*dx-window/2
            rp2=x*x+y*y
            if (rp2.gt.radius*radius) then
               mir(i,j)=0
            else
               phase=2*pi*rp2*curvature/lambda
               mir(i,j)=ci*reflect*dcmplx(dcos(phase),-dsin(phase))
            endif
         enddo
      enddo
      return
      end
```

A reflection will then be carried out by a simple term to term product of the amplitude by the mirror :

```
      do i=1,n
         do j=1,n
            aref(i,j)=ain(i,j)*mir(i,j)
         enddo
      enddo
```

The main example we shall study is the case of a Fabry-Perot cavity. In fact, round trips in the cavity will be implicitly replaced by a direct propagation through a series of thin lenses. It is remarkable that propagation looks just like a lens (see below) in the Fourier space, so that diffraction and refraction processes are exchanged by the FT. As an example we can make the following numerical experiment. Start from the TEM_{00} amplitude at its waist (this situation exists at the corner mirrors of the Virgo cavities, with $w_0 \simeq 2$ cm, then propagate over the distance $L = 3$ km, first by using the analytical formula giving the continuous paraxial result, second by using one step of the FFT scheme, with a sampling of n×n over a grid of size s. Then make numerically the two wave interfere. For measuring the distance between two complex amplitude arrays, we use the Hilbert Space metrics :

$$d(e_1, e_2) = \left[\frac{s}{n} \sum_{i,j=0}^{n-1} |e_{2,ij} - e_{1,ij}|^2 \right]^{1/2}$$

which maybe interpreted as the square root of the total power in the interference of the two waves. The following table gives an idea of the convergence of the discrete world towards continuous.

Sample size	*Optimal window*	*rms error*
32	34 cm	$4.9\ 10^{-5}$
64	49 cm	$1.6\ 10^{-8}$
128	70 cm	$4.9\ 10^{-15}$
256	81 cm	$8.3\ 10^{-15}$

Finding the field reflected off a resonant cavity

Assume an amplitude $E_{in}(x, y)$ is entering a Fabry-Perot cavity. The intracavity field $B(x, y)$ obeys the following equation :

$$B = t_1\ E_{in} - r_1 r_2\ M_1 \times P_L \times M_2 \times P_L\ B \tag{70}$$

(see Fig.2 for notation). $M_{1,2}$ are the reflection operators on the two mirrors respectively, for instance given by (56). Mathematically, this is an implicit linear equation, and it could be in principle solved by matrix inversion. But for a n×n sampling grid, the linear operator P_L is a $\mathrm{n}^2 \times \mathrm{n}^2$ rank operator, which, for large n would lead to invert huge matrices. It is therefore much more convenient if possible, to solve eq.(70) by successive iterations, provided some inital guess of the intracavity field. For instance, if we study small geometrical defects of the mirrors surfaces, the initial guess could be the ideal TEM_{00} mode $\Phi_{00}(x, y)$, with the correct surtension coefficient :

$$B_{guess}(x, y) = < E_{in}, \Phi_{00} > \frac{t_1}{1 - r_1 r_2} \Phi_{00}(x, y)$$

But other choices are possible. The speed of the convergence to the solution depends on the finesse of the cavity. Once the intracavity field $B(x, y)$ is found, the reflected amplitude is obtained by

$$E_{out} = M_1^* E_{in} + P_L \times M_2 \times P_L B$$

. It is possible to bring into evidence the various eignemodes of a parabolic Fabry-perot cavity. In the following numerical experiment we try to scan the different resonances of a VIRGO type cavity by adding a varying phase $\Phi \in [0, 2\pi]$ to the propagator, in order to simulate the fine tuning of the cavity. The inital tuning is assumed to correspond to a TEM_{00}. If the input field is a pure TEM_{00}, we see only the resonances of the fundamental (Fig38), other modes being orthogonal to the input field are never excited. In order to excite higher order modes, we have to take an input field not strictly orthogonal to the TEM_{00}. For instance, adding terms in x, y^2 , x^3, ... to the phase of the

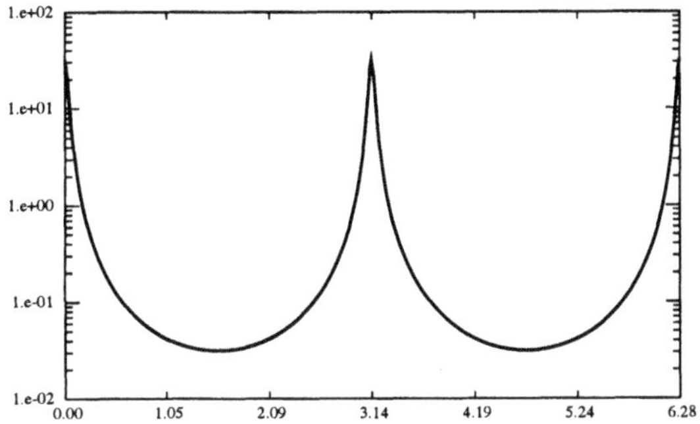

Figure 38: Fabry-Perot cavity : resonances of the fundamental mode

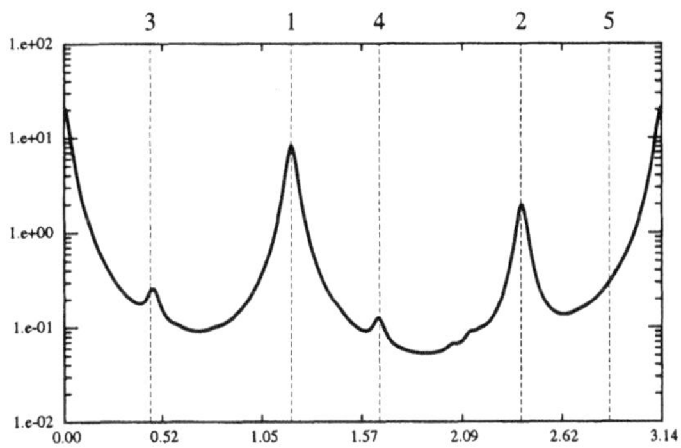

Figure 39: Fabry-Perot cavity : excitation of the $TEM_{m,n}$ modes. Dashed lines correspond to the first theoretical resonances, and are labeled by $m + n$.

input wave allows resonances of TEM_{mn} up to $m + n = 4$, as can be seen on fig.39. A more accurate study shows that the resonances are slightly different from their theoretical values. This is a consequence of the discretization of the field, of the mirrors, of the propagator. The discrete world has different rules. By increasing the order of the calculation (the preceding was carried out on a 64×64 grid), this small discrepancy vanishes. In order to drive the cavity exactly at resonance, it is necessary to achieve a fine tuning of the propagator. The brute force method would consist in computing the power in the intracavity field for various values of the varying phase precedingly introduced, and search for the maximum, but this would result in a very time costly code. It is better to use the following scheme. Let us denote by $\mathcal{C}$ the cavity operator :

$$\mathcal{C} = M_1 \times P_L \times M_2 \times P_L$$

and ϕ an arbitrary phase representing the fine tuning of the cavity. The intra-cavity field B obeys the implicit equation

$$B = t_1 A + e^{i\phi}\mathcal{C} B \tag{71}$$

where A is the input field. If the input field is the fundamental mode TEM_{00} with an amplitude a, and if for the sake of simplicity we note $\{\psi_p \; ; \; p = 1, 2, \ldots, \infty\}$ (with a unique index) the basis of TEM modes, we can write :

$$B = b_0\psi_0 + \sum_{p>0} b_p\psi_p$$

For a small perturbation, the coefficients $\{b_p \; ; \; p > 0\}$ are first order quantities. By taking the scalar product of eq.71 with the fundamental, we get

$$b_0 = t_1 a + e^{i\phi} \left[b_0 < \psi_0 \, , \, \mathcal{C} \, \psi_0 > + \sum_{p>0} b_p < \psi_0 \, , \, \mathcal{C} \, \psi_p > \right]$$

At the lower order we get

$$b_0 = \frac{t_1 \, a}{1 - e^{i\phi} < \psi_0 \, , \, \mathcal{C} \, \psi_0 >}$$

making clear that the value of ϕ which corresponds to resonance is

$$\phi = -\mathrm{Arg}\left[< \psi_0 \, , \, \mathcal{C} \, \psi_0 >\right]$$

The one way propagator must therefore be corrected by the phase factor

$$e^{i\phi/2}$$

The phase discrepancy ϕ of discrete vs continuous eigenmodes of the cavity is given in the following table.

Sample size	*Optimal window*	*phase corr.*
32	34 cm	-7.4 10^{-8} Rd
64	49 cm	-4.5 10^{-15} Rd
128	70 cm	-6.1 10^{-16} Rd

The Michelson Interferometer

It is easy to model a Michelson interferometer having two cavities as arms, for instance is order to study the field reflected off a cavity having imperfect mirrors, we take a perfect reference cavity and recombine the two reflected fields. Denoting by E_1 , E_2 the amplitides reflected by the two cavities, the total output field is

$$E_{\text{tot}} = E_1 + e^{i\phi} E_2$$

where ϕ represents the differential optical path between the two arms. It is necessary to adjust this phase to obtain the darkest field. In the ideal case, the phase is π and the resulting field is zero. In the general case, let

$$< E_1 , E_2 > = |E_1|.|E_2| \, e^{i\alpha}$$

it is easy to obtain the following equation for the power ;

$$P_{\text{tot}} = \left(\sqrt{P_1} - \sqrt{P_2}\right)^2 + 4\sqrt{P_1 P_2} \sin^2[(\phi - \alpha)/2]$$

so that it is clear that we must take $\phi = \alpha$, i.e.

$$e^{i\phi} = \frac{< E_1 , E_2 >}{|E_1|.|E_2|}$$

As an example, we consider a Michelson having the same parameters as Virgo, and in which one spherical mirror has a wrong curvature radius (1% error): After computation, we read that the relative power on the dark fringe is $1.6\,10^{-3}$, the intensity field having the structure shown on Fig.3.2. The interference between the two slightly differently curved wavefronts gives a series of rings of which one is visible in the non zero zone of the globally gaussian intensity. Another example corresponds to a misalignment of a mirror. For instance a corner mirror of one cavity has a pointing error of 10^{-8} Rd. The relative power on the dark fringe is 4.810^{-8}. The structure of the fringe (see Fig.3.2) is analogous to the intensity pattern of a TEM_{01} mode.

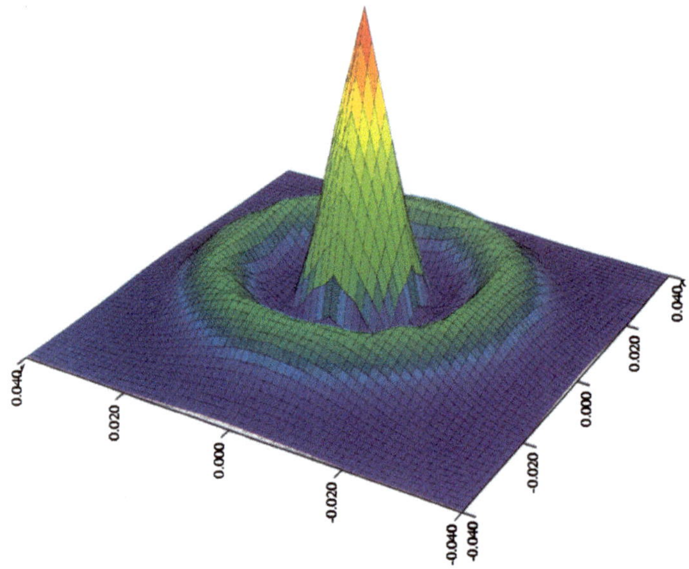

Figure 40: Michelson interferometer : Dark fringe pattern for 1% curvature radius error on a far mirror

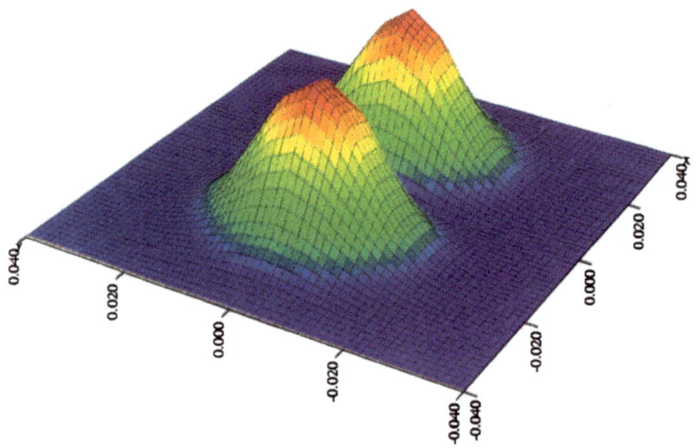

Figure 41: Michelson interferometer : Dark fringe pattern for 10^{-8} Rd pointing error on a corner mirror

On the intrinsic limitation to basic DFT based algorithms

When chosing the spatial window and the rank of the transform, special care must be taken of the angular limitation induced by the finite sampling of optical amplitudes. We know that the maximum spatial frequency, for the rank N and the window F is

$$p_{\max} = \frac{N\pi}{F}$$

This is a limitation on the structure of the admissible optical amplitudes : if their variations are of scale shorter than $\Lambda_{\min} \equiv 2\pi/p_{\max} = 2F/N$, corresponding to the Shannon frequency, aliasing will follow, and the algorithm fails. The Fourier variable p being interpreted as a transverse component of an oblique wave vector, we can set $p_{\max} \equiv k\theta_{\max}$ where θ represents the propagation direction with respect to the main optical axis. We can thus write :

$$\theta_{\max} = \frac{N\lambda}{2F} = \frac{\lambda}{2D}$$

D being the size of the sampling interval. This means that larger divergences are forbidden. For a gaussian beam of amplitude

$$f(r) = \exp\left[- r^2/w_0^2\right]$$

having a Fourier transform given by

$$\tilde{f}(\rho) = \frac{\pi}{w_0}\exp\left[- \rho^2 w_0^2/4 \right]$$

with $\rho^2 = p^2 + q^2$, by substituting $\rho = k\theta$, we get

$$\tilde{f}(\theta) = \frac{\pi}{w_0}\exp\left[- \theta^2 \;/\; \theta_g^2 \right]$$

where $\theta_g \equiv \lambda/\pi w_0$ is the gaussian divergence angle of the beam. The spectrum becomes negligible for $\theta_0 = \sigma \times \theta_g$ where σ of the order of 3 or 4. Owing to the condition $\theta_0 < \theta_{\max}$, we get the condition

$$D < \frac{\pi w_0}{2\sigma}$$

On the other hand, the window must be significantly larger than w_0, say $F > \sigma\, w_0$, so that we get the new limitation

$$D > \frac{\sigma w_0}{N}$$

The compatibility of these two conditions requires that

$$N > \frac{2\sigma^2}{\pi}$$

which is easily met. However, we would have serious troubles if after diffraction, the intensity distribution were out of, or larger than the computation window.

Hankel Transform methods

The 2D Fourier Transform of a function $f(x, y)$ is

$$\tilde{f}(p, q) = \int_{\mathbf{R}} e^{ipx} e^{iqy} f(x, y) dx\, dy$$

In polar coordinates, defined by :

$$\begin{aligned} x &= r\cos\phi, \quad y = r\sin\phi \\ p &= \rho\cos\psi, \quad q = \rho\sin\psi \end{aligned}$$

the transform is :

$$\tilde{f}(\rho, \psi) = \int_0^{2\pi} d\phi \int_0^{\infty} r\, dr\; e^{i\rho r\cos(\phi-\psi)}\; f(r, \phi)$$

If the initial function is axially symmetrical, i.e. independent of ϕ, then its transform is also axially symmetrical in the Fourier space, the ϕ integration can be carried out, and we obtain :

$$\tilde{f}(\rho) = 2\pi \int_0^{\infty} J_0(\rho r)\; f(r)\; r\; dr$$

Integrals of this type, involving a Bessel function, are called Hankel transforms. The inverse Fourier transform reads

$$\tilde{f}(r) = \frac{1}{2\pi} \int_0^{\infty} J_0(\rho r)\; \tilde{f}(\rho)\; \rho\; d\rho$$

It is cumbersome to keep these factors of 2π throughout all foregoing calculations. It is much more convenient to use here a different convention in the definition of the Fourier Transform :

$$\tilde{f}(p) = \frac{1}{\sqrt{2\pi}} \int_{-\infty}^{\infty} e^{ipx} f(x)\; dx \tag{72}$$

In this case, the 2D Fourier transform of $f(r)$ is simply the Hankel transform :

$$\tilde{f}(\rho) = \int_0^\infty J_0(\rho r)\; f(r)\; r\; dr \tag{73}$$

and its inverse is

$$f(r) = \int_0^\infty J_0(\rho r)\; \tilde{f}(\rho)\; \rho\; d\rho \tag{74}$$

We are dealing with special physical solutions of the wave equation, i.e. amplitudes of finite spatial extension (or almost), such as gaussian waves, which are practically zero for $x^2 + y^2 > 10w_0^2$, and of finite extension in the Fourier space (again gaussian waves). There exists a circle of radius a outside of which the amplitude is negligible, and there exists a circle of radius b in the Fourier plane outside of which the Fourier transform of the amplitudes is negligible. It is well known that there exists a family of orthogonal functions on the disk $\mathcal{D}_a \equiv \{r < a\}$, i.e. the functions

$$\varphi_\beta(r) = J_0(\zeta_\beta r/a), \quad \beta = 1, 2, \ldots, \infty$$

where the ζ_β, $\beta = 1, 2, \cdots, \infty$ are the zeros of $J_1(z)$. The orthogonality relation is :

$$\int_0^a \varphi_\beta(r)\; \varphi_\alpha(r)\; r\; dr = \frac{a^2}{2} J_0^2(\zeta_\alpha)\; \delta_{\alpha\beta} \tag{75}$$

Let us note

$$p_\alpha = \frac{a^2}{2} J_0^2(\zeta_\alpha)$$

Obviously, if $\mathcal{D}_b \equiv \{\rho < b\}$ is the disk in the Fourier space, it admits a corresponding family of functions :

$$\psi_\beta(\rho) = J_0(\zeta_\beta \rho/b), \quad \beta = 0, 1, \ldots, \infty$$

with the notation

$$q_\alpha = \frac{b^2}{2} J_0^2(\zeta_\alpha)$$

we have the orthogonality relation :

$$\int_0^b \psi_\beta(\rho)\; \psi_\alpha(\rho)\; \rho\; d\rho = q_\alpha\; \delta_{\alpha\beta} \tag{76}$$

We call B_a the set of all functions of r negligible outside $\mathcal{D}_a$, and B_b the set of functions of ρ negligible outside $\mathcal{D}_b$. We can assume in the formulas of the Hankel Transform (73,74), that $\tilde{f} \in B_b$, and expand it on the basis. Such an expansion is called Dini expansion :

$$\tilde{f}(\rho) = \sum_{\beta=0}^{\infty} f'_\beta \psi_\beta(\rho) = \sum_{\beta=0}^{\infty} f'_\beta \; J_0(\zeta_\beta \rho/b)$$

by substituting in (74) we get :

$$f(r) = \sum_{\beta=0}^{\infty} f'_\beta \int_0^\infty J_0(\rho r) \; J_0(\zeta_\beta \rho/b) \; \rho \; d\rho$$

We can sample the values of $f(r)$ by choosing an elementary distance in the plane. It is convenient to take $\delta r = 1/b$ as the distance element and sample the radii according to

$$r_\alpha = \zeta_\alpha/b$$

so that there is a strong link between the coefficients c_β introduced in the Dini expansion of $\tilde{f}$ and the samples $f(r_\alpha)$:

$$f_\alpha \equiv f(r_\alpha) = \sum_{\beta=0}^{\infty} f'_\beta \int_0^\infty J_0(\rho \zeta_\alpha/b) \; J_0(\zeta_\beta \rho/b) \; \rho \; d\rho$$

which owing to the orthogonality relation gives :

$$f'_\alpha = \frac{f_\alpha}{q_\alpha}$$

by substituting in the expansion of $\tilde{f}$ we get :

$$\tilde{f}(\rho) = \sum_{\beta=0}^{\infty} \frac{f_\alpha \; J_0(\zeta_\beta \rho/b)}{q_\beta}$$

We can now also sample the values of $\tilde{f}(\rho)$ by chosing the elementary frequency as $\delta\rho = 1/a$, and the spectrum samples will be evaluated at $\rho_\alpha = \zeta_\alpha/a$. The preceding expression, after sampling becomes :

$$\tilde{f}_\alpha \equiv \tilde{f}(\rho_\alpha) = \sum_{\beta=0}^{\infty} \frac{1}{q_\alpha} \; f_\alpha \; J_0(\zeta_\beta \zeta_\alpha/ab)$$

This is the discrete expression of the Hankel Transform, a linear relation between the vector $\tilde{f}_\alpha$ and the vector f_α :

$$\tilde{f}_\alpha = \sum_{\beta=0}^{\infty} H_{\alpha\beta}^{(+)} f_\beta$$

The direct transform is thus represented by the matrix

$$H_{\alpha\beta}^{(+)} = \frac{2J_0(\zeta_\alpha\zeta_\beta/ab)}{b^2 J_0^2(\zeta_\beta)} \tag{77}$$

Obviously, a similar treatment can be carried out for the inverse transform. The previously introduced function $f \in \mathcal{D}_a$ admits a Dini expansion of the form

$$f(r) = \sum_{\beta=0}^{\infty} \tilde{f}'_\beta \, J_0(\zeta_\beta \, r/a)$$

so that the expression (73) of the continuous direct Hankel Transform becomes :

$$\tilde{f}(\rho) = \sum_{\beta=0}^{\infty} \tilde{f}'_\beta \int_0^a J_0(\rho r) \, J_0(\zeta_\beta \, r/a) \, r \, dr$$

Now, in the Fourier plane, we can sample the conjugated variable ρ according to $\rho_\alpha = \zeta_\alpha/a$, so that we have the discrete version :

$$\tilde{f}_\alpha \equiv \tilde{f}(\rho_\alpha) = \sum_{\beta=0}^{\infty} \tilde{f}'_\beta \int_0^a J_0(\zeta_\alpha \, r/a) J_0(\zeta_\beta \, r/a) \, r \, dr$$

from the orthogonality relation we obtain:

$$\tilde{f}'_\alpha = \frac{\tilde{f}_\alpha}{p_\alpha}$$

the Dini expansion of $f(r)$ is now determined, and we have :

$$f(r) = \sum_{\beta=0}^{\infty} \frac{1}{p_\beta} \tilde{f}_\beta \, J_0(\zeta_\beta \, r/a)$$

Sampling of the values of f according to $r_\alpha = \zeta_\alpha/b$ leads to the inverse Hankel Transform :

$$f_\alpha = \sum_{\beta=0}^{\infty} H_{\alpha\beta}^{(-)} \tilde{f}_\beta$$

with

$$H_{\alpha\beta}^{(-)} = \frac{2J_0(\zeta_\alpha\zeta_\beta/ab)}{a^2 J_0^2(\zeta_\beta)} \tag{78}$$

The fact that the studied function f is in the set B_a implies that it takes negligible values for $r > a$. The sampling r_α can therefore stop at $r = a$, and we have :

$$\zeta_\alpha /b \leq a \Rightarrow \zeta_\alpha < ab$$

The same result is obviously obtained by considering the Fourier space. If the transformed function $\tilde{f}$ takes negligible values for $\rho > b$, then the sampling of ρ must stop at b :

$$\zeta_\alpha /a \leq b \Rightarrow \zeta_\alpha < ab$$

We can freely decide the size of the computation window a. Then we can still freely decide the maximum number N of zeros we shall take into account in the infinite sums encountered in the expressions of the Discrete Hankel Transformn (DHT). This being done, we have

$$\alpha = 0, 1, 2, \ldots, N$$

$$ab = \zeta_N \Rightarrow b = \zeta_N/a$$

The radial sampling is thus :

$$r_\alpha = a\,\zeta_\alpha/\zeta_N$$

and the spatial frequency sampling :

$$\rho_\alpha = \zeta_\alpha/a$$

The expressions for the Transforms (FDHT) are finite rank matrix algebra :

$$\tilde{f}_\alpha = \sum_{\beta=0}^{N} H_{\alpha\beta}^{(+)} f_\beta \ , \quad \alpha = 0, 1, 2, \ldots, N$$

with $H^{(\pm)}_{\alpha\beta}$ having the following definitions :

$$H^{(+)}_{\alpha\beta} = \frac{2a^2 J_0(\zeta_\alpha \zeta_\beta / \zeta_N)}{\zeta_N^2 J_0^2(\zeta_\beta)} \tag{79}$$

$$H^{(-)}_{\alpha\beta} = \frac{2 J_0(\zeta_\alpha \zeta_\beta / \zeta_N)}{a^2 J_0^2(\zeta_\beta)} \tag{80}$$

The Hermitian scalar product of two functions $f(r)$, $g(r)$ is defined in the direct Hilbert space by

$$< f \, , \, g > = \int_0^\infty f(r)^* \; g(r) \; r \; dr$$

if moreover, $f, g \in \mathcal{D}_a$, then the integral can be stopped at $r = a$ and f , g may be replaced by their Dini expansions on the φ_α, so that, using the orthogonality,

$$< f \, , \, g > = \sum_\alpha \tilde{f'}^*_\alpha \; \tilde{g'}_\alpha \; p_\alpha$$

or,

$$< f \, , \, g > = \sum_\alpha \tilde{f}^*_\alpha \; \tilde{g}_\alpha \; / p_\alpha$$

The Hermitian scalar product is invariant by a Fourier Transform, so that we have as well :

$$< f \, , \, g > = < \tilde{f} \, , \, \tilde{g} >$$

it is easily seen that

$$< \tilde{f} \, , \, \tilde{g} > = \sum_\alpha f^*{}_\alpha g_\alpha / q_\alpha$$

these formula provide the way of computing the scalar product either in the spatial or in the frequency space. The power carried by a given amplitude f is now

$$P(f) \; = < f \, , \, f >$$

and we have a distance in the Hilbert space, defined for two functions f, g by

$$\mathrm{d}(f,g) \; = \; \sqrt{P(f-g)}$$

If efficient FFT routines are available in all mathematical computer libraries, this is not the same for DHT, this is the reason why we give here the basic ideas for building specific libraries. All preceding formulas deal with the two numerical tables of the zeros of J_1 (including 0), ζ_α and $J_0(\zeta_\alpha)$. It is easy to obtain the table ζ_α by the following scheme. An initial guess of ζ being given, a better estimate is found by the Newton formula :

$$\zeta_{new} = \zeta_{old} - \frac{J_1(\zeta_{old})}{J_1'(\zeta_{old})}$$

which can be iterated until a given accuracy is met. Now, using the well known relation

$$J_1'(z) = J_0(z) - \frac{1}{z} J_1(z)$$

this is

$$\zeta_{new} = \zeta_{old} \left[1 - \frac{J_1(\zeta_{old})/J_0(\zeta_{old})}{\zeta_{old} - J_1(\zeta_{old})/J_0(\zeta_{old})} \right]$$

The problem reduces to the calculation of $J_1(z)/J_0(z)$. This can be done using a well known algorithm based on the recursion formula for Bessel functions, namely

$$J_n(z) = \frac{2(n+1)}{z} J_{n+1} - J_{n+2}$$

The recursion begins by taking arbitrarily $J_M = 0$ and $J_{M-1} = 1$, and then descending to J_1 and finally J_0 by the preceding formula. M must be chosen sufficiently large depending on the argument. If during the recursion the J_n's become to large, leading to a possible overflow, all the terms of the recursion can be divided by a common arbitrary renormalization constant. Finally, the ratio of the two last terms gives the value of $J_1(z)/J_0(z)$. The calculation of $J_0(z)$ is identical, except that the last term of the recursion gives $J_0(z)$ only after normalization. This normalization is done using the well known relation

$$1 = J_0(z) + 2 \sum_{n=1}^{\infty} J_{2n}(z)$$

Finally, the determination of all $\{\zeta_\alpha \; ; \; \alpha = 0, 1, \ldots, N\}$ is done by taking the first ones from any mathematical handbook :

$$\zeta_0 = 0 \,, \; \zeta_1 = 3.83 \,, \; \zeta_2 = 7.02$$

(values already encountered in the diffraction problem for a uniform circular aperture) as initial guesses for initializing the Newton refinement process. Then for all higher indices, the initial guess for ζ_α is $\zeta_{\alpha-1} + \pi$. One can easily imagine routines providing at the same time the two families ζ_α and $J_0(\zeta_\alpha)$.

The question "is the inverse HT actually the algebraic inverse of the direct HT ?" must be considered, because it is not manifest that

$$\sum_{\sigma=0}^{N} H^{(\pm)}_{\alpha\sigma} H^{(\mp)}_{\sigma\beta} = \delta_{\alpha\beta} \ (?)$$

In fact this is not true. What is true is that the linear operator $H^{(-)}\ H^{(+)}$ is a projector on B_a , and $H^{(+)}\ H^{(-)}$ a projector on B_b, for N infinite. Practically, this means that for a given function f negligible outside $\mathcal{D}_a$, and for N large enough, we have

$$f_\alpha \simeq \sum_{\beta=0}^{N} \sum_{\sigma=0}^{N} H^{(-)}_{\alpha\sigma} H^{(+)}_{\sigma\beta}\ f_\beta$$

and the corresponding formula in the Fourier space, with an accuracy depending on the window size a and the rank N. To be more specific, it is possible to reach the limit accuracy of the computer ($\simeq 10^{-15}$) in `double` , or `REAL*8` , by suitably chosing the window a. We conclude that the situation is theoretically different from the 2D DFT, in which the product $\mathrm{DFT}^{-1} \times \mathrm{DFT}$ is exactly the unity operator, regardless of the rank of the transform or the window size, but practically, the window being correctly chosen, the numerical accuracy is the same. If we consider a gaussian wave at its waist ($w_0 = 2$ cm) and compare it with its double DHT, we obtain the following results :

Sample size	*Optimal window*	*rms error*
10	8 cm	$5.2\ 10^{-7}$
20	11 cm	$1.8\ 10^{-13}$
50	17 cm	$1.0\ 10^{-15}$

Thus, with N about 50 , there is no significant numerical discrepancy between the DFT and the DHT

The correspondance between an initial field distribution $e_0(r)$ and the field $e_z(r)$ diffracted at a distance z is represented by a matrix that can be computed explicitly. The paraxial propagator, as seen previously, is expressed in the Fourier variables p , q, by

$$\widetilde{G}(p,q,z) = \exp\left[-\ i\ \frac{z}{2k}\left(p^2\ +\ q^2\right)\right]$$

with $\rho^2 = p^2 + q^2$, and using the sampling $\rho_\alpha = \zeta_\alpha / a$, we have

$$\widetilde{G}_\alpha = \exp\left[-\ i\ \frac{\lambda z}{4\pi a^2}\ \zeta_\alpha^2\right]$$

The Fourier tranform of the initial field is :

$$\tilde{e}_{1,\alpha} = \sum_{\beta=0}^{N} H^{(+)}_{\alpha\beta} \, e_{1,\beta}$$

The Fourier Transform of the final field is

$$\tilde{e}_{2,\alpha} = \widetilde{G}_\alpha \times \tilde{e}_{1,\alpha}$$

And the final field itself is

$$e_{2,\alpha} = \sum_{\beta=0}^{N} H^{(-)}_{\alpha\beta} \, \tilde{e}_{2,\beta}$$

All this can be summarized by the simple linear operation

$$e_{2,\alpha} = \sum_{\beta=0}^{N} P_{\alpha\beta} \, e_{1,\alpha} \tag{81}$$

Where the matrix P is :

$$P_{\alpha\beta} = \sum_{\sigma=0}^{N} H^{(-)}_{\alpha\sigma} \, \widetilde{G}_\sigma \, H^{(+)}_{\sigma\beta} \tag{82}$$

In order to compare with the 2D DFT, we do the same numerical experiment, and we propagate a normalised TEM_{00} from its waist ($w_0 = 2$cm) over a distance $z = 3$km. We compare in the following table the numerically propagated wave with the theoretical.

Sample size	*Optimal window*	*rms error*
10	13 cm	$6.3\ 10^{-3}$
20	18 cm	$3.1\ 10^{-5}$
50	29 cm	$9.5\ 10^{-13}$
100	40 cm	$1.8\ 10^{-15}$

The intensity of the field can be represented on a radial plot (see Fig.3.2). The sampling has been represented by small spots. The red profile corresponds to the initial gaussian wave, the blue to the diffracted one.

The optical system being assumed axially symmetrical, the mirror surfaces are sampled along a radius. A mirror will be defined by the diagonal operator

$$M_\alpha = i \, r_P \, \exp\left[i \, k \, \frac{a\zeta_\alpha^2}{2R_c \, \zeta_N^2} + i \, k \, f(a\zeta_\alpha/\zeta_N) \right]$$

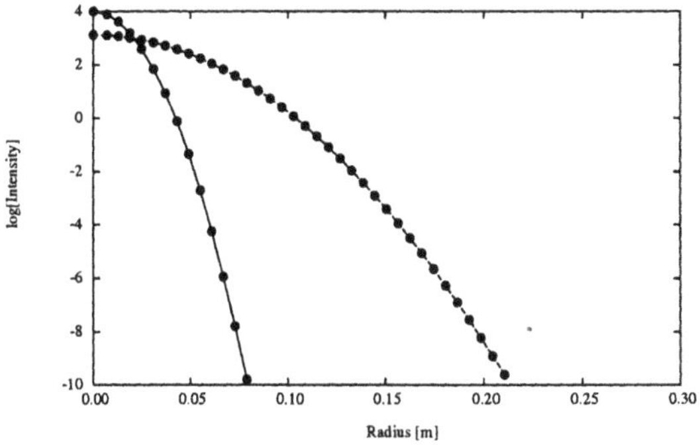

Figure 42: Diffraction of a gaussian wave, N=50 samples, window = 30 cm

where r_P is the photometric reflection coefficient, and R_c the curvature radius. The function $f(r)$ represents possible geometrical defects of the mirror. Action on an optical amplitude, giving A_{ref} from A_{in} is :

$$A_{\mathrm{ref},\alpha} = M_\alpha A_{\mathrm{in},\alpha}$$

It follows that to any optical system involving distances and mirrors (or thin lenses) can be associated a single matrix which represents explicitly the optical transfer function. An example is treated below.

Finding the field stored in a resonant cavity is especially convenient in the DHT scheme because it can be done by one matrix inversion. Recall that if the input amplitude is A, the intracavity field B obeys the implicit equation

$$B = t_1 \, A + r_1 r_2 \, \mathrm{e}^{i\phi} \, [\, M_1 \, P \, M_2 \, P \,] B$$

where the phase ϕ determines the tuning of the cavity, and P is the diffraction operator described above. Calling $\mathcal{C} = M_1 \; P \; M_2 \; P$ the cavity operator, representing the effect of a round trip, we see from the preceding subsection that $\mathcal{C}$ is an explicitly known matrix. The solution of the preceding equation can therefore be found as

$$B = \left[\mathcal{I} - r_1 r_2 \mathrm{e}^{i\phi} \mathcal{C}\right]^{-1} \, t_1 A$$

where $\mathcal{I}$ is the identity matrix. Instead of a series of iterations, as in the DFT scheme, we may now solve the problem by a single $N \times N$ matrix inversion. In case of moderate finesse ($r_1 r_2$ not too close to 1), this is a huge benefit. If the finesse is high, the iteration scheme in the FFT method converges very slowly, but the matrix inversion may become problematic too, due to the very small diagonal elements. An idea of the accuracy of the algorithm can be drawn fron the following experiment : We consider a resonant cavity of finesse 50, tuned on the fundamental mode, and we check the relative accuracy on the solution of the implicit cavity equation, i.e :

$$\epsilon = \| B - t_1 A + r_1 r_2 e^{i\phi} C \|$$

Sample size	*Optimal window*	*rms error*
20	17 cm	4.8 10^{-5}
50	29 cm	1.7 10^{-12}
100	39 cm	3.0 10^{-15}

For the reflected field, we have

$$A_{\text{ref}} = M_1^\dagger \, A \, + \, t_1 \, B$$

so that the reflection off the cavity reduces to a matrix product :

$$A_{\text{ref}} = \left[M_1^\dagger \, + \, t_1 \left(\mathcal{I} \, - \, r_1 r_2 e^{i\phi} C \right) \, t_1 \right]^{-1} A$$

Modal representation

The set of Hermite-Gauss functions $HG_{mn}, \quad m, n = 0, 1, \ldots, \infty$ is a complete set. Thus any optical amplitude admits a unique expansion of the type

$$E(r) = \sum_{m,n} E_{mn} \, HG_{z;m,n}(x, y)$$

This kind of representation has the key advantage that, dealing with its eigenfunctions, the propagation operator has a diagonal matrix representation. In fact, in a perfect system where all mirrors are matched, the propagation problem completely decouples in independent scalar equations, one for each mode. linear coupling of these modes are caused by perturbations of the optical elements. In all what follows, we take the normalized $HG_{m,n}$ functions as the basis. This means that at all optical element, there is a basis of functions

$$\phi_{z;m,n}(x, y) = c_{mn} \, H_m \left(\sqrt{2} x / w(z) \right) H_n \left(\sqrt{2} y / w(z) \right) \times$$

$$\times \exp \left[- \frac{x^2 + y^2}{w(z)^2} \right] \exp \left[i \, \frac{\pi (x^2 + y^2)}{\lambda R(z)} \right]$$

having a $w(z)$ parameter equal to what diffraction imposes. The normalisation constant is :

$$c_{mn} = \sqrt{\frac{2}{\pi w(z)^2}\frac{1}{2^{m+n}\, m!\, n!}}$$

In other words, we have one Hilbert space per element, and all these Hilbert spaces are connected by the function $w(z)$. For instance, at the waist, we have the basis $\phi_{w_0;m,n}(x,y)$, and at the end mirror, at a distance L, the basis $\phi_{w_1;m,n}(x,y)$, where $w_1 = w_0\sqrt{1+L^2/b^2}$

The optical elements we shall consider as examples are weak curvature matched mirrors having defects or wrong location. In general, the effect of such a mirror on an incoming wave $E_{mn,\ in}$ will be represented by a matrix operation

$$E_{mn,\text{out}} = ir\sum_{p,q} R_{mn,pq}E_{pq,\text{in}}$$

where r is the photometric coefficient, and where the coefficients $R_{mn,pq}$ are given by

$$R_{mn,pq} = c_{mn}c_{pq}\iint H_m\left(\sqrt{2}\frac{x}{w}\right)H_n\left(\sqrt{2}\frac{y}{w}\right)\times$$
$$H_p\left(\sqrt{2}\frac{x}{w}\right)H_q\left(\sqrt{2}\frac{y}{w}\right)\exp\left(-2\frac{x^2+y^2}{w^2}\right)\exp\left(2ikf(x,y)\right)dx\,dy$$

The total curvature phase factor vanishes, because if we represent the incoming wave's by $e^{i\phi}$, the matched mirror is $e^{-2i\phi}$, and the reflected wave $e^{-i\phi}$. Then we take the scalar product of this wave with the inversely directed wave which contains $e^{-i\phi}$, and eventually, ϕ disappears. The function $f(x,y)$ represents consequently only the departure of the surface with respect to the ideal paraboloid. We take the opportunity to remark that in kilometric FP cavities, curved mirrors can be considered as well as spherical or parabolic. The apex equation of a sphere of radius R_c osculating the plane $z=0$ is :

$$z = R_c - \sqrt{R_c^2 - x^2 - y^2}$$

The expansion for large R_c gives

$$z = \frac{x^2+y^2}{2R_c^2} - \frac{\left(x^2+y^2\right)^2}{8R_c^3} + \mathcal{O}(a^6/R_c^5)$$

(a being the radius of the mirror). The first term is the parabolic approximation, and the second is bounded by

$$\delta z_{\max} = \frac{a^4}{8R_c^3}$$

For typical values, $a = 17.5$ cm, and $R_c = 1.45$ km, we have $\delta z_{\max} \simeq 4. \times 10^{-14}$ m, so that there is no significant difference between the paraboloid and the sphere. The scalar product can be written

$$R_{mn,pq} = \frac{w^2}{2} c_{mn} c_{pq} \iint H_m(X) H_n(Y) H_p(X) H_q(Y) \times \tag{83}$$

$$\exp(-X^2) \exp(-Y^2) \exp\left(2ikf(wX/\sqrt{2}, wY/\sqrt{2})\right) dx\, dy$$

The first case we shall examine is the rotation of a mirror. Obviously we consider small rotation angles. The general apex equation of a parabolic mirror, as already said, is, in the X, Y, Z coordinates

$$Z = \frac{X^2 + Y^2}{2R_c}$$

Suppose that the X, Y, Z frame is rotated by an azimutal angle ϕ and a colatitude angle θ from the reference frame x, y, z. We need the apex equation in that reference frame. We have

$$\begin{aligned} X &= x \cos\theta \cos\phi + y \cos\theta \sin\phi - z \sin\theta \\ Y &= -x \sin\phi + y \cos\phi \\ Z &= x \sin\theta \cos\phi + y \sin\theta \sin\phi + z \cos\theta \end{aligned}$$

By substituting in the apex equation, we get a second order equation in z, whose solution is

$$z = \frac{x^2 + y^2}{2R_c \cos\theta} - u \tan\theta + \frac{u^2 \sin^2\theta}{2R_c \cos^3\theta} + \mathcal{O}(\theta^3 a^3 / R_c^2)$$

where we have set

$$u = x \cos\phi + y \sin\phi$$

Consider now orders of magnitude. A sophisticated numerical model is useful if the mirrors are aligned enough to allow some interference in the system. If not, geometrical optics models are quite sufficient to describe what happens to the light. If the waves interfere, it means that the rotation angles allow some light off a mirror to reach the opposite one. In other words, the colatitude angle θ is less than $\theta_{\max} = a/L$. This is less than 10^{-4} Rd. With 3.45 km curvature we see that the neglected terms were less than 10^{-21} m. Now the third kept term is less than 10^{-13} m. It is therefore possible to take, with a very good accuracy,

$$z = \frac{x^2+y^2}{2R_c} - \theta\,(x\,\cos\phi + y\,\sin\phi)$$

as the rotated mirror apex equation. It means that the function $f(x,y)$ introduced above is simply

$$f(x,y) = \theta\,(x\,\cos\phi + y\,\sin\phi)$$

In order to compute the rotation matrix, we have to calculate integrals of the form

$$I_{mp} = \int_{-\infty}^{\infty} H_m(X)\,H_p(X)\,\mathrm{e}^{-X^2}\,\mathrm{e}^{i\rho X}\,dx \qquad (84)$$

with $\rho = -\sqrt{2}kw\theta\cos\phi$ then $\rho = -\sqrt{2}kw\theta\sin\phi$. It is easy to compute the integral 84 by using the translation formula 61. We can write indeed :

$$I_{mp} = \mathrm{e}^{-\rho^2/4}\int_{-\infty}^{\infty}\mathrm{e}^{-(X-i\rho/2)^2}H_m(X)\,H_p(X)\,dX$$

By considering a closed loop in the complex plane and using the Cauchy theorem, it can be immediately seen that

$$I_{mp} = \mathrm{e}^{-\rho^2/4}\int_{-\infty}^{\infty}\mathrm{e}^{-X^2}H_m(X+i\rho/2)\,H_p(X+i\rho/2)\,dX$$

now, with the translation formula this is

$$I_{mp} = \mathrm{e}^{-\rho^2/4}\sum_{k=0}^{m}\sum_{l=0}^{n} C_n^k\,C_p^l\,(i\rho)^{m-k+p-l}\int_{-\infty}^{\infty}\mathrm{e}^{-X^2}H_k(X)\,H_l(X)\,dX$$

and with the orthogonality relation of the Hermite polynomials, this gives

$$I_{mp} = \mathrm{e}^{-\rho^2/4}\sum_{k=0}^{m}\sum_{l=0}^{n} C_n^k\,C_p^l\,(i\rho)^{m-k+p-l}\,\sqrt{\pi}2^k k!\delta_{kl}$$

or, finally

$$I_{mp} = \sqrt{\pi}\,i^{m+p}\,\mathrm{e}^{-\rho^2/4}\sum_{k=0}^{\min(m,p)}(-2)^k\,\frac{m!\,n!}{k!\,(m-k)!\,(n-k)!}\,\rho^{m+n-2k}$$

It is convenient to introduce the polynomials

$$Q_{mp}(x) = \sum_{k=0}^{\min(m,p)} (-2)^k \frac{m!\,p!}{k!\,(m-k)!\,(p-k)!} x^{m+p-2k} \tag{85}$$

so that the result is

$$I_{mp} = \sqrt{\pi}\, i^{m+p}\, \mathrm{e}^{-\rho^2/4}\, Q_{mp}(\rho)$$

The first Q_{mp} polynomials are listed below

	0	1	2	3
0	1	x	x^2	x^3
1	x	x^2-2	x^3-4x	x^4-6x^2
2	x^2	x^3-4x	x^4-8x^2+8	x^5-12x^3+24x
3	x^3	x^4-6x^2	x^5-12x^3+24x	$x^6-18x^4+72x^2-48$

The rotation matrix takes on the form

$$R_{mn,pq} = \frac{i^{m+n+p+q}}{\sqrt{2^{m+n+p+q} m!n!p!q!}} \times \tag{86}$$

$$Q_{mp}\left(-\sqrt{2}kw\theta\cos\phi\right) Q_{nq}\left(-\sqrt{2}kw\theta\sin\phi\right) \mathrm{e}^{-k^2w^2\theta^2/2}$$

Remark that

$$\frac{k\theta w}{2} = \frac{\theta}{\theta_g}\frac{w}{w_0}$$

where $\theta_g = \lambda/\pi w_0$ is the divergence of the gaussian beam.

Consider now a parallel displacement of the beam. The incoming beam was

$$E_{\mathrm{in}} = \sum_{mn} E_{mn}\phi_{mn}(x,y)$$

and the translation operator T acts as

$$T.E\mathrm{in} = E_{\mathrm{in}}(x+\Delta x, y+\Delta y)$$

The operator matrix is :

$$T_{mn,pq} = \frac{w^2}{2} c_{mn} c_{pq} \times$$

$$\times \int_{-\infty}^{\infty} dX \int_{\infty}^{\infty} dY \exp\left[-\frac{(X+\Delta X)^2+(Y+\Delta Y)^2}{2}\right] \exp\left[-\frac{X^2+Y^2}{2}\right] \times$$

$$H_m(X)\, H_n(Y)\, H_p(X+\Delta X) H_q(Y+\Delta Y)$$

with $\Delta X \equiv \sqrt{2}\Delta x/w$. A change of variables leads to

$$T_{mn,pq} = \frac{w^2}{2} c_{mn}c_{pq} \int_{-\infty}^{\infty} dX \int_{-\infty}^{\infty} dY \exp\left[-\frac{(X+\Delta X/2)^2+(Y+\Delta Y/2)^2}{2}\right] \times$$

$$\exp\left[-\frac{(X-\Delta X/2)^2+(Y-\Delta Y/2)^2}{2}\right] \times$$

$$H_m(X-\Delta X/2)\, H_n(Y-\Delta Y/2)\, H_p(X+\Delta X/2) H_q(Y+\Delta Y/2)$$

By using again the translation formula, we get successively

$$T_{mn,pq} = \frac{w^2}{2} c_{mn}c_{pq}\, (-1)^{m+n}\, \pi\, Q_{mp}(\Delta X)\, Q_{nq}(\Delta Y) \exp\left[-\frac{\Delta X^2+\Delta Y^2}{4}\right]$$

then

$$T_{mn,pq} = \frac{(-1)^{m+n}}{\sqrt{2^{m+n+p+q} m!n!p!q!}} Q_{mp}(\sqrt{2}\Delta x/w)\, Q_{nq}(\sqrt{2}\Delta y/w) \times \quad (87)$$

$$\exp\left[-\frac{\Delta x^2+\Delta y^2}{2w^2}\right]$$

We often need a fast simulation code for a dynamical model of misaligned and detuned interferometer, in order for instance to study the global control of the system. If the angles are small compared to the divergence of the beam, it is possible to limit at 2d order in the colatitude angle the reflection operators for each mirror. Moreover, we see fom the preceding table that the modes of order m, n are coupled to the TEM_{00} at the $m+n$ order. A consistent 2d order expansion will thus involve the only first 6 modes, and we shall see that it is possible to carry out all calculations using a 2d order, rank 6×6 matrix algebra that we call A266. If we limit the expression to the second order in θ, the expression of the mirror operator becomes, after expansion of the exponential, and with the notation

$$p = \sqrt{2}kw\theta\cos\phi, \quad q = \sqrt{2}kw\theta\sin\phi$$

	00	10	01	20	11	02
00	$1-\frac{p^2+q^2}{4}$	$i\frac{p}{\sqrt{2}}$	$i\frac{q}{\sqrt{2}}$	$-\frac{p^2}{2\sqrt{2}}$	$-\frac{pq}{2}$	$-\frac{q^2}{2\sqrt{2}}$
10	$i\frac{p}{\sqrt{2}}$	$1-\frac{3p^2+q^2}{2}$	$-\frac{pq}{2}$	ip	$i\frac{q}{\sqrt{2}}$	0
01	$i\frac{q}{\sqrt{2}}$	$-\frac{pq}{2}$	$1-\frac{p^2+3q^2}{2}$	0	$i\frac{p}{\sqrt{2}}$	iq
20	$-\frac{p^2}{2\sqrt{2}}$	ip	0	$1-\frac{5p^2+q^2}{8}$	$-\frac{pq}{\sqrt{2}}$	0
11	$-\frac{pq}{2}$	$i\frac{q}{\sqrt{2}}$	$i\frac{p}{\sqrt{2}}$	$-\frac{pq}{\sqrt{2}}$	$1-\frac{3p^2+3q^2}{4}$	$-\frac{pq}{\sqrt{2}}$
02	$-\frac{q^2}{2\sqrt{2}}$	0	iq	0	$-\frac{pq}{\sqrt{2}}$	$1-\frac{p^2+5q^2}{4}$

The parallel displacement operator, at second order in $\Delta x/w$ has a similar expression. The free propagation along the optical axis is represented by the diagonal operator

$$P_{mn,pq} = \exp\left[-i(m+n)\arctan\left(\frac{L}{b}\right)\right]\delta_{mp}\delta_{nq}$$

where L is the propagation distance and b the Rayleigh Range. It is therefore clear that all operators involved in A266 are of the form

$$M = M_0 + \epsilon M_1 + \epsilon^2 M_2$$

where ϵ is a scaling parameter of the order of the normalized displacements (θ/θ_g or $\Delta x/w$). Moreover, each partial operator has a special structure. We note $\mathcal{O}_3$, $\mathcal{O}_{12}$ and $\mathcal{O}_{18}$ the sets of operators having these structures. Namely, $\mathcal{O}_3$ is the set of 6×6 operators of the form

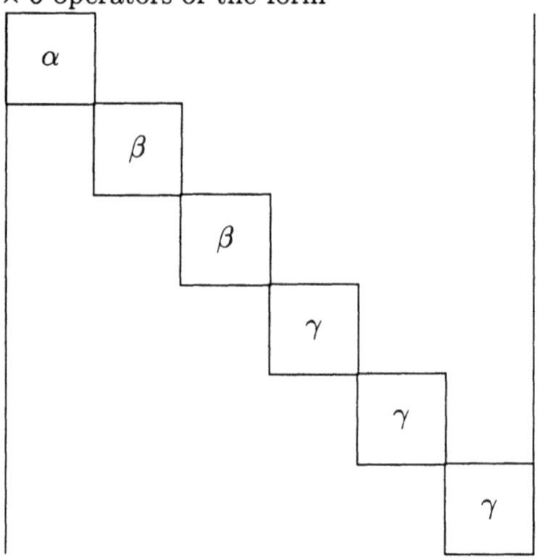

to which obviously belongs the propagation operator, and the zeroth order of any operator, $\mathcal{O}_{12}$ is the set of 6×6 operators of the form

$$\left|\begin{array}{cccccc} & \zeta_1 & \zeta_2 & & & \\ \zeta_7 & & & \zeta_3 & \zeta_4 & \\ \zeta_8 & & & & \zeta_5 & \zeta_6 \\ & \zeta_9 & & & & \\ & \zeta_{10} & \zeta_{11} & & & \\ & & \zeta_{12} & & & \end{array}\right|$$

to which belongs the first order part of the operators. $\mathcal{O}_{18}$ is the set of 6×6 operators of the form

$$\left|\begin{array}{cccccc} \xi_1 & & & \xi_2 & \xi_3 & \xi_4 \\ & \xi_5 & \xi_6 & & & \\ & \xi_7 & \xi_8 & & & \\ \xi_9 & & & \xi_{10} & \xi_{11} & \\ \xi_{12} & & & \xi_{13} & \xi_{14} & \xi_{15} \\ \xi_{16} & & & & \xi_{17} & \xi_{18} \end{array}\right|$$

to which belongs the second order part of the operators. This kind of storage requires $3 + 12 + 18 = 33$ places instead of 36 in the general case : There is no waste of memory. The global structure is stable by the elementary algebraic operations. More specifically, it is obvious that if $A, B \in \mathcal{O}_3$, then $AB \in \mathcal{O}_3$. if $A \in \mathcal{O}_3$ and $B \in \mathcal{O}_{12}$ resp $\mathcal{O}_{18}$ then $AB \in \mathcal{O}_{12}$ resp $\mathcal{O}_{18}$. What is more remarkable is the following property which is the basis of A266 : if $A, B \in \mathcal{O}_{12}$

then $AB \in \mathcal{O}_{18}$. The separation in three partial operators is therefore stable, and any algebraic operation reduces to trivial multiplications

- The sum is trivially defined by

$$(A+B)_0 = A_0 + B_0, \quad (A+B)_1 = A_1 + B_1, \quad (A+B)_2 = A_2 + B_2$$

- The product of two operators is defined by

$$(AB)_0 = A_0 B_0, \quad (AB)_1 = A_0 B_1 + A_1 B_0 \quad (AB)_2 = A_0 B_2 + A_1 B_1 + A_2 B_0$$

- The inverse of an operator is defined recursively by

$$\begin{aligned} (A^{-1})_0 &= A_0^{-1} \\ (A^{-1})_1 &= -(A^{-1})_0 A_1 (A^{-1})_0 \\ (A^{-1})_2 &= -(A^{-1})_0 \left[A_1 (A^{-1})_1 + A_2 (A^{-1})_0\right] \end{aligned}$$

- The square root X of an operator A is defined recursively by the

$$\begin{aligned} X_0 &= \sqrt{A_0} \\ X_{1,ij} &= \frac{A_{1,ij}}{X_{0,ii} + X_{0,jj}} \\ X_{2,ij} &= \frac{A_{2,ij} - (X_1^2)_{ij}}{X_{0,ii} + X_{0,jj}} \end{aligned}$$

3.3 Monte-Carlo methods

If the system in which light propagates has a complex geometry, and if the information carried by the phase is not essential, it is possible to represent light by particles following straight trajectories between reflections or diffusion processes. This is approximately the Newton theory of light. We call "photons" these particles for brevity, though the quantum nature of light is completely ignored in this approach. By launching randomly a large number of such photons, statistics can tell us where the light goes and how we can forbid certain areas to it (e.g. stray light studies). What is interesting is that the diffraction phenomena can be represented up to a certain extent by this particle description, leading to realistic models of light propagation.

Spatial spectra, plane waves and photons

We consider on an initial plane, a given complex amplitude of light $A(x,y)$. We can interpret the normalized square modulus of the amplitude as a probability density for a photon to be launched :

$$\frac{dP'}{ds}(x,y) = |A(x,y)|^2$$

This doesn't tell us the direction of the photon. We know that the angular information on the angles can be extracted from the Fourier transform of the amplitude. The 2D Fourier transform of the amplitude is nothing but an expansion in terms of plane waves having for transverse components of $\vec{k}$, the conjugated variables (p,q). If we consider the square modulus of the latter, we have, due to the Parseval-Plancherel theorem :

$$\int_{\mathbf{R^2}} dp\, dq\, |\tilde{A}(p,q)|^2 = 4\pi^2$$

On the other hand, by using the substitution

$$p = k \sin\theta \cos\phi \quad , \quad q = k \sin\theta \sin\phi$$

we obtain

$$\int_{\mathbf{R^2}} dp\, dq\, |\tilde{A}(p,q)|^2 = k^2 \int_0^{2\pi} d\phi \int_0^{\pi} d\theta\, |\tilde{A}(\theta,\phi)|^2$$

and consequently :

$$\frac{1}{\lambda^2} \int_0^{2\pi} d\phi \int_0^{\pi} d\theta\, |\tilde{A}(\theta,\phi)|^2 = 1$$

which shows that we can obtain a probability density for the angular distribution by taking

$$\frac{dP''}{d\Omega}(\theta,\phi) = \frac{1}{\lambda^2}\, |\tilde{A}(\theta,\phi)|^2$$

Propagation of the light described by the complex amplitude $A(x,y)$ will thus be described by the two densities of probabilities : The departure point of a photon will obey the statistics corresponding to dP'/ds, and its direction, the statistics defined by $dP''/d\Omega$.

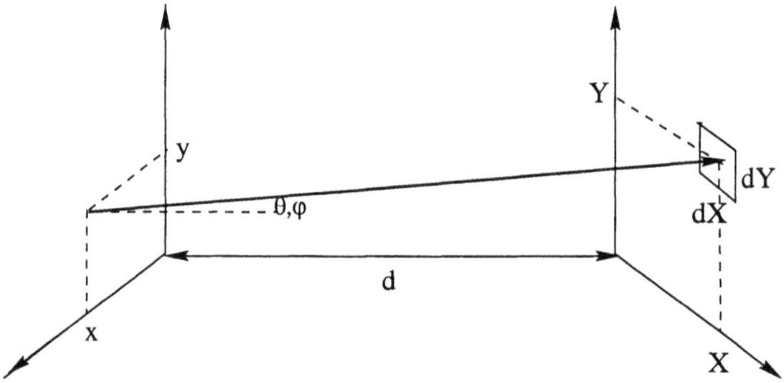

Figure 43: propagation from plane to plane

Propagation

After having chosen the departure point (x, y) and the direction of the photon (θ, ϕ), its trajectory is defined, and it is possible to compute the point (X, Y) at which it hits the plane $z = d$ (see Fig.43). It is even possible to compute the probability density dP'/dS of the arrival point. The conditonal probability for a photon starting from (x, y) at $z = 0$ to hit the small target of area $dXdY$, is

$$\frac{dP'}{dS}(X,Y,x,y)dXdY \;=\; \frac{dP''}{d\Omega}(\theta,\phi)d\Omega$$

where $d\Omega \;=\; dXdY/\rho^2$ is the elementary solid angle corresponding to the target seen from the initial point, $\rho(x, y, X, Y)$ being defined by

$$\rho(x,y,X,Y) \;=\; \sqrt{(x-X)^2+(y-Y)^2+d^2}$$

For the full density, we get thus :

$$\frac{dP'}{dS}(X,Y) \;=$$

$$\int_{\mathbf{R}^2} dx\,dy\,\frac{1}{\rho^2(x,y,X,Y)}\,\frac{dP''}{d\Omega}(\theta(x,y,X,Y),\phi(x,y,X,Y))\,\frac{dP'}{ds}(x,y)$$

This is the integral expressing the transfer of the probability density from $z = 0$ to $z = d$. If we adopt the paraxial approximation, we can replace ρ by d, and θ by $\sqrt{(x-X)^2+(y-Y)^2}/d$, so that :

$$\frac{dP'}{dS}(X,Y) \;=$$

$$\frac{1}{d}\int_{\mathbf{R}^2} dx\,dy\,\frac{dP''}{d\Omega}\left(\sqrt{(x-X)^2+(y-Y)^2}/d,\phi(x,y,X,Y)\right)\;\frac{dP'}{ds}(x,y)$$

We can show on a very simple example how it works. Take a TEM_{00} normalized mode defined by the amplitude :

$$A(x,y) \;=\; \sqrt{\frac{2}{\pi w^2}}\,\exp\left(-\,\frac{x^2+y^2}{w^2}\right)$$

The Fourier transform is :

$$\tilde{A}(p,q) \;=\; \sqrt{2\pi w^2}\,\exp\left(-\,\frac{w^2(p^2+q^2)}{4}\right)$$

We have thus :

$$\frac{dP'}{ds}(x,y) \;=\; \frac{2}{\pi w^2}\,\exp\left(-\,\frac{2(x^2+y^2)}{w^2}\right) \tag{88}$$

and

$$\frac{dP''}{d\Omega}(\theta,\phi) \;=\; \frac{2\pi w^2}{\lambda^2}\,\exp\left(-\,\frac{k^2w^2\theta^2}{2}\right) \tag{89}$$

And the transfer equation is :

$$\frac{dP'}{d\Omega}(X,Y) \;=$$

$$\frac{4}{\lambda^2 d^2}\int_{\mathbf{R}^2} dx\,dy\,\exp\left(-\,\frac{2(x^2+y^2)}{w^2}\right)\,\exp\left(-2\,\frac{\pi^2w^2[(x-X)^2+(y-Y)^2]}{\lambda^2d^2}\right)$$

This can be calculated either directly or by Fourier transform (being a convolution product) and the result is

$$\frac{dP'}{dS}(X,Y) \;=\; \frac{2}{\pi w'^2}\,\exp\left(-2\,\frac{X^2+Y^2}{w'^2}\right)$$

with

$$w' \;=\; w\sqrt{1+\left(\frac{\pi w^2}{\lambda d}\right)^2}$$

exactly as in paraxial wave optics. See Fig.44 for an example based on the Virgo parameters : The initial waist is w_0 =2 cm, the photons are assumed emitted at a point given by the 2D gaussian random variable (x,y) at $z=0$, of parameter w_0, the joint probability density being given by (88), then the

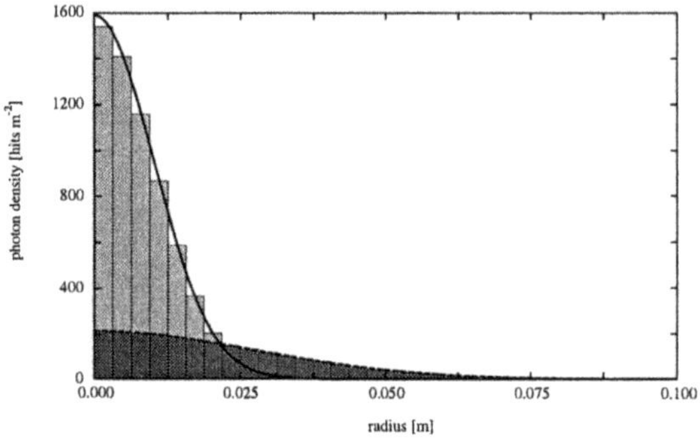

Figure 44: Diffraction spot at 3 km of a 2 cm waist initial TEM mode

direction of flight is given by a new 2D gaussian random variable (ξ, η) of parameter $\theta_g = \lambda/\pi w_0$, the joint probability density being given by (89). and the hit point at $z = d$ is calculated as

$$X = x + d\xi \quad , \quad Y = y + d\eta$$

We see the agreement between the wave optics theory (lines) and the Monte-Carlo result (histograms)

3.4 Diffraction patterns

It is not always possible to know the Fourier transform of the incoming amplitude. This is generally possible at the initial plane, where a clean source is assumed installed, but after propagation in a complex system, photons randomly emitted propagating as in a billiard, and reaching a given plane, do not allow to reconstruct the complex amplitude necessary to determine $dP''/d\Omega$. A very interesting procedure has been proposed in [16]. As they pass near the edge of any aperture, the photons are scattered at random angles, the standard deviation being inversely proportional to the distance at the edge. The heuristic argument is borrowed from quantum mechanics : a particle being at the distance Δx of the edge of a screen may be seen as having its location determined with accuracy Δx. Consequently, the accuracy on its momentum is $\Delta p_x = h/4\pi\Delta x$ (h is Planck's constant). Now, the momentum of a photon is known to be $p = hk/2\pi$, The relation $\Delta p_x/p = \tan\theta^*$ allows then to compute the standard deviation θ^* corresponding to the uncertainty Δp_x :

$$\theta^* = \arctan\left(\frac{\lambda}{4\pi\Delta x}\right)$$

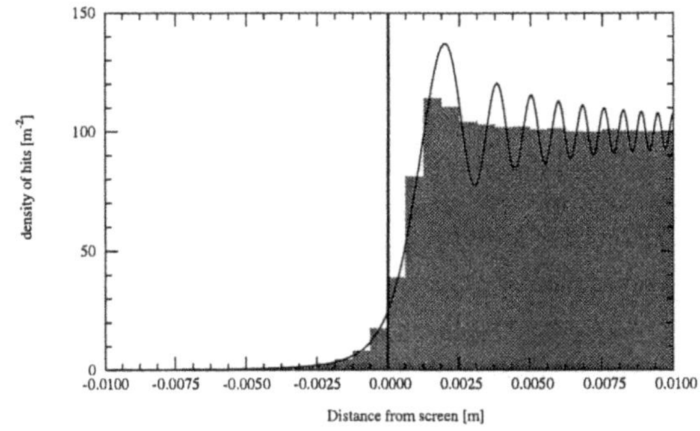

Figure 45: Diffraction by a half-plane. The screen is at $d = 5$m, the wavelength is $\lambda = 1.06\,\mu$m. Number of photons launched : 10^6

We can test the procedure on the well-known problem of diffraction by a half plane. We assume a wide and uniform light beam centered at $x = y = 0$ falling on a black screen masking half of the plane ($x < 0$). If we consider a screen at $z = d$, the intensity of the light is given (see[6]) by

$$I(x) = I_0 \times \frac{1}{2}\left[\left(\frac{1}{2} + C\left(\sqrt{\frac{2}{\lambda d}}x\right)\right)^2 + \left(\frac{1}{2} + S\left(\sqrt{\frac{2}{\lambda d}}x\right)\right)^2\right] \quad (90)$$

Where $C(x)$ and $S(x)$ are the Fresnel functions (see[17]). We assume a uniform random law for launching photons at x off the interval $[-x_M, x_M]$, then, x being randomly chosen, if $x < 0$ the process stops and a new photon is launched. if $x > 0$, its direction θ is drawn as a random deviate knowing its standard deviation θ^* (a gaussian deviate works). The hit point on a screen at distance d is $X = x + \theta d$. The statistics is reported on Fig.45 (histogram), the line represents the wave optics theory with $I_0 = 1/2x_M$ (Eq.90). Remark the excellent agreement in the shadow region, and the averaged behavior in the fringes region, due to the loss of information about the phase.

References

1. Jean-Yves Vinet, Brian Meers, Catherine Nary Man and Alain Brillet Optimization of long-baseline interferometers for gravitational-wave detection,
   Phys. Rev. D , Vol **38**, N.2 (1988) p. 433
2. Brian J. Meers
   Recycling in laser-interferometric GW detectors
   s *Phys. Rev.* D , Vol **38**, N.2 (1988) p. 2317
3. Ronald Drever
   in *Gravitational Radiation*
   edited by N. Deruelle and T. Piran (North Holland 1983) p.321
4. F. Pegoraro, L¿A¿ RAdicati, Ph. Bernard and E. Picasso
   Phys. Lett. **68** A (1978) p.165
5. Jun Mizuno in
   Comparison of optical configurations for laser-interferometric gravitational-wave detectors
   PHD Thesis, Max-Planck-Institut für Quqntenoptik (MPQ 203) 1995
6. Max BORN & Emil WOLF
   Principles of Optics
   Pergamon Press 1980
7. Fritz Oberhettinger
   Tables of Bessel Transforms
   Springer 1972
8. Fritz Oberhettinger
   Tabellen zur Fourier Transformation
   Springer 1957
9. J.W. Cooley and J.W. Tukey
   An algorithm for the machine calculation of complex Fourier series
   Math. Comput. **19**, pp.297-301 (1965)
10. Anthony E. Siegman
    Lasers
    Univ. Sc. Book (Oxford University Press) 86
11. H. Kogelnik, Tingye Li
    Laser beams and resonators *Appl. Opt.* **5**, p.1550 (1966)
12. E. Sziklas, A. E. Siegman
    Mode calculation in unstable resonators with flowing saturable gain I : Hermite-Gaussian expansion
    *Appl. Opt.***13**, p.2775 (1974)
    Mode calculation in unstable resonators with flowing saturable gain II :

FFT methods
Appl. Opt. **14**, p.1874 (1975)

13. Michael Hercher
The spherical mirror Fabry-Perot interferometer
Appl. Opt. **7**, p.951 (1968)
14. Patrice Hello
Modèle physique et simulation de l'antenne interférométrique gravitationnelle Virgo
Thèse de Doctorat en Sciences, Université Paris-Sud (1990)
15. G.D. Bergland
A guided tour of the fast Fourier Transform
IEEE spectrum, July 1969
16. B.K. Likeness
Stray light simulation with advanced Monte-Carlo techniques
Proc. Soc. of Photo-optical Instrumentation Engineers
April 18-21, 1977, Reston (Virginia/USA)
17. M. Abramowitz and I. Stegun
Handbook of mathematical functions
Dover

OPTICS MANUFACTURING AND TESTING FOR INTERFEROMETRIC GRAVITATIONAL-WAVE DETECTORS

VINCENT LORIETTE
ESPCI Laboratoire d'Optique, CNRS UPR5, 10 rue Vauquelin 75005 Paris, France
E-mail: loriette@optique.espci.fr

The core of all interferometric detectors of gravitational radiation is a set of very basic optical components, mirrors and beamsplitters, that form combinations of resonant optical cavities. The purpose of all the shells that surrounds these objects, seismic isolation, locking, detection,... being to stabilize and measure the relative distance between those optical components. From this information, error signals of the stabilization loops, or direct measurement of the distances variations, gravitational waves can be shown off. The core optics of these detectors are quite unique because they fulfill quality requirements that certainly make them the most perfect optical components ever made. The aim of this text is to give an overview of the challenge that represents the fabrication and control of such objects.

1 Optics manufacturing

In this section we want to discuss the definitions of some of the various parameters that characterize an optical component. We are mainly interested in describing, using a few scalar values, the modification of an incident electromagnetic field as it interacts with a component. Those modifications can be for example changes of the polarization state, of the wavefront geometry, of the field amplitude. The modifications can be homogeneous or not across the wavefront.

1.1 Laser beams and definition of optical parameters

All the interferometric detectors of gravitational radiation share the same optical scheme. They are basically giant Michelson type interferometers with Fabry-Perot arms. The set core optics is composed of one beamsplitter, two Rmax mirrors and two to four other mirrors depending of the recycling scheme that is used. The main properties of Virgo core optics are summarized in table 1 (data took from[1]). Interferometric gw detectors work with laser sources.

The profile of those laser beams has to be purely TEM_{00} everywhere in the optical cavities for any detector to work in its optimal configuration. The wavefront amplitude of a TEM_{00} propagating along the z axis is found by

	beamsplitter	input mirrors	end mirrors
substrate material	suprasil 311sv	suprasil 312sv	herasil 1top
subs. diameter (mm)	230	350	350
subs. thickness (mm)	55	100	200
radius of curv. 1 (m)	∞	∞	3450 ± 100
radius of curv. 2 (m)	∞	∞	∞
incidence (°)	45	0	0
wedge (mrad)	1 ± 0.07	0	0
coating diam. (mm)	200	100	280
reflectance (1)	0.5	0.88	$L < 50\,\text{ppm}$ $T > 10^{-5}$
reflectance (2)	$< 10^{-3}$	$< 5 \times 10^{-4}$	10^{-3}
total losses coat	<5 ppm	<5 ppm	<5 ppm
total losses subs.	<1 ppm/cm	<1 ppm/cm	
surface figure (1)	$\lambda/70$ rms $\emptyset = 7\,\text{cm}$	$\lambda/120$ rms $\emptyset = 4\,\text{cm}$	$\lambda/120$ rms $\emptyset = 10\,\text{cm}$
surface figure (2)	$\lambda/70$ rms $\emptyset = 12\,\text{cm}$	$\lambda/70$ rms $\emptyset = 4\,\text{cm}$	no spec.

Table 1. Some Virgo optical components specifications. From the may 1997 final design report, some numbers may be obsolete.

solving the time independent wave equation in the paraxial approximation:

$$A_{00}(r,z) = A_{00}\frac{w_0}{w(z)}\exp\left(-\frac{r^2}{w(z)^2}\right)\exp\left(-i\left(kz - \tan^{-1}\left(\frac{z}{z_0}\right) + \frac{kr^2}{2R(z)}\right)\right) \tag{1}$$

where $k = 2\pi/\lambda$ is the wavenumber, w_0 is the beam waist, $w(z)$ is the beam size at a distance z from the beam waist position, z_0 is the Rayleigh range and $R(z)$ is the wavefront radius of curvature. At a distance z from the beam waist position, and for a given wavelength, the laser beam scalar amplitude is fully characterized by three parameters : $w(z)$, $R(z)$, A. The surface properties of interest in our case are the ones that will modify any one of those parameters upon reflection or transmission. Moreover, the TEM_{00} mode is not the only one that can propagate inside the detector. Any beam of the form:

$$A_{mn}(r,z) = A_{mn} H_m \left(\frac{x\sqrt{2}}{w(z)}\right) H_n \left(\frac{y\sqrt{2}}{w(z)}\right) \frac{w_0}{w(z)} \exp\left(-\frac{r^2}{w(z)^2}\right) \times \exp\left(-i\left(kz - (1+m+n)\tan^{-1}\left(\frac{z}{z_0}\right) + \frac{kr^2}{2R(z)}\right)\right) \quad (2)$$

where H_m and H_n are hermite polynomials of order m and n is a solution of the same propagation equation. The so called Hermite-Gauss functions TEM_{mn} $(m \geq 0, n \geq 0)$ form a complete orthogonal basis. Any modification of one of the parameters $w(z)$, $R(z)$, $[A_{mn}]$ will not only attenuate the TEM_{00} beam, but will also lead to an energy transfer towards other TEM_{mn} modes. Optical components, even the simplest ones, can be described in a variety of ways, using different sets of numbers that all contain the same information. One problem is to find the most useful set of parameters to handle. We have to understand to what purpose we need a description of optical parameters and find the most adapted description. In the case of gw detectors optics characterization is used in two different contexts, as was mentioned in the preface. Optics characterization is of course used to develop and improve the manufacturing processes, but the data is also used in numerical simulations of gw antennas.

1.2 Properties of optical components

In most textbooks on the subject the only effects that an interface can have, in terms of energy, on an incident light beam is either to reflect it, to absorb it or to transmit it. In this picture the interface is seen as a boundary between two half-spaces. All the energy that remains in the same half-space as the incident light is said to be reflected, all the energy that passes to the second half-space is transmitted, and the rest is absorbed by the interface. This definition is very simple but does not suit our needs because it does not make any distinction between *useful* light and *lost* light.

The definition of reflectance and transmittance depends strongly on the use of these quantities. The most versatile definition seems to be: *the reflectance of an optical component is the ratio of the usable reflected power over the total incident power.* The same definition can be adopted for transmittance by replacing *reflected* by *transmitted.* The word *usable* is very important because it distinguishes between light that is 'reflected' but cannot be used after reflection, this is the case of scattered light, and that has to be counted as lost by the system, and light that can be used after the reflection has occurred, which is truly what we call *reflected light.* Using this as a definition of reflectance and transmittance we see that their value must depend on the way the components are used, and in particular on the incident beam shape and on the wanted reflected and transmitted beam shapes. Let the incident amplitude be $A_{inc}(x, y)$, and let the mirror reflection factor be of the form

$$r(x, y) = \rho(x, y) \exp(i\phi(x, y)) \tag{3}$$

(x, y) are spatial coordinates orthogonal to the incident beam direction of propagation. The reflected amplitude is then

$$A_{ref.}(x, y) = \rho(x, y) \exp(i\phi(x, y)) \times A_{inc}(x, y) \tag{4}$$

The fraction of the incident beam in the TEM_{mn} mode is

$$a_{inc}^{mn} = \langle TEM_{mn} \mid A_{inc} \rangle = \iint TEM_{mn}^{*}(x, y) A_{inc}(x, y)\, dxdy \tag{5}$$

In this modal representation the reflection factor is in fact an operator **R** which transforms the incident amplitude vector $\{a_{inc}^{mn}\}$ into a reflected vector. The reflection factor $r_{mn \to pq}$ between two modes TEM_{mn} and TEM_{pq} is thus

$$r_{mn \to pq} = \iint TEM_{mn}^{*}(x, y)\, TEM_{pq}^{'}(x, y)\, \rho(x, y) \exp(i\phi(x, y))\, dxdy \tag{6}$$

and the reflectance is

$$R_{mn \to pq} = r_{mn \to pq}\, r_{mn \to pq}^{*} \tag{7}$$

In the same way we have defined the reflectance $R_{mn \to pq}$ we define the transmittance $T_{mn \to pq}$ and the losses:

$$L_{mn \to pq} = 1 - R_{mn \to pq} - T_{mn \to pq} \tag{8}$$

It should be sufficient to define the reflectance and eventually the transmittance of a component, because all the energy that is not reflected nor transmitted is lost (for the experiment) and it may seem useless to describe the lost part of energy in detail. In fact we need a precise description of the various sources of losses in order to detect them, analyze them, and find technical ways to lower them. In what follow we will discuss on the various possible sources of losses, that when added together give the total losses L of the component.

n	l	$Z_{n,l}(r,\theta)$	$Z_{n,l}(x,y)$	aberration
0	0	1	1	piston
1	1	$r\sin\theta$	x	tilt about y axis
1	-1	$r\cos\theta$	y	tilt about x axis
2	2	$r^2\sin 2\theta$	$2xy$	astigmatism, $\pm 45°$ axis
2	0	$2r^2-1$	$-1+2x^2+2y^2$	focus shift
2	-2	$r^2\cos 2\theta$	$-x^2+y^2$	astigmatism, $0°$ or $90°$ axis
3	3	$r^3\sin 3\theta$	$3xy^2-x^3$	triangular astigmatism
3	1	$\left(3r^3-2r\right)\sin\theta$	$-x+3xy^2+3x^3$	3rd order coma along x axis
3	-1	$\left(3r^3-2r\right)\cos\theta$	$-y+3y^3+3x^2y$	3rd order coma along y axis
3	-3	$r^3\cos 3\theta$	$-3x^2y+y^3$	triangular astigmatism

Table 2. First Zernike polynomials

1.3 Surface figure and roughness

The distinction between surface figure and reflectance is not as sharp as could be thought, because of the definition we have adopted for reflectance. If one thinks of a metal mirror and has in mind the simple definition of reflectance based on the Fresnel parameters, then reflectance has nothing to do with surface figure and vice-versa. In this picture, the reflectance is only

$$R(x,y)=\rho^2(x,y) \tag{9}$$

and the surface figure

$$h(x,y)=\lambda\frac{\phi(x,y)}{2\pi} \tag{10}$$

but as we have seen in Eq. (3), surface figure is a part of reflectance. By using a Fizeau interferometer we obtain directly this phase on a cartesian grid $\phi(x,y)$. This phase information is translated in equivalent height and usually projected on the Zernike functions basis $\{Z_{t,u}(x,y)\}$ which is a complete orthogonal basis on a disc of radius 1.

$$z(x,y)=\sum_{t,u}\varrho_{t,u}Z_{t,u}(x,y) \tag{11}$$

Table (2) gives the Zernike polynomials in cylindrical and cartesian representation, as well as the classical aberration associated. The number of Zernike functions that are used to represent the surface figure is limited to a few tens, usually 36, because the weight of a given Zernike function decreases rapidly with its order. This is not a property of the Zernike functions, but of any real optical surface which exhibits a surface figure frequency spectrum (Fourier transform of the surface figure) having a power law behavior that spread over many decades of spatial frequencies, with a rather large exponent value close to -1.7. This value, that comes from experimental verifications, states that the most influent deformations of any optical surface are the ones with the lowest spatial frequencies, or the largest period. Curvature mismatch is usually the most influent surface defect, followed by the usual optical aberrations: coma, astigmatism,... and higher order aberrations are usually less influent. It is very important for gw detectors to characterize surface figure and to minimize the mismatching between the surface figures and the wavefront shapes because bad surface figure may be a source of considerable losses, and also it is very important, in order to have a good contrast at the output of the interferometer, that only the TEM_{00} mode propagates, because an important property of gaussian beams is that their phase shifting during propagation depends on their mode number. So if an interferometer is set up to work at the dark fringe of mode TEM_{00} it will usually not be at the dark fringe of other modes that will be able to leak out towards the detectors. When using a limited number of Zernike functions it is relatively easy to compute the losses of the TEM_{00} mode, induced by surface figure L_{sf}, once the local reflectance has been measured or evaluated, with the help of equations (6) and (11) :

$$
\begin{aligned}
L_{sf} &= 1 - \left\| r_{00\to 00} \right\|^2 \\
L_{sf} &= 1 - \left\| \begin{array}{c} \iint TEM_{00}^{*}(x,y)\, TEM_{00}^{'}(x,y)\, \rho(x,y) \\ \times \exp\left(\frac{2i\pi}{\lambda} \sum_{t,u} \varrho_{t,u} Z_{t,u}(x,y) \right) dxdy \end{array} \right\|^2
\end{aligned}
\qquad (12)
$$

The distinction between roughness and surface figure is purely artificial. Both terms refer to the modification of the wavefront shape of an incident beam upon reflection or transmission. Surface figure usually designates low spatial frequency deformations while the term roughness, or *surface finish* is used when speaking about high spatial frequencies. Roughness, like surface figure, is a source of loss, but it seems very cumbersome to compute it using equation (12) with an arbitrarily large number of Zernike functions. Moreover it is very interesting, from a practical point of view, to have a different number for characterizing the losses induced by roughness, because the control of surface roughness happens at a different level of polishing than the control of surface figure. Roughness is characterized by a remarkable property: it can be treated statistically, because the characteristic spatial period p_{sca} under which roughness is present is much smaller than the component diameter. We can thus obtain statistical results on roughness, independently of the large scale surface defects. In the case where the height distribution is a gaussian function with a white spectrum, and its standard deviation σ is much smaller than the wavelength,the losses can be approximated by

$$L_{sca} = TIS = \left(\frac{4\pi\sigma\cos\theta_i}{\lambda}\right)^2 \tag{13}$$

which is called *total integrated scattering* or TIS. Today the surface roughness in the spatial frequency range $[1\mu m^{-1}, 1mm^{-1}]$ has dropped down to $\sigma = 1$ rms (General Optics data). In the low spatial frequency region things are less clear because the quality of surface figure depends very strongly on the surface diameter, and it seems very difficult to obtain good reproducibility of performance. The best polishing results in surface quality between $\lambda/50$ and $\lambda/100$.

1.4 Absorption losses in coatings

Imperfections of the optical coatings may result in absorption of a fraction of the incident light. Absorption may be induced by the presence of impurities inside the materials of the coatings, or because of bad stoichiometry, or by contaminations at the interfaces. Absorption can be enhanced because of geometrical defects of the dielectric layers, be it gaps between the layers, structures inside the layers, or wrong thickness, that may change the electromagnetic field geometry. In the case where absorption is homogeneous over the whole surface it is represented by a scalar L_{abs} just like scattering losses. Absorption inside the coatings is a limitation to the ultimate value of reflectance that can be achieved.

1.5 Optical thickness

Optical thickness is the transmission counterpart of the surface figure. The transmission factor of a mirror can be written like its reflection factor of Eq. (3) as

$$t(x,y) = \tau(x,y)\exp(i\theta(x,y)) \tag{14}$$

The optical thickness is:

$$e(x,y) = \lambda\frac{\theta(x,y)}{2\pi} \tag{15}$$

The phase $\theta(x,y)$ depends on the thickness of the material that form the component, and on the value of the refraction index which can be inhomogeneous inside the substrate.

1.6 *Birefringence*

The most important source of loss due to polarization coupling comes from birefringence inside the coatings and we suspect mostly inside the substrates. If the direction of incident polarization does not coincide at each point (x, y) of the incident wavefront to the local direction of the principal axes, polarization coupling occurs and some energy is transferred for one polarization to the other. In the Virgo interferometer the P polarization is not used for detection, so the losses induced because of birefringence are given by

$$L_{biref}(x,y) = 1 - \cos^4(\gamma) + \sin^4(\gamma) + 2\cos^2(\gamma)\sin^2(\gamma)\cos\frac{\Delta}{2} \tag{16}$$

the worst case is when $\gamma = \pi/4$ in which case we have

$$L_{biref}(x,y) = \frac{1}{2}\left(1 - \cos\frac{\Delta}{2}\right) \approx \frac{\Delta^2}{8} \tag{17}$$

1.7 *Absorption and scattering losses*

Absorption and scattering losses are present inside the substrates as well as in the coatings. Coating scattering losses are mainly due to surface roughness while volume scattering involves static fluctuations of the material optical properties. As long as we do not deal with point defects, that we will treat later, the main mechanism of scattering, in glass substrates at the YAG wavelength, is Rayleigh scattering. Scattering as well as absorption losses can be simply represented by scalars that represent the fraction of energy that is lost by an incident beam after propagation through a thickness l of material:

$$P(z+l) - P(z) = (1 - \exp(-\alpha l))\,P(z) \tag{18}$$

The parameter α quantifies the losses that take place inside the substrate. Usually absorption and scattering losses occur anywhere in the material, losses of these kinds are very homogeneous, so we do not have to discuss about different losses for the different modes of the incident light and one scalar value is sufficient to represent any of those losses as long as they are small (as long as the first order Taylor expansion of the exponential in (18) is valid).

$$L_{abs} = \alpha_{abs} l \tag{19}$$

$$L_{sca} = \alpha_{sca} l \tag{20}$$

where l is the component thickness. The source of absorption losses is manifold in optical materials. Depending on the nature of the material the energy transfer between the incident beam and thermal agitation can have different causes. There are some well identified causes but in the overall many aspects of absorption are unknown, especially in materials like sapphire, or YAG, which have a more complex structure than fused silica, and which have been much less studied. In the case of fused silica it has been found that the main cause of absorption is the presence of OH^- ions and that the loss coefficient α depends linearly on the OH concentration.

1.8 Point like defects

Point defects are very local modifications of the components that cannot be handled in a statistical way like scattering, because they are accidents in the global structure of the component, and because they can be located everywhere, on the coating as well as inside the substrate. Point defects can be for example scratches on the coatings, dust particles on the surface or even inside the coating structure. Because they usually have a spatial extend much smaller than the incident wavefront, they couple the modes of the incident beam with very high order modes, so that the energy lost because of the presence of point defects can usually be treated as scattered light. But because they are few, the amount of scattered power strongly depend on their geometry, on their structure and on their position inside the component. Those losses, even if they are seen as scattering losses, depend on the beam geometry. If there are p defects in a component, the losses will be:

$$L_{def} = \frac{1}{P_i} \sum_{j=1}^{p} P(x_j, y_j) \, TIS_j \tag{21}$$

where P_i is the total incident power, $P(x_j, y_j)$ is the energy density at point (x_j, y_j) position of the defect and TIS_j is the total integrated scattering of the defect j. The value of TIS is a function of defect refraction index and geometry (especially scattering cross section).

2 Coatings manufacturing

Since the end of the forties, all-dielectric multilayer coatings are broadly used to make reflective as well as anti-reflective coatings. The development of laser boosted the need for high reflectivity mirrors, but the most stringent requirements came during the 70s with the laser gyroscope. This application demanded interference coatings with no scattering and negligible losses (in the 10^{-5} range). In order to respond to this challenging need, opticians developed a new fabrication method, called ion beam sputtering (*IBS*), which is used today, in a more elaborate form, to make the gw detectors optical coatings.

Thanks to IBS, between the middle of the 70's to the middle of the 90's the reflection factor of Rmax interference coatings has raised from 99% up to 99.9998%. Interferometric gw detectors need a wide range of reflectances, from the lowest anti reflective coatings to the highest Rmax mirrors. They also need reflectances in the intermediate range, for example 50% for the beamsplitter, that must be made with a great precision in order to maintain the symmetry of the interferometer. The design of all those coatings is a complex task but luckily gw detectors are made to work with monochromatic light, which simplifies a lot the problem (at least for high refection coatings) and makes the choice of multilayer coatings the most evident.

Transparent optical materials (in the visible and near infrared) have index of refraction in the range 1.35 (cryolite) to 2.45 (TiO_2). With monolayers it would be impossible to make efficient AR coatings with $R \ll 0.1\,\%$, or Rmax coatings with $R \gg 99.9\,\%$. The idea of multilayer coating is to enhance the effect of the index discontinuities by stacking many interfaces with as large as possible index discontinuities. In the most general case each layer can be made of a different material, and with a different thickness, so the calculation of reflectance and transmittance can be rapidly made very cumbersome. Mathematical tools based on the concepts of *optical* impedance Y and admittance have been developped since the beginning of the advent of interference coatings to make those calculations easy to handle.

2.1 Quarterwave stacks

The formulation of Y takes a very simple form when all the δ_x are set equal to $\pi/2$. In this case it is straightforward to show that the maximum reflectance of a multilayer stack is (when the only loss source is transmittance)

$$R_{\max} = \left(1 - \frac{n_H^2}{n_s}\left(\frac{n_H}{n_L}\right)^{2p}\right)^2 \left(1 + \frac{n_H^2}{n_s}\left(\frac{n_H}{n_L}\right)^{2p}\right)^{-2} \tag{22}$$

It is important to obtain the required high reflectivity with a small number of layers, mainly because each interface is a source of losses, and so the smaller the number of layers the smaller the scattering and absorption losses. Today we can say that the maximum acheivable reflectance is $R_{\max} \approx 1 - 10^{-6}$.

In order to realize anti-reflection coatings we need to make Y as close as possible to one. As we have seen, with a single layer we can already obtain a good AR coating. In order to decrease the reflectance below one tenth of a percent one possibility is to use two quarterwave layers. In this case we have

$$Y = \left(\frac{n_2}{n_1}\right)^2 n_s \tag{23}$$

In order to make a good AR coating we need to find two materials that match the equality

$$\frac{n_1}{n_2} = \sqrt{n_s} \tag{24}$$

There are in fact many ways to make AR coatings: we can use more layers, or non-quarterwave layers, or even graded index coating. Because of the lower number of layers in an AR coating compared to a high reflection coating, the performance of AR coatings are more sensitive to the thickness of the layers. It is important to keep in mind that it is thus much more difficult to make a good AR coating that to make a good reflector. The best AR coatings that can be made today do not have reflectances below a few 10^{-5}.

2.2 Halfwave layers

When the δ of one layer is set equal to π its admittance is simply the identity matrix Thus adding a halfwave plate in a multilayer coating has no visible effect on the reflectance. This property is used to strengthen coatings: the last layer of an interference filter is a halfwave plate, which is used to protect the coating from physical damage. Another very interesting property of this top layer, that we will discuss in more details later on in this course, is that even if its thickness is not exactly half of a wavelength it does not significantly change the reflectance of the coating, but it can change a lot the surface figure.

2.3 Coating process

Many different methods are currently used for making multilayer coatings, each one having its own set of advantages and drawbacks. In the case of gw detector optics the first goal is to obtain low total losses. The most suitable methods are based on the use of ion beams because they create layers having nearly the same optical properties as the bulk material. There are two basic ways to use ion beams. In the first case called *ion assisted deposition* or IAD, the material is evaporated and vapor deposits on the substrate. The ion source is present to correct the stoichiometry of the growing layer and to make it more compact. The second way to use on ion beam is to direct it towards the target to extract atoms of the species to be deposited, giving then a large kinetic energy, this method is called *ion beam sputtering.* The combination of those two methods gives *dual ion beam sputtering.* The process that has been chosen for coating manufacturing of gravitational wave detectors is ion beam sputtering with an assisted source, known as dual ion beam sputtering, or DIBS. This process produces the most efficient layers with the following properties:

- dense layers without any voids, no possibility of contamination
- perfect stoichiometry, the optical properties can be optimized thanks to the second beam
- not geometrical flaws, like columns
- good adhesion

Measurements show that DIBS produces bell shaped layers on a still substrate. In order to obtain homogeneous thickness of the layers it is necessary to move the substrate in front of the arriving atoms, so that each point of the substrate receives the same average number of atoms. This is achieved by placing the substrate on a system of two or three rotation stages with different rotation axes.

2.4 Recent improvements of the coating process

Both the amplitude $\rho(x, y)$ of the reflection factor and its phase $\phi(x, y)$, or equivalently its surface figure, depend on the geometry of the multilayer stack: the number of layers, the thickness of each of them, as well as their optical and mechanical properties. It seems impossible to control independently ρ and ϕ and what is done usually is to make a coating for a given value of ρ and rely on the shape of the underlying substrate to give ϕ. This approach strongly limits the control one can have on the surface figure. Once the coating has been deposited, its contribution on the phase can change a lot the surface figure from the original one given by the surface figure of the bare substrate. Inhomogeneous thickness of the layers, or internal stresses, can modify the radius of curvature of the mirror. The only way to correct this is to clean out the coating, repolish the substrate to a new shape, a then re-coat it, with the possibility that the new coating will give a different contribution to ϕ than the first one. A solution to this problem has been developped by Jean-Marie Mackowski.

2.5 Corrective coating, Mackowki's approach

It was emphasized that the half wave plate at the top of a multilayer coating has not effect on the reflection factor amplitude. Nevertheless its effect on the reflected phase may be important. The idea is thus to adapt locally the thickness of this last half-wave layer in order to control the phase of the reflection factor. This procedure requires two steps: first the mirror is made for a given value of ρ and only about 50% of the half-wave layer is deposited; then its surface figure is measured (with a Fizeau for example) and in a second step extra material is sputtered on the half-wave to correct the surface shape. This idea is in fact not new but Jean-Marie Mackowski is the first to have fully developped it and made $\lambda/100$ mirrors with this technique.

3 Optics testing

3.1 Surface figure and roughness

Surface figure testing is the oldest and most developped branch of optical metrology, for the reasons just mentioned before. It is not exaggerated to say that today 99% of surface figure testing is done using interferometers,be it with Newton, Fizeau, Twyman-Green or other setups (see references [2] and [3] for a review of those instruments). Fizeau interferometers seem to have the leading place in the multitude of different possible optical schemes, and in fact Ligo and Virgo mirrors will be, or have been tested at least with commercial Fizeau interferometers. One of the main advantages of these instruments is their ease of use, speed, and very high sensitivity. On the other hand Fizeau interferometers cannot perform absolute measurement of surface figure, and have thus a limited precision. Some instruments, not based on interferometry, have been developped to get rid of this difficulty.

Fizeau interferometers are among of the simplest and yet most powerful tools available to characterize surface figure as well as optical thickness. Phase shifting interferometry is a major improvement of classical laser interferometry that took place in the 80s. The mainframe of phase shifting interferometers is globally the same as that of a classical laser Fizeau. The main difference is that a phase modulator is fixed to the reference surface and is used to move the reference plane axially by a fraction of the laser wavelength. The reference/sample distance L is modulated by a PZT transducer and the recorded interferogram is time dependent. This is a fundamental feature of phase shifting interferometry, and a valuable advantage over standard interferogram analysis. In the phase shifting technique there is no need to study the fringe pattern of the interferogram since the result at a point is the arctangent of the ratio of intensities measured at that point. Even if no fringe is present in the interference field the wavefront phase difference can still be computed.

3.2 Absolute surface figure testing

The capability to perform absolute surface figure measurements is essential for using the corrective coating technique, but there is a gap between the performances of the standard surface figure testing tool, the Fizeau interferometer, and the required performances for Virgo. As an order of magnitude the best measurements that can be done with a Fizeau lead to a precision of $\lambda/50$ and a sensitivity of $\lambda/200$. The measurement precision is much worse than the sensitivity because this kind of interferometric measurement is based on the comparison of a reference surface and a sample surface, and it is quasi impossible to realize a reference surface with surface state specified to better than $\lambda/50$. Three solutions are: three flats test with a Fizeau interferometer; standard Fizeau measurement with a liquid flat, and profilometry with a reference direction. We will describe only the third method.

We can impose, since the beginning as a rule, the absence of any reference flat. The idea is to represent a surface, not by its height as a function of position, like was done in the first part of this course, but by its local gradient $\partial_{x,y}z$. The same information is contained in the gradient and in the topography apart from the average height and tilt that are useless. Instead of directly measuring the surface height $z\,(x,y)$ with respect to an arbitrary ideal mathematical plane, we measure the surface gradient. So instead of needing a reference surface we only need a reference direction, which is much easier to obtain. The setup scheme is presented on figure 1.

The light source is a 10 mW cw TEM_{00} Nd-YAG laser. This laser is used at the same time as a probe beam and as the reference direction. The beam is directed on the sample surface through a pentaprism mounted on a long translation stage. We use a pentaprism in order to be insensitive to any tilt induced by the mechanical translation stage. After reflection the beam is directed on an angle (a plane parallel Fabry-Perot interferometer) detector which is only sensitive to the beam tilt in the (x,z) plane. With this scheme we only measure the projection of the local surface gradient on this (x,z) plane and have to perform line scans with various orientations of the sample in order to obtain enough information to finally integrate the data and find the sample surface figure. One typical measurement on a two inches flat mirror is shown on figure 2. The main advantage of this method over the precedings is of course the absence of reference flat, but there are also lots of drawbacks.

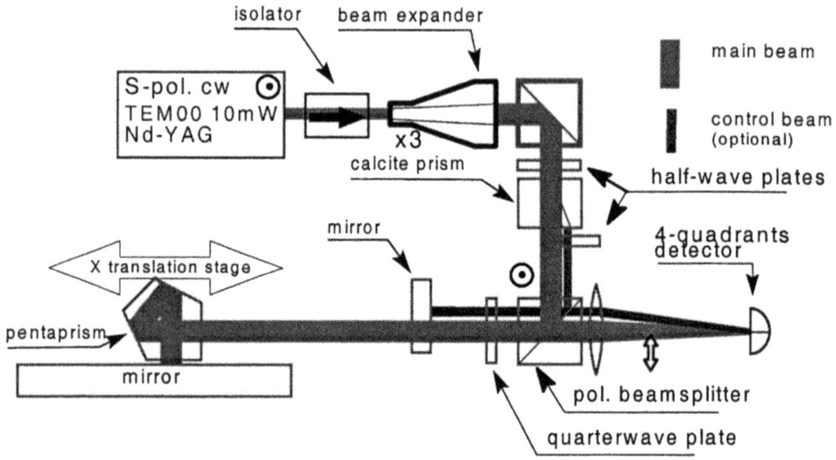

Figure 1. Scheme of the surface figure testing bench. The sample is holded on a θ_z rotation stage and on a y-translation stage. The bench is no more than a modified linear profilometer.

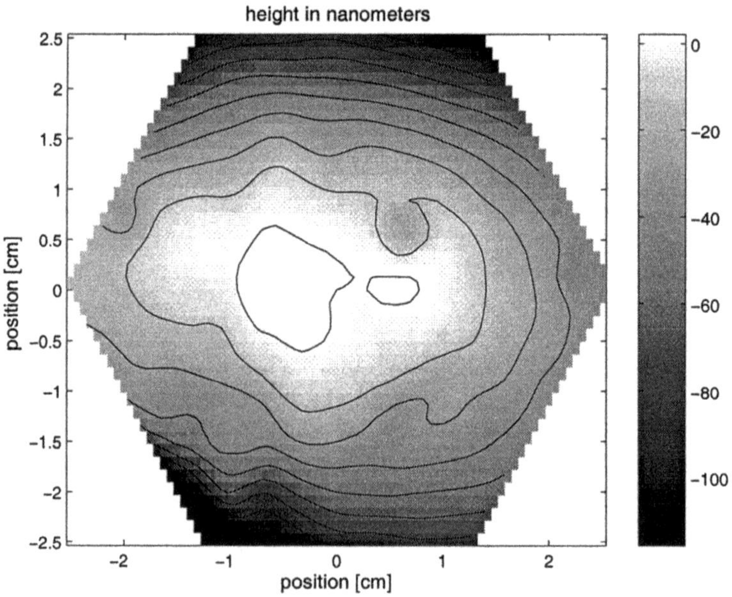

Figure 2. Surface figure of a 2" diameter Rmax flat mirror (exhibiting curvature error). The grey scale is in nanometers.

The main drawback is that the measurement at one point influences the whole reconstructed map. Contrarily to Fizeau measurements where at each point of the interferogram the height is determined by the field intensity at this particular location, gradient measurements require an iterative integration step to obtain the surface topography. This allows any measurement error to spread throughout the whole surface.

3.3 Absorption

Absorption of energy in coatings and substrates cannot be directly evaluated by measuring the intensity decrease of a light beam interacting with a mirror. Absorption factors and coefficients are so small (typically below the part per million range in the best fused silicas, and below a hundred part per million in sapphire or YAG) that it would require instruments with sensitivities and dynamical ranges impossible to achieve. The only way to measure absorption is to create a signal directly proportional to the amount of energy absorbed. There are different ways to perform this, and we will present the most classical one, namely the mirage bench. Even this setup can be modified in a variety of ways, the Virgo team and Ligo team both use tools based on mirage effect but their principle is very different. We will describe the most basic arrangement that can be used to measure either coating absorption factors, or substrate absorption coefficients.

It is well known that when a light beam crosses a medium inside which the index of refraction is not homogeneous, it does not follow a straight line but a trajectory that minimizes the optical path $n(\vec{s})\,l(\vec{s})$, where $\vec{s}$ is the curvilinear abscissa on the path l. This behavior is responsible for the mirages induced by a vertical gradient of the air index of refraction above the sun-heated ground. The temperature dependence of the index of refraction also exists in optical materials. The principle of the mirage technique is thus to heat a sample using a high power laser beam, to create a gradient of the material index of refraction, and to probe this gradient by measuring the deviation of a second laser beam that crosses the heated region. The deviation angle is proportional to the amplitude of the index gradient, which is in turn proportional to the temperature rise, which is proportional to the absorption properties of the material. As in this measurement we do not compute power ratios, the result depends very much on the setup geometry, on the beam powers and geometrical parameters, and on a number of thermal properties of the material under investigation. Because of all these parameters it is nearly impossible to perform accurate absolute measurements, and in practice the measurements are always done versus a reference sample of the same material.

Starting with the Fourier equation one can write the temperature rise induced by surface optical absorption as:

$$T_{s,(m)}(r,z) = T_{amb} + \varepsilon \frac{P_0}{2\pi} \int_0^\infty \frac{e^{-\left(\frac{\lambda^2 w_0^2}{8} + q_{s,(m)}(\lambda)|z|\right)}}{\kappa_s q_s(\lambda) + \kappa_m q_m(\lambda)} J_0(\lambda r)\, \lambda\, d\lambda \qquad (25)$$

with P_0 the total pump beam power, w its beam waist size and ε the absorption factor of the surface $(z = 0)$, $\kappa_{s,(m)}$ $\left(\kappa_{s,(m)} = D_{s,(m)} \rho_{s,(m)} C_{s,(m)}\right)$ are the thermal conductivities of the substrate (s) and the external medium (m). Then using the temperature dependence of the index of refraction

$$n_{s,(m)}(T) \approx n_{s,(m)}(T_{amb}) + \left.\frac{\partial n_{s,(m)}}{\partial T}\right|_{T=T_{amb}} (T - T_{amb}) \qquad (26)$$

in the eikonal equation

$$\frac{d}{ds}\left(n(\vec{r},t) \frac{d\vec{r_\perp}}{ds}\right) = \nabla_\perp n(\vec{r},t) \qquad (27)$$

one ends with

$$\frac{d}{ds}\left(n(\vec{r},t) \frac{d\vec{r_\perp}}{ds}\right) == \frac{1}{n_0} \int_{path} \frac{\partial n}{\partial T} \nabla_\perp T(\vec{r},t)\, ds \qquad (28)$$

The sample is irradiated by an intensity modulated focussed light beam. For the surface absorption measurement we use a Q-switched Nd-YAG laser delivering 100 ns pulses, 2.5 kW peak, at a frequency of 2 kHz. For volume absorption measurements the pump beam is a 3O W cw Nd-YAG laser modulated by an external mechanical chopper at 120 Hz. The pump beam is focused either on the surface or inside the mirror substrate. The periodic local temperature rise induces a thermal lens whose edges are probed by a second laser beam (4 mW cw He-Ne).

The iris prevents the pump beam from damaging the detector in case of bad alignment of the beams, and the interference filter stops any spurious beam or scattered light from the pump. The pump beam focusing length is placed on a motorized translation stage so as to change the pump-probe distance. The periodic deviation induced by the refractive index gradient is easy to measure with a silicon bicell and sensitivity of the order of $10^{-10} \sim 10^{-11}\mathrm{rad}/\sqrt{\mathrm{Hz}}$ can be achieved with 4 mW of probe beam power (shot noise). This non destructive approach is used to qualify raw test materials as well as final parts.

The experimental setup has been improved these last years by taking care of mechanical effects and of damping of air currents. Typical sensitivity at

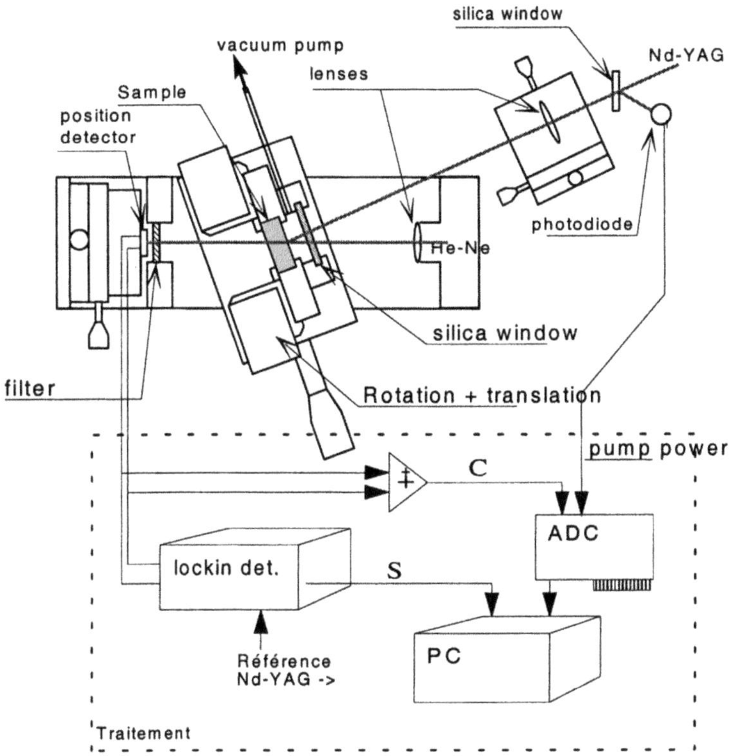

Figure 3. Mirage bench setup for surface measurements.

1.064 μm is 0.01 ppm for coating absorption and 0.2 ppm/cm for volume absorption in fused silica. Today HERAEUS company is able to prepare large samples whose absorption coefficient went down for 5 to 0.8 ppm/cm, whereas IPNL/SMA makes multidielectric coatings with 0.5 ppm losses.

A typical bulk measurement on silica is shown on figure 5 : the scan of the heated zone within the bulk of the silica sample show us the signature of the induced thermal lens. The setup is fully automatized it allows us to control the homogeneity of the absorption losses which is an important parameter for the VIRGO mode degradation.

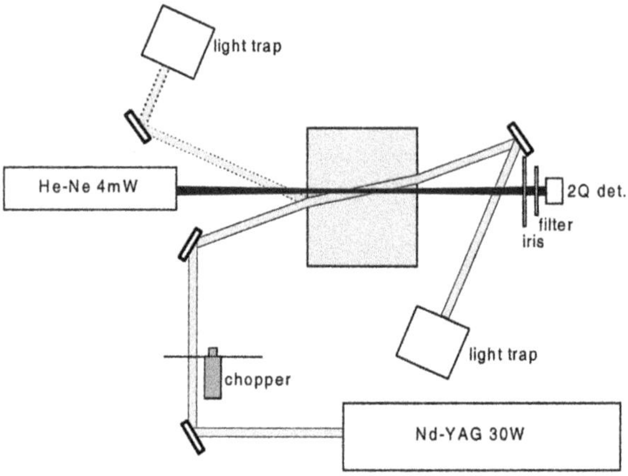

Figure 4. Optical setup of the mirage bench for volume absorption measurement.

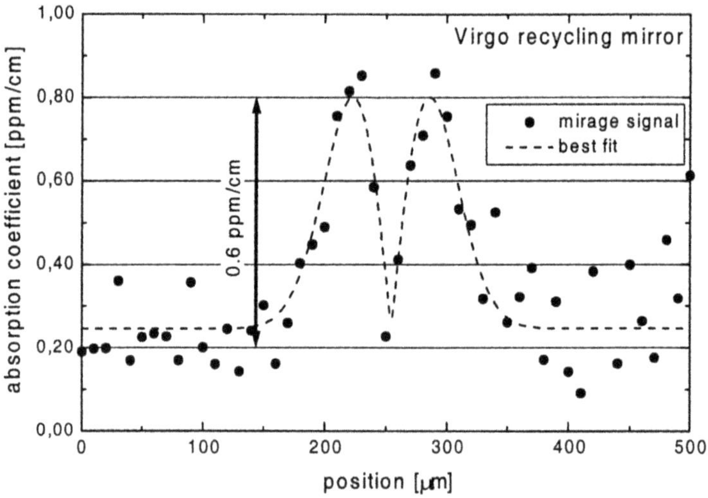

Figure 5. Volume absorption measurement in the Virgo power recycling mirror.

3.4 Reflection and total losses

Optical interfaces are usually characterized using average absorption, scatter, reflection and transmission factors. However, this set of parameters is not sufficient in experiments where the homogeneity of those interfaces is a limiting factor. This is the case for large interferometric gw detectors where the signal to noise ratio is degraded by inhomogeneities of the reflection factor of the various optical coatings. Virgo for example requires a relative homogeneity $\delta R/\overline{R}$ better than 10^{-4} on the reflectivity of beamsplitter, R=0.88, to R>0.99995 mirrors. It is very difficult to think of a bench capable at the same time to be sensitive to total losses $1-\overline{R}$ of Rmax mirrors like the F-P far mirrors, and sensitive to their homogeneity $\delta R/\overline{R}$. The simplest way is to use two different benches.

Measuring the reflectivity of a component is a classical test of optical metrology and there exist many efficient configurations that are simple to use. Measuring power ratios with more or less sophisticated power meters allows to measure reflection factors between a few per cent up to ninety-nine percent with an uncertainty of a fraction of a percent. As soon as the reflection factor exceeds ninety-nine percent it becomes uneasy to obtain a good precision this way because the value of transmitted power may be eventually comparable to the noise in the measurement of incident power. A common way of testing of high reflectivity components is to use a Fabry-Perot cavity and measure with an arbitrary method its transfer function [4,5]. One can for example measure the step response of a cavity and detect the exponential decay of transmitted intensity , or its response to a intensity modulated incident light and detect the phase lag between incident and transmitted light. The difficulty of those methods is that they require the cavity to stay resonant during the measurement [6], which may be uneasy to obtain with cavities having finesses higher that one hundred thousands (a common value today [7]). Nevertheless one can remark that, even if the complex transmission factor of a Fabry-Perot depends on the cavity microscopic length and its eventual offset from resonance, the storage time of energy inside the cavity does not, and only depends on the macroscopic length of the cavity. So by directly evaluating the storage time of energy in a Fabry-Perot cavity we have access to the average reflection factor of the mirror pair. We consider for simplicity the case of a plane parallel cavity , with a length L . By using incident exponential pulses of the form:

$$A_i\left(t\right)=A_0 H\left(t\right) e^{\frac{-t}{2\tau_i}} e^{i(2\pi\nu_{opt}t+\varphi)} \tag{29}$$

A_0 being the peak amplitude, H the heaviside step distribution, τ_i the pulse decay time, and ν_{opt} the optical frequency, and after some easy algebra, we

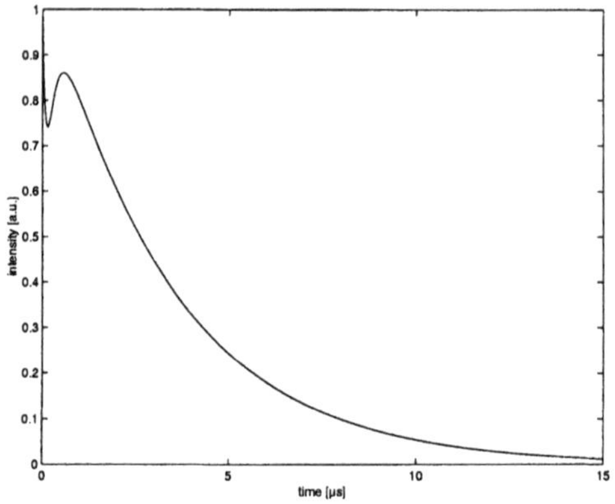

Figure 6. incident $\tau_i = 100ns$ exponential pulse and transmitted pulse shape for $F = 20000$, $\overline{R} = 0.9997$, $L = 30cm$

calculate the transmitted power as :

$$I_{tr}(t) = H(t)\, I_0 e^{\frac{-\tau_c}{2\tau_i}} T_1 T_2 \begin{bmatrix} e^{-\frac{t}{\tau_i}}\left(\frac{1}{2} + F^{-1}\left(\frac{1}{\chi^2} - \frac{1}{2}\right)\right) + \dots \\ +e^{-\frac{t}{\tau_p}}\left(\frac{1}{2} + F^{-1}\left(\frac{1}{\chi^2} - \frac{1}{2}\right)\right) - e^{\frac{-t}{2}\left(\frac{1}{\tau_i}+\frac{1}{\tau_p}\right)} \frac{2F^{-1}}{\chi^2} \end{bmatrix} \tag{30}$$

where F is the finesse of the cavity, τ_p is the storage time of the cavity, $\chi = \left(\tau_i^{-1} - \tau_p^{-1}\right) \times L/c$, and $T_{1,2}$ are the transmission factors of the two mirrors. This equation holds when the cavity is *not* kept on resonance but on the contrary when its length is left free to fluctuate. Figure (6) shows a typical pulse shape. The bump at the left side of the transmitted pulse is created by the filling of the cavity when it is resonant. Once τ_p is measured the calculation of the (geometrical) average of the two reflection factors is straightforward :

$$\sqrt{R_1 R_2} = \overline{R} = \exp\left(\frac{-L}{c\tau_p}\right) \tag{31}$$

The figure (7) shows such a measurement performed on a high finesse cavity. The precision of the measurement depends strongly on the cleanliness of the mirrors. With manual cleaning it is difficult to obtain reflection factors higher than 99.997%. For low reflection factors the precision is limited by the measurement of the cavity length L.

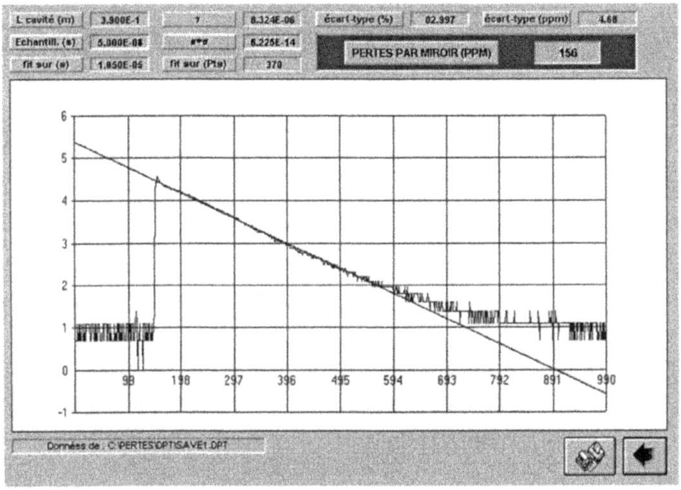

Figure 7. Transmitted pulse shape at the output of a $L = 39\,cm$ cavity with $\overline{R} = 1 - 156 \times 10^{-6}$, in log scale

For mirrors having a reflectivity lower than 99.99% the sensitivity limitation is dictated by the short storage time in the cavity and in practice it is difficult to evaluate τ_p when $\overline{R}$ is lower than 99.9%. On the contrary, when $\overline{R}$ is higher than 99.999% the sensitivity is limited by the signal level because of the T_1T_2 terms. Using a Nd-YAG TEM_{00} laser delivering $\tau_i = 100\,\mathrm{ns}$ 1kW peak pulses at a frequency of 1kHz on a 30 cm long half-confocal cavity, the best sensitivity ever obtained with this method was:

$$\sigma(1 - \overline{R}) = 10^{-6} \tag{32}$$

In order to measure the homogeneity of reflection factors over large (> 100 cm^2) surfaces, the use of a Fabry-Perot cavity is not the most clever choice. When testing very high reflectivity mirrors the Fabry-Perot alignment is not an easy task and should be left to local measurements, not to large mapping tests. As was mentioned earlier it is often more important to measure the variations of a parameter over the component surface, than to know precisely its average value. This is particularly true for reflectivity. Keeping this fact in mind we have developped a way to measure the relative variations of reflectivity of components with arbitrary reflectivity [8,9]. At this point it is important to define clearly what we call reflection factor of a surface. When we deal with laser sources, as it is the case here, we usually define the reflection factor as the ratio between the reflected power in a certain mode of the laser beam, over the incident power in this same mode, so that we should write :

$$R \equiv \frac{\langle U_i, U_r \rangle^2}{\langle U_i, U_i \rangle^2} \tag{33}$$

where Ui is the mode of the incident beam and Ur is the reflected field. The scalar product on the numerator can be written as:

$$\langle Ui, Ur \rangle = K \int_{surface} U_i^* U_i \rho(x,y)\, e^{i\phi(x,y)} dxdy \tag{34}$$

It depends on the local value of the reflection factor $\rho(x,y)$, but also on the surface figure $\phi(x,y)$. What is measured here is the local reflection factor $\rho(x,y)$, the bench is insensitive to the surface figure of the sample.

The bench is presented on figure 8. The light source is a stabilized 6 mW TEM_{00} Nd-YAG laser with a built-in optical isolator. The beam passes through a Glan-Thompson polarizing prism that defines the S-polarization plane. It then crosses a quarterwave plate with axes at forty-five degrees with respect to the S plane. After this quarterwave plate the light is thus circularly polarized. It enters the photoelastic modulator used as a dynamic quarterwave plate with the same axes as the previous one, so that after the modulator the polarization state is switched between S and P at the modulation frequency of 50 kHz. The polarizing beamsplitter cube is oriented so that the S-polarized part of the beam emerges at right angle to the incident beam, and the P-polarized part passes straight through the cube. The S-polarized beam is reflected by the reference mirror, the P-polarized beam by the mirror under test. The beam spot diameter on the mirrors is half a millimeter. Two quarterwave plates with axes at forty-five degrees with respect to the S plane are used to switch the two beam polarizations after two passes, and after a second pass through the cube, the two beams are focused on a photodiode. A lock-in amplifier detects the difference of intensity of the two beams. This signal is normalized by the average dc signal to give the reflection coefficient difference between the reference mirror and the sample. In order to avoid any spurious interference effect on the detector, and any Fabry-Perot effect inside the modulator, the beamsplitter cube, the modulator and the reference mirror are slightly tilted by a few degrees with respect to the beam direction. Interferences, if present, being the main noise source. The sample is fixed in a holder on a motorized translation stage. From a practical point of view it is important to emphasize that no component orientation is critical to the measurement so that the setting is very easy.

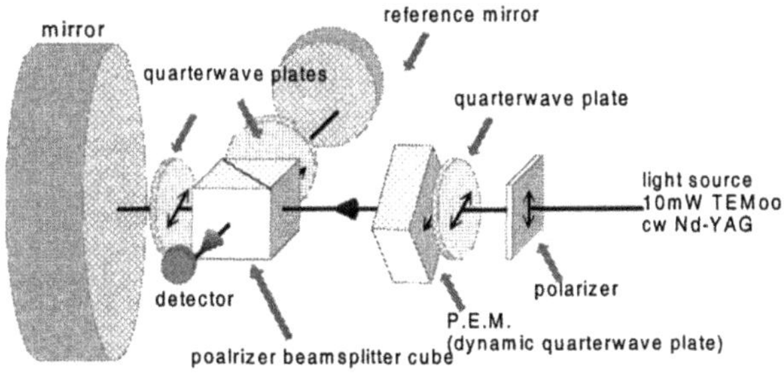

Figure 8. Optical setup of the reflectometer. The detection part (lock-in amplifier and computer) and the translation stages moving the sample are not represented.

The light intensity on the detector is given by

$$\begin{aligned} I_d &= I_L \left(\frac{(1-\varepsilon) R_{ref} + R(x,y)}{2} \right) - I_L \left(R_{ref} - R(x,y) \right) J_1(\psi) \sin(\Omega t) \\ &= I_0 + I_\Omega \sin(\Omega t) \end{aligned} \tag{35}$$

where ψ is the modulation depth and Ω is the modulation frequency and

$$\begin{aligned} R_{ref} &\equiv r_{ref}^2 \\ R(x,y) &\equiv r^2(x,y) \end{aligned} \tag{36}$$

So by defining $\delta R(x,y)$ by

$$\delta R(x,y) \equiv R(x,y) - \overline{R} \tag{37}$$

where $\overline{R}$ is the average reflection factor of the sample. In practice when testing mirrors one can expect $\delta R(x,y) \ll \overline{R}$ so that

$$\begin{cases} I_0 \approx \rho \left(\frac{R_{ref}+\overline{R}}{2} \right) \\ I_\Omega(x,y) \approx \rho \left(R_{ref} - \overline{R} - \delta R(x,y) \right) J_1(\psi) \end{cases} \tag{38}$$

The ratio between the dc term and the Ω term contains two parts

$$\frac{I_\Omega(x,y)}{I_0} \approx J_1(\psi) \frac{\Delta R}{\overline{R}} - J_1(\psi) \frac{\delta R(x,y)}{\overline{R}} \tag{39}$$

The second term contains the information on the sample relative reflection coefficient fluctuations

$$\frac{\delta R(x,y)}{\overline{R}} \approx \frac{\overline{I_\Omega} - I_\Omega(x,y)}{J_1(\psi) I_0} \tag{40}$$

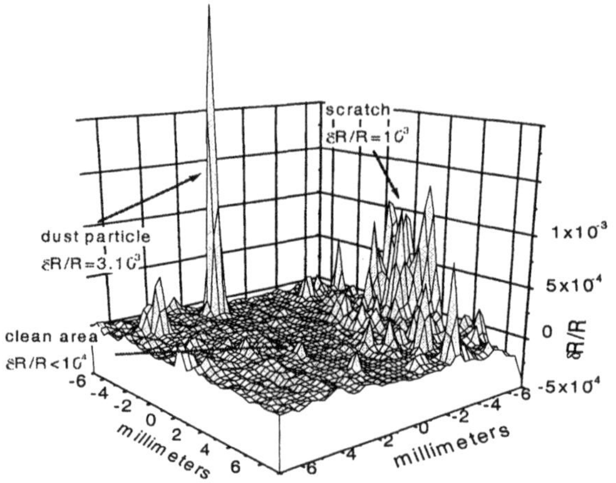

Figure 9. Homogeneity of the reflectance of a multilayer coating.

by using the lock-in amplifier we measure $S_{\Omega} = I_{\Omega}/\sqrt{2}$, the dc signal S_0 is simply measured with the help of a low pass filter and an adc converter

$$\frac{\delta R(x,y)}{\overline{R}} \approx \sqrt{2}\frac{\overline{S_{\Omega}} - S_{\Omega}(x,y)}{J_1(\psi)\, S_0} \tag{41}$$

Figure 9 shows a typical one inch long linear scan on a test Rmax mirror provided by SMA-IPNL. When interferences do not disturb the measurement the two main noise sources are beam wander in front of the sample, chiefly induced by air turbulence, and thermal drifts of the polarization components.

3.5 Point defects

One of the main sensitivity limitations of kilometer-scale gravitational wave detectors [10,11], will be the generation of scattered light, created by the microroughness of the various optical interfaces, and by point defects either situated on the surfaces or buried inside the substrates of the optical components. Scattered light may bounce on the interferometer arms walls, acquire phase noise because of the walls movements induced by seismic displacements, and eventually recombine with the main beam [12,13]. For the detectors not to be limited by this noise source, scattering losses should not exceed a few parts per million of the incident power on each optical component. It is thus a crucial test to characterize the scattering properties of the coatings and bulk

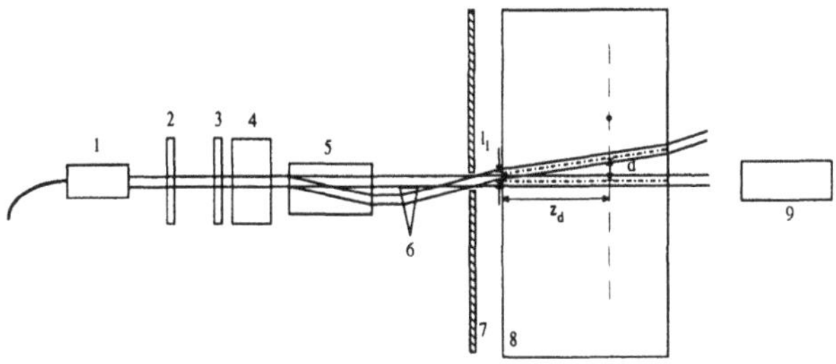

Figure 10. Optical setup: laser beam collimator (1), polarizer (2), quarterwave plate with axes at 45° with respect to the polarizer orientation(3), photoelastic modulator with the same axes orientation as the quarterwave plate (4), calcite beam displacing prism (5), small angle prism (6), light shield (7), sample (8), photomultiplier tube (9). The electronic part, lock-in amplifier and computer, as well as the translation stages are not shown.

materials. Surface microroughness can be investigated using standard scatterometers and other techniques. Scattered light created by mean surface roughness can thus be easily evaluated and today the scattering losses of optical coatings in the range of spatial frequencies between $1\,\mu m^{-1}$ and $1\,cm^{-1}$ have dropped down to the part per million level. This level of scattering losses will not degrade the sensitivity of detectors, unless roughness is not the only source of scattering. The presence of point defects, on the contrary, is difficult to detect, because one can not use a statistical approach. The optical properties of a quasi perfect component can be ruined by the presence of a single point defect, be it a scratch on the surface or an air bubble inside the substrate. On substrates that exhibit scattering losses of the order of one tenth of a ppm in defect free regions, one single defect can raise the total scattering losses by more than a factor ten to one hundred.

In order to find the three coordinates of the point defects we use a photomultiplier tube and detect laser light scattered by the defects. This measurement is insensitive to the nature and shape of the defect, only the precision is affected by the amount of energy that can be collected. We use an optical scheme with two laser beams as shown on figure 10.

In order to scan the whole volume, the component under test moves in the x direction and the bench is translated along the y axis. The light source is a 1 watt continuous argon laser ($\lambda_{Ar} = 514\,nm$) sent into a multimode fiber. At the output of the fiber the beam is collimated and S-polarized. The beam then passes through a system of a photoelastic modulator and a quarterwave plate with axes at 45°with respect to the incident polarization. The modulation depth is chosen so that the modulator behaves as a dynamic quarterwave plate. Thus the polarization plane of the transmitted beam is periodically switched between S and P. A calcite beam displacing prism separates the two polarization components and a small BK7 wedge ($\alpha_p = 11°22'$) prism gives a small tilt in the x-z plane to one of the two beams. At the output of this optical arrangement we thus obtain two coplanar linearly polarized intensity modulated beams that enter the component under test. The intensity modulations are 180°phase shifted. The calcite prism and small angle prism can be replaced by a Wollaston prism but it is easier for practical reasons to use two components instead of a single one. The beams diameter $2w$ can be chosen between 600 μm and 3 mm by changing the collimator at the output of the fiber. Light scattered from any beam by a point defect is collected by a photomultiplier tube and the signal is sent to a lock-in amplifier. The signal phase indicates which beam is affected by the presence of the defect. If a defect is present in the plane defined by the two beams then it will cross one beam at one particular position of the component, and then cross the second beam at a different position. Because of the angle between the two beams the distance between the two signals is proportional to the depth of the defect inside the component. If l_1 is the distance between the two beams on the sample front surface then the depth z_d of the defect is related to the distance d between the two signals by :

$$z_d = \frac{n\,(d - l_1)}{(n_p - 1)\,\alpha_p} \tag{42}$$

where n is the index of refraction of the sample, and n_p the index of refraction of the wedge. The value of l_1 is evaluated experimentally by using a silica reference sample with a bad surface quality. With this reference sample we obtain a lot of signals coming either from the front or back surface.

The figure 11 is an example of a measurement performed on the reference with 600

The measured value is $l_1 = 4.70 \pm 0.03\,\mathrm{mm}$. Using the following numerical values : $n_{p@514\mathrm{nm}} = 1.52049, n_{@514\mathrm{nm}} = 1.46156, \alpha_p = 198.39\,\mathrm{mrad}$ one obtains

$$\frac{z_d}{1\,\mathrm{mm}} = 14.154\left(\frac{d}{1\,\mathrm{mm}} - 4.70\right) \tag{43}$$

Using equation (42) we see that the z resolution depends on the d resolution :

$$\Delta z = z\Delta\alpha_p + \frac{n\,(\Delta l_1 + \Delta d)}{(n_p - 1)\,\alpha_p} \tag{44}$$

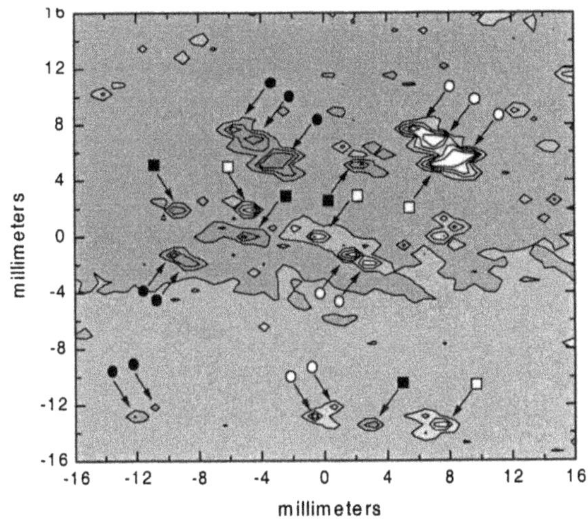

Figure 11. Defect detection on a bad surface quality reference plate. Each couple of peaks can be used to calculate the beams separation on the plate front surface. The peaks pointed by square symbols are created by defects on the front surface while the ones pointed by circles are on the back surface. The symbol color distinguish between peaks created by the untilted beam (white) or by the rotated beam (black).

in any case we have a depth resolution $\Delta z_{ref} < 500\,\mu$m in the reference. The resolution obtained here is a compromise between performance and measurement speed for 30 centimeter diameter samples, but it can be adapted to any need, and small samples can be scanned with a much better sensitivity.

Once the value of l_1 the sample is placed in the holder, with its front face at the same z coordinate as the reference front surface. The precision of this placement decreases the precision of z_d evaluation by $500\,\mu$m . The depth precision in the sample is thus:

$$\Delta z = 1\,\text{mm} \tag{45}$$

unless there is a defect that can be recognized as a surface defect, in which case the sample can also be used as the reference and $\Delta z = \Delta z_{ref}$.

In figure 13 we present the detection of a point defect inside a 10 cm thick transparent sample with a good polishing on both faces and no apparent surface defects. The depth of the defect, as calculated using equation (43) is 24.06 ± 1.2 mm, the origin of the coordinate system is arbitrarily defined by the position of the untilted beam at the center of the map.

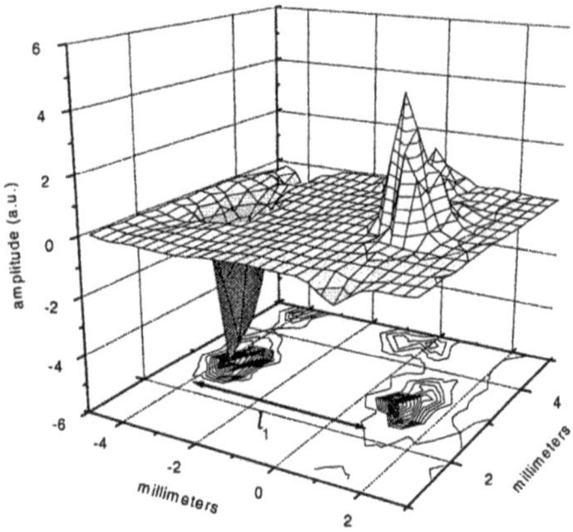

Figure 12. One couple of peaks is used to calculate the distance l_1 between the two beams on the reference front surface.

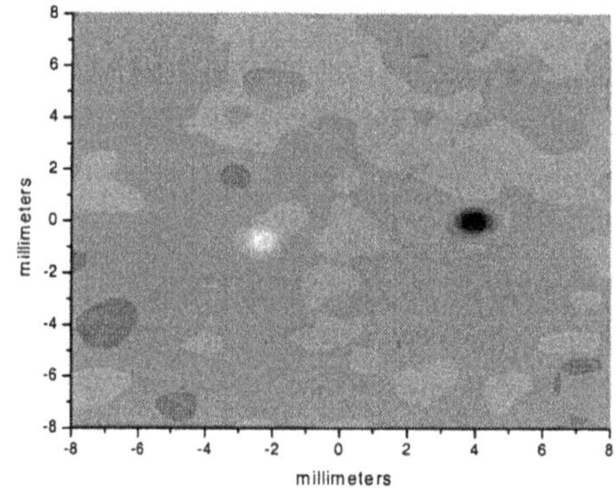

Figure 13. Detection of a point defect inside a superpolished synthetic fused silica substrate. The depth accuracy is 1.2 mm.

References

1. Virgo final design report, 1997.
2. *Selected Papers on Optical Shop Metrology*, SPIE Milestone Series vol. MS18, edited by D. Malacara, SPIE editor, 1990 .
3. Daniel Malacara, *Optical shop metrology*, John Wiley & Sons editors,

1992..

4. J.M. Herbelin, J.A. McKay, A. Kwok, R.H. Ueunten, D.S. Urevig, D.J. Spencer and D.J. Benard, "Sensitive Measurement of Photon Lifetime and True Reflectances in an Optical Cavity by a Phase-Shift Method," Appl. Opt. **19,** 144–147 (1980).
5. D.Z. Anderson, J.C. Frisch and C.S. Masser, "Mirror Reflectometer Based on Optical Cavity Decay Time," Appl. Opt. **23,** 1238–1245 (1984).
6. M. Billardon, M.E. Couprie, J.M. Ortega and M. Velghe, "Fabry-Perot Effects in the Exponential Decay and Phase Shift Reflectivity Measurements Methods," Appl. Opt. **30,** 344–350 (1991).
7. G. Rempe, R.J. Thompson, H.J. Kimble and R. Lalezari, "Measurement of Ultralow Losses in an Optical Interferometer," Opt. Lett. **17,** 363–365 (1992).
8. V. Loriette, "Test of Mirror Pair Reflectivity Coefficient Equality and Homogeneity for Interferometric Gravitational Wave Detectors," J. Optics **26,** 35–38 (1995)
9. V. Loriette and R. Nahoum, "High Sensitivity Reflectometer at 1.064 μm for Testing the Homogeneity of the Optical Components of Gravitational Wave Antennas," J. Opt **29,** 268–274 (1998)
10. A. Abramovici, W.E. Althouse, R.W.P. Drever, Y. Gürsel, S. Kawamura, F.J. Raab, D. Shoemaker, L.Sievers, R.E. Spero, K.S. Thorne, R.W.Vogt, R. Weiss, S.E. Whitcomb, and M.E. Zucker, Science **256,** 325 (1992).
11. C. Bradashia, R. Del Fabbro, A. Di Virgilio, A. Giazotto, H. Kautzsky, V. Montelaticci, D. Passuello, A. Brillet, O. Cregur, P. Hello, C.N. Man, P.T. Manh, A. Marraud, D. SChoemaker, J.Y. Vinet, F. Barone, L. Di Fiore, L. Milano, G. Russo, J.M. Aguirregabiria, L. Bel, J.P. Duruisseau, G. Le Denmat, Ph. Tourrenc, M. Capozzi, M. Longo, M. Lops, L. Pinto, G. Rotoli, T. Damour, S. Bonazzola, J.A. Marck, Y. Gourgoulhon, L.E. Holloway, F Fuligni, V. Iafolla, and G. Natale, Nucl. Instrum. Methods Phys. Res. A **289,** 518 (1990).
12. Jean-Yves Vinet, Violette Brisson and Stefano Braccini, "Scattered light noise in gravitational wave interferometric detectors: Coherent effects," Phys. Rev. D **54,** 1276 (1996).
13. Jean-Yves Vinet, Violette Brisson, Stefano Braccini, Isidoro Ferrante, Laurent Pinard, François Bondu and Eric Tournié, "Scattered light noise in gravitational wave interferometric detectors : A statistical approach," Phys. Rev. D **56,** 6085 (1997).

RESONANT BAR GRAVITATIONAL WAVE DETECTORS

M. VISCO
Istituto di Fisica dello Spazio Interplanetario, CNR, Via del Fosso del Cavaliere, 00133 Roma
E-mail: visco@ifsi.rm.cnr.it

L. VOTANO
Laboratori Nazionali di Frascati, INFN, Via Enrico Fermi, 00044 Frascati
E-mail: votano@lnf.infn.it

The resonant bar detectors have been operating reliably for many years. We describe their basic principles and review highlights of the design of the main detector subsystems. We report the current status of art, performances and sensitivity of the present running antennas. Some hints on future spherical detectors are also given.

1 Introduction

The first experiment to detect gravitational waves was conceived forty years ago by J. Weber [1]. The first pioneering detectors were cylindrical aluminum bars at room temperature, whose fundamental modes of oscillation at 1.6 kHz, excited by a gravitational wave, were monitored using piezoelectric ceramics. In 1969 Weber reported an anomalous number of "pulses" in coincidence between two detectors 1000 km apart. Furthermore the pulses appeared correlated with the sidereal time in a way which suggested a gravitational source at center of the galaxy.

Since then, many groups of scientists have devoted their effort to developing experiments with resonant mass detectors. The increase in energy sensitivity have been of many orders of magnitude and many detectors have been operating almost continuously for years, but no one has ever found results similar to those found by Weber.

The sensitivity of a resonant detector can be further increased both by improving the readout system of present detectors and constructing larger antennae of spherical shape.

2 Interaction between gravitational waves and bar detectors

Two freely falling masses make up the simplest system for measurements of a gravitational field. The two test bodies, during their motion, will follow two geodetics and the detection of a gravitational field will be possible by observing their relative acceleration.

If the two test masses at a distance η^m in a space crossed by a gravitational wave characterized by a metric perturbation h and there is no other force but the gravitational field, the time evolution is described by the so-called geodetic deviation equation. This equation for velocities small compared to that of light, in a local galileian system can be written in this form:

$$\frac{d^2\eta^i}{dt^2} = \frac{1}{2}\frac{\partial^2 h^i_k}{\partial t^2}\eta^k \tag{1}$$

This is the basic equation for any detector of gravitational waves.

If one of the two test bodies is at the origin of the coordinates and we consider displacements ξ^n small compared to the original position ($\eta^n \equiv \xi_0, \xi_x, d + \xi_y, \xi_z$), for a gravitational wave travelling along the x-axis in the TT gauge, Eq.(1) becomes:

$$\begin{cases} \ddot{\xi}_y = \frac{d}{2}\ddot{h}_+(t) \\ \ddot{\xi}_z = \frac{d}{2}\ddot{h}_\times(t) \end{cases} \tag{2}$$

The two bodies will exhibit a relative acceleration in a plane orthogonal to the propagation direction (transversal waves). The two polarization states h_+ and $h_\times$ will produce effects respectively along y and z axes. The acceleration will be proportional to the initial distance d of the masses.

2.1 Resonant detector

A resonant detector of gravitational waves is a system of two or more masses which interact with an elastic force and are free to move with respect to one another. In real experiments, the two free falling masses connected by a spring are substituted by a metallic bar, the longitudinal vibration of which is left free by hanging the bar on its center of gravity.

The free longitudinal displacement of a bar of length L made of a material having a sound velocity v_s is characterized by an infinite number of normal modes having resonance frequencies:

$$\omega_n = (2n+1)\omega_0 \qquad \text{with } \omega_0 = \frac{\pi}{L}v_s \quad \text{and} \quad \text{n} = 1,2.... \tag{3}$$

If we consider only the first longitudinal mode at frequency ω_0, a bar with length L and mass M is completely equivalent to an oscillator having two masses $m = \frac{M}{2}$ at a distance $d = \frac{4}{\pi^2}L$ connected by a spring with elastic constant $k = m\omega_0^2$. Moreover, in order to model the mechanical dissipations present in the bar, a frictional force proportional to $m\frac{\omega_0}{Q}$ must be introduced as a damper between the two masses.

The effect of a gravitational wave on a two-mass system can be modeled by introducing the tidal acceleration $\frac{d}{2}\ddot{h}$ given by (2). The equation describing the displacement ξ with respect to the resting position is the following:

$$\ddot{\xi} + \frac{\omega_0}{Q}\dot{\xi} + \omega_0^2\xi = \frac{d}{2}\ddot{h} \tag{4}$$

The solution of this equation can be found by means of the Fourier transform

$$\xi(t) = \frac{1}{2\pi}\int_{-\infty}^{\infty} H(\omega)T(\omega)e^{i\omega t}d\omega \tag{5}$$

where $T(\omega)$ is the transfer function of the oscillator and $H(\omega)$ is the Fourier transform of $h(t)$

$$T(\omega) = \frac{d}{2}\frac{\omega^2}{(\omega^2 - \omega_0^2) - i\frac{\omega_0}{Q}\omega} \tag{6}$$

Different solutions can be find for different shapes of h. Let us consider a signal δ pulse $h(t) = H_0\delta(t)$, which has been the primary aim for resonant gravitational wave detectors. The displacement of the oscillator is the following:

$$\xi(t) = -H_0\frac{d}{2}e^{-\frac{\omega_0 t}{2Q}}\omega_0 sin(\omega_0 t) \tag{7}$$

It is also interesting to consider a monochromatic excitation at the bar resonance $h(t) = -h_0 sin(\omega_0 t)$. In this case the response is given by:

$$\xi(t) = -h_0\frac{d}{2}Q sin(\omega_0 t) \tag{8}$$

These results, obtained for a two-mass system, can be extended to the bar in the limit previously given.

¿From (7) and (8) it is evident that a larger amplitude of oscillation is expected for a bar of larger dimension and the mechanical quality factor Q also has an important role in increasing the response.

2.2 Energy absorbed by the antenna

The total energy carried per unit time across a unit area by a gravitational wave can written as:

$$E_{wave} = \frac{1}{2\pi} \int_{-\infty}^{+\infty} f(\omega) d\omega \tag{9}$$

where $f(\omega)$ is the spectral energy density

$$f(\omega) = \frac{c^3}{16 \cdot \pi \cdot G} \left| \omega H(\omega) \right|^2 \tag{10}$$

In the analysis of the interaction between the antenna and the gravitational wave, it is fundamental to calculate the fraction of the energy carried by the wave (represented by $f(\omega)$), that becomes vibrational energy E_{vib} in the detector. A cross-section σ is usually introduced to model this process:

$$E_{vib} = \int_{-\infty}^{+\infty} \sigma(\omega) f(\omega) d\omega \tag{11}$$

An explicit expression for σ can be easily found[3] using (5) and (11). If we consider a plane-polarized wave impinging on the detector, $\sigma(\omega)$ is given by:

$$\sigma(\omega) = 2\pi \frac{G}{c^3} M v_s^2 \frac{\frac{\omega_0}{Q}}{(\omega - \omega_0)^2 + \frac{\omega_0^2}{Q^2}} sin^4\theta \; cos^2 2\phi \tag{12}$$

where θ and ϕ are the angles between the bar axis and the propagation direction and the polarization axis, respectively, of the wave.

This expression means that, whatever the shape of the gravitational wave, in order to have a stronger interaction with the bar, the detector must have the largest possible mass and should be made of a material having high speed of sound (i.e., once the frequency for the detector is fixed, its linear dimensions must be as large as possible).

The cross section is a function of the frequency which is sharply peaked at the resonant frequency of the bar. Because of this characteristic many people believe that resonant detectors are very narrow band apparatuses. This is wrong as σ represents the response to the signal while the bandwidth of a detector can only be calculated in comparison with the noise, which has also a resonant behavior (see par.4).

3 Detection apparatus

In the preceding paragraph we have analyzed the interaction between the gravitational wave and the antenna and calculated the mechanical energy produced in the bar. This vibration is transformed into a mechanical signal by the transducer, then amplified and sent through an ADC to a computer for the analysis.

3.1 The transducer

An electro-mechanical transducer transforms the mechanical vibrational energy of the antenna into electrical energy. The working principle of a transducer is very simple: the mechanical vibration modulates one parameter of an electromagnetic circuit, converting the signal from one form of energy into another one.

The electromagnetic energy to be modulated can be stored in the electric circuit in different forms:

- As an electric field in capacacitive and piezoelectric transducers.
- As magnetic field inside a coil in inductive transducers.
- As electromagnetic energy in optical transducers.

Another great distinction can be drawn between different classes of transducers on the basis of the energy gain $\beta = \frac{E_{el}}{E_{vib}}$ between the input and the output of the transducer.

- Active transducer if $\beta > 1$. Usually the energy gain is obtained by parametric amplification biasing the electromagnetic circuit with a time variable field.
- Passive transducer if $\beta < 1$. In this case the electrical circuit is biased by a dc field.

Most transducers are bi-directional devices, which means that they operate also backwards, transforming electrical signal in mechanical oscillations (although some are non-linear in this respect). This characteristic must be taken in account when the noise is considered. In particular, the transducer brings the electric noise of the amplifier back to the antenna, producing the so-called back-action.

As an example, we will describe the behavior of a capacitive transducer. This device consists of a variable capacitor, with an electrode rigidly connected to the antenna face, the motion of which produces a change of the capacitance C and consequently a modulation of the high electric field present in the gap. Let us consider a plane capacitor that has two electrodes with surface S at a distance D. If the transducer is biased by an electric field $E_0 = \frac{V_0}{D}$, the motion of the antenna $x(t)$ will change the initial distance D between the two faces producing a change of the potential:

$$V(t) = \frac{Q}{\epsilon_0 S}(D + x(t)) = V_0 + E_0 x(t) \tag{13}$$

With some elementary calculations one can get the coefficient β for a capacitive transducer:

$$\beta = \frac{\frac{1}{2}CV(t)^2}{\frac{1}{2}m\omega_0^2 x(t)^2} = \frac{CE_0^2}{m\omega_0^2} \tag{14}$$

¿From (14) it easy to see that to obtain a greater energy transfer it is necessary to increase the electric field in the gap and the capacitance. So we are interested in having gaps D as small as possible. Nowadays, transducers with gaps $D \approx 10\ \mu m$ are used.

3.2 Resonant transducers and readout circuit

It is well known that when a signal is transferred from a device to another, the energy at the connection point is in part transferred and in part reflected. The energy transfer can be maximized by a careful matching between the impedances. Actually, in a more correct approach one should analyze the condition separately for signal and noise, trying to maximize their ratio.

On the basis of these considerations, a transducer must be designed with an input mechanical impedance and an output electrical impedance matched to the antenna and the amplifier used.

Transformers are used for electrical matching and something similar is also possible in a mechanical circuit. The transducer is not connected directly to the bar, but through a small mass resonating at the same frequency of the antenna. The energy flows from the bar to this resonator that, having a smaller mass, oscillates with a larger amplitude. In this way the energy is conserved, but the mechanical impedance is changed. A transducer made as above is called a "resonant transducer".

A system made by a resonant transducer and an antenna can be modeled as two coupled oscillators; the transfer function presents two resonances at the frequencies of the mechanical normal modes (see Fig. 1).

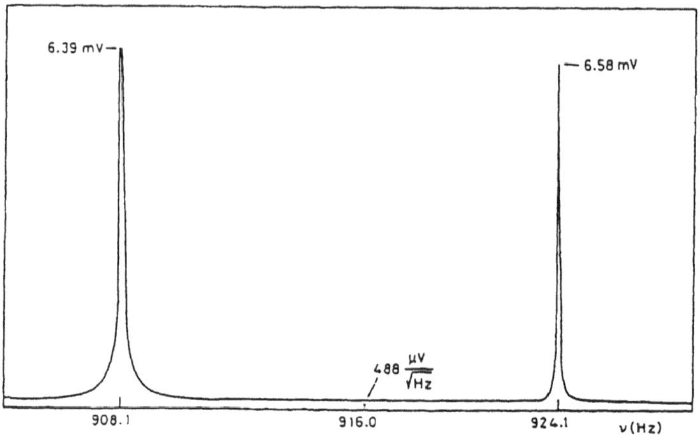

Figure 1. Output power spectrum of a bar detector equipped with a resonant transducer.

In order to reach optimum performance, the readout circuit must be designed according these principles. Fig.2 shows, as an example, the scheme of the readout circuit used for the experiments NAUTILUS and EXPLORER.

One face of the capacitive transducer is a mass of the order of $m_t = 0.1\ kg$ (the antenna equivalent mass is $m_a = \frac{2300}{2}\ kg$). The transducer capacitance is biased with a constant charge by two high value resistances R_p; the capacitor C_d avoids the discharge of the transducer.

The capacitive resonant transducer is connected through a low dissipation transformer to a dc SQUID (direct current Superconducting QUantum Interference Device) amplifier. A dc SQUID is a device based on two Josephson junctions. By its nature it is a very sensitive magnetometer. Introducing an input coil it can be used as electronic amplifier. At frequencies around $1\ kHz$ it has a noise energy, which approaches the quantum limit. Any linear device, in fact, is affected by a noise that, by quantum considerations, cannot be smaller than $T_n^{min} = \frac{h\nu}{K_B ln(2)}$.

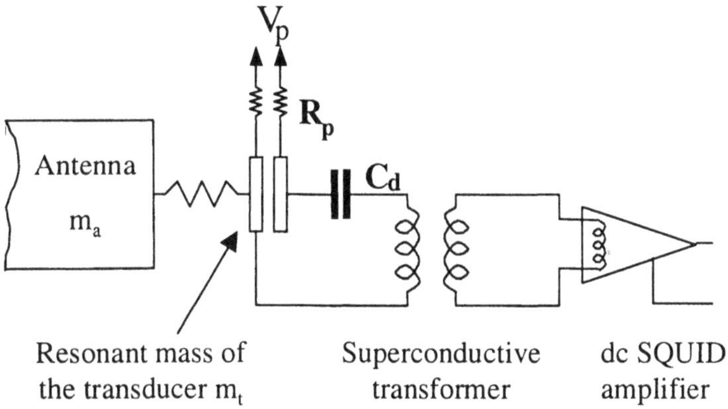

Figure 2. The readout used in EXPLORER and NAUTILUS experiments

4 Noise analysis

We can divide the sources of noise in two categories: external to the experimental apparatus and intrinsic to the detector design. The first class includes the noise produced by human activity and seismic noise, while the second one includes the thermal noise of the detector and the electronic noise of the amplifier used.

External noise in general is not stationary, non-Gaussian and cannot be easily modeled. Its effect can only be attenuated by mechanical filters and by a careful insulation of the apparatus.

On the other hand, the intrinsic noise is generally Gaussian and can be carefully modeled to optimize the design of the detection apparatus.

4.1 Thermal noise

Thermal noise is produced in the bar due to the chaotic motion of the atoms in the thermal bath at temperature T. It can be modeled considering the oscillator affected by a stochastic force with a mean square amplitude proportional to the temperature T, to the dissipation (represented by the mechanical quality factor Q) and to the mass m of the oscillator:

$$S_F(\omega) = 4\frac{\omega_0}{Q} m K_B T. \tag{15}$$

where K_B is the Boltzmann constant.

To reduce the brownian noise it is necessary to have a high mechanical Q value and to cool the detector. In the most recent experiments the resonant bar is cooled to temperatures lower than 100 mK.

4.2 Electronic noise

The noise of the amplifier can be expressed by two parameters: T_n and R_n. T_n is the minimum temperature noise reachable with a source impedance equal to R_n. In this case the power spectrum of the potential is given by $4K_BT_nR_n$.

T_n and R_n can be easily expressed in terms of voltage and current equivalent-noise generators at the input of the amplifier:

$$T_n = \frac{\sqrt{V_n^2 I_n^2}}{K_B} \ ; \ R_n = \sqrt{\frac{V_n^2}{I_n^2}} \tag{16}$$

The amplifier produces at its input a random electrical potential having a constant spectrum given by:

$$S_v(\omega) = K_BT_n\,|Z_{out}|\left(\lambda + \frac{1}{\lambda}\right) \tag{17}$$

where $\lambda = \frac{R_n}{|Z_{out}|}$ is the ratio between its optimum impedance R_n and the transducer output impedance Z_{out}. To obtain the best performance the value of λ should be around unity.

Due to the bi-directionality of the transducer, the amplifier noise also produces on the antenna a stochastic force with mean square amplitude given by:

$$S_f(\omega) = m\omega_0\beta\frac{K_BT_n}{\lambda} \tag{18}$$

4.3 Total noise

The total random force acting on the antenna is given by the sum of the back-action and of the brownian noise and can be written as:

$$S_f^{tot}(\omega) = 4K_Bm\frac{\omega_0}{Q}T_e \tag{19}$$

where $T_e = T(1 + \frac{\beta QT_n}{2\lambda T})$ is an equivalent temperature higher than the thermodynamic one.

The potential power spectrum (17), using the antenna transfer function (6), can be expressed as a random force on the antenna to be summed up to (19):

$$S_f^{el}(\omega) = \frac{K_B T_n \left(\lambda + \frac{1}{\lambda}\right) m\omega^2}{\beta\omega_0} \left[\left(1 - \left(\frac{\omega_0}{\omega}\right)^2\right)^2 + \left(\frac{\omega_0}{\omega}\right)^2 \frac{1}{Q^2} \right] \tag{20}$$

This force power spectrum has a narrow ($\Delta\omega = \omega_0/Q$) anti-resonance for $\omega = \omega_0$, while the brownian spectrum (19) is white.

Nevertheless we want to stress that when the noise is analyzed at the amplifier input, the electronic noise has a white power spectrum and the brownian noise has a narrow resonance ($\Delta\omega = \omega_0/Q$) at $\omega = \omega_0$.

The total noise force at the input of the antenna, obtained by adding (19) and (20), can finally be expressed in terms of the spectral density S_h of h,

$$S_h(\omega) = \frac{4K_B T_e \omega_0}{\frac{d}{2} m Q \omega^4} \left\{ 1 + \Gamma \left[Q^2 \left(1 - \frac{\omega^2}{\omega_o^2}\right)^2 + \frac{\omega^2}{\omega_0^2} \right] \right\} \tag{21}$$

where $\Gamma = \frac{T_n(\lambda + \frac{1}{\lambda})}{2\beta Q T_e}$ is a useful constant given by the ratio, at the resonant frequency ω_0, between electronic and brownian noise.

$S_h(\omega)$ has an anti-resonance at $\omega = \omega_0$ of which the half-height bandwidth with $\Gamma << 1$ and $Q >> 1$ is:

$$\Delta\omega \approx \frac{\omega_0}{Q\sqrt{\Gamma}} \tag{22}$$

S_h is equal to the spectral density of a gravitational wave that could be detected with a signal to noise ratio $SNR = 1$.

The expression (21) is not valid for resonant transducers, in this case S_h has two anti-resonances.

In Fig.3 is shown a measurement of S_h made on NAUTILUS detector.

5 Data filtering

If the power spectra for signal and noise are different, the noise to signal ratio can be improved by filtering the response of the detector in such a way to cut the spectral ranges where the noise is dominant over the signal.

The so called "optimum" filter is based on this principle. Its transfer function is given by $F(\omega) = \frac{T(\omega)^*}{S_h|T(\omega)|^2} = \frac{1}{S_h T(\omega)}$.

If an optimum filter is applied, for short bursts of gravitational waves $H(\omega) = H_0$, the SNR is given by:

$$SNR = H_0 \int_{-\infty}^{\infty} \frac{1}{S_h(\omega)} d\omega \tag{23}$$

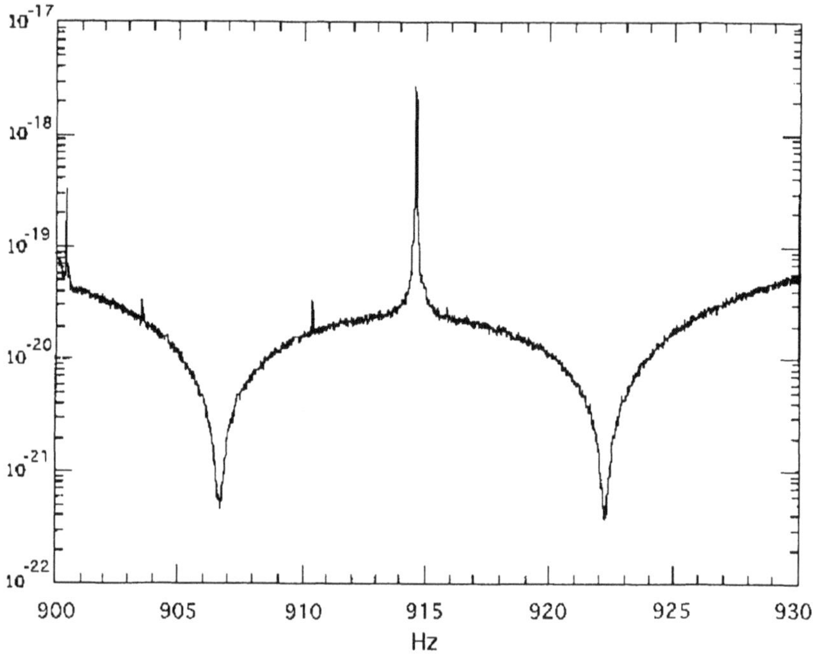

Figure 3. $\sqrt{S_h}$ measured on NAUTILUS experiment. The best sensitivity is reached at the frequencies of the two normal modes. The central peak is a calibration signal.

Since a real pulse cannot have a power spectrum constant over all frequencies, the integral (23) must be limited to a smaller interval dependent on the pulse length. If we limit the integration to the bandwidth (22), we obtain:

$$SNR = \frac{\frac{d^2}{2} m \omega_0^4 H_0^2}{K_b T_{eff}} \tag{24}$$

where $T_{eff} = 4T_e\sqrt{\Gamma}$.

The numerator of (24) is the energy released in the detector and therefore the denominator $K_B T_{eff}$ represents the noise energy of the antenna for short pulse detection.

T_{eff} is commonly used in the bar detector community to express the sensitivity of experiments.

Using the definition of T_{eff}, the bandwidth of the detector in the case of SNR=1, (22) becomes:

$$\Delta\omega_0 = 4\frac{\omega_0}{Q}\frac{T_e}{T_{eff}} \tag{25}$$

The bandwidth, as one might expect, is therefore connected to sensitivity, mechanical quality factor and temperature. For a detector with $T_{eff} = 10\ \mu K$ and $Q = 10^6$, working at a frequency $\omega_0 = 2\pi\ 900\ Hz$ and cooled at $T = 50\ mK$, which is a reasonable target, the expected bandwidth is around $100\ Hz$.

6 Status of present experiments

After many years of development, presently there are five cryogenic resonant detectors with comparable sensitivity. In Tab.1 the main characteristics of these detectors are reported.

The experiments use different transduction schemes. ALLEGRO[4] has a superconductive transducer and a dc SQUID as an amplifier. AURIGA[5], EXPLORER[6] and NAUTILUS[7] have capacitive resonant transducers and dc SQUID's as amplifiers. NIOBE[8] is equipped with a superconductive microwave parametric transducer. Two of these experiments, AURIGA and NAUTILUS, belong to the more recent generation of resonant detector with bars cooled to around $100\ mK$.

Table 1.

Experiment name	Location	Antenna mass (kg)	Thermodynamic temperature (K)	Sensitivity $\sqrt{S_h}$ $\left(\sqrt{Hz}\right)$
ALLEGRO	Baton Rouge Louisiana - US	2296	4.2	10^{-21}
AURIGA	Legnaro Italy	2230	0.25	$2 \cdot 10^{-22}$
EXPLORER	CERN Switzerland	2270	2.6	$6 \cdot 10^{-22}$
NAUTILUS	Frascati Italy	2260	0.1	$2 \cdot 10^{-22}$
NIOBE	Perth Australia	1500	5	$8 \cdot 10^{-22}$

It is fundamental to have more than one experiment contemporarily working, since local disturbances could produce a few signals on a single detector that could be considered as candidates for gravitational wave pulses. Even if these events are quite rare, there is no way to eliminate this background, other than to make a coincidence analysis between more than one detector.

All the resonant mass detector groups formed in 1997 the International Gravitational Event Collaboration (IGEC)[9] and recently, preliminary results on the coincidence analysis using data from the five detectors have been reported[10].

In the next paragraphs we will present the experiment NAUTILUS and some results obtained with this detector.

7 The Nautilus Experiment

A schematic layout of the experimental apparatus of NAUTILUS is shown in fig.4.

The antenna is an aluminum bar, 2350 kg in mass, 3 m long, and 0.6 m in diameter. It is suspended from a massive copper ring by a high-conductivity copper cable wrapped around its central section. The bar is cooled through this cable by a special ^{3}He-^{4}He dilution refrigerator with a cooling power of 1 mW at 100 mK. Six shields suspended from one another surround the antenna and have two roles: they provide the proper thermal insulation and constitute the different stages of mechanical filters with an overall attenuation of -260 dB at 900Hz.

In addition to the refrigerator, part of the apparatus is cooled by a reservoir of 2000 l of liquid helium. The antenna and the refrigerator are kept at a pressure of 10^{-7} mb.

The cool-down of the detector goes through different steps; 3 weeks and 8000 liters of liquid nitrogen are necessary to reach 77 K; one week is needed to achieve 4.2 K using about 5000 liters of liquid helium. Afterwards, in few days the antenna can be cooled below 100 mK by means of the dilution refrigerator. The lowest temperature achieved was about 90 mK on the bar end face and 45 mK on the mixing chamber of the refrigerator.

The antenna is equipped with a capacitive resonant transducer and a dc-SQUID amplifier. The proper impedance matching is provided by a superconducting transformer (see Fig.2). The output signal is monitored around the two frequencies $\nu_- = 908.34\ Hz$ and $\nu_+ = 923.81\ Hz$. The output of the dc SQUID, amplified by a FET amplifier is sampled at 220 Hz and sent to a computer. The data are analyzed using three different techniques: the standard Wiener filter with fixed parameters, an adaptive Wiener filter which

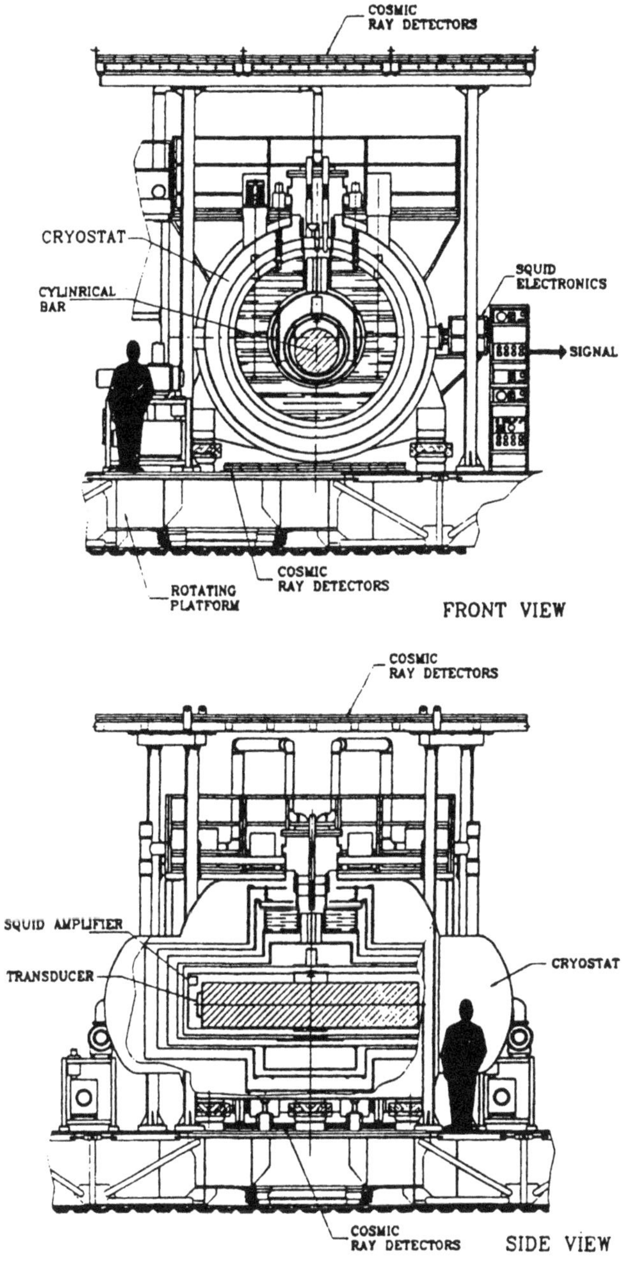

Figure 4. Schematic layout of the NAUTILUS experiment

improves the overall sensitivity by taking account the non-stationary nature of the noise, and an adaptive matched filter.

NAUTILUS is also equipped with a cosmic ray detector[11]; this is necessary because Extensive Air Showers (EAS) or energetic single particles (muons or hadrons) interacting in the antenna may produce detectable signals whose rate increases with increasing sensitivity of the antenna

The cosmic ray detector system of NAUTILUS consists of seven layers of streamer tubes for a total of 116 counters. Three superposed layers, each one of 36 m^2 are located over the cryostat. Four superimposed layers are under the cryostat, each one with an area of 16.5 m^2. The signal from each counter is digitized to measure the charge which is proportional to the number of particles. For an Extensive Air Shower (EAS) the efficiency is close to 100 %, but the systematic error on the absolute number of particles crossing the apparatus is of the order of 30%. In addition, saturation begins to show for multiplicities greater than 100 $particles/m^2$.

7.1 Measurements in coincidence with the cosmic ray detector

High energy cosmic-ray events could excite mechanical vibrations in the bar. The mechanical vibrations originate from the local thermal expansion caused by the warming due to the energy lost by the particles crossing the material. The effect depends on the ratio between the thermal expansion coefficient and the specific heat of the material, called the Grüneisen coefficient.

Calculations[12,13,14,15,16] made by several authors agree in predicting the following formula, for the vibrational energy in the fundamental mode of vibration of an aluminum cylindrical bar:

$$\epsilon = 7.64 \cdot 10^{-9} W^2 f[K] \tag{26}$$

where W(in GeV) is the particle energy dissipated in the bar and f is a geometrical factor of the order of unity.

The analysis[17] performed using the NAUTLUS data collected during a total time of 47.7 days has shown for the first time a clear correlation between the passage of a high-multiplicity EAS in an ultra-cryogenic antenna and the mechanical excitations of the antenna. In this analysis, a threshold of multiplicity $M \geq 10^4$ in the lower detector has been imposed, which corresponds to a signal of the order of 1 mK expected in the antenna NAUTILUS. For each EAS event, 20000 samples of Nautilus data, processed by a matched filter, were taken centered at the time when the number of particles exceeding the selected threshold crossed the cosmic-ray detector. 93 stretches of data during a total time of 47.7 days were selected. The data were averaged superimposing the selected stretches at the same relative time with respect to the EAS event. In this way the variance of noise should decrease with the number of stretches.

In fig.5 the result of the analysis is reported.

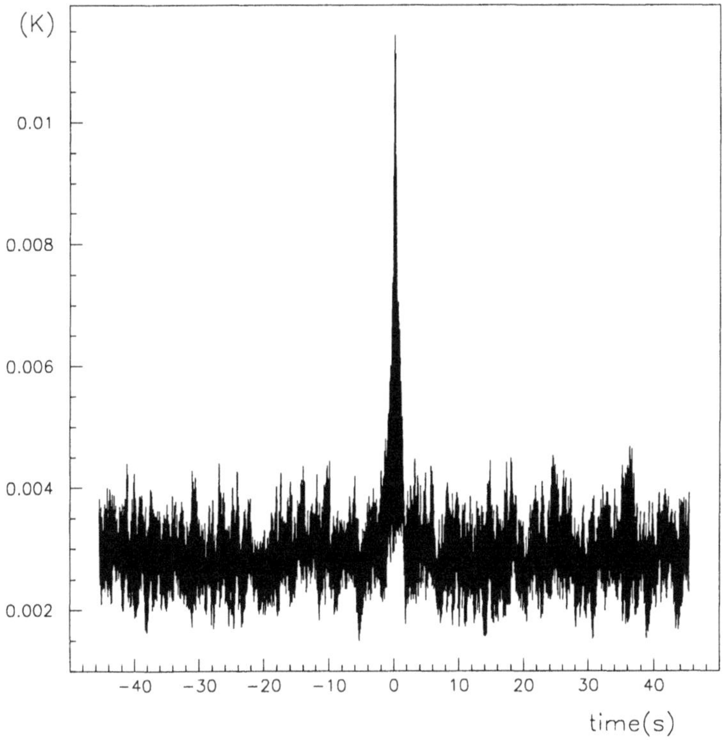

Figure 5. The averaged energy over 92 stretches of NAUTILUS data versus time. The large signal corresponds at the cosmic ray arrival time

The experimental response confirms the calculations on the cosmic ray effect made by various authors, both in terms of rate of occurrence and in terms of amplitude of the cosmic ray interaction with the resonant detector. Also very important are the experimental demonstration of the overall well-functioning of the apparatus under physical impulsive excitations, and in particular, the verification of the performance of the filtering algorithm.

8 Future resonant detectors

Recently some groups[19] have started to study the possibility of building large spherical detectors with diameters of 3 *m* and masses as large as 40 *tons*. A single sphere exhibits 5 modes of oscillation with the same resonance frequency, that are excited by waves coming from any direction and with any polarization. Therefore, a single sphere acts as many cylindrical bars oriented in different directions. Moreover the equivalent mass of the 5 degenerate modes of a sphere is larger by an order of magnitude than that of a cylinder working at the same frequency.

So a spherical detector is much more sensitive and powerful than a bar detector.

Some experiments have confirmed the possibility of using spherical detectors to reconstruct the propagation direction of a gravitational wave[20].

Spherical detectors are therefore very promising apparatus and complementary to large laser interferometers.

Acknowledgments

We would like to thank all our colleagues of ROG group and particularly E.Coccia, G.Modestino and F.Ronga.

References

1. J.Weber,*Phys. Rev.* **117**, 306 (1960)
2. G.Pizzella, *Fisica Sperimentale del Campo Gravitazionale*, La Nuova Italia Scientifica, Roma, 1993.
3. C. W. Misner, K.S.Thorne, J A. Wheeler, *Gravitation*, W.H.Freeman and Company, New York, 1973.
4. E.Mauceli *el al.*, *Phys. Rev.* D **54**, 1264 (1996).
5. G.Prodi *el al.*, in *Gravitational Wave, Proceedings of the 2nd Amaldi Conference*, Editors. E.Coccia, G.Pizzella and G.Veneziano, World Scientific, Singapore,1998.
6. P.Astone *el al.*, *Phys. Rev.* D **47**, 2 (1993).
7. P.Astone *et al.*, *Astroparticle Phys.* **7**, 231 (1997).
8. D.G.Blair *el al.*, *Phys. Rev. Lett.* **74**, 1908 (1995).
9. Information on IGEC at http://igec.lnl.infn.it/igec.

10. Presented at the 4th Gravitational Wave Data Analysis Workshop, Roma (Italy), 2-4 december 1999.
11. E.Coccia, A.Marini, G.Mazzitelli, G.Modestino, F.Ricci, F.Ronga and L.Votano, *Nucl. Instrum. Methods* A **333**, 555 (1994).
12. A.M.Allega and N.Cabibbo, *Lett. Nuovo Cimento* **38**, 263 (1983).
13. C.Bernard, A.De Rujula and B.Lautrup, *Nucl. Phys.* B **242**, 93 (1984).
14. A.De Rujula and S.L. Glashow, *Nature* **312**, 734 (1984).
15. E.Amaldi and G.Pizzella, *Nuovo Cimento* **9**, 612 (1986).
16. G.Liu and B.Barish, *Phys. Rev. Lett.* **61**, 271 (1988).
17. P.Astone *et al.*, *Phys. Rev. Lett.* in print.
18. E.Coccia, in *Proceedings of the 14th International Conference on General Relativity and Gravitation*,Editors M.Francaviglia, G.Longhi, L.Lusanna, E.Sorace, World Scientific, Singapore,1997.
19. *Proceedings of the first international workshop on Omnidirectional gravitational radiation observatory*, Editors Velloso W.F., Aguiar O.D. and Magalhães N., World Scientific, Singapore, 1997.
20. S.M.Merkowitz and W.W.Johnson, *Phys. Rev.* D **53**, 5377 (1996).

AN OPTICAL TRANSDUCER FOR BAR DETECTORS

F. MARIN, M. DE ROSA
Dipartimento di Fisica, Università di Firenze, LENS, and INFN, Sez. di Firenze, L.go E. Fermi 2, I-50125 Firenze, Italy
E-mail: : marin@lens.unifi.it

L. CONTI, G. PRODI, S. VITALE
Dipartimento di Fisica, Università di Trento, and INFN Gruppo Coll. Trento, I-38050 Povo (Trento), Italy

M. CERDONIO, L. TAFFARELLO, J.P. ZENDRI
INFN Sez. di Padova and Dipartimento di Fisica, Università di Padova, Via F. Marzolo 8, I-35100 Padova, Italy

We present a new kind of transducer for gravitational waves bar detectors. The bar vibrations are measured by means of a mechanically coupled optical cavity whose length is compared with that of a reference cavity by means of a laser system. We describe in details the transduction chain, we analyze the noise sources and the achievable sensitivity and present some experimental steps for the system implementation.

1 Introduction and working principle

Ultracryogenic resonant detectors of gravitational waves (g.w.) have reached a very high peak spectral sensitivity, thanks to the extremely low level of thermal noise [1]. In order to further increase the burst sensitivity and fully exploit the detection capabilities an improved transduction chain must be implemented for translating the weak bar excitation into a readable signal with the smallest added wideband noise.

The possibility to implement an optical signal extraction chain has already been explored in the past [2]. In the framework of the AURIGA collaboration we are developing a full readout for an ultracryogenic resonant bar detector. The AURIGA detector is based on a 3 m long, 0.6 m diameter cylinder made with Al5056, an alloy which permits to obtain an high mechanical quality factor (Q) at low temperatures. The bar is cooled down to about 0.1 K, where it is characterized by a vibration frequency (first longitudinal mode) of 920 Hz and a Q of about 10^7.

The basic idea for an optical transduction system is to build a Fabry-Perot cavity by means of a mirror attached to the bar end face and a second mirror attached to a resonant mechanical transducer. The relative motion of the two

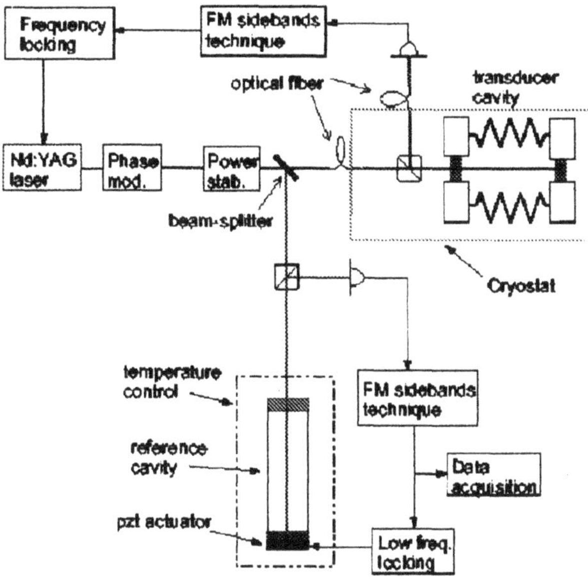

Figure 1. Scheme of the opto-mechanical readout for g.w. resonant detectors.

mirrors is then converted into a frequency shift of the cavity optical resonance frequency. The mechanical resonator is an oscillator formed by a load on an elastic plate, made from a single piece of Al5056. The system composed by the bar and the resonant transducer can be viewed as two coupled harmonic oscillators. If their oscillation frequencies are the same, the bar vibration amplitude is amplified by a factor equal to the square root of the oscillators mass ratio.

The complete scheme of the readout chain is shown in Fig. 1 [3]. It is based on the use of two Fabry-Perot cavities which we will call transducer and reference cavity. The former is installed on a bar end face, as already described, while the latter is kept in an acoustically isolated and thermally stabilized environment.

A Nd:YAG laser is frequency locked to the transducer cavity by means of a FM sidebands technique [4] and its frequency fluctuations are monitored using the same technique with the reference cavity, which acts as frequency reference. A single-mode polarization maintaining optical fiber is used to carry the laser radiation into the bar vacuum chamber. The beam reflected by the transducer cavity, after an optic circulator, is collected by a second fiber and detected outside the bar chamber.

2 Noise sources and sensitivity

The bar and readout system can be considered as a transduction amplifier with a frequency response corresponding to the mechanical transfer function of the coupled oscillators, i.e., with two resonances $\omega_{\pm}$ around 1 kHz and an high Q.

For the calculation of the sensitivity the noise sources can be divided into two classes: narrow-band and wide-band noise. The former corresponds to noise sources acting at the level of the bar or of the resonant transducer. This noise is amplified roughly like the signal, according to the system transfer function, and is seen at the output as a narrow-band noise. This kind of noise limits the peak spectral sensitivity, measured at the $\omega_{\pm}$.

Far away from these frequencies, the narrow-band noise (and the signal) measured at the output become lower and eventually reaches the plateau given by the wide-band noise. This letter class of fluctuations is due to the noise sources acting after the resonant components of the transduction chain and consequently are linearly amplified. The wide-band noise level determines the useful bandwidth of the detection system. Indeed, the spectral signal-to-noise ratio remains roughly the same (limited by the wide-band noise) until the wide-band plateau is reached.

The total strain sensitivity is obtained by integrating the spectral sensitivity and thus depends on both narrow-band and wide-band noise. The performances of the g.w. detector can be given in terms of the equivalent strain noise amplitude Shh, considered at the detector input (i.e., at the level of the bar), that would produce the observed output noise power spectrum (see the contributions by L. Votano and by M. Visco in this volume).

In the case of the optical transduction chain, the narrow-band noise is given by the thermal noise of the bar and of the resonant mechanical transducer and by the back-action. The thermal noise can only be reduced by decreasing the temperature and increasing the Q. Even if in principle the readout chain does not influence the thermal noise, on must take great care not to reduce the Q with the mechanical components installed on the bar and not to heat up the system with the laser radiation. These requirements give probably the most difficult technical problems.

The other narrow-band noise source is the back-action: the amplitude fluctuations of the laser radiation stored in the transducer cavity are transformed into displacement noise by the radiation pressure effect on the cavity

Cavity	Finesse	Lenght (cm)	Laser power (mW)
Transducer	300000	1	2
Reference	40000	20	5

mirrors. In order to reduce this effect the laser intensity noise must be minimized (for frequencies around 1 kHz). The lower limit, for a classical laser source, is given by the radiation shot noise.

Different sources of 'technical' noise contribute to the wide-band fluctuations: acoustic and thermal noise in the reference cavity, phase fluctuations induced by Doppler effect in the alignment mirrors and optical components, fluctuations of the refraction index of the fiber and of the air, etc. These effects can in principle be reduced to negligible levels. A more basic limitation comes from the laser amplitude noise which reduces the sensitivity in the measurement of the detuning between laser and Fabry-Perot cavities. Again, the fundamental limit is given by the laser shot-noise. We stress that while for the back-action what is important is the amplitude noise around 1 kHz, for the detection sensitivity the important frequency is determined by the FM sidebands and is typically in the rf region.

If a shot-noise limited detection is obtained, the mirrors displacement sensitivity is proportional to $F\sqrt{P}$, where F is the cavity Finesse and P is the laser power, while the laser frequency displacement sensitivity is proportional to $F\sqrt{P}/L$ where L is the cavity length. For the transducer cavity, the Finesse is limited by present mirror construction technologies. Together with an high reflectivity, a low absorption is necessary for avoiding heating up the mirrors. The usable laser power is limited by the same thermal considerations. In the case of a future technical improvement, leading to higher F and P, the back-action must also be considered.

For the transducer cavity a lower Finesse is sufficient, provided that it is compensated (or, better, overcompensated) by a longer cavity and an higher laser power. Approximate parameter values for our apparatus are reported in Table 2.

With these values, the displacement sensitivity is about $4 \times 10^{-21}\mathrm{m}/\sqrt{\mathrm{Hz}}$, corresponding to a laser frequency change of about 10^{-4} $\mathrm{Hz}/\sqrt{\mathrm{Hz}}$.

The complete system sensitivity, calculated with a temperature of T = 0.1 K, with Q = 5×10^6 for both the bar and the transducer and for a transducer effective mass of 10 kg, gives an expected sensitivity at the level of $\sqrt{S_{hh}} \approx 10^{-22}/\sqrt{\mathrm{Hz}}$. The bandwidth is about 50 Hz, with a minimum detectable strain of $h_0 = 2 \times 10^{-20}$. These figures allow an improvement of more than one order of magnitude with respect to present sensitivities.

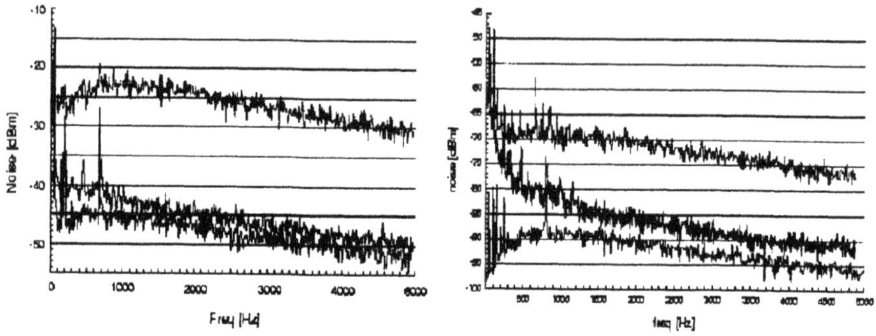

Figure 2. Intensity noise power spectrum. Upper curve: free-running laser; middle curve: intensity stabilized laser; lower curve: shot-noise reference. Left graph: for a laser power of 14 mW impinging on the photodiode. Right graph: for 15 mW impinging on the photodiode, after propagation through a 2 m optical fiber.

3 Implementation of the transduction chain

3.1 The laser system

The laser is a commercial Nd:YAG source, with an output power of 100 mW. Its intensity noise must be actively reduced in the spectral region around 1 kHz. At this purpose, the laser beam passes through an electro-optical modulator (EOM) whose crystal axes are slightly rotated with respect to the laser polarization direction. The EOM is followed by a polarizer and a 50on the voltage applied to the EOM crystal we can change the beam polarization and these changes are transformed into amplitude variations by the polarizer. The beam reflected by the beam-splitter is detected by a photodiode whose amplitude, properly amplified, is fed back to the EOM in order to realize a closed feedback loop. The transmitted beam is used for the experiment.

The results of the active stabilization are shown in Fig. 2 (left). The laser amplitude noise is reduced from 23 dB above shot-noise level (SNL) down to 5 dB, in agreement with the theoretical calculations.

The transmission through the optical fiber deteriorates this performance, as shown in Fig. 2 (right) which refers to a transmitted power of 15 mW. The measured noise is 8 dB above SNL in the frequency range of interest. However, the amplitude noise approaches SNL due to the effect of attenuation or losses. Scaling the measurement of Fig. 2 to the 2 mW of laser power which we plan to couple to the transducer cavity, we deduce an amplitude noise 2 dB higher than SNL. This corresponds to a back-action noise at least 5 dB lower than the transducer thermal noise, i.e., to a negligible effect.

The phase modulation must be performed in a frequency region where the laser amplitude noise is nearly at SNL and, at the same time, the detection can still be performed with high efficiency, low electronic noise, high enough saturation level. The optimum region is $10 \div 15$ MHz.

We phase modulate the laser with the optimum modulation depth (about 1 rad) using a resonant EOM. Even if the phase modulation also produces an amplitude modulation, we have checked that the we do not introduce any additional noise, around 1 kHz, in the signal recorded and de-modulated. This is true also after the propagation through the fiber.

3.2 The cavities

The reference cavity must have the following characteristics:

- high stability (i.e., low mirrors displacement noise) around 1 kHz, where it must act as frequency reference;
- it must follow the laser locked to the transducer cavity for slow frequency variations, e.g., for times longer than ~ 0.1 s.

The cavity is made by a 20 cm long, 10 cm diameter zerodur spacer with optically contacted mirrors. It is supported by a double cantilever and kept under vacuum.

The coarse tuning is obtained by changing the vacuum chamber temperature in the range $20 \div 100$ °C. The temperature is then stabilized with residual fluctuations lower than 0.001 °C in 24 hours. The fine tuning ($\sim$ 30 kHz) is obtained by compressing the spacer with piezoelectric ceramic disks.

The transducer cavity has been realized and tested on a room temperature bar, showing a satisfactory Q. We have performed preliminary measurements of the thermal noise excited vibrational modes of the bar and transducer system using a simplified optical transduction chain.

3.3 The locking

The free-running laser frequency noise is about 300 $\mathrm{Hz}/\sqrt{\mathrm{Hz}}$ in the region around 1 kHz. When locked to the transducer cavity, the laser frequency fluctuations with respect to this cavity must be reduced down to a level lower than the intrinsic transducer sensitivity, as already discussed. At this purpose, we have implemented a feedback loop with a gain of 130 dB at 1 kHz. When the laser is locked to the reference cavity, the in-loop fluctuations are below the SNL.

4 Conclusions

We have described the principles of operation and some experimental steps for the implementation of an optical transduction chain for g.w. bar detectors. The complete realization of such a transducer would allow a huge improvement in the performances of present detectors.

Acknowledgments

We thank M. Inguscio for his useful and encouraging support to this research.

References

1. C Jarlskog in *CP Violation*, ed. C Jarlskog (World Scientific, Singapore, 1988).
2. L. Maiani, *Phys. Lett.* B **62**, 183 (1976).
3. Cerdonio M.et al., Cryogenic resonant detectors of gravitational waves: current operation and prospects. In Gravitation and Relativity: at the Turn of the Millennium ed. by Dadhich N. and Narlikar J. (Pune, India, 1998) pp. 211- 230.
4. Richard J. P., J. Appl. Phys. 64 (1988) pp. 2202-2205.
5. Conti L. et al., Rev. Sci. Instr. 69 (1998) pp. 554-558.
6. Drever R. W. P. et al, Appl. Phys. B 31 (1983) pp. 97-105.

THE VIRGO PROJECT

ADALBERTO GIAZOTTO
I.N.F.N. - Pisa Section, I-56010
S.Piero a Grado, Pisa, Italy
e-mail: giazotto@pi.infn.it

A short description of Virgo Interferometric Gravitational Detector is here presented, pointing out working principle, foreseen antenna sensitivity, planned development.

1 Introduction

Virgo is a French-Italian project, approved in 1993. The goal of this experiment is to detect gravitational waves in a large bandwidth using a 3 km armlength Michelson interferometer, which is being built in Càscina (near Pisa, Italy). The collaboration, funded by CNRS (France) and by INFN (Italy), involves eleven Laboratories: five in France (LAL, Orsay - LAPP, Annecy - IPN, Lyon - ESPCI, Paris - OBSERVATOIRE de la COTE D'AZUR, Nice) and six in Italy (INFN Sections of: Pisa - Florence/Urbino - Rome - Perugia - Neaples; INFN National Lab of Frascati).

2 Working Principle

Virgo aims to detect gravitational waves by measuring the small effects induced on a Michelson interferometer with suspended optical components (see figs 1 - 2). A gravitational wave, impinging on the plane of the suspended interferometer, "stretches" one of its arm and "compresses" the other one, inducing the opposite effect half cycle later. The amplitude of the mirrors displacement is approximately given by the product between the arms length L and the dimensionless amplitude of the gravitational wave (usually indicated with $h(t)$). A detection of the gravitational wave can be provided by a standard interferometric measurement of the phase shift between the interfering beam. The phase shift to be detected is

$$\Delta\Phi(t) = \frac{4\pi h(t)L}{\lambda}$$
$$\lambda \gg L$$

where λ is the laser wavelength. Since the small phase shift to be detected is proportional to the length of the arms, an interferometer with very long arms is desired for the purpose, but the light storage time should always be kept smaller than G.W. period.

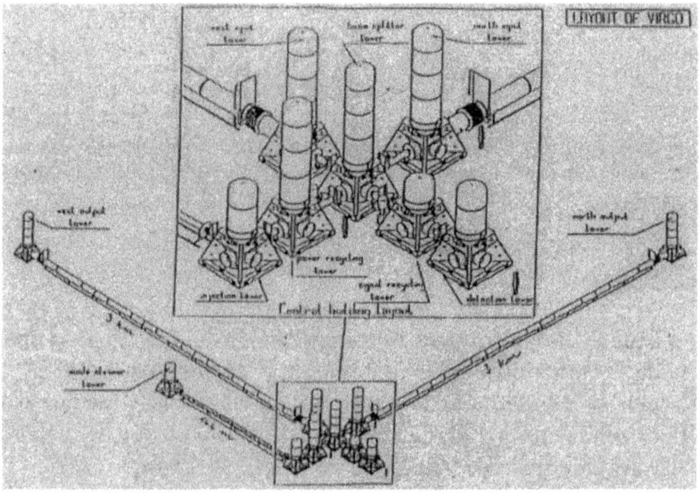

Figure 1. Layout of VIRGO.

In order to evaluate the sensitivity that VIRGO has to have, let us consider the typical signal emitted by a supernova in the Virgo cluster. This signal, at the Earth, is a pulse of a few milliseconds with dimensionless amplitude h ranging from 10^{-22} to 10^{-20} . If we refer to a 3 km long arms interferometer (as VIRGO) the passage of such a signal should induce a pulse displacement of the mirrors of about 10^{-19} - 10^{-17} m and therefore a phase shift of 10^{-12} - 10^{-10} radians (where a laser wavelength of about 1 μ m has been considered). It is clear that the VIRGO interferometer has to be characterized by an extremely high sensitivity.

A few interferometric gravitational wave antennae are under construction around the world with the aim to reach a sufficient sensitivity to detect gravitational waves emitted by astrophysical sources in the frequency range from few tens of Hz to few kHz. At lower frequencies the detection is hindered by seismic noise, namely by continuous vibrations of the ground which induce spurious motions of the optical components and mask the small displacements induced by gravitational waves. Since many astrophysical sources, such as pulsars and coalescing binaries, are expected to emit mainly low frequency gravitational waves (from a fraction of a hertz to a kHz), it is of great

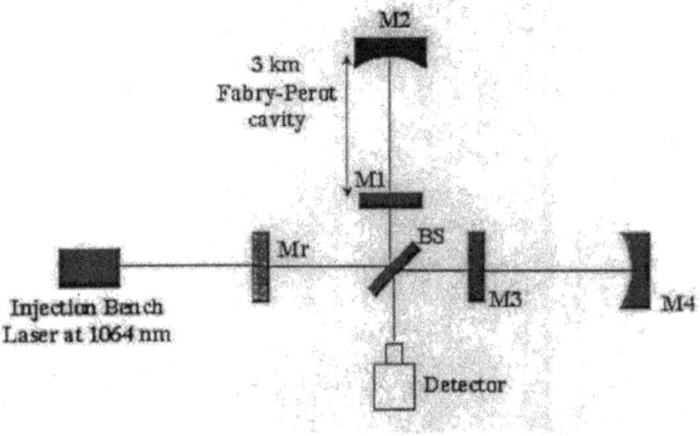

Figure 2. Optical components of VIRGO : $M_{1,2}$; $M_{3,4}$ = Mirrors of Fabry-Pérot cavities. BS = Beam Splitter. Mr = Mirror for power recycling.

importance to lower the frequency detection threshold as much as possible.

VIRGO is currently the only ground based interferometric antenna that aims to lower the detection band down to a few Hz. This is accomplished by suspending each optical component of the interferometer from a special attenuation chain (superattenuator: see fig.3) which reduces the transmission of seismic noise.

3 Sources of gravitational waves

General relativity predicts that an accelerated system of masses could emit gravitational waves. The main difference with respect to the case of electromagnetism is that the lowest order radiation is now quadrupolar and not dipolar. The strongest gravitational signals which can reach an observer are expected to be produced by astrophysical processes in which high accelerated coherent motions of large masses take place. These astrophysical sources can be classified as belonging to three main categories: burst sources, periodic sources, and stochastic background. The expected theoretical behaviors are crucially depending upon the adopted models: some of them are indeed affected by rather large uncertainties.

Burst sources produce gravitational signals for a short time only. This category includes coalescing compact binaries (of black holes and neutron stars), infalls of massive objects into black holes, supernovae, and so on. A typical supernova in Virgo cluster is expected to radiate a gravitational pulse of a few milliseconds, having a dimensionless amplitude h, at the Earth, of 10^{-22} - 10^{-20} .

Periodic sources, such as neutron stars, are expected to emit gravitational waves continuously. The periodicity of these signals could favor their detection despite the involved very low amplitudes. Some billions of neutron stars in our galaxy are expected to produce signals in the frequency range from a fraction of Hz to a few hundreds of Hz, with dimensionless amplitudes h, at the Earth, ranging from 10^{-24} to 10^{-30} or less.

Stochastic background is a superposition of random signals, coming from random directions at all frequencies. This background is produced by the radiation originated in the early evolution of the universe, and by all the gravitational signals which come from unknown sources. Its amplitude is expected to be well below the detection threshold of single ground-based detectors: so, coincidence measurements between more detectors are necessary.

4 Design of the VIRGO interferometer

In VIRGO the technique employed to enhance the phase shift induced by the gravitational signal is to increase the optical path of the light in the two arms of the interferometer. This result can be attained by allowing the light to bounce back and forth several times along the two arms before the recombination on the beam splitter. In this way, the phase shift to be detected ($\Delta\Phi$) is magnified by the number of reflections made by the light in each arm. This increase of the optical path can be obtained by replacing each arm of the interferometer with a Fabry-Pérot cavity, resonating at the laser frequency. In this way each photon goes back and forth in each arm an average number of times related to the Finesse F of the cavity (in VIRGO a Finesse of 50 has been chosen). By the use of this technique the relevant phase shift can be amplified by a factor F/π .

Several noise processes can generate spurious signals in the antenna (see fig. 4), masking the effect induced by the gravitational wave. Some of these processes, such as seismic noise and thermal noise, induce displacements of the mirrors ("displacement noise"), while others, such as the noise induced by the frequency fluctuations of the laser, affect the phase of the optical rays even if a real movement of the mirrors is not present ("phase noise"). In order to define the sensitivity of the antenna it is necessary to compare the real signal $h_{GW}(t)$ to each relevant fake signal $h_{Noise}(t)$. This comparison is usually expressed in terms of the so called "linear spectral density" which

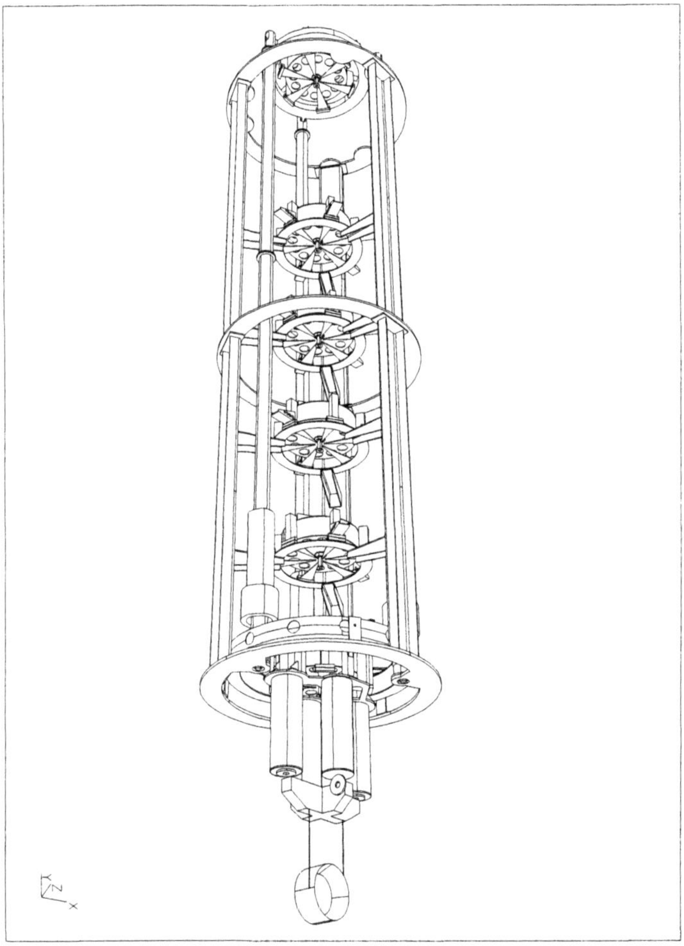

Figure 3. The superattenuator: it has a structure of a compounded pendulum; each stage is a filter for vertical oscillations. Higher suspension point is sustained by a (pre-isolation) inverted pendulum; active control can be exerted on different point of the chain, with relation to different frequency range. The mirror is suspended at the end of the pendulum through a more complex structure constituted by a particular filter (historically called filter seven) and a "marionetta", a rigid body of particular shape in order to steer the mirror. All the components are enclosed in a vacuum tower.

is defined as the square root of the power spectrum of the signal. In the case of dimensionless amplitudes, the linear spectral density can be therefore expressed in units of $\sqrt{\text{Hz}}$. The sensitivity curve of VIRGO (see fig. 5) is plotted by adding in an incoherent way the linear spectral density of all the fake signals associated to spurious processes. We briefly review principal noise sources.

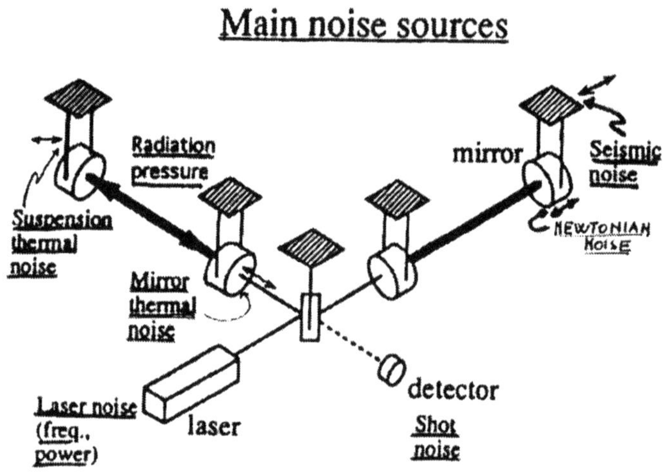

Figure 4. Main noise sources in an interferometric, ground-based antenna.

Poisson statistical fluctuations of the number of photons in the light beam generate shot-noise on the interference signal; the linear spectral density of this noise decreases like the square root of the beam power: this last can be increased by "recycling" the light which is back-reflected from the interferometer. This is obtained by inserting a semi-transparent mirror in order to build a resonating cavity between this mirror and the whole interferometer. With this technique, and with a very high power laser source (in VIRGO a 10 W Nd:Yag laser is used), an impinging power of 1 kW on the beam splitter can be obtained. In this condition shot-noise is a white noise on the interference phase corresponding to a linear spectral density of the wave amplitude close to $10^{-23}\sqrt{\text{Hz}}$. Any other source of noise which affects the antenna is normally compared to this level of noise, called "shot-noise limit".

The static path difference between the two arms of the interferometer generates noise through frequency fluctuations of the laser source. In fact, because of an unavoidable asymmetry between the two arms, the fields which interfere at the beam splitter are originated from the laser at different times, and a fluctuation of the laser frequency is converted into a fluctuation of the phase difference $\Delta\Phi(t)$. In order to keep this noise below the shot-noise limit, it is adequate to develop a very efficient system for the stabilization of the laser frequency. The goal is to achieve a stability of the frequency ν of $\delta\tilde{\nu}(f) < 10^{-6}\mathrm{Hz}/\sqrt{\mathrm{Hz}}$.

Fluctuations of the gas density in the two arms generate a variation in the refractive index, and therefore differences in the speed of light in the two cavities. Since these fluctuations are not correlated, they will produce a fluctuating phase difference and thus a spurious signal. For this reason the 3 km arms of the interferometer must be kept under vacuum: a pressure of 10-8 mbar in the pipe surrounding each arm of the interferometer is required.

By the improvement of the techniques described above it is possible to develop an antenna of very high sensitivity, able to detect several gravitational signals at medium/high frequencies (above a few tens of Hz). At lower frequencies the detection is hindered by seismic noise which can cause spurious vibrations of the optical components of the interferometer. Measurements of the spectrum of the seismic noise have been performed by many groups. The linear spectral density of the seismic displacement of a point on the ground is found to be of the same order of magnitude in all the directions and to vary, to a good approximation, as A/f^2 (where A ranges from 10^{-6} to $10^{-9}\mathrm{mHz}^{3/2}$, depending on the location of the measurement). In order to keep the mirrors spurious motion well below the displacements generated by gravitational signals, it is necessary to design a suspension system able to attenuate the propagation of the seismic vibrations. VIRGO is the unique interferometric gravitational wave detector which plans to incorporate sophisticated suspension system (superattenuator) for the optical components that will be able to isolate the antenna from seismic noise. The VIRGO superattenuator is essentially a eight meters high six stages pendulum where each mass is replaced by a special mechanical filter designed to reduce the transmission of seismic vibrations in all the degrees of freedom. The goal of the VIRGO superattenuators is to attenuate the seismic noise transmitted to the optical components by a factor larger than 10^{10}, starting from frequency value of few Hz upwards. Recently the performances of the apparatus have been tested, obtaining an attenuation larger than 10^{10} in all degrees of freedom, starting from about 4 Hz. We can obtain in this way a mirror displacement sensitivity of about $3\ 10^{-8}\ \mathrm{m}/\sqrt{\mathrm{Hz}}$, limited by the thermal noise on optical components. This corresponds to a gravitational wave sensitivity (in the low frequency region) of $\tilde{h} \approx 10^{-21}/\sqrt{\mathrm{Hz}}$.

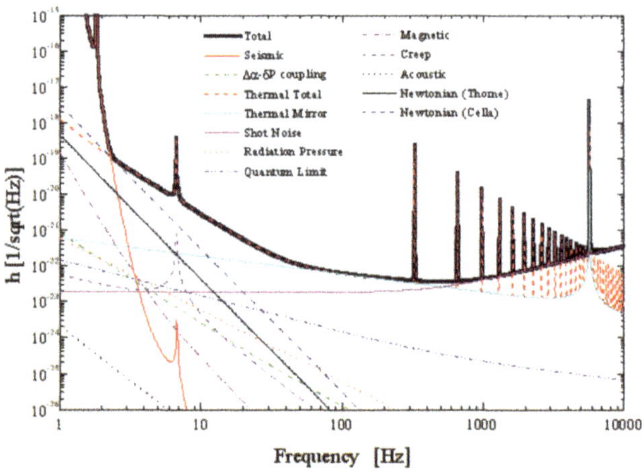

Figure 5. The VIRGO sensitivity curve.

5 Conclusions

In fig. 5 the sensitivity curve of VIRGO is reported. As discussed in the previous paragraphs, in the low frequency range (below few tens of Hz) the VIRGO sensitivity is limited by the thermal noise of the pendulum suspension. Between few tens of Hz and few hundreds of Hz, the dominant mechanism is the thermal noise of the mirrors internal modes. At higher frequencies the VIRGO sensitivity curve is limited by shot-noise. With this sensitivity curve, VIRGO should be able to detect several high frequency gravitational signals coming from supernovae in the Virgo cluster (few events per week). Furthermore it is clear the possibility for Virgo, owing high seismic isolation reached by superattenuators, to detect several periodic or quasi-periodic signals in the low frequency range (galaxy pulsars and coalescing binaries). The central part of the interferometer will be concluded by the end of 2000 and some preliminary test will be performed in the next 2 years. In parallel, the construction of the two 3 km long arms (tunnels and vacuum pipes) is beginning. VIRGO should start to take data for the gravitational wave detection in 2002.

References

1. Abramovici A. et al: Science, 256, 325 (1992)
2. Braccini S. *et al*: Design of the Superattenuators for VIRGO construction - VIRGO int.Rep.,VIR-TRE-PIS-4600-134 (1997)
3. Brillet A.: Ann. De Phys., 10, 219 (1985)
4. Cattuto C. *et al*: The Virgo sensitivity curve, Virgo Int.Rep., VIR-NOT-PER-1390-51 (1999)
5. Giazotto A.: Phys.Reports, 182 (6),365 (1989)
6. Misner C.W., Thorne K.S., Wheeler J.A.: Gravitation, W.H.Freeman, S.Francisco (1973)
7. Saulson P.R.: Fundamentals of Interferometric Gravitational Wave Detectors, World Scientific, Singapore (1994)
8. Vinet J.Y. *et al*: Phys. Rev. D,38, 433 (1988)
9. VIRGO Collaboration: Virgo Final Design Report - VIR-TRE-DIR-1000-13 (1997)

LOW FRICTION MATERIALS FOR HIGH SENSITIVITY GRAVITATIONAL WAVE DETECTORS

C. CATTUTO[a,b], L. GAMMAITONI[*a,b] J. KOVALIK[a],
F. MARCHESONI[a,c] AND M. PUNTURO[a]

[a]*Ist. Naz. Di Fisica Nucleare–Sezione di Perugia, VIRGO Project, I-06100 Perugia ITALIA*
[b]*Dipartimento di Fisica, Universita di Perugia, I-06100 Perugia ITALIA*
[c]*Dipartimento di Fisica, Universita di Camerino, I-62032 Camerino ITALIA*

Thermal noise in mechanical suspension systems is presently the most severe limitation to the sensitivity of the new generation of interferometric gravitational wave detectors, like VIRGO, in the frequency range between 4 and 500 Hz.

A gravitational wave interferometer must detect very small displacements of free masses which are hung as mechanical pendulums. Once seismic noise is filtered, the dominant noise source between 10 and 50 Hz is likely to be the thermal noise of the pendulum motion. One way to reduce this noise source is to construct the pendulum with as high a mechanical Q as possible in order to concentrate the thermal motion in a narrow band around the pendulum resonant frequency.

In designing a high Q pendulum, one must take into account the ratio of energy stored in the pendulum to the energy dissipated in one cycle of the pendulum oscillation. Since most of the energy is stored in the gravitational field which is lossless, while all of the energy dissipated is in the wires, the highest pendulum, Q_p, occurs when the wires are as thin as possible. The quality factor Q_p of an individual pendulum depends upon the internal losses of the suspension wires and for a pendulum suspended by two wires (or a single loop of wire) is given by the formula [1]:

$$Q_p = l\sqrt{\frac{2Mg}{EI}}\frac{1}{\phi_w} \tag{1}$$

where l is the length of the pendulum wires, M is the suspended mass, E is the Young's modulus of the suspension wire material, and $I = \pi d^4/64$ is the moment of inertia of the cross section of a cylindrical wire with diameter d.

Thus, the optimum wire is one that is made from a material with high breaking stress and low internal friction. Thus it is clear that an important task is to identify materials with low internal friction and to understand their

*Author who presented the comunication

frequency dependence. The mechanical properties of these materials, especially their breaking strength, have also to be examined in view of possible applications to the present generation of gravitational wave detectors.

Such a study has been carried out in few laboratories and the internal losses for different wire materials have already been measured (see below). The conclusion is that both fused silica (synthetic fused silicon dioxide) and fused quartz (natural fused silicon dioxide) show inherent losses that are about two orders of magnitude lower than metals[2,3,4]. For these reasons, fused silica suspensions are currently under investigation for costructing high sensitivity gravity wave interferometers.

In the following we summarize some recent experimental results collected at the Virgo Perugia laboratory [4]. We measured the loss angle of clamped-free thin wires of many different materials. The wires were excited by means of an electrostatic actuator that was a copper plate placed a few millimetres from the wire. Voltages of typically one kilovolt were used. The wires were first excited by white noise in order to find the resonant frequencies. The wires were then excited at their resonant frequencies. The driving signal was switched off and the amplitude of the oscillation free decay was recorded by a personal computer. The slope γ of the exponential function which describes the decay, is related to the quality factor, Q_i, of the mode with frequency f_i by

$$Q_i = \frac{\pi f_i}{\gamma} \tag{2}$$

The internal friction of the wire is the inverse of its Q_i, that is $\phi_w(f_i) = Q_i^{-1}$

Different wire materials were tested: C85 steel with a diameter of $200\mu m$ from Wagner, Marval (maraging) steel wire (prepared by De Salvo of INFN, Pisa), sapphire with a diameter of $500\mu m$ from Saphikon, and fused silica with a diameter between $200\mu m$–$300\mu m$(pulled by the Glasgow group of GEO600 from a 3 mm diamter rod from Heraeus). The silicon carbide sample (supplied by Goodfellow) had a diameter of $100\mu m$ with a $10\mu m$ tungsten core. The remaining metals: niobium, Invar (Fe64/Ni36), Cu-Be (Cu98/Be2), Chromalloy (Fe75/Cr20/Al5), Aluchrom (Fe70/Cr25/Al5), titanium, and tungsten were all of diameter $250\mu m$ and ordered from Goodfellow. The metal samples and the sapphire were all clamped in the tool steel. The 3 mm head of the fused quartz sample was clamped between a V-shaped groove and a flat piece of aluminum.

The fused silica, sapphire and silicon carbide all formed a straight wire. The fused silica was pulled from a straight rod and always kept straight. The sapphire and silicon carbide were spooled and might have been affected by

Table 1. Internal friction parameters for various samples. The relaxation strengths and frequencies have been extracted from our data by means of the fitting law (2) and compared with Zener's theoretical prediction reported in the text. Note that for both Aluchrom and Chromalloy a Young's Modulus could not be provided by the manufacturer so that a typical value for a metal of $1.5 \times 10^{11} N/m^2$ was used.

material	Δ theor. ($\times 10^{-3}$)	f_r theor. (Hz)	ϕ_0 expt.	Δ expt.	f_r expt. (Hz)
C85 Steel	2.2	648	$1.9 \pm 0.1 \times 10^{-4}$	$1.9 \pm 0.4 \times 10^{-3}$	480 ± 20
Marval Steel	2.2	648	$6.0 \pm 0.1 \times 10^{-5}$	$1.36 \pm 0.04 \times 10^{-3}$	340 ± 10
Aluchrom	2.0	131	$2.1 \pm 0.6 \times 10^{-4}$	$1.6 \pm 0.2 \times 10^{-3}$	130 ± 10
Cu-Be	3.5	889	$5 \pm 3 \times 10^{-5}$	$2.57 \pm 0.08 \times 10^{-3}$	650 ± 30
Chromalloy	3.1	168	$9.0 \pm 5 \times 10^{-5}$	$2.0 \pm 0.1 \times 10^{-3}$	135 ± 8
niobium	1.15	807	$1.1 \pm 0.1 \times 10^{-4}$	$6.7 \pm 0.4 \times 10^{-4}$	680 ± 50
titanium	1.1	322	$7 \pm 0.1 \times 10^{-4}$	$1.5 \pm 0.2 \times 10^{-3}$	420 ± 40
tungsten	0.74	233	$1.7 \pm 0.1 \times 10^{-4}$	$7.8 \pm 0.4 \times 10^{-4}$	1200 ± 100

the coiling process.

All the metals were delivered on a spool and may have also been affected by being coiled. The softer metals (Marval, Invar, Chromalloy, Aluchrom, titanium) could all be straightened by pulling the wire with 10-50 N of force. The remaining stiffer wires were straightened by weaving them through a grating and then pulling them through. All the wires were wiped clean with ethanol before being measured. Table 1 lists both the calculated and fitted parameters for the metal samples where the data appears to follow the model. Note that the metal Invar does not follow this model. This is due to the fact that its thermal expansion coefficient, α, is an order of magnitude smaller than that of usual metals so that the relaxation strength of the thermoelastic damping (which depends upon α quadratically) is negligible compared to the inherent material loss. The magnitude of the damping strength, Δ, agrees quite well (within 20%) for C85 steel, Aluchrom, titanium and tungsten. For Marval, Cu-Be, Chromalloy and niobium, the measured Δ is actually smaller than the calculated one, thus suggesting that the usual parametrization for its relaxation strength may be inadequate.

The relaxation frequencies f_r in general do not agree well with the relevant predictions of Zener's model [6]. Since the relaxation frequency does depend quadratically on the wire diameter, variations in the wire cross-section (especially where the wire is tightly clamped) may have caused this discrepancy.

It is important to note the mechanical properties of the samples. The C85 steel has by far the best yield strength (3 GPa) and handling properties. It has a yield point near 90% of its breaking point and can thus be highly loaded before behaving in an anelastic manner. The Marval steel has a breaking strength that is about 70% that of C85 steel. The other materials have several undesirable properties: titanium, niobium, Invar, Chromalloy and Aluchrom are all very soft and stretch plastically when loaded near 50% of their breaking point; Cu-Be and tungsten are both brittle; tungsten tends to crack when squeezed too tightly, while Cu-Be tends to break when bent too sharply; silicon carbide is also brittle as a thin wire and, finally, sapphire and fused quartz are extremely fragile.

In view of the fluctuation-dissipation theorem, the constant low-frequency loss angle $\phi_w \simeq \phi_0$ can be interpreted as an experimental evidence for a mechanical $1/f$ noise in the lattice structure. However, the microscopic mechanisms underlying such a phenomenon prove elusive [8]. Moreover, the very existence of a frequency independent structure damping has been questioned, for instance, by A. V. Granato [9]. He pointed out that very long time-scale relaxation processes might affect internal friction experiments at low frequency, so that their outcome is likely to depend critically on the thermal history of the samples and the preparation procedures followed at each stage of the experiment.

While the losses for the metals do follow the phenomological model Eq.(2) quite well, the actual magnitudes of the parameters calculated from handbook values for the materials do not necessarily agree with the experimentally determined parameters. Since the frequency independent structure damping, ϕ_0, must be measured by experiment, it is clear that the only way to actually obtain good parameters is to fit experimental data in the frequency range of interest. Once this is done, one obtains a good frequency dependent loss function, $\phi_w(f)$, that can be used in calculations for the thermal noise spectrum of a high Q_p pendulum, for example.

It is also clear that all metals have inherently the same order losses (10^{-3}–10^{-4}) at frequencies on the order of one to a few tens of Hz. Fused silica is a promising material because of its low loss. Some studies suggest that if the surface of a thin fused silica fibre is damage free, it can have a breaking stress better than that of carbon steel. Further research is being carried out to investigate this phenomenom.

References

1. P. R. Saulson, Phys. Rev. **D42** 2437 (1990); J. Kovalik and P. R. Saulson, Rev. Sci. Instrum. **64** 2942 (1993), G. Cagnoli, L. Gammaitoni, J.

Kovalik, F. Marchesoni and M. Punturo, Rev. Sci. Instrum. 1999, in print).
2. V. B. Braginsky, V. P. Mitrofanov and and O. A. Okhrimenko, Phys. Lett. A **175** 82 (1993)
3. S. Rowan, R. Hutchins, A.C. McLaren, N. A. Robertson, S.M. Twyford and J. Hough, , Phys. Lett. A **227** 153 (1997)
4. G. Cagnoli, L. Gammaitoni, J. Kovalik, F. Marchesoni and M. Punturo, Phys. Lett. A 255, 230-235, (1999).
5. A. S. Nowick and B. S. Berry, *Anelastic Relaxation in Crystalline Solids* (Academic, New York, 1972)
6. C. Zener, *Elasticity and Anelasticity of Metals* (University of Chicago Press, Chicago, 1948)
7. P. M. Morse, *Vibration and Sound* (McGraw-Hill,New York, 1948)
8. F. Marchesoni and M. Patriarca, Phys. Rev. Lett. **72** 4101 (1994)
9. A. V. Granato, private communication.

AN INTRODUCTION TO FEEDBACK CONTROL SYSTEMS

L. BENVENUTI
Dip. Informatica e Sistemistica, Università degli Studi di Roma "La Sapienza"
Via Eudossiana 18, 00184, Roma, Italy
E-mail: benve@dis.uniroma1.it

M. D. DI BENEDETTO
Dip. Ingegneria Elettrica, Università degli Studi di L'Aquila
Poggio di Roio, 67140, L'Aquila, Italy
Dept. of EECS, University of California at Berkeley, CA 94720
E-mail: dibenede@ing.univaq.it

1 Concept of Control

Automatic control is at the root of many of the most significant advances in engineering of the past and continues to be an essential discipline. The basic idea is to add to a given system (the ***plant***), representing either a physical system or a design, a component (***the controller***) capable of modifying the behavior of the original system to achieve a set of desired properties. To control the behavior of a system we must be able to represent and analyze it in mathematical form. It is assumed that the variables describing the interaction of the system with its environment are in a cause-effect relationship so that we can talk about input and output of a system (see Figure 1). Control is the process of causing the system variable $y(t)$ (plant output) to conform to some desired value (reference input) $r(t)$ by means of a signal $u(t)$ affecting the system and called control input. Other signals $d(t)$ (disturbance) affect the output of the system adversely.

An *open–loop* control system utilizes a controller in order to obtain the desired response, as shown in Figure 2. In contrast to an open–loop control

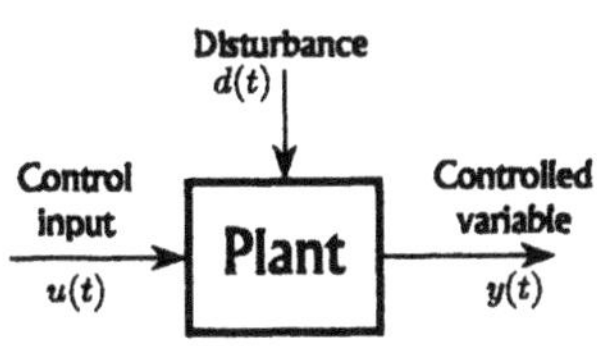

Figure 1: Plant to be controlled.

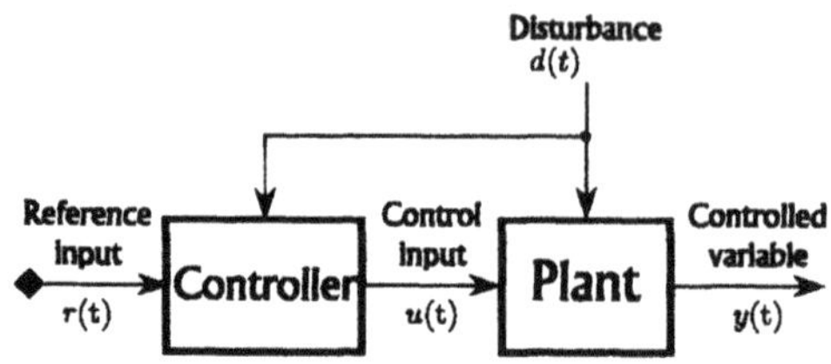

Figure 2: Open–loop control system.

system, a closed–loop control system utilizes an additional measure of the actual output to compare the actual output with the desired output response. The measure of the output is called the *feedback* signal. A simple closed–loop control system is shown in Figure 3. In an open–loop control system the output is not compared with the reference input. Thus to each reference input there corresponds a fixed operating condition; as a result, the accuracy of the system depends on calibration. In the presence of disturbances an open–loop system will not be able in general to perform the desired task. Open–loop control can be used, in practice, only if the relationship between the input and output is known with good accuracy and if there are no disturbances. A closed–loop control system has the advantage that the use of feedback makes the system response relatively insensitive to external disturbances and internal variations in system parameter. It is thus possible to use relatively inaccurate and inexpensive components to obtain the accurate control of a given plant, whereas doing so is impossible in the open–loop case. The importance of closed-loop control cannot be overemphasized. However, the design of feedback controllers

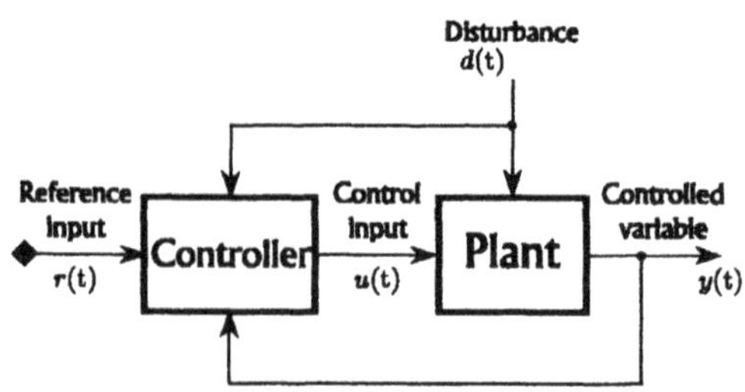

Figure 3: Closed–loop feedback control system.

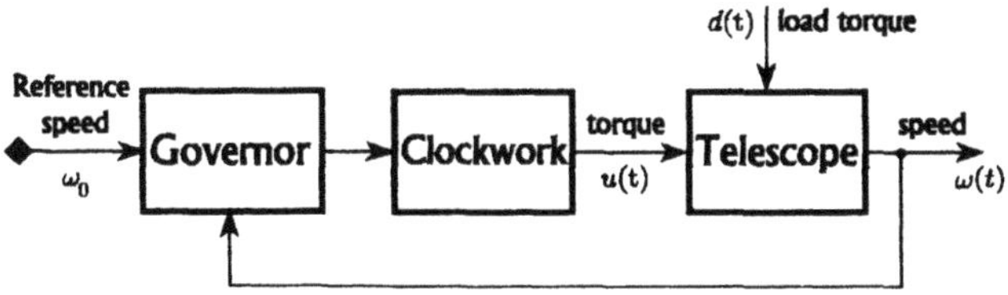

Figure 4: Block diagram of the system.

is not always straightforward. It involves careful analysis and design.

The Watt–type governor is one of the first examples of the use of closed-loop control. This kind of governor is due to James Watt (1736–1819), developer of the steam engine, who seems to have adapted the governor traditionally used in windmills to one of his own engine in 1788. He described his governor in the following way: "it consists of a perpendicular axis, turned by the engine. To a joint near the top of this axis are suspended two iron rods carrying heavy balls of metal at their lower ends, in the nature of pendulums. When this axis is put in motion by the engine, the balls recede from the perpendicular by the centrifugal force, and by means of a combination of levers fixed to their upper end, raise the end of a lever which acts upon the spanner of the throttle valve, and shuts it more or less according to the speed of the engine, so that as the velocity augments, the valve is shut, until the speed of the engine and the opening of the valve come to a maximum and balance each other.[a]".

The use of this kind of governor to astronomy by George Biddel Airy (1801–1892) professor of mathematics and astronomy at Cambridge University uncovered the most troublesome aspect of closed-loop control: dynamic instability. In 1834 Airy became involved with the design of governors able to regulate the speed of astronomical telescopes. The control problem is as follows: a star seen through a fixed telescope appears to cross the field of view within a few seconds, owing to the rotation of the earth; in order to track the star it is thus necessary to keep the telescope rotating slowly at a uniform speed round its polar axis. Clockworks were used to drive the telescope and different type of governors used to control the speed. Airy drew attention to the phenomenon of dynamical instability of a feedback system: "I constructed a machine for uniform motion. It went exceedingly well till inequality began

[a]Watt envisaged the system variables as changing monotonically so that it seems that he had no concept of overshoot or similar dynamical behavior.

to be perceptible; this inequality, though still of an oscillatory character, increased rapidly, and finally became so great as entirely to destroy the character of the previous motion; and the machine became perfectly wild".

Stability is possibly the most studied problem in automatic control. Techniques to avoid instability must be considered at all times. Note that feedback control can be also used to stabilize intrinsically unstable systems such as modern combat airplanes.

The topics covered in this chapter are intended to give a simple and basic introduction to automatic control issues and results. Given the tutorial nature of the paper, we have borrowed material from the textbooks [3, 5, 8, 13, 14]. The chapter is organized as follows: In Section 2 we present mathematical models of systems with particular emphasis on linear systems since the analysis and design of linear control systems is the most well understood branch of control theory and design. In Section 3, we introduce the concept of stability of a dynamical system and show how a general non-linear system can be analyzed using a linear approximation around stable trajectories. In Section 4 the structural properties of a system are introduced and in Section 5 we present transfer functions and block diagrams as mathematical representations of linear systems that can be used to simplify the analysis of their properties. Finally in Section 6 we present the foundations of feedback-control system design.

2 Dynamic Models

A system is called ***static*** if its current output depends only on its current input, otherwise it is known as ***dynamic***. The output of a static system remains constant if the input does not change. In a dynamic system, the output changes with time even if the input does not change, if it is not in a state of equilibrium.

Any attempt to design a system must begin with a prediction of its performance, which is usually based on a mathematical description of the system's dynamic characteristics, called ***mathematical model***. Although there are endless examples of dynamic situations arising in a spectrum of areas, the number of corresponding general forms for mathematical models is relatively small. Most commonly, dynamic systems are represented mathematically in terms of either differential or difference equations. These representations correspond, respectively, to whether the behavior is viewed as occurring in continuous or discrete time. A model of a given system can be obtained usually by applying physical laws and may include some unknown parameters, which must then be evaluated through actual tests. Sometimes, however, the physical laws governing the behavior of a system are not completely defined so that an experimental modeling process is needed. In this process the system is subjected to a set of known inputs, and its outputs are measured; then a mathematical model is derived from these input–output relationship.

In the sequel, for the sake of simplicity, we will mainly consider systems which can be described by linear differential equations with constant coefficient. The relationship between the input $u(t)$ and the output $y(t)$ is then of the form

$$a_N \frac{d^N y}{dt^N} + a_{N-1} \frac{d^{N-1} y}{dt^{N-1}} + \ldots + a_1 \frac{dy}{dt} + a_0 \, y = u(t) \tag{1}$$

The differential equation (1) has order N and can be always expressed as a system of N first-order differential equations defined in terms of N variables $x_1(t), \ldots, x_N(t)$ that are functions of the continuous variable t. More generally, using a matrix notation, we will consider Single–Input Single–Output (SISO) Linear Time–Invariant (LTI) systems of the form

$$\begin{cases} \dot{x}(t) = A\,x(t) + b\,u(t) \\ y(t) = c\,x(t) \end{cases} \tag{2}$$

for $t \geq t_0$ and initial conditions $x(t_0) = x_0$. $x(t)$ is the $n \times 1$ state vector, $u(t)$ is the input (forcing) of the system, and $y(t)$ is the output of the system. A is a $n \times n$ constant matrix usually referred to as system matrix, b is the $n \times 1$ input vector, and c is the $1 \times n$ output vector.

The solution $x(t)$ of system (2) in terms of the initial state x_0 and the input $u(t)$ is

$$\begin{aligned} x(t) &= e^{A(t-t_0)} x_0 + \textstyle\int_{t_0}^{t} e^{A(t-\tau)} b\, u(\tau) d\tau \\ y(t) &= c e^{A(t-t_0)} x_0 + \textstyle\int_{t_0}^{t} c e^{A(t-\tau)} b\, u(\tau) d\tau \end{aligned} \tag{3}$$

where

$$e^{At} - I + At + A^2 \frac{t^2}{2!} + \cdots = \sum_{k=0}^{\infty} A^k \frac{t^k}{k!}$$

The first and second terms of the right hand side of the first (second) of equations (3) are referred to as the ***state (output) free evolution*** and the ***state (output) forced evolution*** of the system, respectively. The output free evolution is then composed of linear combinations of the entries of the matrix e^{At} which are called ***natural modes.***

Moreover, the output evolution $y(t)$ can be also decomposed in ***steady–state*** and ***transient*** response. The steady–state response $y_{ss}(t)$ to the input $u(t)$ is defined as

$$y_{ss}(t) = \lim_{t_0 \to -\infty} \left(c e^{A(t-t_0)} x_0 + \int_{t_0}^{t} c e^{A(t-\tau)} b\, u(\tau) d\tau \right)$$

when the above limit does exist and is independent of the value of x_0. The transient response $y_t(t)$ is then defined as

$$y_t(t) = y(t) - y_{ss}(t)$$

Unfortunately, the equations describing the behavior of a physical system are usually nonlinear. Nonlinear equations are much more difficult to solve than linear ones and the kinds of possible behaviors resulting from a nonlinear model are much more difficult to analyze. Nonlinear systems are described by equations of the following general form

$$\begin{cases} \dot{x}(t) = f(x(t), u(t)) \\ y(t) = h(x(t)) \end{cases} \tag{4}$$

where $f(\cdot)$, $h(\cdot)$ are appropriate functions and $f(\cdot)$ is such to guarantee the existence and uniqueness of the solution of equations (4).

3 Stability and Linearization

In many situations the natural rest points of a dynamic system $\dot{x} = f(x, \overline{u})$ under constant input $u(t) = \overline{u}$ are as much of interest as the mechanism of change. In order to study this situation, we set in this Section $u(t) = \overline{u}$ for all $t \geq 0$. Throughout the paper we will assume, without loss of generality, $t_0 = 0$. The rest points are handled by the following definition:

Definition 1 *A state $x_e \in \mathbb{R}^n$ is an* equilibrium point *of a dynamic system if it has the property that once the system state is equal to x_e it remains equal to x_e for all future time, i.e. $f(x_e, \overline{u}) = 0$.*

If the system state is initiated at an equilibrium point, then by definition it will never move. Stability properties characterize how a system behaves if its state is initiated close to, but not precisely at, a given equilibrium point. Roughly speaking, an equilibrium point is stable if, whenever the state is initiated near that point, the state remains near it, perhaps tending toward it as time increases. An equilibrium point is unstable if it is not stable, that is when the state initiated near that point moves away as time increases. More precisely:

Definition 2 *An equilibrium point x_e of a dynamic system described by the differential equation $\dot{x} = f(x, \overline{u})$ is*

- stable *if, for any given $\epsilon > 0$, there is a $\delta > 0$, depending on ϵ, such that for every x for which $|x - x_e| < \delta$ the solution $x(t)$ of $\dot{x} = f(x, \overline{u})$ through x at $t = 0$ satisfies the inequality $|x(t) - x_e| < \epsilon$ for all $t \geq 0$.*

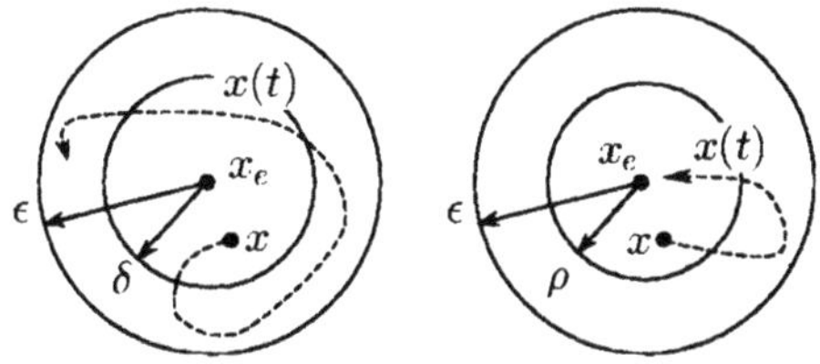

Figure 5: Definition of stability (left) and asymptotic stability (right).

- asymptotically stable ***whenever it is stable and, in addition, there is a*** $\rho > 0$ ***such that the solution*** $x(t)$ ***of*** $\dot{x} = f(x, \overline{u})$ ***through*** x ***at*** $t = 0$ ***tends to*** x_e ***as time increases for all*** x ***satisfying*** $|x(t) - x_e| < \rho$.
- marginally stable ***if it is stable but not asymptotically stable.***
- unstable ***if it is not stable.***

The definitions are illustrated in figure 5.

According to the basic definitions, stability properties depend only on the behavior of the system near the equilibrium point. Therefore, to analyze the stability properties of a nonlinear dynamic system, it is convenient to replace its full description by a simpler one that describes in an approximate way the behavior of the system near the equilibrium point. A linear approximation is often (but not always) sufficient to reveal stability properties. The linearization of a nonlinear system $\dot{x} = f(x, u)$ is based on linearization of the nonlinear function $f(x, u)$ near an equilibrium point $(x_e, \overline{u})$:

$$f(x_e + \delta x, \overline{u} + \delta u) = f(x_e, \overline{u}) + \left.\frac{\partial f}{\partial x}\right|_{\substack{x=x_e \\ u=\overline{u}}} \delta x + \left.\frac{\partial f}{\partial u}\right|_{\substack{x=x_e \\ u=\overline{u}}} \delta u + o(\delta x, \delta u)$$

where δx and δu are the deviations of the state x from the equilibrium point x_e and of the input u from the constant value $\overline{u}$, respectively. Since the equilibrium point is such that $f(x_e, \overline{u}) = 0$, then it follows that

$$\dot{\delta x} = \dot{x} \simeq A\,\delta x + b\,\delta u$$

where

$$A = \left.\frac{\partial f}{\partial x}\right|_{\substack{x=x_e \\ u=\overline{u}}} = \begin{bmatrix} \frac{\partial f_1}{\partial x_1} & \frac{\partial f_1}{\partial x_2} & \cdots & \frac{\partial f_1}{\partial x_n} \\ \vdots & \vdots & & \vdots \\ \frac{\partial f_n}{\partial x_1} & \frac{\partial f_n}{\partial x_2} & \cdots & \frac{\partial f_n}{\partial x_n} \end{bmatrix}_{\substack{x=x_e \\ u=\overline{u}}}$$

is referred to as the *Jacobian matrix* of f and

$$b = \left. \frac{\partial f}{\partial u} \right|_{\substack{x=x_e \\ u=\overline{u}}} .$$

The behavior of the nonlinear system for $u(t) = \overline{u}$ is then close to the behavior of the LTI homogeneous system $\dot{\delta x} = A\delta x$ when $x(t)$ is close to x_e. Therefore, it is useful to analyze stability properties in the case of homogeneous LTI systems. The system $\dot{x} = Ax$ always has the origin as an equilibrium point. If the matrix A is singular, corresponding to A having zero as an eigenvalue, then there are other equilibrium points that are all in the null space of A. On the other hand, the origin is the unique equilibrium point whenever A is nonsingular. Moreover, the following holds:

Proposition 3 *An equilibrium point x_e of a LTI dynamic system is* asymptotically stable *if and only if the eigenvalues of A all have negative real part. If at least one eigenvalue has positive real part, then the point is unstable.*

Because the stability of an equilibrium point of an LTI system depends only on the structure of the system matrix A, it is common to refer to stability of the system itself, or stability of the matrix A.

On the basis of the previous results, the stability properties of the original nonlinear system can be inferred from the linearized system using the following:

Proposition 4 *The equilibrium point x_e of the nonlinear system $\dot{x} = f(x, \overline{u})$ is*

- asymptotically stable *if all the eigenvalues of A have negative real part.*
- unstable *if at least one eigenvalue of A has a positive real part.*

If no eigenvalues of A have positive real part, but at least one has a zero real part, then x_e may be either stable, asymptotically stable, or unstable.

The essence is that, for small deviations from the equilibrium point, the performance of the system is approximately governed by the linear terms. These terms dominate and thus determine stability, provided that the linear terms do not vanish. If there are eigenvalues with zero real part, then the linearized systems gives no information on the stability properties of the original system. In this *critical case*, a different analysis is required which can be carried out using, for example, the Lyapunov theory [11] or the center manifold theory [1].

Example. The Watt fly-ball governor used by Airy is shown in Figure 6 where m is the mass of each ball, l the length of the rods and ω the speed of the telescope. θ is the angle which the suspending rod makes with the vertical. At steady–state and speed ω_0, the angular position of the rods θ_0 is such that

$$|F_1| = m\omega_0^2 l \sin\theta_0 \cos\theta_0 = |F_2| = mg\sin\theta_0$$

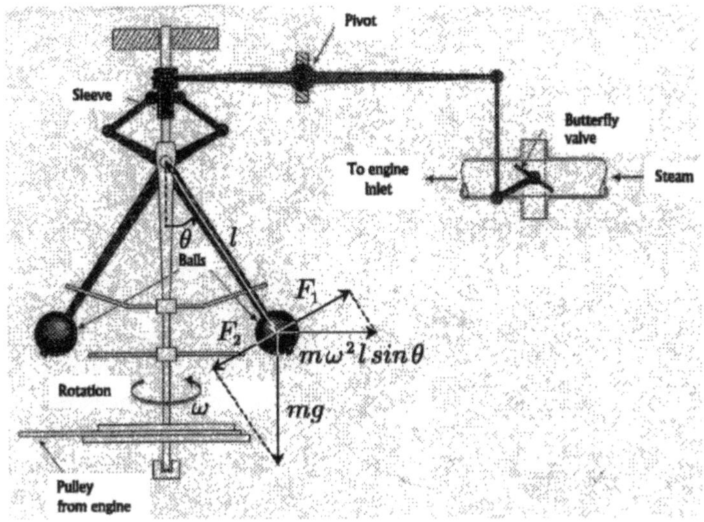

Figure 6: The fly–ball governor.

holds. Then, the relationship between θ_0 and ω_0 is the following:

$$\theta_0 = \arccos\left(\frac{g}{l}\frac{1}{\omega_0^2}\right)$$

The equation which describes the angular motion $\theta(t)$ of the rods with respect to the speed of the telescope $\omega(t)$ can be obtained writing the momentum acting on each ball

$$\left(ml^2\right)\ddot{\theta} = m\omega^2 l^2 \sin\theta\cos\theta - mgl\sin\theta - f\dot{\theta}$$

where f is the friction coefficient of the force acting on the top joint. The linear approximation of the system about the equilibrium point can be easily found considering small deviations $\delta\theta = \theta - \theta_0$ and $\delta\omega = \omega - \omega_0$. The resulting linear differential equation is

$$\ddot{\delta\theta} + \frac{f}{ml^2}\dot{\delta\theta} + \frac{g}{l}\frac{\sin^2\theta_0}{\cos\theta_0}\delta\theta - \left(2\omega_0\sin\theta_0\cos\theta_0\right)\delta\omega = 0 \tag{5}$$

Then, setting $x = \begin{pmatrix} \delta\theta & \dot{\delta\theta} \end{pmatrix}^T$ and $u = \delta\omega$, the previous equation can be written in matrix form as

$$\dot{x}(t) = \begin{pmatrix} 0 & 1 \\ -\frac{g}{l}\frac{\sin^2\theta_0}{\cos\theta_0} & -\frac{f}{ml^2} \end{pmatrix} x(t) + \begin{pmatrix} 0 \\ 2\omega_0 \sin\theta_0 \cos\theta_0 \end{pmatrix} u(t)$$

◁

4 Structural Properties

This section deals with two basic concepts of system theory, controllability and observability, which allow the analysis of the structural relationship between input and output variables and the state variable. The controllability concept is related to the existence of a control strategy $u(t)$ to achieve a particular given objective in the state space. The observability concept is related to the possibility of deriving the state of a system from observation of its output. This is important when the state of the system cannot be measured directly. In fact the behavior of the system depends on the state of the system and, if we only have access to the output, then we must make sure that the state can be mathematically derived from the available measurements. In particular, a state $\tilde{x}$ is said to be *reachable* for the system (2) with $x_0 = 0$ if there exist a time $\tilde{t}$ and an input $\tilde{u}(t)$ steering the origin to $\tilde{x}$ at time $\tilde{t}$, i.e.

$$x(\tilde{t}) = \int_0^{\tilde{t}} e^{A(\tilde{t}-\tau)} b\, \tilde{u}(\tau) d\tau = \tilde{x}$$

On the other hand, a state $\tilde{x}$ is said to be *unobservable* for the system (2) if the output free evolution originated by $\tilde{x}$ is identically zero, i.e.

$$y(t) = ce^{At}\tilde{x} = 0 \quad \text{for all } t \geq 0$$

A system is said to be *reachable* if every state is reachable and *observable* if there is no nonzero unobservable state.

Note that controllability involves statements about input and states, while observability involves output and states. Controllability of a system is closely linked to the question of existence of optimal control and observability stands in a similar position to state estimation. The present form of the definitions of controllability and observability and the recognition of duality between them were worked out by R. E. Kalman. Figure 7 shows the simple structural role of these two properties. Every LTI dynamic system can be in fact decomposed into four parts as shown in the figure. These properties play an important part in theory; roughly speaking, only the controllable and observable part of a system can be identified from input–output data.

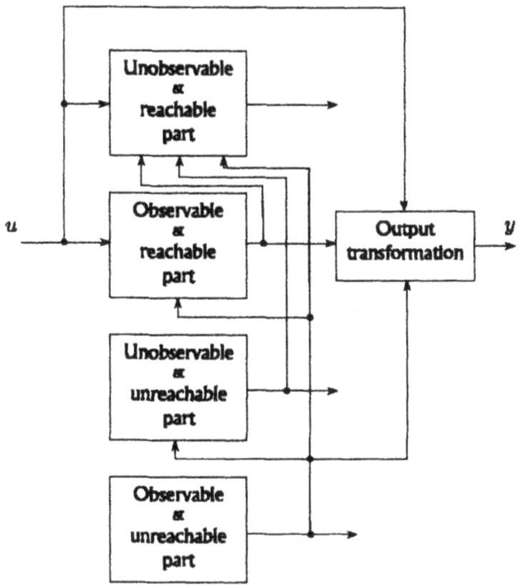

Figure 7: Kalman decomposition.

5 Transfer Functions and Block Diagrams

While the state equation representation is perfectly adequate and contains all the necessary information about the behavior of the system, an input-output relation can be conveniently used to design control systems and to assess important properties with simple tests. For LTI systems, the Laplace transform technique can be applied to equations (2) to obtain an input–output relation in the complex frequency domain s, referred to as ***transfer function.*** The Laplace transform of (2) yields

$$\begin{cases} sX(s) - sx_0 = A\,X(s) + b\,U(s) \\ Y(s) = c\,X(s) \end{cases} \tag{6}$$

where $X(s)$, $Y(s)$ and $U(s)$ are the Laplace transforms of $x(t)$, $y(t)$ and $u(t)$, respectively. Assuming $x_0 = 0$ and eliminating $X(s)$ from equations (6) one has

$$Y(s) = c(sI - A)^{-1}b\,U(s) = F(s)\,U(s) \tag{7}$$

where $F(s)$ is the ***transfer function*** of the system and carries all the information necessary to determine the output in terms of the input (when $x_0 = 0$). The transfer function $F(s)$ of a system (2) with n states is a strictly proper rational function of degree no greater than n. In fact, the poles of $F(s)$ (i.e. the roots of the denominator of $F(s)$) are equal to the eigenvalues of the observable and reachable part of the system.

The transfer function $F(s)$ of a system can be described in the frequency domain by the relation

$$F(j\omega) = F(s)|_{s=j\omega} = Re[F(j\omega)] + j\, Im[F(j\omega)] \tag{8}$$

Alternatively, the transfer function can be represented by a magnitude $|F(j\omega)|$ and a phase $\phi(\omega)$ as

$$F(j\omega) = |F(j\omega)|e^{j\phi(\omega)} \tag{9}$$

where

$$|F(j\omega)|^2 = Re[F(j\omega)]^2 + Im[F(j\omega)]^2 \qquad \phi(\omega) = \tan^{-1}(Im[F(j\omega)]/Re[F(j\omega)])$$

The graphical representation of the frequency response of the system $F(s)$ can utilize either equations (8) or (9). The ***polar (or Nyquist) plot*** representation is obtained by using equations (8). The coordinates of the plot are the real and imaginary parts of $F(j\omega)$ and ω is the current parameter over the plot.

On the other hand, equations (9) are used in the ***logarithmic (or Bode) plot*** representation. In this representation, the plot of logarithmic magnitude in dB [b] versus ω and the plot of the phase $\phi(\omega)$ versus ω are plotted on two different sets of axes. Usually, for convenience, a logarithmic scale of frequency is used.

An important characteristic of the transfer function is that the transfer function of a large structured system can be easily written in terms of the transfer functions of each subsystem. Consider the basic structures depicted in Figure 8. The transfer function $W(s)$ of the series combination (a) is equal to the product of the transfer functions of the individual subsystems, i.e. $W(s) = F_1(s)F_2(s)$ while the transfer function $W(s)$ of the parallel combination (c) is equal to the sum of the transfer functions of the individual subsystems, i.e. $W(s) = F_1(s) + F_2(s)$. Finally, the transfer function $W(s)$ of the feedback structure (b) is $W(s) = F_1(s)/(1 + F_1(s)F_2(s))$ [c].

Example. Revisiting the Watt governor example, by Laplace transforming

[b] The logarithmic magnitude is defined as $20\log_{10}|F(j\omega)|$ where the units are decibels (dB).

[c] It is worth noting that the connection of observable and reachable individual subsystems may destroy the observability and/or controllability properties of the overall system. For example, this happens to be the case when two parallel subsystems have some common poles or when there is a pole–zero cancellation in a series combination. Moreover, if such common or canceled pole has a positive real part, then the overall system displays an unobservable and/or uncontrollable unstable part so that stability cannot be achieved anymore.

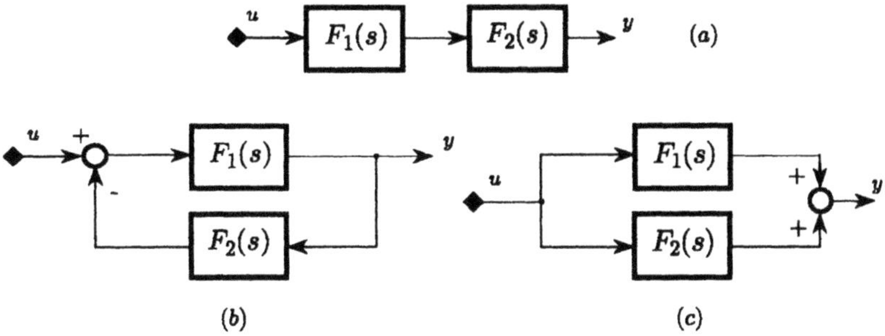

Figure 8: Interconnected systems.

equation (5) one obtains the following transfer function for the governor:

$$\frac{\delta\theta(s)}{\delta\omega(s)} = \frac{k_g}{1 + \frac{2\zeta}{\omega_n}s + \frac{1}{\omega_n^2}s^2}$$

where

$$\omega_n^2 = \frac{g}{l}\frac{\sin^2\theta_0}{\cos\theta_0} \qquad \zeta = \frac{f}{2ml^2} \qquad k_g = \frac{2l\omega_0\cos^2\theta_0}{g\sin\theta_0} > 0$$

The relation between small deviation of θ and small deviation of the torque u driving the telescope can be assumed linear, i.e.

$$\delta u = k_m \delta\theta \qquad \text{with } k_m < 0$$

Note that, k_m is assumed to be negative since the governor acts in such a way to reduce the torque acting on the telescope whenever the speed increases and vice versa. Finally, the speed of the telescope depends on its moment of inertia J and the following holds

$$J\dot{\omega} = u - d$$

where d is the load torque acting on the telescope. Then the transfer function between small deviations of the net driving torque and small deviations of the speed of the telescope is

$$\frac{\delta\omega(s)}{\delta u(s) - \delta d(s)} = \frac{1}{Js}$$

The overall block diagram of the system is shown in Figure 9.

◁

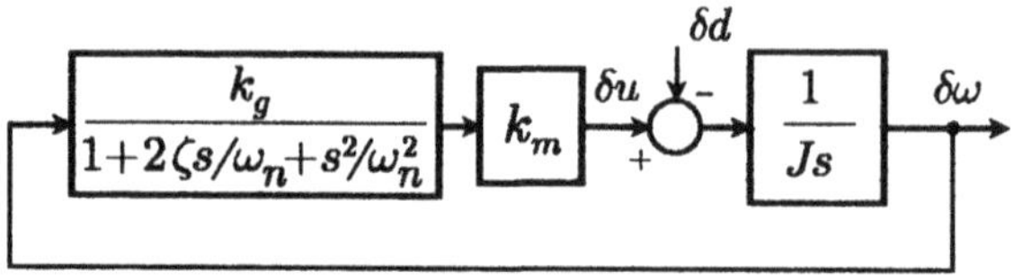

Figure 9: Block diagram of the overall system.

6 The Design of Feedback Control Systems

The design of feedback control systems consists of determining the feedback controller that modifies the behavior of the plant to satisfy, possibly in an optimal way, a set of specifications and constraints. Procedures to determine a controller in the general case of nonlinear, time-varying systems, when available, are at best very complex. As before, we present some considerations about the case of LTI systems that is much easier to tackle and for which synthesis techniques are available. In this case, the design problem is to determine the parameters of an LTI controller. This amounts to the determination of poles, zeros and gain of the transfer function of the controller.

The specifications are in general grouped into two main categories:

- Asymptotic, e.g., stability and steady-state response;
- Transient, e.g., overshoot, settling time and rise time.

Asymptotic specifications can be satisfied by a unique choice of some parameters of the controller, while for transient ones, different solutions may lead to satisfactory results. Hence, in general, a design procedure is recommended that begins by satisfying the asymptotic specifications and then selects, often by a trial-and-error process, the remaining part of the controller to satisfy the transient specifications. The techniques that can be used to satisfy transient specifications are mainly based either on frequency-domain representations of the system or on complex-domain representations (transfer functions).

In the sequel, without loss of generality, we will consider negative unity feedback system as the one shown in Figure 10. We denote by $F(s)$ the transfer function of the open–loop system and by $W(s)$ the transfer function of the overall closed–loop system.

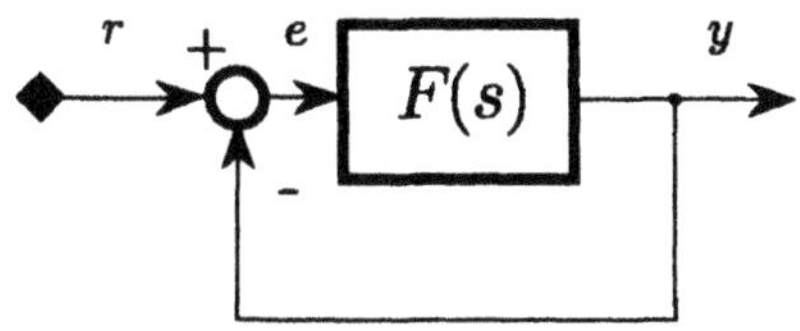

Figure 10: Feedback system.

6.1 Performance Specifications

Time–domain Performance Specifications

Stability is the primary requirement of a feedback control system. In fact, an high–gain controller may tend to over-correct errors thus causing oscillations of larger and larger amplitude, leading to instability. In addition, the response of the system is required to show some characteristics such as a reasonable damping and a reasonably fast speed. Moreover, a control system is required to be able of reducing the error between output and reference to zero or some small tolerable value and to make the output as much as possible independent of the disturbance. Because control systems are inherently dynamic systems, performances are usually specified in terms of both ***steady–state*** error and time ***transient response*** for a specific standard set of test input signals $r(t)$. The steady–state error is usually measured for test input signals $r(t)$ such as step, ramp, and parabola input, that is for signals of the form $r(t) = t^k/k!$ (see Figure 10). For such reference inputs, when the system is asymptotically stable, the steady–state output is well defined and the following definition is given:

Type k system. *A system is referred to as* type *k if it has a constant nonzero steady–sate error e for a reference input* $r(t) = t^k/k!$.

Control system are usually described in terms of their type number and the error constants e.

Standard transient performance measures are usually defined in terms of the step transient response of a system as shown in Figure 11. The swiftness of the response is measured by the rise time t_r and the similarity of the response with regard to the reference step is measured by the overshoot $\hat{s}$ and the settling time t_s. Necessarily, a compromise between the swiftness of the response and the allowable overshoot is needed.

Example. Consider the system shown in Figure 9; in order to make the speed of the telescope as much as possible independent of the disturbance torque d, the steady–state deviations $\delta\omega$ of the speed must be as much as

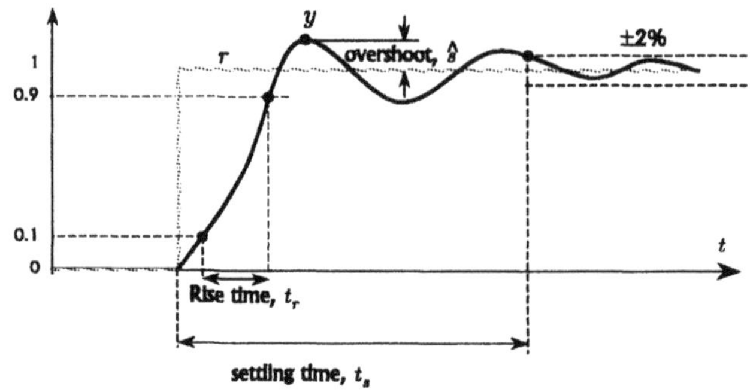

Figure 11: Step response y of a control system.

possible independent of δd. Since the steady–state output value $(\delta\omega)_{ss}$ for a step disturbance δd is

$$(\delta\omega)_{ss} = \frac{1}{k_g k_m}\delta d$$

then the larger the value of $k_g k_m$, the more the system rejects the disturbance. Unfortunately, as will be clear in the sequel, a large value of $k_g k_m$ undermines the stability of the closed–loop system.

◁

Frequency–domain Performance Specifications

In the time–domain, transient performances of a control system have been described in terms of overshoot, settling time and rise time. In the frequency domain, the transient response can be specified in terms of the resonant peak M_r and the bandwidth frequency ω_{BW}. The first parameter is defined as

$$M_r = \frac{\max_\omega |W(j\omega)|}{|W(0)|}$$

and it is usually related to the overshoot to a step input by the relation $0.85 \cdot M_r \cong 1 + \hat{s}$. The second parameter corresponds to the maximum frequency at which the output of a system will track an input sinusoid in a satisfactory manner and is defined as the frequency ω_{BW} for which

$$|W(j\omega_{BW})| = 0.707 \cdot |W(0)|$$

It is related to the speed of the transient response and usually the relation $\omega_{BW} \cdot t_r \cong 3$ does hold.

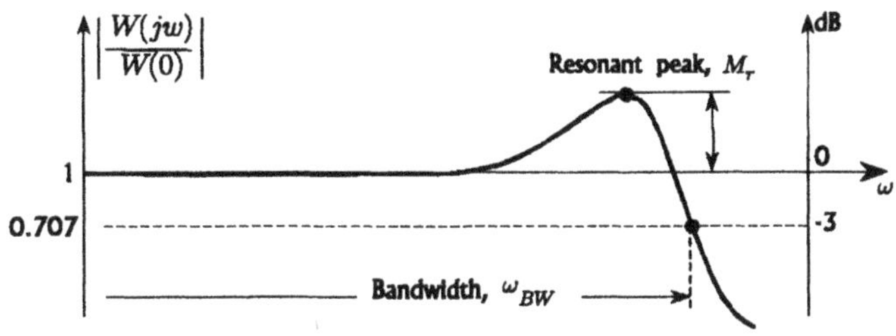

Figure 12: Magnitude characteristic of a control system $W(s)$.

6.2 Stability

As already pointed, stability is the most important specification for a closed-loop system. In this section, we present two important criteria to assess the stability of a closed-loop system, one based on transfer functions and one based on frequency domain responses.

The Routh–Hurwitz Stability Criterion

The Routh–Hurwitz stability method provides an answer to the question of stability by considering the characteristic equation $\varphi(s)$ of the system, that is the denominator of the transfer function $W(s)$ of the overall system,

$$\varphi(s) = a_n s^n + a_{n-1} s^{n-1} + \ldots + a_1 s + a_0 = 0$$

The method is based on ordering the coefficients of $\varphi(s)$ into an array as follows

$$\begin{array}{c|cccc} n & a_n & a_{n-2} & a_{n-4} & \cdots \\ n-1 & a_{n-1} & a_{n-3} & a_{n-5} & \cdots \end{array}$$

Then, further rows of the array are completed as follows

$$\begin{array}{c|cccc} n & a_n & a_{n-2} & a_{n-4} & \cdots \\ n-1 & a_{n-1} & a_{n-3} & a_{n-5} & \cdots \\ n-2 & b_{n-1} & b_{n-3} & b_{n-5} & \cdots \\ n-3 & c_{n-1} & c_{n-3} & c_{n-5} & \\ \vdots & \vdots & \vdots & \vdots & \\ 0 & h_{n-1} & & & \end{array}$$

where

$$b_{n-1} = \frac{-1}{a_{n-1}} \begin{vmatrix} a_n & a_{n-2} \\ a_{n-1} & a_{n-3} \end{vmatrix} \qquad b_{n-3} = \frac{-1}{a_{n-1}} \begin{vmatrix} a_n & a_{n-4} \\ a_{n-1} & a_{n-5} \end{vmatrix}$$

and

$$c_{n-1} = \frac{-1}{b_{n-1}} \begin{vmatrix} a_{n-1} & a_{n-3} \\ b_{n-1} & b_{n-3} \end{vmatrix}$$

and so on.

The Routh-Hurwitz criterion. *If no entry in the first column of the array is zero, then the system is asymptotically stable if and only if all the entries of the first column have the same sign.*

In the case of unstable systems, the method states that the number of roots with positive real part is equal to the number of changes in sign of the first column. When an element of the first column is zero, then suitable modifications of the array calculation procedure allow to compute the number of roots with positive real part.

Example. In order to investigate the stability of the closed–loop system depicted in Figure 9, one can consider the characteristic equation $\varphi(s)$ of the closed–loop system and apply the Routh stability method. The characteristic equation of the closed–loop system is

$$\varphi(s) = \left(1 + \frac{2\zeta}{\omega_n}s + \frac{1}{\omega_n^2}s^2\right) Js - k_g k_m = \frac{J}{\omega_n^2}s^3 + \frac{2\zeta J}{\omega_n}s^2 + Js - k_g k_m$$

and the corresponding Routh–array is

$$\begin{array}{c|ll} 3 & J/\omega_n^2 & J \\ 2 & 2\zeta J/\omega_n & -k_g k_m \\ 1 & (J/\omega_n^2)(k_g k_m + 2\zeta J\omega_n) & \\ 0 & -k_g k_m & \end{array}$$

Since $k_g k_m < 0$, then the closed–loop stability is achieved if $2\zeta/\omega_n > -k_g k_m/(J\omega_n^2)$ i.e., for a sufficiently large value of the damping coefficient $2\zeta/\omega_n$ of the fly–ball transfer function. It is worth noting that the larger the value of $k_g k_m$, i.e. the more the closed–loop system rejects the disturbance δd, the larger the damping coefficient has to be in order to ensure stability. To this end, Airy proposed the use of a hydraulic damper.

◁

Stability Criteria in the Frequency Domain

The frequency response of a system represents the sinusoidal steady–state response of a system and, in the case of asymptotically stable systems, can be readily obtained experimentally by exciting the system with sinusoidal input signals. It provides sufficient information for the determination of the stability of the system so that it can be utilized to investigate the stability of the system when the system parameter values have not been determined and the Routh-Hurwitz criterion cannot be applied.

A frequency-domain stability criterion was developed by H. Nyquist in 1932 and remains a fundamental approach to the investigation of the stability of linear control systems. The criterion is as follows:

The Nyquist Criterion. *The feedback system of Figure 10 is asymptotically stable if and only if, for the polar plot of $F(j\omega)$ with $\omega \in (-\infty, +\infty)$, the number of counterclockwise encirclements of the $(-1, 0)$ point is equal to the number of poles of $F(j\omega)$ with positive real part.*

The Nyquist criterion is defined in terms of the $(-1, 0)$ point on the polar plot. Clearly, the proximity of the $F(j\omega)$–polar plot to this stability point is a measure of the robustness of the stability of the system. It can be expressed by means of the ***phase margin*** or ***gain margin***. The phase margin φ_m is defined as the phase angle through which the $F(j\omega)$ plot must be rotated in order that the unity magnitude point $|F(j\omega_t)| = 1$ passes through the $(-1, 0)$ point. The gain margin k_m is defined as the reciprocal of the gain $|F(j\omega_\pi)|$ at the frequency ω_π at which the phase angle reaches -180°; it is the factor by which the system gain would have to be increased for the $F(j\omega)$ plot to pass through the $(-1, 0)$ point. The phase and gain margins can also be defined in terms of the 0 dB, -180° point on the logarithmic plot. The above definitions are illustrated in Figure 13 on the polar and logarithmic plots.

Sometimes, it is possible to apply the Nyquist criterion directly on the Bode plots of $F(j\omega)$, and this gives the following criterion:

The Bode Criterion. *When $F(s)$ has no poles with positive real part and the magnitude plot of $F(j\omega)$ has only one intersection with the 0 dB axis, then the feedback system of Figure 10 is asymptotically stable if and only if, the gain of $F(s)$ and the phase margin are positive.*

Example. In order to investigate the stability of the closed–loop system depicted in Figure 9, the magnitude (in dB) and the phase of the open–loop system

$$F(s) = \frac{-k_g k_m}{\left(1 + \frac{2\zeta}{\omega_n}s + \frac{1}{\omega_n^2}s^2\right) Js}$$

are depicted in Figure 14 for a small and a large value of the damping coefficient

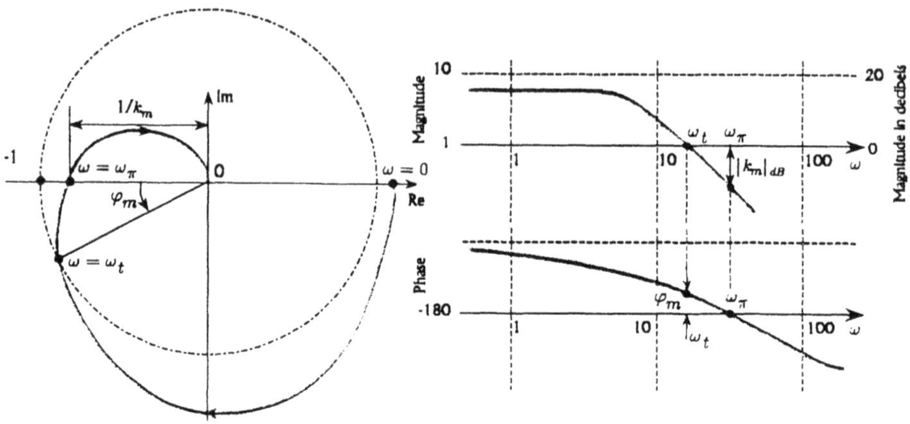

Figure 13: Phase and gain margin definitions.

$2\zeta/\omega_n$. The phase margin is equal to 0 when $F(j\omega) = -1$ that is when $\omega = \omega_n$ and $2\zeta/\omega_n = -k_g k_m/(J\omega_n^2)$. Then the closed–loop system is asymptotically stable when $2\zeta/\omega_n > -k_g k_m/(J\omega_n^2)$ as already known. ◁

6.3 Synthesis Techniques: a Sketch

In the case of a design based on the frequency domain representation of the system, it is important to relate the graphical representations of the open-loop $F(j\omega)$ and of the closed-loop $W(j\omega)$, so that the transient response characteristics can be translated in terms of the characteristics of the frequency response of the open-loop system. Nichols charts yield a very popular graphical procedure that allows this mapping of closed-loop specifications into open-loop specifications. Then, the synthesis procedure is based on modifying the open-loop frequency response $F(j\omega)$ in order to satisfy the open-loop specifications. The simplest controllers that can be used for this purpose are the so-called Proportional-Integrator-Differentiator (PID) controllers that are characterized by a very simple transfer function that contains a constant term, a term in s and one in $1/s$. Hence the design problem is to determine the values of the coefficients of these terms so that the composition of the PID controller and of $F(j\omega)$ satisfies the specifications. Lead-lag compensators are elementary controllers that can be used in more general situations.

In the case of transfer function based synthesis techniques, the quality of the transient response can be easily related to the location of the poles of the

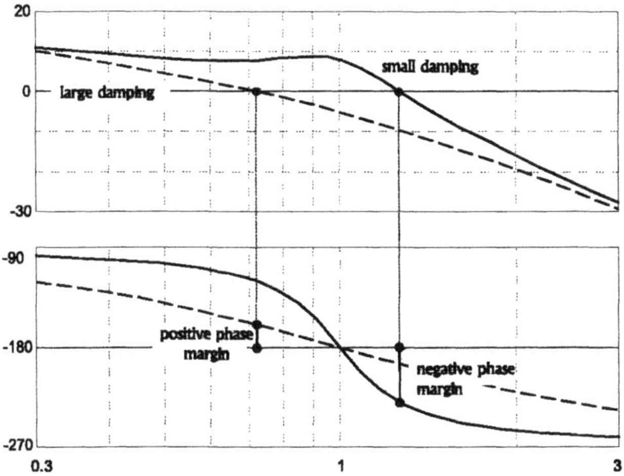

Figure 14: Bode plots of $F(s)$ for a small and a large value of the damping coefficient.

transfer function $W(s)$ on the complex plane. The problem here is then to correlate the values of poles, zeros and gain of the closed-loop system with the ones of the open-loop system. A very much used graphical procedure to do so is the so-called *root-locus technique*.

7 Conclusions

In this chapter, we briefly presented the basic foundations of control theory and design. Given the space limitations, this review can only be useful as a reference and a taxonomy of issues and results in control. We point the readers to the textbooks in the references for a complete treatment of the subject.

Among the issues we have not covered, we recommend the interested reader to look at:

- State-space methods, based on the structural concepts of controllability, observability and minimal realizations, introduced by Kalman in his seminal papers[10].

- Digital control, which covers the theory and design of control systems based on digital hardware (mostly programmable processors). Digital controllers are used in most of the modern control applications due to their flexibility, availability, cost and computational power. Several textbooks and papers (see e.g. [6, 15]) have been published on this important topic.

- Control techniques for Multi–Input Multi–Output (MIMO) systems, such as eigenvalues assignment [9], Linear Quadratic Regulator (LQR) Control [12], H_2-H_∞ control [4], l_1 control [2], Loop Transfer Recovery [16].
- Analysis and design of nonlinear control systems [7], [11].

References

1. J. Carr, *Applications of Centre Manifold Theory*, Springer–Verlag, 1981.
2. M. A. Dahleh and I. Diaz-Bobillo, *Control of Uncertain Systems: A Linear Programming Approach*, Prentice-Hall, 1995.
3. R. C. Dorf, *Modern Control Systems*, 6th Ed., Addison–Wesley, 1992.
4. J. Doyle, B. Francis and A. Tannenbaum, *Feedback Control Theory* , Macmillan, 1992.
5. G. F. Franklin and J. D. Powell, *Feedback Control of Dynamic Systems*, 3rd Ed., Addison–Wesley, 1994.
6. G. F. Franklin, J. D. Powell and M.L. Workman, *Digital Control of Dynamic Systems*, 2nd Ed., Addison–Wesley, 1990.
7. A. Isidori, *Nonlinear Control Systems*, 3rd Ed., Springer–Verlag, 1995.
8. A. Isidori, *Sistemi di controllo*, Seconda Edizione, Vol.1 e 2, Siderea, 1992 (in Italian).
9. T. Kailath, *Linear Systems*, Prentice–Hall, 1980.
10. R. E. Kalman, P. Falb and M. A. Arbib, *Topics in Mathematical System Theory*, McGraw-Hill, 1967.
11. H. K. Khalil, *Nonlinear Systems*, 2nd Ed., Prentice-Hall, 1996.
12. H. Kwakernaak and R. Sivan, *Linear Optimal Control Systems*, John Wiley & Sons, 1972.
13. D. G. Luenberger, *Introduction to Dynamic Systems*, John Wiley & Sons, 1979.
14. K. Ogata, *System Dynamics*, 2nd Ed., Prentice–Hall, 1992.
15. K. Ogata, *Discrete-Time Control Systems*, Prentice-Hall, 1987.
16. A. Saberi, Ben M. Chen and Peddapullaiah Sannuti, *Loop Transfer Recovery: Analysis and Design*, Springer-Verlag, 1993.

INTRODUCTION TO THE MECHANICAL SIMULATION OF SEISMIC ISOLATION SYSTEMS

ANDREA VICERÉ
INFN, Sezione di Pisa, Via Vecchia Livornese 1291, I-56125 Pisa, Italy
Email: Andrea.Vicere@pi.infn.it

In these few lessons we introduce some of the basic tools needed to develop models of mechanical systems; we have no ambition of generality, and focus instead on the the modeling of the attenuation elements used in interferometers. Because of their function, they present some peculiarities, which make their simulation different from the usual engineering applications of structural analysis, and require ad-hoc solutions, which we try to present in a self contained manner.

1 Motivations

The optical elements of ground based interferometers require a suspension system capable of isolating from the ground seismic motion and of steering the optical elements, to keep the interferometer in the *locked* state. Among the interferometric experiments VIRGO is the most ambitious in requiring that the seismic noise is not a limiting factor already at 4 Hz. To this end, the VIRGO suspensions (the super-attenuators, SAs) are composed of several attenuating elements, cascaded in a multiple oscillator with many low frequency resonances: at frequencies larger than those of the resonances we have a steep transfer function from the soils motion to the suspended element, and therefore a good attenuation. The price to pay is the complication of the system, which has to isolate in all the degrees of freedom, while in turn making it possible to control the position of the mirror, through feedback loops, displayed in Fig. (1) which receive informations from the interferometer itself, or from some local sensors. To predict its behavior, a numerical model is needed.

2 System analysis

An attenuation system, as displayed in Fig. (1) is characterized by the alternance of relatively rigid elements (masses, filter bodies, ...), and comparatively elastic elements (suspension wires, blades ...).

This means that rigid parts can truly be considered elementary, that is described only in terms of their rigid translations and rotations.

These elements are connected by the elastic ones, whose internal mechan-

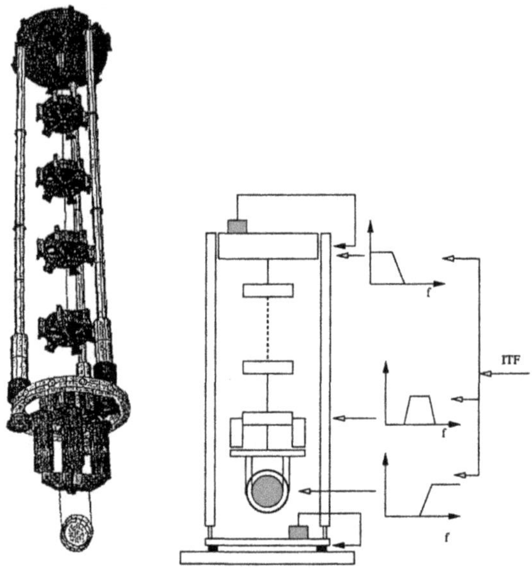

Figure 1. A VIRGO super-attenuator, and its control scheme.

ical stresses (together with the gravity!) give rise to the recalling forces which make the system behave as a multiple oscillator.

There are inputs, which may be forces applied, like those needed to steer the elements, or noises, like the seismic noise motion of the base of the suspension; and there are interesting outputs, for instance read-outs of accelerometers placed on some stage, or the position or velocity of some element.

The questions typically asked to the simulation software are to evaluate the transfer functions between arbitrary inputs and outputs, and to provide a model of direct use in simulating the control loop of the system. The first question has meaning because the typical amplitudes of the motion are so small[a] that a linear approximation suffices, and the transfer function

$$TF_{i \to o}(f) \equiv \frac{\tilde{o}(f)}{\tilde{i}(f)} \tag{1}$$

is well defined.

[a]With a proviso: there are effects which require a non-linear treatment, like the so called up-conversion effects.

The representation in terms of transfer functions is sufficient for many purposes[b], but for controlling the system it is more effective to build a state space model, which is a first order in time evolution equation for the system:

$$\frac{dQ}{dt} = A \cdot Q + B \cdot U, \qquad Y = C \cdot Q + D \cdot U\,; \tag{2}$$

in a mechanical system, Q is a vector containing both positions and velocities of the mechanical elements. Which elements are actually contained in Q depends on the accuracy of the simulation: for instance in a certain range of frequencies one may require more elements (say, more finite elements) to catch certain resonances which would go otherwise unnoticed.

The vector U contains the inputs, noises or external forces applied, while the vector Y contains the interesting outputs, for instance the coordinates of the mirror and the read-outs of accelerometers or other position sensors.

3 Potential methods

If the mechanical elements can be divided in "masses", where the potential and kinetic energy is accounted for by the overall motion, and "links", where the internal degrees of freedom are relevant, a further simplification is sometimes possible: we can neglect altogether the mass of the elastic elements.

In this case the equation of motion for the internal DOFs of the "link" do not depend on time and reduce to the minimization of the potential energy, at fixed boundary conditions.

To be more precise, given a potential energy depending on a continuum of internal degrees of freedom $x(\xi)$ and on external coordinates q, we have

$$U(q) = \min_x \int U(x(\xi), q)\, d\xi \tag{3}$$

note that even if the internal degrees are continuum, the resulting potential depends on a finite number of variables, the ones which determine the boundary: at fixed boundary the elastic massless element relaxes "instantaneously" to the internal shape with minimum energy.

In practice one assumes that $x(\xi)$ takes the shape dictated by the minimum energy limit, and therefore its time dependence is dictated by the boundary conditions $q(t)$: that is we have $x(\xi, t) = x(\xi\, q(t))$.

[b] An advantage of the transfer matrix representation is that it may not require to build a time-domain model, thus allowing some simplifications: for instance, the use of structural damping.

Now if we are studying the system at frequencies well below the internal resonance frequencies, we expect that the internal mass of the "link" is not very important. However this is not quite true: there is an effect even at low frequencies, due to the fact that the kinetic energy

$$K(\dot{q}) = \int K(\frac{dx(\xi, q(t))}{dt})\, d\xi \tag{4}$$

depends on time through $q(t)$. Taking into account this effect one can extend the range of validity of the approximation, as will shall see in the following, without paying any price in terms of the number of degrees of freedom.

A few examples will make everything completely clear.

3.1 Compressed wire

Consider the longitudinal motion of a wire, having a linear density ρS and a Young modulus E, described by a coordinate $x(\xi)$, where $\xi \in [0,\, L]$: we have

$$K = \frac{1}{2}\int_0^L \rho S\, [\dot{x}(\xi, t)]^2\, d\xi, \qquad U = \frac{1}{2}\int_0^L ES\, [x'(\xi, t)]^2\, d\xi \tag{5}$$

The state is specified, in the potential approximation, by the boundaries $q = [x(0), x(L)]$, and the condition of minimal energy is $x''(\xi, t) = 0$, hence

$$x(\xi, t) = x(0, t)(1 - \frac{\xi}{L}) + x(L, t)\frac{\xi}{L} \tag{6}$$

and the potential energy is immediately given by

$$U(x(0, t), x(L, t)) = \frac{1}{2}\frac{ES}{L}\, [x(L, t) - x(0, t)]^2 \tag{7}$$

which is as expected the potential of a simple spring. But we can go beyond the simple spring and compute also the kinetic energy, in the above described low frequency approximation:

$$K = \frac{1}{2}\int_0^L \rho S\, [\dot{x}(\xi, t)]^2\, d\xi = \frac{1}{2}\rho S L \frac{1}{3}\left[\dot{x}(0, t)^2 + \dot{x}(0, t)\dot{x}(L, t) + \dot{x}(L, t)^2\right] \tag{8}$$

which is some sort of average of the boundary values.

Assume to drive one end of the wire with a prescribed $x(0, t)$ and measure on the other side $x(L, t)$, where a mass M has been connected; this is what is needed to compute the *transfer function.* Then the equation of motion is

$$\rho\ddot{x} - Ex'' = 0 \tag{9}$$

subject to the assigned $x(0,t)$ at one extremum, and to the condition $ESx'(L,t) = M\ddot{x}(L,t)$ at the other extremum, which means the effect of the inertia of the mass [c]. We have in the frequency domain the exact solution

$$x(\xi\,\omega) = x(0,\omega)\left[\cos\kappa\xi - \frac{\cos\kappa L + \frac{\rho S}{M}\sin\kappa L}{\sin\kappa L - \frac{\rho S}{M}\cos\kappa L}\sin\kappa\xi\right] \tag{11}$$

where $\kappa \equiv \frac{\omega^2\rho}{E}$. This solution, computed in $\xi = L$, gives[d]

$$x(L,\omega) = x(0,\omega)\frac{1}{\cos\kappa L - \frac{M}{\rho S}\sin\kappa L} \tag{13}$$

that can now be directly compared with the potential approximation

$$x_0(L,\omega) = x(0,\omega)\frac{\frac{ES}{ML}}{\frac{ES}{ML} - \omega^2} \tag{14}$$

and also with the solution obtained including the correction in Eq. (8)

$$x_1(L,\omega) = x(0,\omega)\frac{\frac{ES}{ML} + \omega^2\frac{\rho SL}{6M}}{\frac{ES}{ML} - \omega^2\left(1 + \frac{\rho SL}{3M}\right)} \tag{15}$$

We plot in Fig. (2) the transfer functions, defined as the ratio $\left|\frac{x(L,\omega)}{x(0,\omega)}\right|$, for particular values of the numerical parameters[e]. The system behaves at low frequencies as a simple spring, and thanks to the small ratio $\frac{\rho SL}{M}$ only at large frequencies the internal structure becomes important. A better approximation, than a simple spring, is obtained taking into account the internal mass as in Eq. (8). The structure displayed in Fig. (2), despite the simplicity of the example, will repeat itself when considering other kinds of elements.

---

[c] A force F at the extremum can be represented by a term $Fx(L,t)$ in the potential energy: when taking the variation of the Lagrangian

$$\int_0^L \rho S\delta\dot{x}(\xi)\dot{x}(\xi)d\xi - \int_0^L ES\delta x'(\xi)x'(\xi) - F\delta x(L) = 0 \tag{10}$$

the integration by parts gives the bulk equation Eq. (9) and the equation at the extremum $ESx'(L,t) = F$. Now one can put for F the inertia $-M\ddot{x}$ of the attached mass.

[d] Note the limit $M \to 0$

$$x(\xi,\omega) = x(0,\omega)\frac{\cos(\kappa(L-\xi))}{\cos(\kappa L)} \tag{12}$$

[e] The parameter used are $E = 1.87\times 10^9, M = 100, L = 2, \rho = 7000, S = \pi/10$.

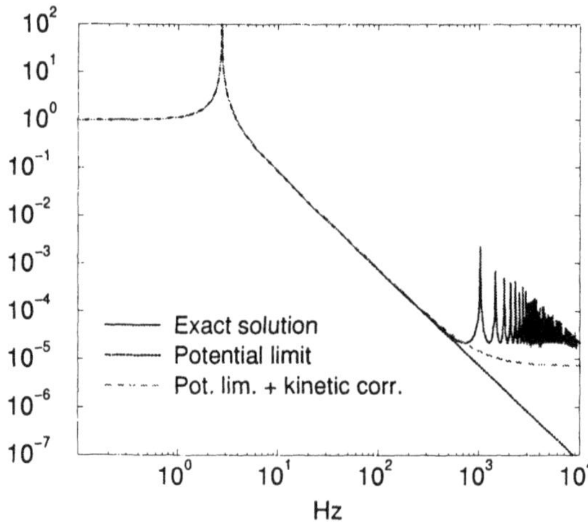

Figure 2. Different approximations of the transfer function of a wire spring oscillator.

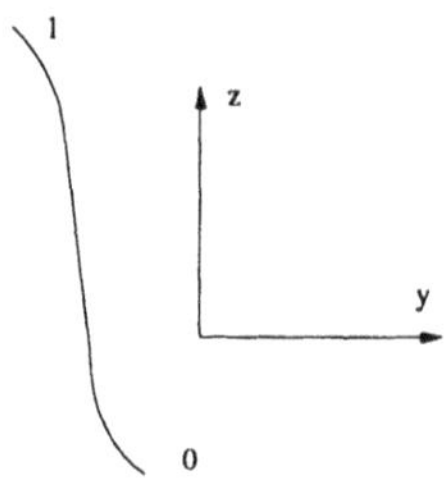

Figure 3. Coordinates for the simple beam in two dimensions.

3.2 Beam in tension (or compression)

At least in the VIRGO suspension the basic element is the simple pendulum, realized with steel wires, which in the language of structural mechanics are just beams under a tension T. Also they are important the beams under compression (in our notation, negative tension) for instance for describing the flex joints of an inverted pendulum.

Also in this case it is immediate [1] to write down the distributed form of the kinetic and potential energy: making reference to Fig. (3) and considering for simplicity just the motion in the $x = 0$ plane, we have

$$K = \frac{1}{2}\int_0^L \rho S \ddot{y}^2 dz, \qquad U = \frac{1}{2}\int_0^L \left[EI\left(y''\right)^2 + Ty'^2\right] dz \tag{16}$$

where $I \equiv \frac{\pi r^4}{4}$. We redefine the units, introducing $z = \zeta L$, and obtain

$$U = \frac{EI}{2L^3}\int_0^1 \left[y''^2 + k^2 y'^2\right] d\zeta = \frac{TL}{2}\frac{1}{k^2 L^2}\int_0^1 \left[y''^2 + k^2 y'^2\right] d\zeta \tag{17}$$

where $k^2 \equiv \frac{TL^2}{EI}$ has no dimensions. As before, we can derive approximations to the potential or the solution of particular motion equations.

Potential limit

Neglecting the internal mass the motion equation *in the bulk* is

$$EIy'''' - Ty'' = 0 \tag{18}$$

which can be solved in terms of 4 boundary conditions, for instance the position and first derivative in $0, L$: let us assemble the boundaries in the vector

$$\mathbf{y}^T = (y(0),\ y'(0),\ y(1),\ y'(1)) \tag{19}$$

. The most general solution of Eq. (18) is a linear combination as

$$y(\zeta) = (\sinh k\zeta,\ \cosh k\zeta,\ k\zeta,\ 1)^T \cdot \mathbf{c} \tag{20}$$

where $\mathbf{c}^T = (c_1,\ c_2,\ c_3,\ c_4)$ is a vector of coefficients.

Then the potential energy can be written in "matrix" form, noting that

$$y''(\zeta)^2 = k^4 \mathbf{c}^T \cdot \mathbf{Y}_2(\zeta) \cdot \mathbf{c} \tag{21}$$

where the non zero entries of the matrix $\mathbf{Y}_2$ are

$$\mathbf{Y}_2[1:2, 1:2] \equiv \begin{bmatrix} \sinh(k\zeta)^2 & \sinh k\zeta \cosh k\zeta \\ \sinh k\zeta \cosh k\zeta & \cosh(k\zeta)^2 \end{bmatrix} . \tag{22}$$

With analogous formulas for $y'(\zeta)$ one defines the integrand of U and then

$$U = \frac{TL}{2}\frac{k^2}{L^2}\mathbf{c} \cdot \mathbf{A} \cdot \mathbf{c}, \quad \mathbf{A}[1:3, 1:3] = \begin{bmatrix} \frac{\sinh(2k)}{2k} & \frac{\sinh(k)^2}{k} & \frac{\sinh(k)}{k} \\ \frac{\sinh(k)^2}{k} & \frac{\sinh(2k)}{2k} & \frac{2\sinh(\frac{k}{2})^2}{k} \\ \frac{\sinh(k)}{k} & \frac{2\sinh(\frac{k}{2})^2}{k} & 1 \end{bmatrix} . \tag{23}$$

Then we impose on $\mathbf{c}$ the boundary conditions $\mathbf{x}$

$$\mathbf{B}\cdot\mathbf{c}=\mathbf{x}, \quad \text{where} \quad \mathbf{B}=\begin{bmatrix} 0 & 1 & 0 & 1 \\ k & 0 & k & 0 \\ \sinh(k) & \cosh(k) & k & 1 \\ k\cosh(k) & k\sinh(k) & k & 0 \end{bmatrix}. \tag{24}$$

It follows that

$$U=\frac{TL}{2}\frac{k^2}{L^2}\mathbf{x}^T\cdot\left[\left(\mathbf{B}^{-1}\right)^T\cdot\mathbf{A}\cdot\mathbf{B}^{-1}\right]\cdot\mathbf{x}\equiv\frac{TL}{2}\mathbf{x}^T\cdot\mathbf{W}\cdot\mathbf{x} \tag{25}$$

where the matrix $\mathbf{W}$ is given by

$$\mathbf{W}=\frac{k}{L^2\left[k-2\tanh(\frac{k}{2})\right]}\times \tag{26}$$

$$\begin{bmatrix} 1 & \frac{\tanh(k/2)}{k} & -1 & \frac{\tanh(k/2)}{k} \\ \frac{\tanh(k/2)}{k} & \frac{1}{k}\left(\frac{1}{\tanh(k)}-\frac{1}{k}\right) & -\frac{\tanh(k/2)}{k} & \frac{1}{k}\left(\frac{\tanh(k)}{k}-\frac{1}{\cosh(k)}\right) \\ -1 & -\frac{\tanh(k/2)}{k} & 1 & -\frac{\tanh(k/2)}{k} \\ \frac{\tanh(k/2)}{k} & \frac{1}{k}\left(\frac{\tanh(k/2)}{k}-\frac{1}{\cosh k}\right) & -\frac{\tanh(k/2)}{k} & \frac{1}{k}\left(\frac{1}{\tanh k}-\frac{1}{k}\right) \end{bmatrix}$$

A similar formula holds for the x coordinate: taking into account that z had been rescaled to $z=L\zeta$, and referring to Fig. (3), the correspondence between derivative x', y' and rotation angles around axes $\mathbf{y}, \mathbf{x}$ are

$$x'(0,1)=+L\theta_y(0,1), \qquad y'(0,1)=-L\theta_x(0,1) \tag{27}$$

Hence the total potential energy of the elastic line is given by

$$U=\frac{TL}{2}\left[\mathbf{y}^T\cdot\mathbf{W}\cdot\mathbf{y}+\mathbf{x}^T\cdot\mathbf{W}\cdot\mathbf{x}\right] \tag{28}$$
$$\mathbf{x}=\left(x(0),\, L\theta_y(0),\, x(1), L\theta_y(1)\right), \quad \mathbf{y}=\left(y(0),\, -L\theta_x(0),\, y(1), -L\theta_x(1)\right)$$

which accounts both for the energy due to the wire's bending (EI term in the Lagrangian) and for the work done against the tension T, for instance, against a gravitational potential originating the tension because of a suspended mass.

Kinetic correction

As in Sec. (3.1) we can write down an approximation to the kinetic energy,

$$K\equiv\frac{1}{2}\int_0^L S\rho\,(\dot{y})^2\,dz \tag{29}$$

substituting for y the solution in terms of the boundary conditions: we quote the result only in the limit of large $k \equiv \frac{TL^2}{EI}$ratio

$$K = \frac{1}{2} SL\rho \dot{\mathbf{y}}^T \cdot \mathbf{W}_{kin} \cdot \dot{\mathbf{y}}$$

$$\mathbf{W}_{kin} = \frac{1}{(k-2)^2} \begin{pmatrix} (\frac{k^2}{3} - k - 1) & -2 & (\frac{k^2}{6} - k + 3) & 1 \\ -2 & 1/3 & -1 & -1/6 \\ \dots & \dots & (\frac{k^2}{3} - k - 1) & 2 \\ \dots & \dots & \dots & \frac{1}{3} \end{pmatrix} .$$

Motivation for the large tension approximation

In the interferometric experiments the suspension is built so as to have almost all the recalling energy due to gravitation: the radius of the wires and therefore the EI term is kept as small as possible. It is therefore sensible to take the large tension limit, in our case the limit of large k. An asymptotic expansion is possible: setting $\tanh(k/2) \simeq 1$ and $\cosh(k) \simeq \infty$ in Eq. (27)

$$\mathbf{W} = \frac{k}{L^2(k-2)} \begin{bmatrix} 1 & \frac{1}{k} & -1 & \frac{1}{k} \\ \frac{1}{k} & \frac{k-1}{k^2} & -\frac{1}{k} & \frac{1}{k^2} \\ -1 & -\frac{1}{k} & 1 & -\frac{1}{k} \\ \frac{1}{k} & \frac{1}{k^2} & -\frac{1}{k} & \frac{k-1}{k^2} \end{bmatrix} ; \tag{30}$$

this expression can be further expanded in $1/k$, obtaining at lowest order the potential of a simple pendulum.

3.3 Blade

The other important element of the VIRGO suspension is the *blade* which is used as a vertical spring: in Fig. (4) we display a blade of length L and thickness h holding a mass M. It is of triangular (actually trapezoidal) shape, and pre-bent so as to assume horizontal position when loaded with the correct mass. The triangular shape (actually $W(0) \gg W(L)$) ensures that the stress remains constant along the blade: indeed the torque due to the mass grows linearly with the distance from the mass. At the wide end the blade is clamped, and therefore $z'(0) = 0$, while at the short end the load is suspended by a thin wire which is equivalent to an *hinged* clamp, therefore no torque is applied and $z''(L) = 0$.

The Landau book [1] provides us with the Lagrangian for the blade

$$\mathcal{L} = \frac{1}{2} \int_0^L dx \left\{ \rho(x)(\dot{z}(x))^2 - \frac{Eh^3 w(x)}{12(1-\sigma^2)} z''(x)^2 \right\} \tag{31}$$

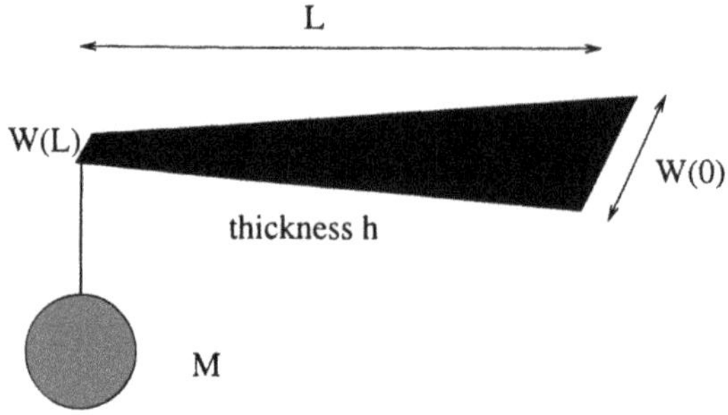

Figure 4. A blade suspending a mass M.

where σ is the Poisson modulus, and one has

$$w(x) = w(0)\left[(1-\frac{x}{L}) + \frac{w(L)}{w(0)}\frac{x}{L}\right] ; \tag{32}$$

we can proceed exactly as before in deriving approximations to the Lagrangian.

Potential limit

Neglecting the mass and taking the variation the equation for the blade is

$$\frac{d^2}{d\,x^2}\left[w(x)\,z''(x)\right] = 0, \qquad \text{plus} \qquad z'(0) = z''(L) = 0\,. \tag{33}$$

Integrating twice we obtain immediately

$$z''(x) = \frac{a_1(x-L)}{w(x)},\; a_1 \equiv \frac{2w(0)(1-\beta)^3(z(0)-z(L))}{L^3(1-4\beta+3\beta^2-2\beta^2\ln\beta)},\; \beta \equiv \frac{w(L)}{w(0)}. \tag{34}$$

One can now directly integrate the potential energy and obtain

$$U = \frac{1}{2}\frac{Eh^3}{12(1-\sigma^2)}\int_0^L w(x)\left[z''(x)\right]^2 dx = \frac{1}{2}\frac{Eh^3}{12(1-\sigma^2)}\int_0^L \frac{a_1^2(x-L)^2}{w(x)}dx \tag{35}$$

$$= \frac{1}{2}\frac{Ew(0)h^3}{4L^3(1-\sigma^2)\alpha}(z(0)-z(L))^2,\; \alpha \equiv \frac{3}{2(1-\beta)}\left[3 - \frac{2}{1-\beta}\left(1+\frac{\beta^2\ln\beta}{1-\beta}\right)\right] ;$$

as expected the potential energy is the one of a simple spring.

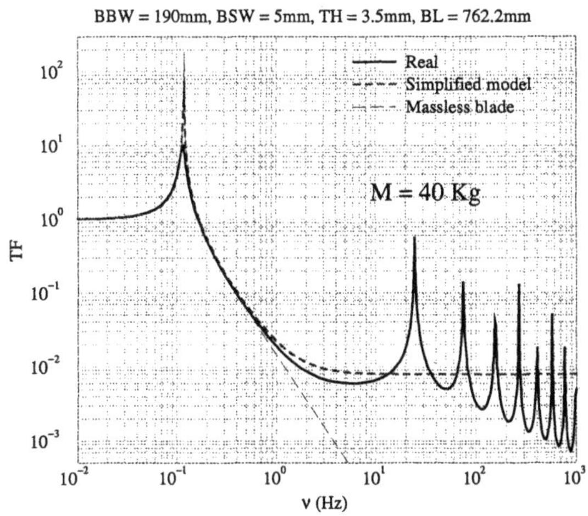

Figure 5. Different approximations to the transfer functions of a single VIRGO blade.

Kinetic correction

Having $z''(x)$ we can integrate to get $z(x)$ and then compute the blade kinetic energy in the low frequency approximation: we have $\rho(x) = \rho h w(x)$ where ρ is the volume density of the metal, and

$$K = \frac{1}{2}\int_0^L dx\, \rho h w(x)\, [\dot{z}(x)]^2 \; ; \tag{36}$$

the computation is direct and the result is a bit cumbersome:

$$\begin{aligned} K = \frac{1}{2}\rho \frac{h w(0) L}{2400\alpha^2(\beta-1)^4} \Big\{ & \dot{z}(0)^2 \left[200\alpha^2(\beta-1)^3(\beta^2-6) + 3(60+8\beta-181\beta^2+93\beta^3) + 20\alpha^2(-30+40\beta-60\beta^3+23\beta^4)\right] \\ & + 3\,\dot{z}(L)^2 \left[60+8\beta-(181+200\alpha+200\alpha^2)\beta^2+(93+400\alpha+600\alpha^2)\beta^3-60\alpha(3+10\alpha)\beta^4+200\alpha^2\beta^5\right] \\ & + 2\dot{z}(0)\dot{z}(L)\left[200\alpha^2(\beta-1)^3\beta^2-3(60+8\beta-181\beta^2+93\beta^3+20\alpha(15-20\beta+2\beta^4)\right] \Big\} \end{aligned}$$

The presence of the mixed term $\dot{z}(0)\dot{z}(L)$ gives rise to a ω^2term at numerator in the transfer function, which is the lowest order effect of the internal mass of the blade in correcting the attenuation properties of the spring. We compare in Fig. (5) the transfer functions in the massless approximation with the one taking into account the lowest order mass contribution, and the exact one: the behavior is similar to the one obtained with a simple spring.

3.4 Damping in the potential approximation

The internal motion in the elastic elements, for instance the bending of the sections of a wire, lead to dissipation and ultimately to the damping of the oscillations.

The simplest possible dissipation model can be derived from a viscous term in the stress-strain relation, as follows

$$\sigma = E\left(\varepsilon + \gamma\dot{\varepsilon}\right) . \tag{37}$$

This model predicts a dissipation increasing with the frequency, which is not observed in the experiments: however such a model can still be used tuning the γ parameter to the measured dissipation at the resonance. It results a modification of the Lagrange equations which can be put in the form

$$\frac{d}{dt}\frac{\partial \mathcal{L}}{\partial \dot{\mathbf{q}}} - \frac{\partial \mathcal{L}}{\partial \mathbf{q}} + \frac{\partial \mathcal{F}}{\partial \dot{\mathbf{q}}} = 0 \tag{38}$$

where $\mathcal{F}$ is the so called Rayleigh dissipation function.

At second order it is possible to write the $\mathcal{F}$ as a quadratic form in $\dot{\mathbf{q}}$,

$$\mathcal{F} = \frac{1}{2}\dot{\mathbf{q}} \cdot \mathbf{F} \cdot \dot{\mathbf{q}} . \tag{39}$$

The problem reduces to the computation of the Rayleigh dissipation function, and we shall provide as an example, again, the beam under tension: to do this, notice that if $\mathbf{F}\cdot\dot{\mathbf{q}}$ is the force due to the dissipation term, the power dissipated $\dot{\mathbf{q}} \cdot \mathbf{F} \cdot \dot{\mathbf{q}}$ gives (apart the 1/2 factor) the Rayleigh dissipation function.

Beam in tension

For a beam in tension, the dissipation is connected only to the elasticity, and not to the origin of the tension (say, the gravity): in other words it is proportional to the work done by the bending moment.

The torque acting on a section of the beam is given by

$$M_y(z) = -EI\left[\frac{\partial^2 y}{\partial z^2} + \gamma\frac{\partial^3 y}{\partial t \partial z^2}\right] ; \tag{40}$$

and on the beam portion between z and $z + dz$ acts the bending moment

$$dM = M(z + dz) - M(z) = \frac{dM}{dz}dz . \tag{41}$$

To compute the dissipation, we need to evaluate the work done by the dissipative portion of this force, or better the power, multiplying by $\dot{\theta}_x = -\frac{\partial \dot{y}}{\partial z}$, that is

$$P = -\gamma EI \int_0^L \frac{\partial^3 \dot{y}}{\partial z^3}\frac{\partial \dot{y}}{\partial z}\, dz = \gamma EI \left\{ \int_0^L \left[\frac{\partial^2 \dot{y}}{\partial z^2}\right]^2 dz - \left[\frac{\partial^2 \dot{y}}{\partial z^2}\frac{\partial \dot{y}}{\partial z}\right]_0^L \right\} \quad (42)$$

where in the last equation we have integrated by parts. The last term which appears is computed at the extrema, and it alters the boundary conditions.

If we make the same approximation as for the potential method, namely that the solution $y(z)$ can be written in terms of the boundaries, disregarding any internal inertia effect, then to be consistent we drop the last term, which would modify the boundary conditions with a time-dependent term, exactly as the internal mass of the wire would do.

Then the dissipation function is simply given by

$$\mathcal{F} = \frac{1}{2}\int_0^L EI\,[\dot{y}'']^2\, dz = \frac{EI}{2L^3}\int_0^1 [\dot{y}'']^2\, d\zeta \quad (43)$$

where $\dot{y}$ is the solution of the beam equation written in terms of the velocities of the boundaries, and we have made the same rescaling $z = L\zeta$ as in computing the potential. We have

$$[\dot{y}''(\zeta)]^2 = k^4 \dot{\mathbf{c}}^T \cdot Y_2(\zeta)\cdot \dot{\mathbf{c}} \quad (44)$$

and the integration in ζ gives

$$\mathcal{F} = \frac{EI}{2L^3} k^4 \dot{\mathbf{c}}^T \cdot \left[\int_0^1 \mathbf{Y}_2(\zeta) d\zeta\right] \cdot \dot{\mathbf{c}} = \frac{TL}{2}\frac{k^2}{L^2}\dot{\mathbf{c}}^T \cdot \mathbf{A}\cdot \dot{\mathbf{c}} \quad (45)$$

where

$$\mathbf{A}[1:2,1:2] = \begin{bmatrix} \frac{\sinh(2k)-2k}{4k} & \frac{\sinh(k)^2}{2k} \\ \frac{\sinh(k)^2}{2k} & \frac{\sinh(2k)+2k}{4k} \end{bmatrix} \quad (46)$$

Following the same steps as in the previous section one easily finds

$$\mathcal{F} = \frac{TL}{2}\dot{\mathbf{x}}^T \cdot \tilde{\mathbf{W}} \cdot \dot{\mathbf{x}} \quad (47)$$

where

$$\tilde{W} = \frac{1}{L^2\left[k\cosh k/2 - 2\sinh k/2\right]^2} \times \mathbf{A}_{ij}$$

where

$$\mathbf{A}_{11} = \mathbf{A}_{33} = \frac{k\,(-k + \sinh(k))}{2}$$

$$\mathbf{A}_{12} = \mathbf{A}_{14} = \mathbf{A}_{21} = \frac{k\,(-k + \sinh(k))}{4}$$

$$\mathbf{A}_{13} = \mathbf{A}_{31} = \frac{k\,(k - \sinh(k))}{2}$$

$$\mathbf{A}_{22} = \frac{\left(-2k + 2k^3 + 4k\cosh(k) - 2k\cosh(2k) - 4\left(1 + k^2\right)\sinh(k) + (2 + k)\sinh(2k)\right)}{16k\,[\sinh(k/2)]^2}$$

$$\mathbf{A}_{23} = \mathbf{A}_{32} = \mathbf{A}_{34} = \mathbf{A}_{43} = \frac{k\,(k - \sinh(k))}{4}$$

$$\mathbf{A}_{24} = \frac{\left(k - \left(2k + k^3\right)\cosh(k) + k\cosh(2k) + \left(2 + k^2\right)\sinh(k) - \sinh(2k)\right)}{8k\,[\sinh(k/2)]^2}$$

$$\mathbf{A}_{42} = \frac{\mathrm{csch}(\frac{k}{2})^2\left(k - \left(2k + k^3\right)\cosh(k) + k\cosh(2k) + \left(2 + k^2\right)\sinh(k) - \sinh(2k)\right)}{8k}$$

$$\mathbf{A}_{44} = \frac{\mathrm{csch}(\frac{k}{2})^2\left(-2k + 2k^3 + 4k\cosh(k) - 2k\cosh(2k) - 4\left(1 + k^2\right)\sinh(k) + \left(2 + k^2\right)\sinh(2k)\right)}{16k}$$

This cumbersome expression can be simplified in the large k limit, but we shall not further elaborate on this.

4 Exact solutions in the frequency domain

We have till now considered methods to evaluate a potential representation for the elastic elements, which is the most economical when one is interested just in the low frequency behavior.

However it may be necessary, for example to assess the attenuation performance, to take into account also the internal modes, exactly if possible. For systems which can be assimilated to chains of elements there is a particularly convenient method, which was introduced by Holzer [2] for the study of torsional vibration, and it has been re-discovered by G.Cella and applied to mechanical system.

The idea is quite simple: each mechanical element in a mechanical chain can be represented as in Fig. (6) where the arrows represent the forces exerted by the element on the boundary (taking conventionally the positive direction as outgoing), while the lines ended by circles represent the coordinates of the boundary of the element.

Each element is represented by a matrix linking input and output:

$$\begin{pmatrix} x_{out} \\ F_{out} \end{pmatrix} = \mathbf{A} \cdot \begin{pmatrix} x_{in} \\ F_{in} \end{pmatrix} . \tag{48}$$

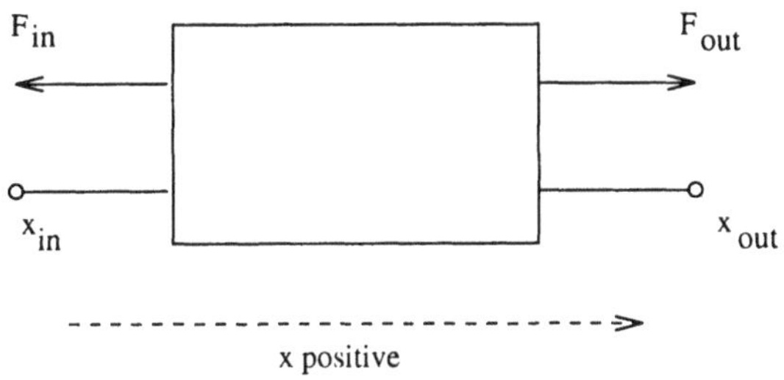

Figure 6. The general scheme.

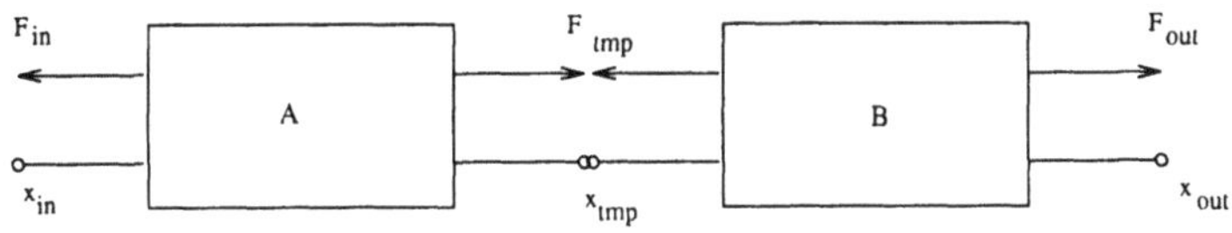

Figure 7. Series connection.

Suppose that this impedance represents an entire chain of elements, and that the last one is "free", that is has no force applied other than those coming from the rest of the chain: then one has $F_{out} = 0$. Now for instance to compute the transfer function one has just to impose $x_{in} = 1$ and solve, for each value of the frequency, the system of equations

$$\begin{pmatrix} x_{out} \\ 0 \end{pmatrix} = \mathbf{A} \cdot \begin{pmatrix} 1 \\ F_{in} \end{pmatrix} , \tag{49}$$

which incidentally gives also the force in input needed to drive the input with unit displacement amplitude.

Joining these elements in series or in parallel allows to build easily complicated TF in the frequency domain: why this method is particularly apt to study chains of elements is made clear by considering the series connection.

Series connection

If two elements are in series, as in Fig. (7), the forces at the boundary have to be compensated, while the coordinates are equal: one has the two equations

$$\begin{pmatrix} x_{out} \\ F_{out} \end{pmatrix} = \mathbf{B} \cdot \begin{pmatrix} x_{tmp} \\ F_{tmp} \end{pmatrix} , \qquad \begin{pmatrix} x_{tmp} \\ F_{tmp} \end{pmatrix} = \mathbf{A} \cdot \begin{pmatrix} x_{in} \\ F_{in} \end{pmatrix} \tag{50}$$

which give

$$\begin{pmatrix} x_{out} \\ F_{out} \end{pmatrix} = \mathbf{B} \cdot \mathbf{A} \cdot \begin{pmatrix} x_{in} \\ F_{in} \end{pmatrix} , \tag{51}$$

simply the product of matrices.

Before considering other kinds of connections, let us make the method more clear with some one-dimensional examples

Simple mass.

The motion equation of a point mass is

$$M\ddot{x} = F_{in} - F_{out} \Rightarrow \begin{pmatrix} x_{out} \\ F_{out} \end{pmatrix} = \begin{pmatrix} 1 & 0 \\ M\omega^2 & 1 \end{pmatrix} \cdot \begin{pmatrix} x_{in} \\ F_{in} \end{pmatrix} \tag{52}$$

Spring

The force relations are

$$F_{out} = F_{in} = -\kappa\,(x_{out} - x_{in}) \Rightarrow \begin{pmatrix} x_{out} \\ F_{out} \end{pmatrix} = \begin{pmatrix} 1 & -\frac{1}{\kappa} \\ 0 & 1 \end{pmatrix} \cdot \begin{pmatrix} x_{in} \\ F_{in} \end{pmatrix} \tag{53}$$

Dashpot

The force relation for a damping element proportional to velocity is

$$F_{out} = F_{in} = -\eta\,(\dot{x}_{out} - \dot{x}_{in}) \Rightarrow \begin{pmatrix} x_{out} \\ F_{out} \end{pmatrix} = \begin{pmatrix} 1 & \frac{i}{\omega\eta} \\ 0 & 1 \end{pmatrix} \cdot \begin{pmatrix} x_{in} \\ F_{in} \end{pmatrix} . \tag{54}$$

Wire

The Lagrangian of a wire with masses at end is given by [1]

$$\mathcal{L} = \frac{1}{2}\int_0^L \left[\rho S\dot{x}^2 - ES\,(x')^2\right] d\xi + \frac{1}{2}\left[M_0\dot{x}(0)^2 + M_L\dot{x}(L)^2\right] \tag{55}$$

where ξ is the unperturbed coordinate along the wire and x is the displacement from equilibrium. The masses are fictitious, to obtain the motion equations

$$\rho\ddot{x} - Ex'' = 0, \qquad ESx'(0) + F_{in} = 0, \qquad ESx'(L) + F_{out} = 0\ ; \tag{56}$$

the general solution is therefore

$$x(\xi) = x_{in}\cos\kappa\xi + \left[x_{out} - x_{in}\cos\kappa L\right]\frac{\sin\kappa\xi}{\sin\kappa L}, \qquad \kappa = \omega\sqrt{\frac{\rho}{E}} \tag{57}$$

which gives $x'(0)$, $x'(L)$ in terms of x_{in}, x_{out} and the transfer matrix

$$\begin{pmatrix} x_{out} \\ F_{out} \end{pmatrix} = \begin{pmatrix} \cos\kappa L & -\frac{\sin\kappa L}{\kappa ES} \\ \kappa ES\sin\kappa L & \cos\kappa L \end{pmatrix} \cdot \begin{pmatrix} x_{in} \\ F_{in} \end{pmatrix} \tag{58}$$

It is now immediate for instance to compute the transfer function of a system composed of a wire and an attached mass: just combine the two matrices in

$$A_{tot} = \begin{pmatrix} \cos\kappa L & -\frac{\sin\kappa L}{\kappa ES} \\ M\omega^2\cos\kappa L + \kappa ES\sin\kappa L & \cos\kappa L - \frac{M\omega^2\sin\kappa L}{\kappa ES} \end{pmatrix} \tag{59}$$

and solve the equation

$$\begin{pmatrix} x_{out} \\ 0 \end{pmatrix} = A_{tot} \cdot \begin{pmatrix} 1 \\ F_{in} \end{pmatrix} \tag{60}$$

which gives

$$x_{out} = \frac{\det A_{tot}}{A_{tot}(2,2)} = \frac{1}{A_{tot}(2,2)} = \frac{1}{\cos\kappa L - \frac{M\omega^2\sin\kappa L}{\kappa ES}} \tag{61}$$

as a consequence of the fact that all the elements we have considered till now have matrices with unit determinant (a sort of unitarity principle).

VIRGO vertical transfer function example

Also the blades can be modeled in the same way, although the computation must be done numerically, for instance resorting to a finite-differencing scheme for computing the response of the elementary blade to the inputs. We shall not go into details here because they would not be much instructive, but we shall just show in Fig. (8) an example of a successful application of this method in a effectively 1 dimensional system, namely the VIRGO super-attenuator chain excited vertically. We see at low frequencies the resonances due to the fundamental blade resonances (lowered by the anti-springs ...), at intermediate frequencies (around 10 Hz) the onset of the mass effect, and then many resonances due to the blades and to the connecting wires.

4.1 Two dimensional examples

We have variables $x_{i,o}$, $\theta_{i,o}$ and forces and momenta $F_{i,o}$, $\mathcal{M}_{i,o}$: this time input and output are linked by the equation

$$\begin{pmatrix} x_o \\ \theta_o \\ F_o \\ \mathcal{M}_o \end{pmatrix} = \mathbf{A} \cdot \begin{pmatrix} x_i \\ \theta_i \\ F_i \\ \mathcal{M}_o \end{pmatrix} , \tag{62}$$

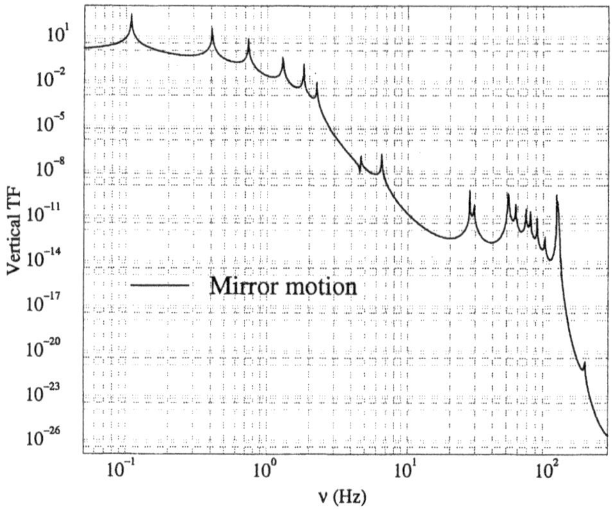

Figure 8. Vertical transfer function for the VIRGO super-attenuator.

with the same sign conventions for forces and momenta as in the previous case. Let us consider again our favorite pendulum examples

Mass in two dimensions

An element with mass M and moment of inertia J is described by

$$M\ddot{x} = F_i - F_o \qquad J\ddot{\theta} = \mathcal{M}_i - \mathcal{M}_o \tag{63}$$

hence by a matrix

$$\mathbf{A} \equiv \begin{pmatrix} 1 & 0 & 0 & 0 \\ 0 & 1 & 0 & 0 \\ M\omega^2 & 0 & 1 & 0 \\ 0 & J\omega^2 & 0 & 1 \end{pmatrix} \tag{64}$$

Beam under tension

In this case the equation of motion in the bulk is

$$EIx''''(z) - T(z)x''(z) = \omega^2 \rho S(z) x(z) \tag{65}$$

while the conditions on forces and torques at the boundaries are

$$\begin{aligned} EIx'''(0) - T(0)x'(0) = F_i, \qquad & EIx'''(L) - T(L)x'(L) = F_o \\ EIx''(0) = -\mathcal{M}_i, \qquad & EIx''(L) = -\mathcal{M}_o \ ; \end{aligned} \tag{66}$$

assuming T and S constant we solve the motion equation in the bulk, setting

$$x(z) = A\cos\lambda_3 z + B\sin\lambda_3 z + C\cosh\lambda_1 z + D\sinh\lambda_1 z \tag{67}$$

where

$$\lambda_{1,3} = \sqrt{\frac{\sqrt{\tau^2 + 4\gamma^4} \pm \tau}{2}}, \quad \tau = \frac{TL^2}{EI}, \quad \gamma^4 = \frac{\omega^2 \rho S L^4}{EI} \ . \tag{68}$$

Then we can solve for the constants A, B, C, D by imposing the conditions at the boundaries, thus obtaining the relations between positions (angles) and forces (torques), which after a bit of work gives

$$\mathbf{A} \equiv \begin{pmatrix} \frac{\lambda_1^2 c_3 + \lambda_3^2 c_1}{\sqrt{4\gamma^4+\tau^2}} & \frac{L(\lambda_3 s_3 + \lambda_1 s_1)}{\sqrt{4\gamma^4+\tau^2}} & \frac{L^3(-(\lambda_1 s_3) + \lambda_3 s_1)}{EI\gamma^2\sqrt{4\gamma^4+\tau^2}} & \frac{L^2(c_3 - c_1)}{EI\sqrt{4\gamma^4+\tau^2}} \\ \frac{\gamma^2(-(\lambda_1 s_3) + \lambda_3 s_1)}{L\sqrt{4\gamma^4+\tau^2}} & \frac{\lambda_3^2 c_3 + \lambda_1^2 c_1}{\sqrt{4\gamma^4+\tau^2}} & \frac{L^2(-c_3 + c_1)}{EI\sqrt{4\gamma^4+\tau^2}} & -\frac{L(\lambda_3 s_3 + \lambda_1 s_1)}{EI\sqrt{4\gamma^4+\tau^2}} \\ \frac{EI\gamma^2\left(\lambda_1^3 s_3 + \lambda_3^3 s_1\right)}{L^3\sqrt{4\gamma^4+\tau^2}} & \frac{EI\gamma^4(-c_3 + c_1)}{L^2\sqrt{4\gamma^4+\tau^2}} & \frac{\lambda_1^2 c_3 + \lambda_3^2 c_1}{\sqrt{4\gamma^4+\tau^2}} & \frac{\gamma^2(\lambda_1 s_3 - \lambda_3 s_1)}{L\sqrt{4\gamma^4+\tau^2}}\} \\ \frac{EI\gamma^4(c_3 - c_1)}{L^2\sqrt{4\gamma^4+\tau^2}} & \frac{EI\left(\lambda_3^3 s_3 - \lambda_1^3 s_1\right)}{L\sqrt{4\gamma^4+\tau^2}} & -\frac{L(\lambda_3 s_3 + \lambda_1 s_1)}{\sqrt{4\gamma^4+\tau^2}} & \frac{\lambda_3^2 c_3 + \lambda_1^2 c_1}{\sqrt{4\gamma^4+\tau^2}} \end{pmatrix}$$

$$c_3 \equiv \cos(\lambda_3), \quad s_3 \equiv \sin(\lambda_3), \quad c_1 \equiv \cosh(\lambda_1), \quad s_1 \equiv \sinh(\lambda_1) \tag{69}$$

The method is powerful but becomes cumbersome when considering higher dimensional systems: also, beyond two dimensions it becomes numerically unstable and cannot be used reliably in systems with an high dynamic range.

5 Finite element method

The potential method does not take into account the internal modes of the elastic elements. These modes are in general in an infinite number and cannot be modeled by a system having a finite number of freedom.

This problem can be circumvented in frequency domain, where we set up a simulation framework which does not require a time-domain representation.

However in solving control problems a time-domain model is useful: moreover, we expect on physical grounds that not all the modes are relevant, but only those at relatively lower frequency. The finite element method is a general approach to the problem which allows to subdivide an elastic element in sub-parts, and introduce therefore as many internal degrees of freedom as subdivisions: the advantage over simply discretizing the motion equations is a greater efficiency in representing the element shape.

Instead of trying to set up a general theory (the interested reader can for instance consult [2]) we go directly through an example.

5.1 FEM applied to beams

We start from the kinetic and potential energy of the beam transverse motion

$$U_{\text{kin}} = \frac{1}{2}\int_0^L dz\, \rho(z)\left[\ddot{y}\right]^2, \quad U_{\text{pot}} = \frac{1}{2}\int_0^L dz\, EI\left[\frac{d^2y}{dz^2}\right]^2 + T\left[\frac{dy}{dz}\right]^2 . \tag{70}$$

The beam will be connected to other elements specifying position and angles at end-points: we have $\frac{dy}{dz} = \theta$ as the local slope of the curve.

The idea is now to divide the beam in a number N of elements, each of length $h \equiv L/N$, separated by *nodes*.

The displacement inside the element j must now be expressed in terms of slopes and displacements at the nodes, introducing a local non-dimensional coordinate $\zeta \equiv \frac{j\,h-z}{h} = j - \frac{z}{h}$ and 4 shape functions $\phi_{1,\dots 4}$, corresponding to the 4 boundary conditions, displacements and slopes at the extrema:

$$y(\zeta) = \phi_1(\zeta)y_{j-1} + \phi_2(\zeta)h\theta_{j-1} + \phi_3(\zeta)y_j + \phi_4(\zeta)h\theta_j \equiv \phi^T\mathbf{y} . \tag{71}$$

Note: $\frac{dy}{dz} = -\frac{1}{h}y'$ where the prime is the derivation with respect to ζ.

We can now choose cubic shape function elements:

$$\phi_i(\zeta) = c_{i1} + c_{i2}\zeta + c_{i3}\zeta^2 + c_{i4}\zeta^3 ; \tag{72}$$

next we impose the conditions on values at end-points

$$\begin{aligned} y(1) &= \phi_1(1)y_{j-1} + \phi_2(1)h\theta_{j-1} + \phi_3(1)y_j + \phi_4(1)h\theta_j = y_{j-1} \\ y'(1) &= \phi_1'(1)y_{j-1} + \phi_2'(1)h\theta_{j-1} + \phi_3'(1)y_j + \phi_4'(1)h\theta_j = -h\theta_{j-1} \\ y(0) &= \phi_1(0)y_{j-1} + \phi_2(0)h\theta_{j-1} + \phi_3(0)y_j + \phi_4(0)h\theta_j = y_j \\ y'(0) &= \phi_1'(0)y_{j-1} + \phi_2'(0)h\theta_{j_1} + \phi_3'(0)y_j + \phi_4'(0)h\theta_j = -h\theta_j \end{aligned} \tag{73}$$

that is

$$\begin{array}{llll} \phi_1(1) = 1 & \phi_2(1) = 0 & \phi_3(1) = 0 & \phi_4(1) = 0 \\ -\phi_1'(1) = 0 & -\phi_2'(1) = 1 & -\phi_3'(1) = 0 & -\phi_4'(1) = 0 \\ \phi_1(0) = 0 & \phi_2(0) = 0 & \phi_3(0) = 1 & \phi_4(0) = 0 \\ -\phi_1'(0) = 0 & -\phi_2'(0) = 0 & -\phi_3'(0) = 0 & -\phi_4'(0) = 1 \end{array} \tag{74}$$

As we have

$$\phi_i(\zeta) = \left[1, \zeta, \zeta^2, \zeta^3\right] \cdot \mathbf{c}_i, \quad -\phi_i'(\zeta) = \left[0, -1, -2\zeta, -3\zeta^2\right] \cdot \mathbf{c}_i \tag{75}$$

we re-arrange the equations in matrix form as $\mathbf{A} \cdot \mathbf{c}_i = \mathbf{e}_i$, and get the coefficients $\mathbf{c}_i$ as the columns of $\mathbf{A}^{-1}$, where

$$\mathbf{A} = \begin{pmatrix} 1 & 1 & 1 & 1 \\ 0 & -1 & -2 & -3 \\ 1 & 0 & 0 & 0 \\ 0 & -1 & 0 & 0 \end{pmatrix}, \qquad \mathbf{A}^{-1} = \begin{pmatrix} 0 & 0 & 1 & 0 \\ 0 & 0 & 0 & -1 \\ 3 & 1 & -3 & 2 \\ -2 & -1 & 2 & -1 \end{pmatrix}. \tag{76}$$

They translate in the shape functions

$$\phi_1 = 3\zeta^2 - 2\zeta^3,\ \phi_2 = \zeta^2 - \zeta^3,\ \phi_3 = 1 - 3\zeta^2 + 2\zeta^3,\ -\zeta + 2\zeta^2 - \zeta^3 \tag{77}$$

which are the so called *Hermite cubics* interpolation functions.

Once the interpolation functions are known, we compute the kinetic and the potential energy in terms of positions and slopes at nodal points. As there is no overlap of the functions, the integrals decompose in a sum of terms

$$U_{\text{kin}} = \frac{1}{2}\dot{\mathbf{y}}^T \cdot \mathbf{M}_j \cdot \dot{\mathbf{y}}, \qquad U_{\text{pot}} = \frac{1}{2}\mathbf{y}^T \cdot \mathbf{K}_j \cdot \mathbf{y} \tag{78}$$

where M_j and K_j are mass and stiffness (sub) matrices.

Assuming constant density, we have

$$U_{\text{kin},j} = \frac{1}{2}\int_{(j-1)h}^{jh} dz \rho(\zeta) \left[\dot{y}(z)\right]^2 = \frac{1}{2}h\rho \int_0^1 d\zeta \left[\dot{y}(\zeta)\right]^2 ; \tag{79}$$

but $\dot{y}(\zeta) = \phi^T \cdot \dot{\mathbf{y}}$ and we can simply rewrite

$$\int_0^1 d\zeta \left[\phi^T \cdot \dot{\mathbf{y}}\right]^2 = \dot{\mathbf{y}}^T \cdot \left[\int_0^1 \phi(\zeta)\phi(\zeta)^T\right] \cdot \dot{\mathbf{y}} \tag{80}$$

hence we finally obtain

$$M_j = \frac{h\rho}{420}\begin{pmatrix} 156 & 22 & 54 & -13 \\ 22 & 4 & 13 & -3 \\ 54 & 13 & 156 & -22 \\ -13 & -3 & -22 & 4 \end{pmatrix}. \tag{81}$$

We next do the same for the potential, starting from

$$U_{\text{pot,j}} = \frac{1}{2}h\int_0^1 d\zeta \frac{EI}{h^4}\left[y''\right]^2 + \frac{T_j}{h^2}\left[y'\right]^2 \tag{82}$$

we obtain

$$K_j = \frac{EI}{h^3}\begin{pmatrix} 12 & 6 & -12 & 6 \\ 6 & 4 & -6 & 2 \\ -12 & -6 & 12 & -6 \\ 6 & 2 & -6 & 4 \end{pmatrix} + \frac{T_j}{30h}\begin{pmatrix} 36 & 3 & -36 & 3 \\ 3 & 4 & -3 & -1 \\ -36 & -3 & 36 & -3 \\ 3 & -1 & -3 & 4 \end{pmatrix} \tag{83}$$

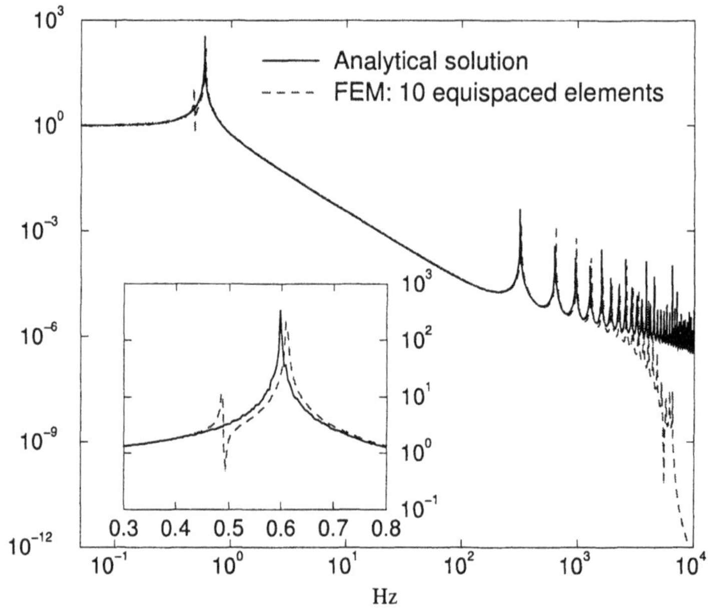

Figure 9. Comparison between analytical and FEM approximations.

Mass and potential matrices need to be connected in a single matrix in order to build the motion equations.

Violin modes example

Let us suspend a mass M to a wire, choosing the typical values for the last suspension elements in the VIRGO chain.

We can compare the solution by the FEM method with the "exact" solution for the transfer function, obtained by analytical methods: we first show in Fig. (9) the result of using 10 elements, all of the same length [f]: We note two features: as expected at high frequencies, say above the fifth violin mode, the FEM approximation fails. But also at low frequencies, as evidenced in the subplot, a spurious zero-pole pair appears!

It is crucial to realize (I thank G.Cella for pointing this out) that this

[f] In this example, we make a relevant simplification, using a single wire, instead of the four wires of the real example; this is a substantial change because the stiffness to angular motions is altered, but is a good example because is somehow "critical".

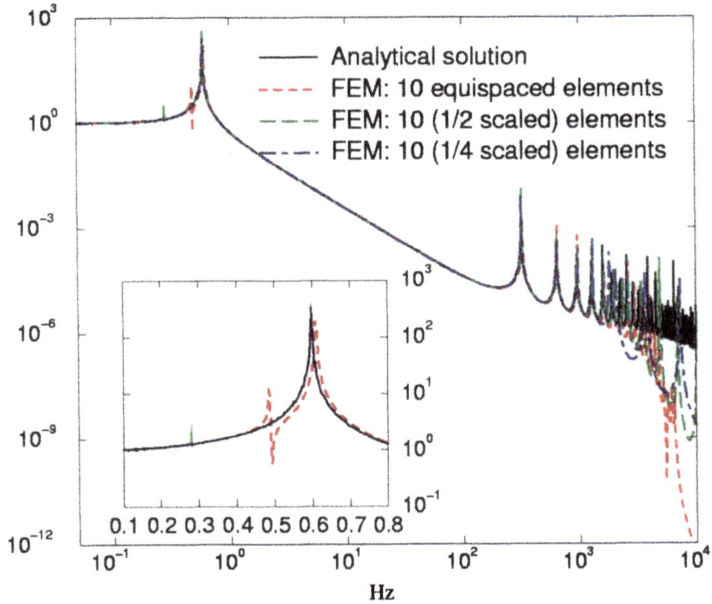

Figure 10. Comparison between analytical and different FEM scalings.

is an effect due to a bad approximation of the wire behavior close to the attachment points: in practice at low frequencies the wire remains essentially rectilinear and bends only at the attachment points, thus requiring a much finer approximation.

We understand that using equispaced elements is an inefficient solution, it is less expensive to use elements of different lengths.

In Fig. (10) we compare the analytical solution with two different element series, still 10 elements but scaled by 1/2 or 1/4 at the extrema, that is

$$[1/2]\ L \times \left(\tfrac{1}{32}, \tfrac{1}{32}, \tfrac{1}{16}, \tfrac{1}{8}, \tfrac{1}{4}, \tfrac{1}{4}, \tfrac{1}{8}, \tfrac{1}{16}, \tfrac{1}{32}, \tfrac{1}{32}\right)$$

$$[1/4]\ L \times \left(\tfrac{1}{512}, \tfrac{3}{512}, \tfrac{3}{128}, \tfrac{3}{32}, \tfrac{3}{8}, \tfrac{3}{8}, \tfrac{3}{32}, \tfrac{3}{128}, \tfrac{3}{512}, \tfrac{1}{512}\right)$$

It is clear that at low frequencies the situation is much improved, at the price of reducing the high frequency accuracy. A compromise is necessary and must be decided on the basis of the band one is interested to simulate.

The lesson is that much care must be put in choosing the right approximation for each element in the system and check different solutions to avoid wasting computer time: an analytical solution is invaluable in checking the correctness of the approximation.

5.2 FEM approximation for a blade

We repeat the analysis for a blade approximated as a thin plate of trapezoidal form, with length L, thickness H and variable transverse section $W(x)$,

$$W(x) = W_{out}\left(1-\frac{x}{L}\right) + W_{in}\frac{x}{L}\ ; \tag{84}$$

and therefore a linear density $\rho(x) = W(x)H\rho_0$. We have

$$U_{kin} = \frac{1}{2}\int_0^L dx\rho(x)\left[\dot{z}(x)\right]^2, \qquad U_{pot} = \frac{1}{2}\int_0^L dx\frac{EH^3W(x)}{12\left(1-\sigma^2\right)}\left[\frac{d^2z}{dx^2}\right]^2 \tag{85}$$

where E is the Young modulus and σ is the Poisson ratio.

We further impose the conditions that in $x = 0$ there is zero derivative, and that at $x = L$ it is applied only a force and not a torque:

$$\frac{dz}{dx}|_{x=0} = 0, \qquad \frac{d^2z}{dx^2}|_{x=L} = 0\ ; \tag{86}$$

this means that the first and the last elements will require a special treatment.

We adopt, for the bulk the same shape functions as before, but in computing kinetic and potential energy we take into account the widths variation:

$$U_{\mathrm{kin},j} = \frac{1}{2}H\rho_0 h\int_{(j-1)h}^{jh} dx\, W(x)\left[\dot{z}(x)\right]\ ; \tag{87}$$

we have

$$\int_0^1 d\xi W(\xi)\left[\phi^T\cdot\dot{\mathbf{z}}\right]^2 = \dot{\mathbf{z}}^T\cdot\left[\int_0^1 W(\xi)\phi(\xi)\phi(\xi)^T\right]\cdot\dot{\mathbf{z}}\,. \tag{88}$$

We are interested in $W(\xi) = W_{j-1}\xi + W_j\left(1-\xi\right)$ linear in ξ; we obtain

$$M_j = \frac{hH\rho_0}{840}\left[W_{j-1}\begin{pmatrix} 240 & 30 & 54 & -14 \\ 30 & 5 & 12 & -3 \\ 54 & 12 & 72 & -14 \\ -14 & -3 & -14 & 3 \end{pmatrix} + W_j\begin{pmatrix} 72 & 14 & 54 & -12 \\ 14 & 3 & 14 & -3 \\ 54 & 14 & 240 & -30 \\ -12 & -3 & -30 & 5 \end{pmatrix}\right] \tag{89}$$

Analogously we have

$$U_{\mathrm{pot},j} = \frac{1}{2}h\int_0^1 d\xi\frac{EH^3W(\xi)}{12(1-\sigma^2)h^4}\left[z''(\xi)\right]^2 \tag{90}$$

hence

$$K_j = \frac{EH^3}{12(1-\sigma^2)h^3}\left[W_{j-1}\begin{pmatrix} 6 & 4 & -6 & 2 \\ 4 & 3 & -4 & 1 \\ -6 & -4 & 6 & -2 \\ 2 & 1 & -2 & 1 \end{pmatrix} + W_j \begin{pmatrix} 6 & 2 & -6 & 4 \\ 2 & 1 & -2 & 1 \\ -6 & -2 & 6 & -4 \\ 4 & 1 & -4 & 3 \end{pmatrix}\right]. \tag{91}$$

The condition at the clamped end of the blade is easily implemented by deleting the second row and column of each matrix for the first element.

At the other end we have to implement a zero second derivative: this requires using a different parameterization and different shape functions. We assume still cubic shape functions

$$z(\xi) = \phi_1(\xi)y_{j-1} + \phi_2(\xi)\,h\theta_{j-1} + \phi_3(\xi)y_j \tag{92}$$

and we obtain

$$M_j = \frac{hH\rho_0}{3360}\left[W_{j-1}\begin{pmatrix} 1155 & 175 & 189 \\ 175 & 35 & 49 \\ 189 & 49 & 147 \end{pmatrix} + W_j \begin{pmatrix} 477 & 113 & 279 \\ 113 & 29 & 83 \\ 279 & 83 & 645 \end{pmatrix}\right]; \tag{93}$$

$$K_j = \frac{EH^3}{12(1-\sigma^2)h^3}\left(\frac{9}{4}W_{j-1} + \frac{3}{4}W_j\right)\begin{pmatrix} 1 & 1 & -1 \\ 1 & 1 & -1 \\ -1 & -1 & 1 \end{pmatrix}. \tag{94}$$

The formulas obtained can be tested considering a typical VIRGO blade:

$$\begin{gathered} E = 1.86 \times 10^11, \quad \sigma = 0.563, \quad \rho_0 = 7874 \\ W_{base} = 0.120, \quad W_{tip} = 0.012, \quad L(\text{length}) = 0.375, \quad H = 0.0035 \end{gathered} \tag{95}$$

and we further load the blade with a mass $M = 54$ concentrated on the tip.

We display in Fig. (11) the resulting transfer function, from the base of the blade to the motion of the mass, in the $[0, 1000]$ Hz range. We compare with a 128 nodes finite difference simulation. It is clear that even a small number of elements is able to capture the main features of the TF.

It is worth noting that a VIRGO blade works at constant curvature, which means that the best strategy is to choose, as we did, equispaced elements.

6 System synthesis and simulation

Both potential methods and FEM methods allow us to define mass $\mathbf{M}_i$ and stiffness matrices $\mathbf{K}_i$, as well as dissipation matrices Γ_i, for all the elements of the mechanical system.

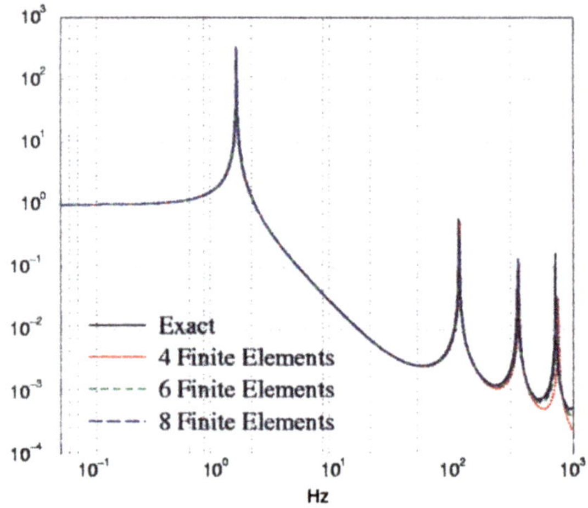

Figure 11. Comparison between finite difference and FEM approximations.

The additivity of energy (or of the work, as is the case of dissipation) allows then to write down the total energy stored in the system, when its positions $\mathbf{x} = \bigcup_i \mathbf{x}_i$ are displaced from the equilibrium. The vector of variables $\mathbf{x}$ include without repetitions the position of rigid elements (say, their CM) and of a certain number of nodes in the elastic elements, sufficient to approximate the behavior of the system in a certain frequency range.

Then we can write down the motion equations as

$$\mathbf{M} \cdot \ddot{\mathbf{x}} + \mathbf{K} \cdot \mathbf{x} + \Gamma \cdot \dot{\mathbf{x}} = \mathbf{F} \tag{96}$$

where $\mathbf{F}$ generically represents the external forces applied. Actually in general we shall not have an external force acting on one and just one internal DOF: if we call $\mathbf{U}$ the external variable, say the currents in some coils or the motion of the soil, the work done on the internal variables can be expressed as

$$W_{ext} = \mathbf{x} \cdot \mathbf{K}_{ext} \cdot \mathbf{U} \, ; \tag{97}$$

and the motion equations become

$$\mathbf{M} \cdot \ddot{\mathbf{x}} + \mathbf{K} \cdot \mathbf{x} + \Gamma \cdot \dot{\mathbf{x}} = \mathbf{K}_{ext} \cdot \mathbf{U} \, . \tag{98}$$

The equations can be rewritten in state space form: one defines

$$\mathbf{Q} = \begin{pmatrix} \mathbf{x} \\ \dot{\mathbf{x}} \end{pmatrix}, \quad \Rightarrow \quad \dot{\mathbf{Q}} = \begin{pmatrix} 0 & 1 \\ -\mathbf{M}^{-1}\mathbf{K} & -\mathbf{M}^{-1}\Gamma \end{pmatrix} \cdot \mathbf{Q} + \begin{pmatrix} 0 \\ \mathbf{M}^{-1} \cdot \mathbf{K}_{ext} \end{pmatrix} \cdot \mathbf{U} ; \tag{99}$$

the two terms define the matrices $\mathbf{A}, \mathbf{B}$ of the state space model. One proceeds similarly for defining the matrices $\mathbf{C}, \mathbf{D}$ but we shall not elaborate on this.

Given the matrices of the state space model, all the standard tools can be applied for the simulation.

6.1 Frequency domain simulation

In frequency domain the equations become

$$(i\omega\mathbf{I} - \mathbf{A}) \cdot \mathbf{Q} = \mathbf{B} \cdot \mathbf{U}, \qquad \mathbf{Y} = \mathbf{C} \cdot \mathbf{Q} + \mathbf{D} \cdot \mathbf{U} \tag{100}$$

and therefore, frequency by frequency, we can compute

$$\mathbf{Q} = (i\omega\mathbf{I} - \mathbf{A})^{-1} \cdot \mathbf{B} \cdot \mathbf{U}, \qquad \mathbf{Y} = \mathbf{C} \cdot \left[(i\omega\mathbf{I} - \mathbf{A})^{-1} \cdot \mathbf{B} + \mathbf{D}\right] \cdot \mathbf{U}; \tag{101}$$

this could be done in several ways. A numerically stable and accurate method is to use LU decomposition: namely

1. cycle over angular pulsations ω and build the matrix $\bar{\mathbf{A}}(\omega) = i\omega\mathbf{I} - \mathbf{A}$.

2. Compute its LU factorization, in the complex domain, obtaining $\bar{\mathbf{A}} = \mathbf{P} \cdot \mathbf{L} \cdot \mathbf{U}$ where $\mathbf{P}$ is a permutation matrix and $\mathbf{L}, \mathbf{U}$ are lower and upper triangular matrices. This factorization costs $O(N^3)$, where N is the order of the matrix A, and must be done once for each frequency.

3. Solve the triangular linear systems and find the different solutions $\mathbf{Q}$ setting in turn one of the entries of $\mathbf{U}$ to 1. This means performing $O(N^2)$ operations for each element of $\mathbf{U}$.

Overall this may become very expensive: one could be tempted to act differently, and apparently also in a smarter way, by performing a modal decomposition of the matrix $\mathbf{A}$: it is not symmetric, therefore it will be possible a decomposition over complex eigenvectors, with complex eigenvalues, so that

$$\mathbf{A} = \mathbf{V} \cdot \Lambda \cdot \mathbf{V}^H, \quad \Rightarrow \quad (i\omega\mathbf{I} - \mathbf{A})^{-1} = \mathbf{V} \cdot (i\omega - \Lambda)^{-1} \cdot \mathbf{V}^H \tag{102}$$

where the matrix Λ is diagonal, and therefore one would simply have

$$Q_k = \sum_{l,m} V_{kl} \frac{1}{i\omega - \Lambda_l} V^*_{ml} U_l \tag{103}$$

which is a considerable saving of time. But this method is terribly inaccurate for system like an attenuator. Indeed, suppose to compute the transfer function between the soil and the mirror: then after n attenuation stage you expect that at high frequency the $TF \propto \frac{1}{\omega^{2n}}$. This behavior is possible, in Eq. (103), only if a delicate cancelation takes place: indeed you shall have

$$Q_k(\omega) = \frac{\sum_{l,m} V_{kl} \prod_{l' \neq l}(i\omega - \Lambda_{l'}) V_{ml} U_l}{\prod_l (i\omega - \Lambda_l)} \tag{104}$$

and in order to have the desired behavior it is necessary that all the $\omega, \omega^2, ...$ terms in the numerator cancel exactly!

On the contrary the procedure based on the LU factorization does not display this problem: so for systems with huge dynamic range in the frequency domain I would recommend not to use a modal expansion[g].

6.2 Time domain simulation

The state space equations Eq. (2) are in continuous time, while a simulation requires a time domain form: one has to translate the equation in discrete time, that is approximate the evolution over discrete steps.

Given the status $\mathbf{Q}(t)$ of the system at time t, the status at time $t + dt$ can be obtained by direct integration: we have

$$\mathbf{Q}(t + dt) = e^{\mathbf{A}dt}\mathbf{Q}(t) + e^{\mathbf{A}(t+dt)} \int_t^{t+dt} e^{-\mathbf{A}\tau}\mathbf{B} \cdot \mathbf{U}(\tau)\, d\tau; \tag{107}$$

this expression is exact, but cannot be computed because we do not have $\mathbf{U}(\tau)$, we just have $\mathbf{U}(t)$: therefore we have to approximate it setting $\mathbf{U}$ constant in steps. In this case the integral over τ can be performed and one obtains

$$\mathbf{Q}(t + dt) = e^{\mathbf{A}dt}\mathbf{Q}(t) + \left[\mathbf{A}^{-1} \cdot \left(e^{\mathbf{A}dt} - \mathbf{I}\right) \cdot \mathbf{B}\right] \cdot \mathbf{U}(t) \tag{108}$$

which defines the discrete form of the state space equations, namely

$$\mathbf{Q}(t+dt) = \mathbf{A}_d(dt) \cdot \mathbf{Q}(t) + \mathbf{B}_d(dt) \cdot \mathbf{U}(t), \qquad \mathbf{Y}(t) = \mathbf{C} \cdot \mathbf{Q}(t) + \mathbf{D} \cdot \mathbf{U}(t) \tag{109}$$

where the matrices

$$\mathbf{A}_d(dt) \equiv e^{\mathbf{A}dt}, \qquad \mathbf{B}_d(dt) \equiv \left[\mathbf{A}^{-1} \cdot \left(e^{\mathbf{A}dt} - \mathbf{I}\right) \cdot \mathbf{B}\right] \tag{110}$$

---

[g] A further observation: one could be also tempted to perform a modal expansion on the motion equation

$$\mathbf{M} \cdot \ddot{\mathbf{x}} + \mathbf{K} \cdot \mathbf{x} + \Gamma \cdot \dot{\mathbf{x}} = \mathbf{F} \tag{105}$$

defining eigenvectors and eigenvalues: however this is not possible [4] because it would require finding a way to simultaneously diagonalize the three matrices $\mathbf{M}, \mathbf{K}, \Gamma$.

Only when the dissipation matrix is a linear combination of the other two

$$\Gamma = a\mathbf{M} + b\mathbf{K} \tag{106}$$

this simultaneous diagonalization is possible: and only in that case one can speak about eigenvectors and eigenvalues, which shall in general have an oscillating and a damping part.

need to be computed only once at the beginning of the simulation.

There is however a practical aspect which must be taken in consideration, that is the computational complexity.

Indeed for most attenuation systems, which are built as chain of objects, the potential and kinetic matrices are very sparse, that is the number of non-zero entries is small, compared to the number of zero entries.

Obviously also the matrix $\mathbf{A}$ will be sparse, but this is not true for the exponentiated form, which in general is full, with heavy consequences on the computational requirements. Indeed, if the system has N degrees of freedom, at each step one shall have to perform $O(N^2)$ operations to perform an update: if N is of the order of 100, and one is interested of simulating with steps of $dt = 1ms$, one is forced to perform $O(10^7)$ operations for a second of simulation: 10 MFlops just for one of the suspensions, say.

A possible way out is suggested in [3], where it is introduced an algorithm which allows to compute in an efficient and numerically reliable way not only the matrix exponential $e^{\mathbf{A}dt}$, but also directly the combination $e^{\mathbf{A}dt} \cdot \mathbf{v}$ without first performing the exponentiation. A naive way to do that would be to expand the exponential in a series

$$e^{\mathbf{A}dt} \cdot \mathbf{v} \simeq \mathbf{v} + dt\mathbf{A} \cdot \mathbf{v} + \frac{1}{2} dt^2 \mathbf{A} \cdot \mathbf{A} \cdot \mathbf{v} + \ldots \tag{111}$$

so that each term of the expansion entails the computation of the matrix-vector product $\mathbf{A} \cdot \mathbf{x}$ with $\mathbf{A}$ sparse matrix: the computational cost of this step is proportional to the number of non-zero elements in $\mathbf{A}$, which may be significantly smaller than N^2. The actual algorithm [3] performs a few ($O(10)$) of these sparse multiplications.

For large systems this method is computationally more efficient, and generally more accurate, although the real efficiency on computer may be different from what one expects just from the operation count: indeed one finds, for intermediate values of N (say, 100), that the two methods require comparable CPU time, or even that the standard method is slightly faster. The reason that on most modern computers are optimized for matrix-matrix and matrix-vector operations, so the greater cost of the full matrix multiplication, in terms of floating point operations, is compensated by a greater efficiency, while the evaluation of the product $e^{\mathbf{A}dt} \cdot \mathbf{v}$ is slowed down by the overheads.

References

1. Lev D. Landau and Evgenij M. Lifsits, *Teoria dell'elasticità*, (Editori Riuniti - Edizioni MIR, 1979).

2. L. Meirovitch, *Principles and techniques of vibrations*, (Prentice -Hall, 1997).
3. R.B.Sidje, *Expokit: a package for computing matrix exponentials*, www.maths.uq.edu.au/expokit
4. H.Goldstein, *Classical mechanics*, (Addison-Wesley, 1975).

ACTIVE CONTROLS IN INTERFEROMETRIC DETECTORS OF GRAVITATIONAL WAVES: INERTIAL DAMPING OF THE VIRGO SUPERATTENUATOR

GIOVANNI LOSURDO
Istituto Nazionale di Fisica Nucleare - Sezione di Pisa
E-mail: losurdo@galileo.pi.infn.it

The operation of an interferometer for gravitational waves detection requires sophisticated feedback controls in many parts of the apparatus. The aim of this lecture is to introduce the types of problems to be faced in this line of research. The attention is focused on the ***inertial damping*** of the test mass suspension of the VIRGO interferometer (the superattenuator): it is a multidimensional local control aimed to reduce the residual motion of the suspended mirror associated to the normal modes of the suspension. Its performance is very important for the locking of the interferometer.

1 Introduction

Operating an interferometer for gravitational waves detection requires the implementation of many active controls on the different parts of the apparatus. For instance, feedback controls are needed to reduce the laser frequency and power fluctuations, to keep the optical cavities in resonance, to maintain the interferometer output on the dark fringe. The basic idea is that the apparatus works at its best only strictly around a well defined working point [1] (laser fluctuations and shot noise at a minimum, optical cavities in resonance, interferometer output on the dark fringe). Internal and external disturbances would swamp the dynamic range of the interferometer. Therefore, it has to be forced to remain in the correct working position, and this is achieved via feedback controls.

In this lecture we examine in details one of the controls needed to reach the final goal of operating the interferometer: the ***inertial damping*** of the VIRGO superattenuator (SA) [3]. The SA can be described as a chain of mechanical "filters", each one acting as a "spring" in 6 d.o.f.. The normal modes frequencies of the SA range between 0.04 and about 2 Hz. At frequencies $f >> 2$ Hz the SA acts as a steep filter of the seismic vibrations of the ground (an attenuation factor of 10^{15}@10 Hz is expected). Therefore, in the interferometer detection band (10 Hz-few kHz), the suspended mirror is "disconnected" from the ground. Beside being a low pass filter for ground vibrations the SA is a tool for actively controlling the mirror position. Forces aimed to move and steer the mirror can be exerted in 3 points of the SA chain: the inverted pendulum (IP) [5], the *marionetta*, the mirror itself (from a suspended *reference mass*).

Keeping the interferometer in the operating position requires the mirror to have a maximum RMS relative motion of less the 10^{-12} m. In the frequency range where the SA is fully effective ($f >> 4$ Hz) the residual motion of the mirror is negligible. On the other hand, the mirror free motion in the region of the SA normal modes is $\sim 100\ \mu$m . Feedback forces acting on the SA must reduce the mirror motion from 100 μm to 10^{-12} m. The dynamic range of a feedback system able to perform this control has to be huge: the control is performed in three steps (hierarchical control [2]). The first step is a damping of the SA normal modes in order to reduce the mirror residual motion to less than 10 μm . This is necessary in order to control the mirror position acting on the lower stages without reinjecting noise into the detection band. The implementation of a high gain and wideband damping of the SA resonances requires a smart approach, that we describe in the following.

2 Experimental setup

The setup (fig. 1) of the experiment is composed by a full scale superattenuator, provided with 3 accelerometers (placed on the top of the IP), 3 LVDT position sensors (measuring the relative motion of the IP with respect to an external frame), 3 coil-magnet actuators. The accelerometers work in the range DC-400 Hz and have acceleration spectral sensitivity $\sim 10^{-9}\ \mathrm{m\,s^{-2}\,Hz^{-1/2}}$ below 3 Hz [6]. The sensors and actuators are all placed in *pin-wheel* configuration. The sensors and actuators signals are elaborated by a computer controlled ADC (16 bit)-DSP-DAC (20 bit) system. The DSP allows to handle the signals of all the sensors and actuators, to recombine them by means of matrices, to create complex feedback filters (like the one of fig. 8) with high precision poles/zeroes placement and to perform all the calculations at high sampling rate (10 kHz). The suspended mirror is also provided with LVDT position sensors to measure its displacement with respect to ground.

3 The approach to the problem of control

The IP, where all the sensors are placed and on which the forces are exerted, has 3 main resonant modes: two translation (X, Y) and the rotation around the vertical axis (Θ). Each sensor is sensitive to all the modes and each actuator can excite all the modes. In the control theory language, such a system is defined MIMO (multiple in-multiple out). Controlling a MIMO system can be very difficult. Our approach has been different: the signals

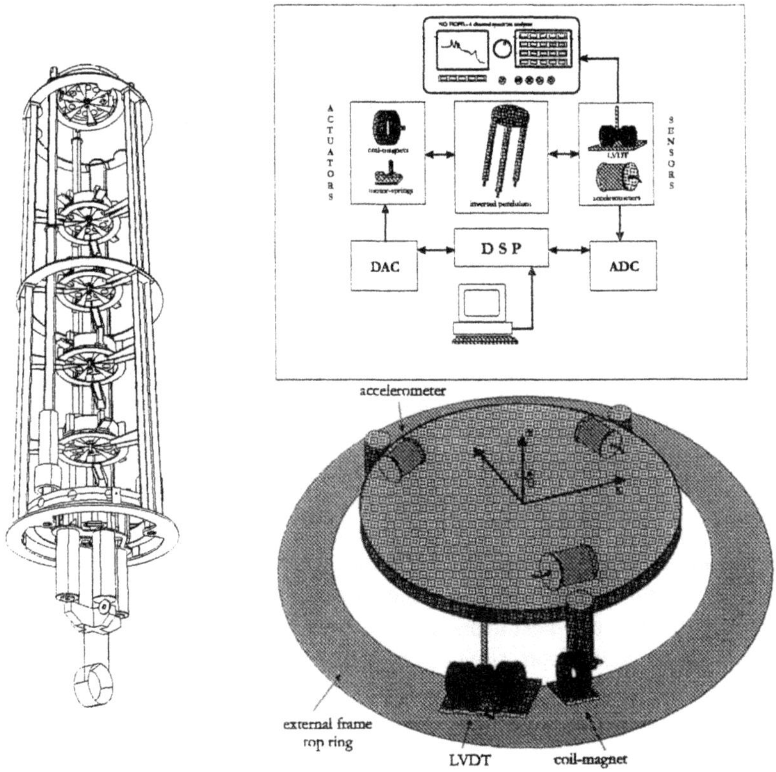

Figure 1. LEFT: the superattenuator; RIGHT TOP: logical scheme of the setup for the local active control; RIGHT BOTTOM: simplified view of the IP top table, provided with the 3 accelerometers. One LVDT position sensors and one coil-magnet actuator are also shown.

of the 3 sensors are digitally mixed (using proper transformation matrices) to build up ***virtual sensors***, sensitive to one mode only and "blind" to the others. At the same time, we build up ***virtual actuators*** able to excite each mode separately. The system is thus uncoupled into 3 SISO (single in-single out) subsystems. In terms of analytical mechanics this means to describe the system in the normal modes basis. The equations of motions take the form:

$$\ddot{x}_k + \omega_k^2 x_k = q_k \quad , \quad k = 1...3 \tag{1}$$

where x_k is the $k-$th normal mode, $\omega_k/2\pi$ is the corresponding resonant frequency and q_k the generalized force on that mode. Let $\mathbf{u} = (u_1, ..., u_3)$ the

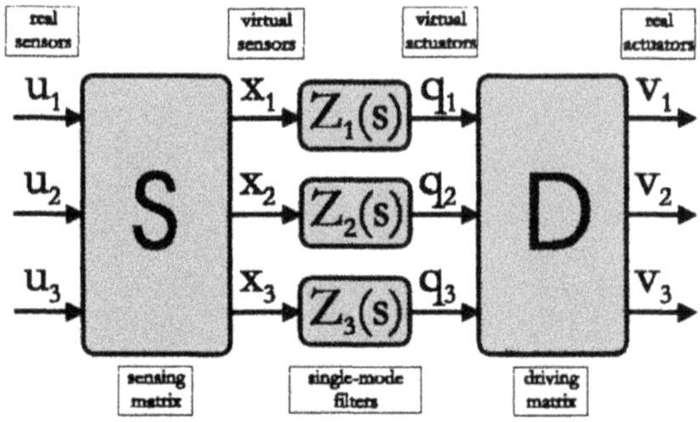

Figure 2. The logic of the diagonalisation: the output u_i of the sensors are linearly recombined by a matrix $\mathbf{S}$ in order to produce 3 *virtual* sensors outputs (x_i), sensitive to pure modes. Three independent feedback filters $Z_i(s)$ are designed for the pure modes and 3 generalized forces q_i are produced. The q_i are turned into real currents (v_i) to feed the actuators via the matrix $\mathbf{D}$.

vector made by the outputs of the 3 sensors and let $\mathbf{x} = (x_1, ..., x_3)$ the vector made by the 3 virtual sensors. Analogously, let $\mathbf{v}$ the vector made by the 3 currents driving the actuators and $\mathbf{q}$ the vector of the 3 generalized forces. The transformation from the system of the physical sensors/actuators to that of the virtual ones is operated by two matrices, such that:

$$\mathbf{v} = \mathbf{D}\mathbf{q} \tag{2}$$

$$\mathbf{x} = \mathbf{S}\mathbf{u} \tag{3}$$

The *sensing* ($\mathbf{S}$) and *driving* ($\mathbf{D}$) matrices are experimentally measured. The measurement procedure is described in details in ref. 9. The result of the measurement is determined by the mechanical characteristics of the system (resonant frequencies, quality factors), the geometry of the sensors and actuators and their calibration, but it is not needed to know them in order to measure the matrices. The logic of diagonalisation is explained by fig. 2.

After the diagonalisation the system can be considered as composed by 3 *uncoupled* oscillators, and the control strategy for each of them has to be defined independently (see fig. 3).

4 Inertial damping: principle

The control we describe here is called *inertial damping* because it is performed by using (mostly) *inertial sensors* (accelerometers). In the following , with

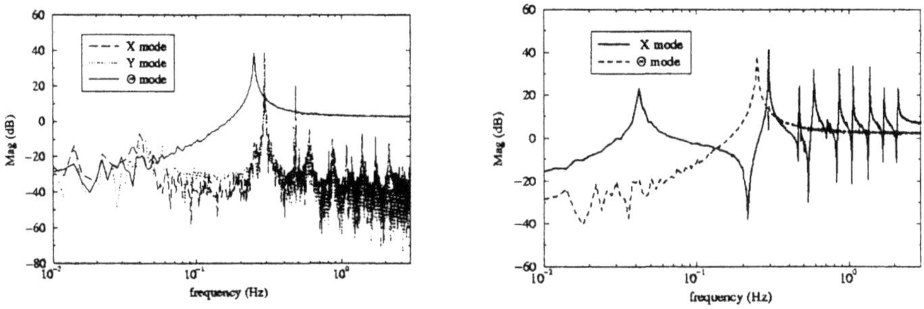

Figure 3. Effect of the digital diagonalisation. LEFT: the output of the 3 virtual accelerometers. Θ is excited. RIGHT: the output of the virtual accelerometers X and Θ are compared. Different feedback strategies are required in the two cases, because X senses all the translational modes of the SA chain.

the help of a simple model, we explain why this is the best choice to achieve an high performance damping.

Let us consider a simple pendulum of mass m and length l. Let x be the abscissa of the suspended mass, x_0 that of the suspension point. Let F_{fb} the external force on the pendulum (i.e. the feedback force to control it). The equation of motion is then:

$$F_{\text{fb}} = m\ddot{x} + \gamma\dot{x} + k(x - x_0) \tag{4}$$

where γ is the viscous dissipation factor and $k = mg/l$. The control loop of such a system is sketched in fig. 4, where $H(s)$ is the mechanical transfer function, $G(s)$ is the compensator and ***out*** is the output of the sensor used. The goal of the control is to damp the pendulum resonance. This can be done easily with a ***viscous*** (theoretical) feedback force:

$$F_{\text{fb}} = -\gamma'\dot{x} \tag{5}$$

Our sensors do not measure x. Their output is:

$$out = \begin{cases} x - x_0 & \text{for displacement sensors} \\ \ddot{x} & \text{for accelerometers} \end{cases} \tag{6}$$

Therefore, the actual "viscous" force that can be built if position sensors are used has the form:

$$F^p_{\text{fb}} = -\gamma'\frac{\mathrm{d}}{\mathrm{d}t}(x - x_0) \tag{7}$$

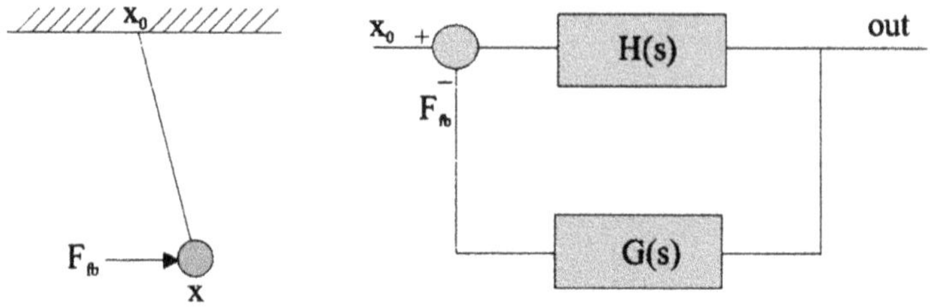

Figure 4. The control scheme for a simple pendulum

It can be easily shown that with such a feedback force the closed loop equation of motion (in Laplace space) reduces to:

$$x(s) = \frac{\omega_0^2 + G_0 s}{s^2 + \omega_0^2 + (\omega_0/Q + G_0)s} \cdot x_0(s) \tag{8}$$

where $G_0 = \gamma'/m$ is a gain parameter[a], measuring the intensity of the viscous feedback force, and Q is the open loop quality factor. When the loop is closed a damping of the resonance is achieved:

$$Q' \stackrel{G>>1}{\longrightarrow} \frac{\omega_0}{G} \tag{9}$$

Nevertheless, as the gain is increased, a larger amount of noise is reinjected off-resonance. This is associated to the term "$G_0 s$" in the numerator of (8) and depends on the fact that the sensor used to build up the feedback force measures the position of the pendulum with respect to ground. Therefore, an infinitely efficient feedback would "freeze" the pendulum to ground (which is seismic noisy), reducing its motion at the resonance, with the drawback of bypassing its attenuation properties above resonance.

The situation is fairly different when an inertial sensor is used. In this case the viscous feedback force is obtained by integrating the accelerometer output, and the output does not depend on x_0:

$$F_{\mathrm{fb}}^a = -\gamma' \int \ddot{x} \mathrm{d}t \tag{10}$$

The closed loop equation of motion is then:

$$x(s) = \frac{\omega_0^2}{s^2 + \omega_0^2 + (\omega_0/Q + G_0)s} \cdot x_0(s) \tag{11}$$

[a] In a real feedback system a frequency dependent gain function $G(s)$ rather than a gain parameter has to be considered.

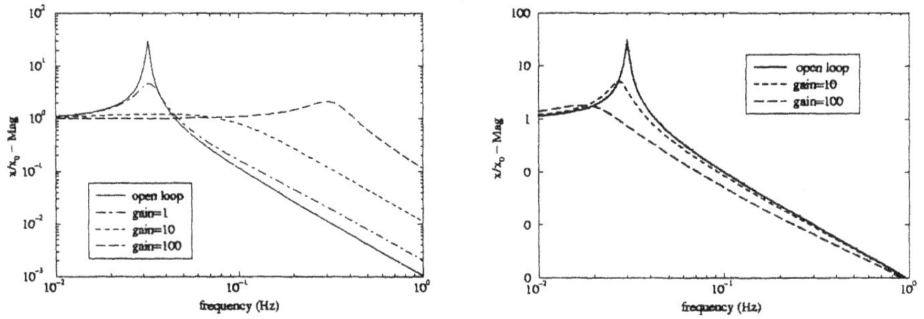

Figure 5. Damping of a simple pendulum: the closed loop transfer function $x(s)/x_0(s)$ (magnitude) when a position sensor is used (LEFT) and when an accelerometer is used (RIGHT).

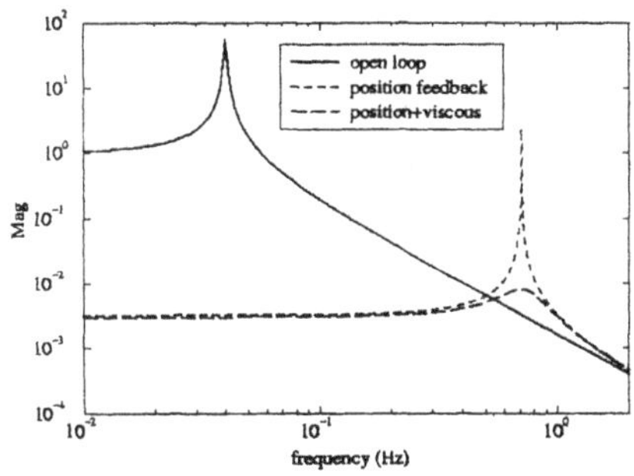

Figure 6. Inertial damping of a simple pendulum when a position feedback is implemented.

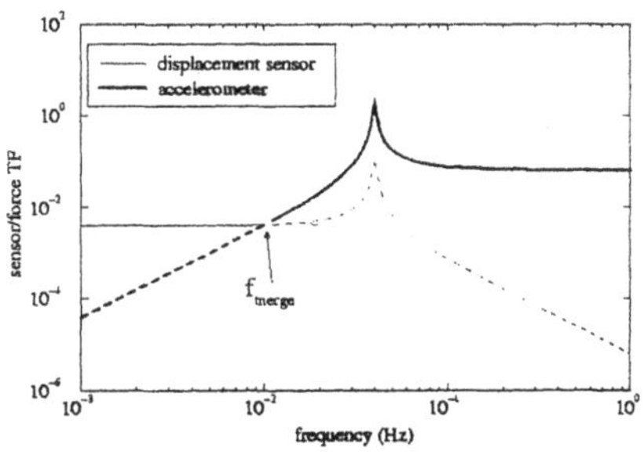

Figure 7. *Merging* of displacement and acceleration sensors (simulation for a simple pendulum).

A damping of the resonance is obtained (exactly as in the previous case) but without reinjection of off-resonance noise. In fig. 5 a simulation of the closed loop transfer function $x(s)/x_0(s)$ is shown in the two cases. Up to now we have considered a simple viscous damping. It is possible to increase the bandwidth of the control if the feedback force contains a term proportional to x (the double integral of the accelerometer signal). The result obtained in this case is shown in fig. 6.

5 Control strategy

In this section we extend the principles of the previous section and describe the strategy to control the SA.

The basic idea of inertial damping is to use the accelerometer signal to build up the feedback force. As a matter of fact, an infinitely efficient feedback using only the inertial sensor information, would null the acceleration of the pendulum, but it would not do anything if the pendulum moves at constant velocity. Such a control would be unstable with respect to ***drifts***. Therefore, if the control band is to be extended down to DC, a position signal is necessary. Our solution was a ***merging*** of the two sensors: the virtual LVDT (position) and accelerometer signals are combined in such a way that the LVDT signal ($l(s)$) dominates below a chosen cross frequency f_{merge} while the accelerometer signal ($a(s)$) dominates above it (see fig. 7 and ref. 7). The feedback force has the form[b]:

$$F_{\mathrm{fb}} = G(s)\left[a(s) + \epsilon l(s)\right] \tag{12}$$

where $G(s)$ is the digital filter transfer function (see fig. 8) and ϵ is the parameter whose value determines f_{merge}. We have chosen $f_{\mathrm{merge}} \sim 10$ mHz (corresponding to $\epsilon \sim 5 \cdot 10^{-3}$). This approach allows to stabilize the system with respect to low frequency drifts at the cost of reinjecting a fraction ϵ of the seismic noise via the feedback.

We describe in the following the feedback design for the 3 d.o.f., starting from the the translational ones. The virtual X and Y sensors show many resonant peaks (the modes of a chain of pendulums) and this requires a more sophisticated feedback strategy. The digital filter used to control the translation modes ($G(s)$) is shown in fig. 8 (LEFT). It shows three main features:

- for $0.01 < f < 2$ Hz the gain is proportional to f^{-2}. This corresponds to the case of fig. 6: the accelerometer signal is integrated twice and the feedback force is proportional to x;
- for $f > 2$ Hz the gain is proportional to f^{-1}. The accelerometer signal is integrated once: the feedback force is proportional to the velocity and a viscous damping is achieved;
- the peaks visible in the filter are necessary to compensate the corresponding dips in the mechanical transfer function ($H(s)$) of fig. 3, in order to make the feedback stable.

[b]Actually, the LVDT signal $l(s)$ is properly filtered in order to preserve the feedback stability at the cross frequency and in order to reduce the amount of reinjected noise at $f > f_{\mathrm{merge}}$.

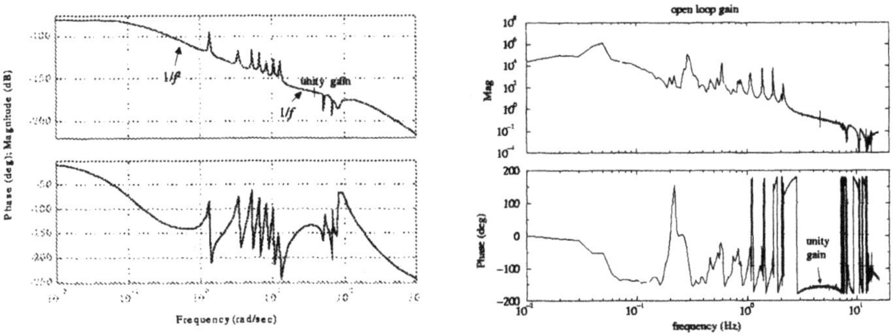

Figure 8. LEFT: Digital filter used for the inertial damping of a translation mode (X). The filter slope is f^{-2} in the range 10 mHz< f <3 Hz, f^{-1} for f >3 Hz. The unity gain is at 4 Hz. The peaks in the digital filter are necessary to compensate the dips in the mechanical transfer function (see the transfer function of the X mode in fig. 3). RIGHT: open loop gain function (measured). The phase margin at the unity gain frequency is about 25°.

Fig. 8 (RIGHT) shows the open loop gain transfer function $G(s)H(s)$.

The damping strategy for the Θ mode is simpler: the Θ virtual sensor (fig. 3, RIGHT) shows one resonance peak only and no dips: no compensation is necessary. Apart from this, the feedback strategy is similar to the ones used for the translational modes.

6 Inertial damping: experimental results

The result of the inertial control (on 3 d.o.f.) is shown in figure 9. The measurement has been performed in air. The noise on the top of the IP is reduced over a wide band (10 mHz - 4 Hz). A gain > 1000 is obtained at the main SA resonance (0.3 Hz). The RMS translational motion of the IP (calculated as $x_{\rm RMS}(f) = \sqrt{\int_f^\infty \tilde{x}^2(\nu)\mathrm{d}\nu}$) in 10 sec. is reduced from more than 30 to 0.3 μm. The closed loop floor noise corresponds to the fraction of seismic noise reinjected by using the position sensors for the DC control and can, in principle, be reduced by a steeper low pass filtering of the LVDT signal at $f > f_{\rm merge}$ and by lowering $f_{\rm merge}$: both this solution have drawbacks and need a careful implementation.

Preliminary measurements of the displacement of the mirror with respect to ground have been performed in air, using an LVDT position sensor. The

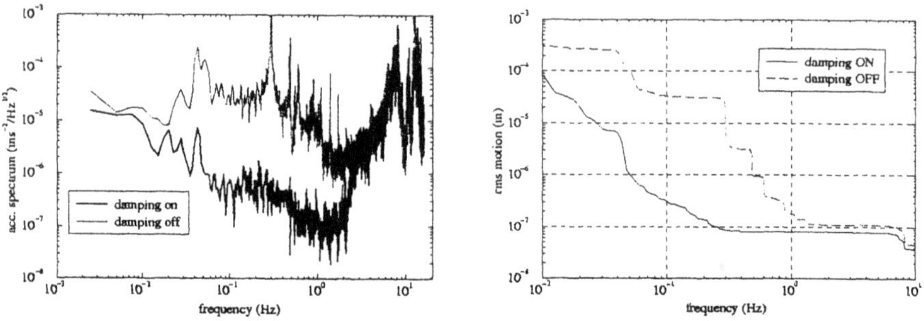

Figure 9. Performance of the inertial control (X, Y, Θ loops closed) of the superattuenuator, measured on the top of the IP: the left plot shows the acceleration spectral density as measured by the *virtual* accelerometer X (translation). The right plot shows the effect of the feedback on the RMS residual motion of the IP as a function of the frequency.

residual RMS mirror motion in 10 sec. is[c]:

$$x_{\rm RMS}(0.1\,{\rm Hz}) \leq 3\,\mu{\rm m}. \tag{13}$$

When the damping is on such a measurement can provide only an upper bound because the LVDT output is dominated by the seismic motion of the ground.

7 Conclusions

In this lecture we have tried to outline how to face the problem of reducing the free motion of the suspended optical components of the VIRGO interferometer, associated to the resonances of the suspension. This is only the beginning: once the motion of the mirrors is reduced to a few microns, the lower control stages can operate to *lock* the interferometer in the correct operation state.

Acknowledgments

The author wishes to thank all the people of the Pisa and Florence VIRGO Groups. Among them, special thanks to Diego Passuello, Alberto Gennai, Andrea Marin.

[c]This number has been obtained with a simpler feedback design, less *aggressive* than the one of fig. 8: the gain raised as $1/f$ (pure viscous damping force), the cross frequency was 30 mHz and no compensation of the dips was needed.

References

1. P.R.Saulson, *Fundamentals of interferometric gravitational wave detectors*, World Scientific (1994).
2. G.Losurdo for Pisa VIRGO Group, Proc. of the *Second E.Amaldi Conference on Gravitational Waves*, CERN, Geneva, 1-4 July 1997, eds. E.Coccia, G.Pizzella, G.Veneziano , World Scientific (1998).
3. F.Frasconi, *et al.*, Performances of the R&D superattenuator chain of the VIRGO experiment, to be published in the proceedings of the *XXXIVth Rencontres de Moriond* (1999).
4. A.Giazotto, this volume.
5. G.Losurdo, *et al.*, *Rev. Sci. Instrum.*, **70** (5), 2507-2515 (1999).
6. S.Braccini, *et al.*, *Rev. Sci. Instrum.*, **66** (3), 2672-2676 (1995).
7. S.J.Richman, J.A.Giaime, D.B.Newell, R.T.Stebbins, P.L.Bender, J.E.Faller, *Rev. Sci. Instrum.*, **69** (6), 2531 (1998).
8. A.Gennai, *et al.*, *A control model for the inverted pendulum*, VIRGO Note VIR-TRE-PIS-4900-102 (1997).
9. G.Losurdo, *Ultra-low frequency inverted pendulum pre-isolator for the VIRGO test mass suspension*, PhD Thesis, Scuola Normale Superiore, Pisa (1998).
10. G.Losurdo, in the Proceedings of the *Third E.Amaldi Conference on Gravitational Waves Experiment*, Caltech, Pasadena, 12-16 July 1999, gr-qc/9911044 (1999).
11. A.Gennai, G.Losurdo, A.Marin, D.Passuello, *Inertial damping of the superattenuator*, VIRGO Note VIR-TRE-PIS-4900-104 (1999).

SIGNAL PROCESSING: ELEMENTS OF DETECTION AND ESTIMATION THEORY

ARMANDO VANNUCCI
Dipartimento di Ingegneria dell'Informazione Università degli Studi di Parma
Viale delle Scienze 181A, Parma, Italy
E-mail: vannucci@tlc.unipr.it

MARIA GABRIELLA DI BENEDETTO
Dipartimento INFOCOM Università degli Studi di Roma "La Sapienza"
Via Eudossiana 18, Roma, Italy
E-mail: gaby@acts.ing.uniroma1.it

The authors wish here to provide a basic introduction to the signal processing tools and techniques necessary for the purpose of detecting a signal corrupted by random noise. This is the core problem of *Detection Theory.* It is typically encountered, among various other scientifical and technological research areas, in the experimental study of Gravitational Waves (GW), where the signal is often of an electrical nature and is provided by a suitable transducer, which can be referred to, in general terms, as an antenna. Since the signal to be detected is rarely completely known, its shape can vary according to many unknown parameters, such as amplitude, frequency, direction of arrival etc. The evaluation of those parameters is thus essential for the correct determination of the GW to be detected. The theory and techniques for a reliable evaluation are referred to as *Estimation Theory.* The fundamental role of noise, as the main drawback in the process of detection and estimation, makes it necessary to introduce appropriate tools for its understanding and modelling: the theory of *Random Processes* provides the background for this purpose and for the successful exploitation of noise models for the detection and estimation problems.
This tutorial paper is organized in three sections: the first is a brief review of the theory of Stochastic Processes (a synonym for Random Processes), of which the reader is assumed to have already a basic knowledge; the second is an introduction to Detection Theory, discussing different optimization criteria for the detection problem, under binary or multiple hypotheses, and leading to the design method of a *receiver* operating in the common case of *Gaussian Noise.* The third section deals with *Estimation Theory*: a survey of the parameters estimation problem under different optimization criteria is given, based on the results of the previous section; some analytical tools for evaluating the reliability of the derived *estimators* are also introduced.
The prerequisites for a full understanding of the following material are: a good knowledge of the Theory of Probability and its main results and theorems, about which the reader is assumed to have taken university level courses, and a knowledge of the theory and techniques of *Fourier Analysis*, with which the reader is also assumed to be familiar.

1 Review of Stochastic Processes

When a physical quantity, such as an electrical field or a gravitational field, is measured for the purpose of gaining a certain amount of *information* on it, it is implicit that its value is, a priori, not known nor predictable. When the same quantity varies in time, the function describing time variations is called a *signal*: in order for a signal to bear some useful information it is thus necessary that it varies in an unpredictable way or, in other words, that it is a *random signal*. Those random variations may be caused by an appropriate device, as happens in *electrical digital communications* where an electrical field is varied by a *modulator* according to some data to be transmitted, or rather be intrinsic in the physical quantity to be measured, as is the case for *gravitational waves* where the evolution of the astrophysical source is responsible for its generations.

Stochastic Processes are the mathematical entity for modelling this kind of signals; one possible point of view is to think of a stochastic process as a whole family of functions of time, each function describing one among all of the possible evolutions of the physical quantity to be observed, thus each function being a *deterministic signal*. The observation is supposed to extend along the whole infinitely long time axis and thus, once the observation is performed, the shape of the unknown signal is completely determined and the process is said to have taken one of its *sample functions* or *realizations*. As depicted in Fig. 1, it is useful to remind the similarity of this approach with the classical description of a *Random Variable* (RV): prior to the execution of a random experiment all possible outcomes S_i of a RV, the so called *sample space* S, is needed to describe the variable itself, while after the observation of the actual outcome, the variable takes a certain value, or, in other words, a particular *determination*.

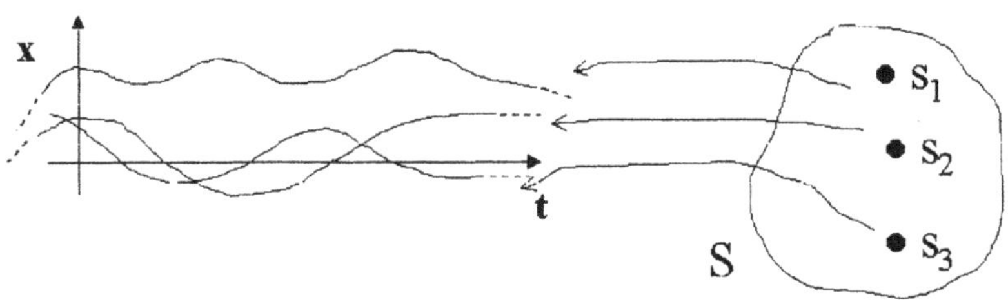

Figure 1. Correspondence between determinations of a random variable and realizations of a stochastic process.

Thus, a stochastic process is a family of signals (its realizations), each describing the same physical quantity (a voltage, a photon flux power ...), where it is not a priori known which signal will experimentally be observed. A process is often denoted by a capital symbol, like $X(t)$, to distinguish it from ordinary functions, or otherwise $X(t; s_i)$ to underline the analogy with RVs, as if the observed signal depended on the outcome s_i of an experiment defined on a probabilistic space S.

Another useful point of view for defining a process is gained by observing its value (i.e. the measured value of the physical quantity that the process represents) at a given time t_1. Prior to the observation, the value $X(t_1; s_i)$ that the process will take depends on which realization will be observed, and is thus a RV X_1 whose outcome depends on s_i. Similarly, by observing the process at an n-tuple of times $(t_1, t_2, ..., t_n)$, a random vector $X=(X_1, X_2, ..., X_n)$ is obtained, as shown in Fig. 2 for $n = 2$. A process can then be tought of as a collection of a noncountable infinity of random variables X_i, each extracted by observing the process at any possible time t_i.

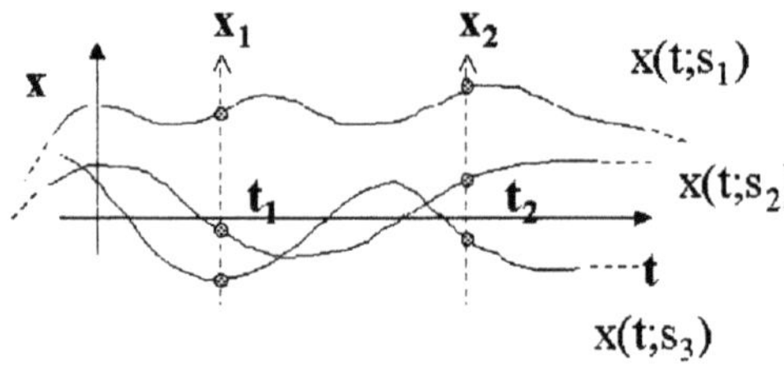

Figure 2. Random variables X_i extracted from the observation of a process; 3 sample functions are shown.

If, on the other hand, in the expression $X(t; s_i)$ it is the second variable s_i that is fixed, the process is reduced to a deterministic signal, i.e. one of its sample functions. If both variables are fixed, $X(t_1; s_i)$ is trivially a number. In the following we will usually omit the dependence on the probabilistic variable s_i, and denote a process by $X(t)$. The probabilistic approach relates any observation $X(t)$ of a process to a RV X: as a random variable, X can be described by the usual probabilistic functions, namely the *Probability Distribution Function* (PDF)

$$F_X(x; t) = P_r\{X(t) \leq x\}$$

where $P_r\{A\}$ denotes the probability of an event A, or equivalently its first derivative, the *probability density function* (pdf)

$$f_X(x;t) = \frac{\partial F_X(x;t)}{\partial x}$$

These functions depend on two variables: the real number x and the observation time t, thus a knowledge of the pdf, or the PDF, for every possible t would be desirable; all these functions, for every t, constitute the first order statistics of the process. Despite the difficulty of practically obtaining all the first order statistics, they would not even be sufficient for a complete probabilistic description of the process $X(t)$: higher order statistics, such as those of second order

$$F_X(x_1, x_2; t_1, t_2) = P_r\{X(t_1) \leq x_1 \cap X(t_2) \leq x_2\}$$

$$f_X(x_1, x_2; t_1, t_2) = \frac{\partial^2 F_X(x_1, x_2; t_1, t_2)}{\partial x_1 \partial x_2}$$

or, in general, n-th order, $f_X(x_1, x_2, ..., x_n; t_1, t_2, ..., t_n)$, for n arbitrary large, would be needed. Given the statistical correlation of the RVs X_i (their outcomes all depend on the same realization $X(t; s_i)$ of the process), n-th order statistics cannot be derived from lower order statistics.

1.1 *Ensemble averages*

Usually, a complete statistical description of a process is not needed: it suffices to know some cumulative values yielding information about the *realizations in the average*. The process mean value, defined as

$$\eta_X t = E[X(t)] = \int_{-\infty}^{+\infty} x f_X(x; t) dx$$

where $E[\,]$ is an ensemble average, i.e. the value of the process X at time t is averaged over all possible realizations, is a deterministic function of time giving the shape of the mean realization. Even though $\eta_X(t)$ could not be one of the possible realizations of the process, it gives the a priori expected value of X at time t. Another fundamental function is the process **autocorrelation**, defined as

$$R_X(t_1, t_2) = E[X(t_1)X(t_2)] = \int_{-\infty}^{+\infty} \int_{-\infty}^{+\infty} x_1 x_2 f_X(x_1, x_2; t_1, t_2) dx_1 dx_2$$

where, for a couple of observation times (t_1, t_2), R_X measures the correlation of the RVs $X_1 = X(t_1)$ and $X_2 = X(t_2)$.

The mean value can be computed if the first order statistics are known; similarly, the autocorrelation requires the second order statistics for its analytical computation: thus mean value and autocorrelation are first and second order ensemble averages, respectively. But, as happens for the average values of random variables, if these ensemble averages are, for some reason, known, it is anyway not possible to derive the first and second order statistics. Another useful second order ensemble average is the **autocovariance**

$$\phi_X(t_1,t_2) = E[(X(t_1)-\eta_X(t_1))(X(t_2)-\eta_X(t_2))] = R_X(t_1,t_2)-\eta_X(t_1)\eta_X(t_2)$$

that can be computed based on the previously introduced mean value and autocorrelation. As for the covariance in the theory of random variables, ϕ_X measure the statistical dependence between the two RVs $X(t_1)$ and $X(t_2)$. The last important average value for a process is the **mean square value**

$$P_X(t) = E\left[|X(t)|^2\right] = R_X(t,t)$$

measuring the average square magnitude of the process at time t. For the computation of all the four mentioned averages, only first and second order statistics would be needed.

The Harmonic process

As an example we shall consider a process $X(t)$ whose realizations are sine waves of fixed amplitude and frequency, having a random initial phase θ. When θ is uniformly distributed over the interval $[0,2\pi]$, the process is called *harmonic*: its general expression is

$$X(t) = a\cos(2\pi f_0 t + \theta)$$

with RV θ uniform in $[0,2\pi]$, and two examples of its realizations are shown in Fig. 3 together with the pdf of θ.

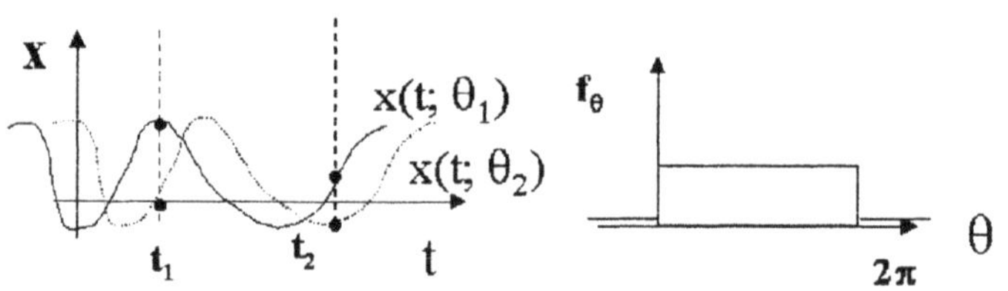

Figure 3. Sample functions of an harmonic process, observed at times t_1 and t_2.

The harmonic process thus represents a sine wave starting at a completely random time and can be practically thought of as the output of a stable oscillator that is turned on at an unknown time. It is one of a wide class of processes, called *parametric processes*, whose peculiarity is the explicit analytic dependence of the process expression on one or more random parameters.

We shall now proceed with the calculation of the previously introduced ensemble averages: the mean value is, by the theorem of expectation, equal to zero

$$\eta_X t = E[X(t)] = \int_0^{2\pi} \frac{1}{2\pi} a \cos(2\pi f_0 t + \theta) d\theta = 0$$

while the autocorrelation, through some trigonometric simplifications, takes the following expression

$$R_X(t_1, t_2) = E[X(t_1)X(t_2)] = \int_0^{2\pi} \frac{1}{2\pi} a \cos(2\pi f_0 t_1 + \theta) \cos(2\pi f_0 t_2 + \theta) d\theta =$$
$$= \frac{a^2}{2} \cos(2\pi f_0 (t_1 - t_2))$$

that only depends on the difference between the chosen observation times. The process autocovariance and mean square value are:

$$\phi_X(t_1, t_2) = R_X(t_1, t_2) \qquad P_X(t) = R_X(t, t) = \frac{a^2}{2}$$

1.2 Stationarity

We have just shown that, for the harmonic process, the mean value is independent of the observation time t (it is constant and equal to zero) and the autocorrelation only depends on $(t_1 - t_2)$. These particular properties shown for the ensemble averages of this process are common to a wider, and extremely important, class of processes: any process with the two mentioned properties is called **Wide Sense Stationary (WSS)**. For these, the mean value η_X is thus constant and the autocorrelation $R_X(\tau)$ is expressed with only one time variable $\tau = t_1 - t_2$.

Another definition is that of **Strict Sense Stationary (SSS)** processes: a process is said to be SSS if its statistical properties are invariant to a shift of the time origin. In other words, it can be said that the sample functions of a SSS process have the same probability of being observed with any time displacement. Therefore, given $X(t)$, its statistics are the same of the process $X(t - t_0)$, whose realizations are a time displaced version of those of $X(t)$, and are equally likely to happen. Having the same statistics implies, for the first order statistics, $f_X(x; t) = f_X(x; t - t_0)$, for any t_0; thus this first order pdf must be independent of the observation time and only depends on the variable

x. Resorting to the definition of mean value, it is easy to show that η_X must be constant and independent of time. A similar reasoning can be applied to the second order statistics too: if the process is SSS, $f_X(x_1, x_2; t_1, t_2) = f_X(x_1, x_2; t_1 - t_0, t_2 - t_0)$ must hold, and then these second order pdf must only depend on the difference of the two time arguments. Through the definition of autocorrelation, also $R_X(\tau)$ will depend only on the difference τ between the observation times.

For any SSS process it is then demonstrated that it obeys the two properties defining the class of WSS processes and thus that the first class includes the second as a subset. SSS is indeed a much more strict definition since it is based on the complete statistics of the process, while WSS only requires some average values to respect some properties. It is then evident that the implication SSS$\Rightarrow$WSS cannot be reversed; anyway it must be noted that the definition of WSS is much easier to verify in practice, as has been done for the harmonic process, than that of SSS. From now on we will concentrate mainly on the important class of WSS processes, showing the meaning and properties of their autocorrelation.

The following equality

$$R_X(-\tau) = E[X(t_1)X(t_1 + \tau)] = E[X(t_2 - \tau)X(t_2)] = R_X(\tau)$$

shows that R_X must be an even function of τ; moreover, since $R_X(0) = E[X^2(t)]$, the autocorrelation must have a positive value on the origin of the τ axis. A third property, that we will not demonstrate, states that the maximum absolute value for R_X must be reached on the origin, analytically $|R_X(\tau)| \leq R_X(0)$. Fig. 4 shows pictorially the implications of these properties on the graphic of the autocorrelation for WSS processes: the second function shown is not allowed since it is not even, the third is even but has a negative value on the origin and the fourth respects the first two properties but has its absolute maximum on the two sidelobes rather than on the origin.

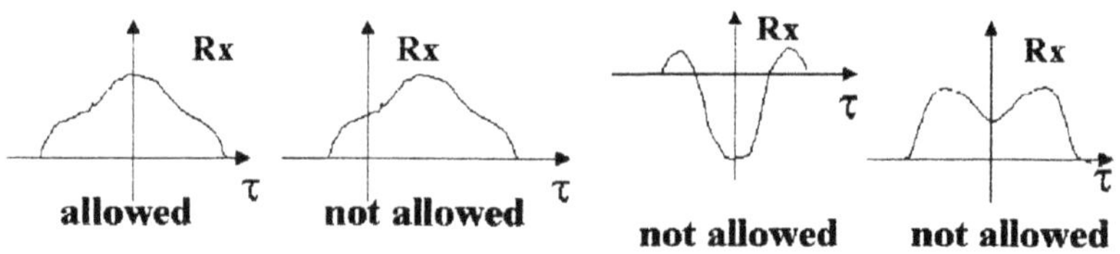

Figure 4. Some graphical properties for the autocorrelation of WSS processes.

As previously outlined, $R_X(\tau)$ is the correlation between the two RVs

$X(t)$ and $X(t+\tau)$ and correlation is, for RVs, a measure of their statistical dependence: if, for instance, RVs $X(t)$ and $X(t+\tau)$ are independent for a certain value of τ, their correlation equals the product of their mean values and thus $R_X(\tau) = \eta_X(t)\eta_X(t+\tau)$, while the autocovariance $\phi_X(\tau)$will be equal to zero. For WSS processes, the value of autocorrelation and autocovariance is the same for any time displacement of the observation times, provided their difference is fixed; then, if the autocovariance has a finite duration on the τ axis, being zero outside a finite interval [-M,M], M can be regarded as the *memory of the process*. The reason for this is that any two observations of the process at times differing by more than M yield two uncorrelated random variables.

Another useful interpretation for the autocorrelation of WSS processes is gained by defining their *power*. Thinking of the process as a random signal conveying energy, if, for instance, $X(t)$ is a voltage or current signal, it can be applied to a resistive load of 1Ω. This way, the instantaneous power dissipated on the load at time t is $|X(t)|^2$, which is a RV whose expectation equals the WSS process mean square value P_X, independent of t. For a single realization, the time average of the dissipated power is

$$P_X(s_i) = \lim_{T\to\infty} \frac{1}{2T}\int_{-T}^{T} |X(t;s_i)|^2 dt$$

which is still a random variable, as underlined by the dependence on s_i. The ensemble average over all realizations of this RV is defined as the **power of the process**. If the process is WSS, this average is constant and equals the previously mentioned average of istantaneous powers P_X:

$$P_X = E[P_X(s_i)] = \lim_{T\to\infty} \frac{1}{2T}\int_{-T}^{T} E[|X(t;s_i)|^2]dt$$

the expectation in the last integral being constant and equal to $R_X(0)$. Thus the power of a WSS process is constant and equal to the value of its autocorrelation for $\tau = 0$. We will further exploit the relationship between a process power and its autocorrelation, after introducing the topic of process filtering.

1.3 Filtering

A filter is defined as a *Linear Time-Invariant* (LTI) system; as usual in systems theory, such systems are characterized by their response to a *Dirac delta* pulse $\delta(t)$ applied at their input: the output is referred to as the impulse response $h(t)$ of the system. The general relationship between the input and the output of a filter is given by the *convolution* of the input signal with the system's impulse response, mathematically defined as

$$y(t) = \int_{-\infty}^{+\infty} x(\tau)h(t-\tau)d\tau = x(t) * h(t)$$

where $x(t)$ is the input signal and $y(t)$ is the output. Signal filtering can as well be performed in the frequency domain, resorting to the properties of *Fourier Transforms*: by the *theorem of convolution*, the output signal's transform is simply given by multiplying the transform of the input signal with the system *transfer function* $H(f)$, which is defined as the Fourier transform of the impulse response, according to the simple equation $Y(f) = X(f)H(f)$.

When the input to a LTI system is a stochastic process instead of a deterministic signal, the output is also a stochastic process whose realizations are the filtered versions of the realizations of the input process, being the system deterministic. It is then useful to compute some ensemble averages for the output process, once this averages are given for the input, in order to explicit how the filter acts on the process as a whole. If $x(t)$ is one possible realization of the input process, then performing convolution with $h(t)$ and averaging with respect to all the realizations gives mean value of the output process.

$$\eta_Y(t) = E\left[\int_{-\infty}^{+\infty} x(\tau)h(t-\tau)d\tau\right] = \int_{-\infty}^{+\infty} \eta_x(\tau)h(t-\tau)d\tau = \eta_x(t) * h(t)$$

Computing the autocorrelation of the output process, similarly yields

$$\begin{aligned} R_y(t_1,t_2) &= E\left[\int_{-\infty}^{+\infty} x(t_1-\tau)h(\tau)d\tau \int_{-\infty}^{+\infty} x(t_2-u)h(u)du\right] \\ &= \int_{-\infty}^{+\infty}\int_{-\infty}^{+\infty} R_x(t_1-\tau,t_2-u)h(\tau)h(u)dud\tau \end{aligned}$$

The two above equations can be further simplified when the input process $X(t)$ is WSS

$$\eta_Y = \eta_X \int_{-\infty}^{+\infty} h(\tau)d\tau \qquad R_Y(\tau) = R_X(\tau) * h(\tau) * h(-\tau)$$

which shows that also the output process obeys the two properties defining wide sense stationarity. It can then be stated that linear filtering preserves WSS; it could be also demonstrated that also strict sense stationarity is preserved. This is a nontrivial result, that does not hold for other kinds of systems, such as nonlinear memoryless which preserve the WSS but not the SSS properties.

1.4 Power Spectral Density

The last equation, relating the autocorrelations of the input and output processes of a filter, is more easily expressed in the frequency domain, where the double convolution is tranformed into a product: $F[R_Y(\tau)] = F[R_X(\tau)]H(f)H(-f)$, where $F[.]$ denotes Fourier transform and $H(-f) = H^*(f)$ for real systems, i.e. for systems having a real impulse response. It is not clear, however, at this stage what such transform could possibly mean. If, however, we introduce the symbol $S_X(f) = F[R_X(\tau)]$, and recalling the properties of Fourier transforms, the mean power of the WSS process $X(t)$ can be obtained through the integration of $S_X(f)$. Moreover, suppose to apply the process $X(t)$ at the input of an ideal passband filter: the situation is depicted in fig.5. Since the filter is real, a multiplication of $S_X(f)$ by the square magnitude of the transfer function $H(f)$ gives $S_Y(f)$

$$S_Y(f) = S_X(f)|H(f)|^2$$

and since $H(f)$ is equal to one in its passband and zero elsewhere, $S_Y(f)$ is nothing but a *slice* of the function $S_X(f)$.

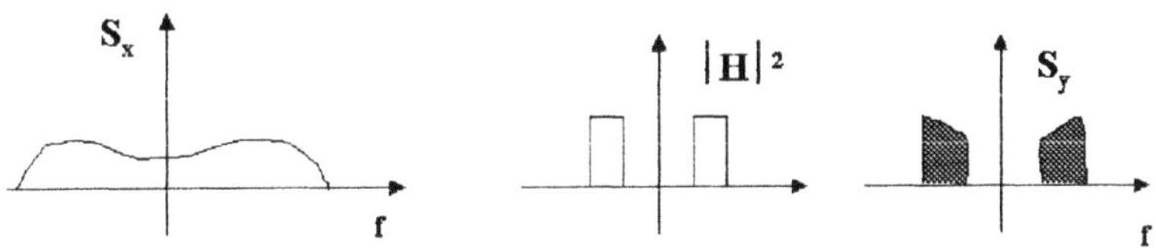

Figure 5. Power spectral density of an ideally bandpass filtered process.

We know that the mean power of the output process is

$$P_Y = R_Y(0) = \int_{-\infty}^{+\infty} S_Y(f)df \geq 0$$

which is the area of the shaded region in Fig. 5. Supposing the filter $H(f)$ has a narrow band centered around the frequency f_0, it suppresses almost all the frequency components of the input process' realizations but those in the neighbourhood of f_0: then it can be said that the mean power of the output process coincides with the power of the input process pertaining those frequencies only. Since, for an extremely narrowband filter, this power is proportional to $S_X(f_0)$, the conclusion is that the introduced function $S_X(f)$ is a description of how the mean power of the input process is distributed on the frequency axis. This conclusion is the celebrated **Wiener-Kintchine**

theorem: $S_X(f)$ is called the **Power Spectral Density (PSD)** of the WSS process $X(t)$ and, from the above reasoning, it is implied that it must be a nonnegative function whose integral on the whole f axis yields the mean power P_X. A similar relationship holds for the so called **covariance spectrum** $\Phi_Y(f)$, which is defined as the Fourier transform of the autocovariance of the process $Y(t)$

$$\Phi_Y(f) = F[\phi_Y(\tau)] = \Phi_X(f)|H(f)|^2$$

Since we stated that the duration of the autocovariance is related to the correlation time, or memory, of the process, processes with short memory, for which their value at time t has little or no statistical dependence on their value at a later time $t+\tau$, with properly large τ, tend to have a covariance spectrum with a wide band.

1.5 Gaussian Processes

Often, the value of a process at a certain time is determined by the superposition of many independent random phoenomena. Since the celebrated *Central Limit Theorem* states that the sum of n independent RVs with finite mean and variance approaches a Gaussian distribution for increasing n, it is intuitively understood how the observation of a process can be a Gaussian RV.

Strictly speaking, a process is called Gaussian if, for any n-tuple of times $\underline{t} = [t_1, t_2, ..., t_n]$, its observed values form a vector $\underline{x} = [x_1, x_2, ..., x_n]$ of jointly Gaussian RVs with joint pdf

$$f_X(\underline{x};\underline{t}) = \frac{1}{\sqrt{(2\pi)^n|C|}} e^{-\frac{1}{2}[(\underline{x}-\underline{\eta})^T C^{-1}(\underline{x}-\underline{\eta})]}$$

where $\eta_i = E[x_i] = E[X(t_i)] = \eta_X(t_i)$ are the elements of the *mean value vector* $\underline{\eta}$ and the *covariance matrix* C has elements

$$C_{ij} = Cov[x_i, x_j] = E[(X(t_i) - \eta_X(t_i))(X(t_j) - \eta_X(t_j)] = \phi_X(t_i; t_j)$$

Since these elements only depend on the mean value and autocovariance functions of the process, statistics of any order can be computed once $\eta_X(t)$ and $\phi_X(t_1, t_2)$ are given. Moreover, if the Gaussian process is WSS, it can be shown, using η_X and $\phi_X(\tau)$, that the statistics of any order are invariant to a shift of the time origin and thus $X(t)$ is also SSS.

These properties make Gaussian processes a very special kind of processes; another fundamental property is that Gaussianity is preserved by linear filtering: writing convolution in a limit form

$$y(t) = \lim_{\Delta t \to \infty} \left[\sum_{i=-\infty}^{+\infty} x(i\Delta t) h(t - i\Delta t) \Delta t \right]$$

the output process at any time t is shown to be a linear combination of the Gaussian RVs $x(i\Delta t)$ and thus Gaussian. If $X(t)$ is Gaussian and WSS then $Y(t)$ is WSS, as stated in the last section, and also gaussian, then both are SSS.

White Gaussian noise

Noise is often the sum of many independent random effects: for example, the thermal agitation of electrons in a conductor at absolute temperature T produces a random voltage signal at the terminals of the conductor; this stochastic process is referred to as **Johnson noise** or **thermal noise**.

Given a complex impedance Z(f), the celebrated **Nyquist theorem** states that thermal noise, i.e. the voltage N(t) measured at the terminals of the impedance, is a stationary Gaussian process N(t) with mean value $\eta_N = 0$ and covariance spectrum $\Phi_n(f) = 2KT\mathrm{Re}\{\mathrm{Z(f)}\}$, where $\mathrm{Re}\{\cdot\}$ stands for the real part and K is the Boltzmann constant.

If Z(f)=R is a pure resistance, then the covariance spectrum and the PSD are constant and the process N(t) is called **white noise** in analogy with pure white light, which contains all frequency components in the same amount. It is evident that white noise is only an abstraction, otherwise its mean power P_N computed integrating the PSD would be infinite, but for the frequencies of interest (about up to 10^{12} Hz) thermal noise can usually be thought of as white; for higher frequencies the actual spectrum will be slowly decreasing. Also, it is common to deal with filtered versions of thermal noise: since filtering alters the PSD, this is referred to as *coloured noise*.

Inverse-transforming the covariance spectrum of white noise yields for the autocorrelation $\phi_N(\tau) = 2KTR\delta(\tau)$, a Dirac delta with zero duration. Since we interpreted this duration as the memory of the process, we can state that a white noise process has the most rapid, istantaneous, random variations since its value at time t is statistically independent of its value at immediately preceding or subsequent times.

We will often use the properties of gaussian processes and white noise in the following chapters, where noise plays a major role in the detection and estimation problems.

2 Elements of Detection Theory

The problem of detecting an information bearing signal arises in many scientifical and technological fields: communication theory, radar theory and experimental study of gravitational waves, just to name a few. The main difficulty is that the information bearing signal may suffer from many impairments while it transits from the *source* that generated it to the destination equipment that is able to detect it: *weakening*, due to dispersion of its energy, *distortion*, due to physical means that supports its transmission, and *noise*, which is always present in any measuring equipment, are the main impairments. The solution to the problem is to design a suitable **receiver** that is able to detect signals which are weakened, distorted and corrupted by random noise. The object of detection theory is to devise the design methodology of such receivers and the assessment of their performance.

There may be different schemes, depending on the application area, in which a detection problem is at hand: in the GW case, for instance, the main question is to decide, at a certain time, if a useful GW signal in the midst of noise has been detected by the measuring equipment or else the measured data consists of noise alone; we talk of **binary hypotheses testing,** in this case. Another scheme is that of *digital communications*, in which it is costumary to transmit one among many, say M, different signals which is in a one-to-one correspondence with a certain *symbol* to be received at destination; we talk of **multiple hypotheses testing**, in this case.

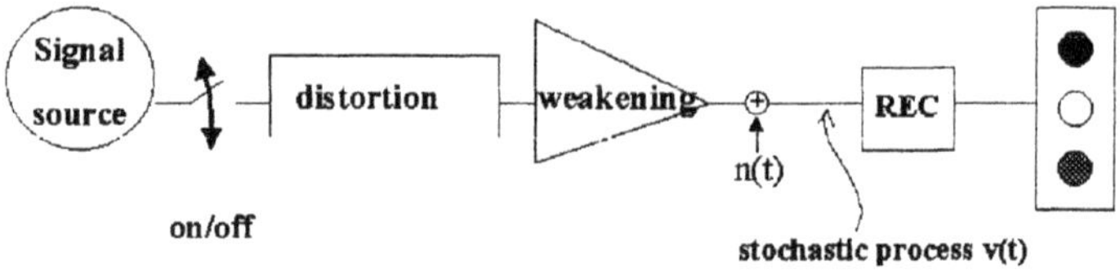

Figure 6. Binary hypotheses testing scheme.

In Fig. 6 a binary hypotheses testing scheme is depicted: the receiver must make a decision whether the source is on or off, and thus turn the traffic light green or red, based on the observation of its input signal $v(t)$. This signal is typically the voltage across the terminals of an antenna and, since the receiver ignores what signal will show up at its input, $v(t)$ is a stochastic process, its randomness being influenced by several factors: in particular, the state of the source (on/off) and the insertion of random noise $n(t)$. The aim of the receiver is, of course, to make mistakes as seldom as possible: this is often a though task since, particularly when dealing with very weak signals, a useful signal plus a strong noise can resemble the shape of noise alone.

2.1 The mathematical approach

Theory of Probability represents a useful mathematical tool for modelling a detection problem: in the general case of multiple hypotheses testing, we can associate the M hypotheses, each representing the detection of a certain signal, to M different *events* H_j. Suppose the receiver observes the signal $v(t)$ in an interval of length T, measuring n data points that are collected in a vector $v = (v_1, v_2, ..., v_n)$: v is a vector of real *samples* extracted from $v(t)$, that lies in the R^n space, R being the set of real numbers. We will discuss later of how the sampling procedure shall be performed; for the moment, the problem can be stated as: choose j, and thus the hypothesis M_j, based on the outcome of the random vector $\underline{v} \in \mathrm{R}^n$. The key to the solution of this problem is the knowledge of the joint conditional pdfs $p_j(v_1, v_2, ..., v_n)$ (we will denote a pdf with the letter p in this chapter), each subscript j denoting conditioning on the event H_j. The problem can thus be restated as: choose which of the M different $p_j(...)$ is believed to characterize the observed data. The receiver must then univocally associate an H_j to any point of the data space, thus dividing R^n in M disjoint regions R_j: any such decomposition is called a **strategy**. Given the strategy, it is determined the rule according to which the receiver decides for a certain hypothesis, having observed the measured data.

2.2 Decision criteria: the Bayes's strategy

To decide which strategy to employ we must seek a decision criterion: it is reasonable to choose that strategy maximizing the probability of correct decision P_c, or equivalently the one minimizing the probability of error $P_e = 1 - P_c$. Assume we know the **prior probabilities** $\Pr(H_j)$ with which the j-th hypothesis is verified, since the receiver chooses H_j whenever $\underline{v} \in \mathrm{R}_j$, we can write, according to the *total probability theorem*

$$P_c = \sum_{j=1}^{M} P_r(H_j) P_r(\underline{v} \in \mathrm{R}_j / H_j) = \sum_{j=1}^{M} P_r(H_j) \int_{R_j} p_j(\underline{v}) d^n v$$

and also the **posterior probabilty** that H_j is true when $\underline{v}$ is observed can be expressed through the Bayes's theorem

$$P_r(H_j/\underline{v}) = \frac{P_r(H_j) p_j(\underline{v})}{p(\underline{v})}$$

with $p(\underline{v}) = \sum_{j=1}^{M} P_r(H_j)p_j(\underline{v})$ the unconditional pdf of the data vector. We can now insert the last equation in the expression of P_c, obtaining

$$P_c = \sum_{j=1}^{M} \int_{R_j} p(\underline{v}) P_r(H_j/\underline{v}) d^n v \quad (p(\underline{v}) \geq 0)$$

Since we wish to maximize P_c, the best strategy, according to the last expression, is to choose the hypothesis with the greatest posterior probability, given the observed data $\underline{v}$. This conclusion was first enunciated by reverend Thomas Bayes in 1763 in his "*An essay on the doctrine of chances*" and is referred to as the **Bayes's rule.**

In the simple case of binary hypotheses, for the detection problem, the receiver can only choose between the presence of noise only in the observed data, which is referred to as the **null hypothesis** and labelled as H_0, and the presence of a useful signal besides noise, labelled H_1 and called the **alternative hypothesis.** Following Bayes's rule, the receiver must evaluate and compare the posterior probabilities: neglecting the unconditional pdf $p(\underline{v})$, which is irrelevant for the comparison,

$$P_r(H_1)p_1(\underline{v}) \mathop{\gtrless}_{H_0}^{H_1} P_r(H_0)p_0(\underline{v}) \qquad \Lambda(\underline{v}) \stackrel{\Delta}{=} \frac{p_1(\underline{v})}{p_0(\underline{v})} \qquad \Lambda_0 \stackrel{\Delta}{=} \frac{\Pr(H_0)}{\Pr(H_1)} \qquad \Lambda(\underline{v}) \mathop{\gtrless}_{H_0}^{H_1} \Lambda_0$$

the first above expression must be evaluated, choosing the appropriate H_j. Equivalently, introducing the **likelihood ratio** $\Lambda(\underline{v})$, as above, it can be compared with a fixed decision threshold, determined through the prior probabilities.

In binary decisions, two kinds of errors can be made: an error of the first kind happens when choosing for the alternative hypothesis when no signal is present: this is called a **false alarm** and denoted by $(\rightarrow H_1|H_0)$. Errors of the second kind happen whenever choosing the null hypothesis while H_1 is at hand: it is called a **false dismissal** and denoted by $(\rightarrow H_0|H_1)$. In practical situations, false alarms and false dismissals may have different costs of operation and even correct decisions, like $(\rightarrow H_1|H_1)$, which is called a **detection**, may have their costs. Following Bayes's rule, the average cost of operation is

$$\bar{C} = \int_{R_0} C(\rightarrow H_0|\underline{v})p(\underline{v})d^n v + \int_{R_1} C(\rightarrow H_1|\underline{v})p(\underline{v})d^n v$$

and, through some calculations, it is seen that it can be minimized again by the same decision rule introduced before, comparing now the likelihood ratio against a modified threshold.

$$\Lambda'_0 = \frac{P_r(H_0)[C(\to H_1/H_0) - C(\to H_0/H_0)]}{P_r(H_1)[C(\to H_0/H_1) - C(\to H_1/H_1)]}$$

This refinement of the Bayes's rule is referred to as the **Bayes's strategy**.

The core of binary hypotheses testing is, as for the general M-ary case, to find the hypersurface D that splits the data space R^n in two disjoint regions, R_0 and R_1, related with the two hypotheses. The probabilities Q_0 of a false alarm and Q_d of correct detection will then read

$$Q_0 \stackrel{\Delta}{=} P_r(\to H_1/H_0) = \int_{R_1} p_0(\underline{v}) d^n v \qquad Q_d \stackrel{\Delta}{=} P_r(\to H_1/H_1) = \int_{R_1} p_1(\underline{v}) d^n v$$

and are referred to as the **size** and the **power** of the test, respectively. Through these, it is easy to calculate the probability of error $P_e = Pr(H_0)Q_0 + Pr(H_1)(1 - Q_1)$ or the probability of a false dismissal, which is (1-Q_d).

2.3 The Neyman-Pearson criterion

In many practical applications, it is difficult, if not impossible, to establish an a-priori value for the probabilities of H_0 and H_1: in the detection of gravitational waves it could even be meaningless to state the probability with which a GW signal is present in the observed data. But, in those situations where H_1 happens "rarely" it can be seen, from its last expression, that the probability of error is dominated by false alarms.

Therefore, an alternative reasonable decision criterion, first established by Neyman and Pearson in 1933, would be "for a given affordable Q_0, seek a strategy maximizing Q_D", or, otherwise stated, **given the size, maximize the power of the test**. We shall now see that the optimal strategy under this criterion again calls for the likelihood ratio, requiring the comparison with a certain threshold λ:

$$\Lambda(\underline{v}) \stackrel{\Delta}{=} \frac{p_1(\underline{v})}{p_0(\underline{v})} \mathop{\gtrless}_{H_0}^{H_1} \lambda \qquad (1)$$

In order to demonstrate it, note that, according to the last expression, the hypersurface D contains data points for which $\Lambda(\underline{v}) = \lambda$, while on the R_0 and R_1 sides of D the likelihood ratio is smaller or larger than λ, respectively.

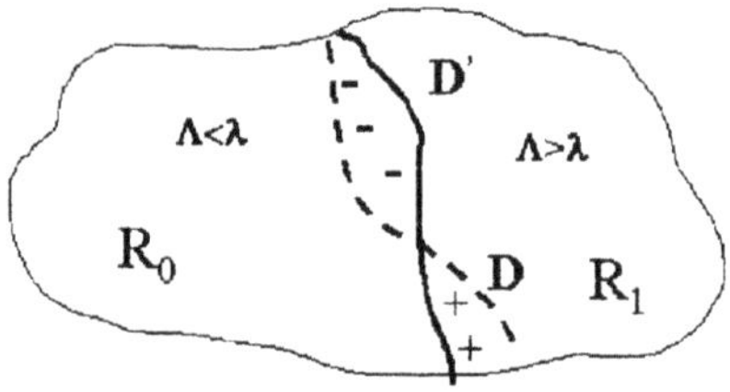

Figure 7. An intuitive demonstration for the Neyman-Pearson decision strategy.

Assume that the strategy is changed by modifying D into another hypersurface D' while keeping Q_0 constant, as sketched in fig.7. Since Q_0 is the integral over R_1 of the conditional pdf $p_0(\underline{v})$, this modification can be accomplished in an infinity of ways, but, whichever may be the new D' (solid line in fig.7), the power

$$Q_d = \int_{R_1} p_1(\underline{v}) d^n v = \int_{R_1} \Lambda(\underline{v}) p_0(\underline{v}) d^n v$$

is easily seen to have decreased. Intuitively, we can see that by modifying D we gave away from R_1 those points, labelled with "-" in the figure, while gaining the points labelled "+" which are of "less value" if measured by $\Lambda(\underline{v})$, as in the above integral.

Since the optimal test is that expressed in (1), how does one set the threshold λ? The first thing to note is that $\Lambda(\underline{v})$ is itself a random variable, depending on the random data, with some conditional pdfs $P_j(\Lambda)$ that can be calculated, under each H_j. The size and the power of the test can be then rewritten as

$$Q_0 = P_r(\Lambda(\underline{v}) > \lambda / H_0) = \int_\lambda^\infty P_0(\Lambda) d\Lambda; Q_d = P_r(\Lambda(\underline{v}) > \lambda / H_1) = \int_\lambda^\infty P_1(\Lambda) d\Lambda$$

and λ be determined from the first integral once the size is fixed, according to the Neyman-Pearson criterion. The same criterion ensures that the second integral yields the maximum possible value. Note that, under this new criterion the threshold is independent of the prior probabilities $\Pr(H_j)$, which can be unknown.

2.4 Sufficient statistics

As seen in the previous sections, both under Bayes's and Neyman-Pearson's criteria, the likelihood ratio $\Lambda(\underline{v})$ embodies all the relevant information, extracted from the data, for performing an optimal decision. Any function of the data points with the latter property is called a **sufficient statistics** for the detection test: besides Λ, other sufficient statistics exist, such as, for instance, any monotone increasing function $G(\Lambda)$ of the likelihood ratio. This should of course be compared with a different threshold G_0, to be determined: in order to set a proper G_0, the conditional pdfs of the new random variable G must be calculated. Once $P_0^G G$ and $P_1^G G$ and are known, one of the two following equations can be used

$$\frac{P_1^G(G_0)}{P_0^G(G_0)} = \Lambda_0 \qquad\qquad Q_0 = \int_{G_0}^{\infty} P_0^G(G)dG$$

Bayes *Neyman – Pearson*

to determine G_0, depending on which criterion is adopted. The most common among these other sufficient statistics is the **log-likelihood ratio**

$$G(\Lambda) - ln\left[\frac{p_1(\underline{v})}{p_0(\underline{v})}\right]$$

of which the usefulness will be evident from the following sections.

2.5 Detection in Gaussian noise

Let us now switch back to the problem of obtaining the data samples $\underline{v}$ from the observed signal $v(t)$. The most celebrated problem in detection theory is to recognize if, observing a noisy signal $v(t)$ in the time interval $0 < t < T$, there is a useful signal $s(t)$ "immersed" in gaussian noise $n(t)$ or else $v(t)$ consists of purely noise. Therefore, this is a binary hypothesis test, for which the alternative hypothesis H_1 is expressed by the equation $v(t) = s(t) + n(t)$, while the null hypothesis H_0 can be written as $v(t) = n(t)$; it is assumed here that the signal $s(t)$ is completely determined while noise is described by its ensemble averages. It is important to recognize that, under both hypotheses, $v(t)$ is a gaussian process: then it can be described by its ensemble averages, computed on the samples of $v(t)$. Instead of using time samples, like $v(t_i)$, assume to use the following original sampling method.

A set of function $\{f_k(t)\}$ is said to be orthonormal on $t \in [0,T]$ if they obey the following equalities

$$\int_0^T f_k(t) f_j(t) dt = 0 \qquad \int_0^T f_k^2(t) dt = 1$$

Some examples of such sets are the Fourier harmonics and the Legendre polynomials. Using an orthonormal set, any function $v(t)$ can be represented on [0,T] by a Fourier-like series using its coefficients v_k, according to the following equations

$$v(t) = \sum_{k=1}^{\infty} v_k f_k(t) \qquad v_k = \int_0^T v(t) f_k(t) dt \qquad \int_0^T v^2(t) dt = \sum_{k=1}^{\infty} v_k^2$$

and the last equation is a Parseval-like theorem expressing the energy of $v(t)$ on the interval [0,T]. The coefficients v_k are easily seen to be Gaussian RVs, since $v(t)$ is Gaussian: their mean value and correlations, under the null hypothesis H_0, can be computed as follows, using the linearity of expectation E[.] and the definition of autocovariance of the noise process $\phi(t,s) = E[n(t)n(s)]$.

$$E[v_k / H_0] = \int_0^T E[n(t)] f_k(t) dt = 0$$

$$\varphi_{jk} = E[v_j v_k / H_0] = \int_0^T \int_0^T f_j(t) E[n(t)n(s)] f_k(s) dt\, ds = \int_0^T \int_0^T f_j(t) \varphi(t,s) f_k(s) dt\, ds$$

One of the reasons for using the samples v_k and not time samples is that if $n(t)$ is white noise the variance of time samples would be infinite.

When setting up the likelihood ratio for the data samples v_k, it is extremely useful to have uncorrelated samples: this task can be accomplished by choosing a particular set of orthonormal functions. Suppose we select those functions which are the eigenfunctions of the following **homogeneous Fredholm integral equation**

$$\lambda_k f_k(t) = \int_0^T \phi(t,s) f_k(s) ds$$

Here the kernel $\phi(t,s)$ acts as a linear operator that "stretches and rotates" functions in an Hilbert space of infinite dimension: the eigenfunctions being those that are stretched but not rotated. It is possible to show that if the kernel is real and symmetrical (or complex but hermitian) the equation has real eigenvalues and orthogonal eigenfunctions, as demonstrated by the following equations

$$\lambda_k \int_0^T f_j(t) f_k(t) dt = \int_0^T \int_0^T f_j(t) \varphi(t,s) f_k(s) ds\, dt$$

$$\lambda_j \int_0^T f_k(t) f_j(t) dt = \int_0^T \int_0^T f_k(t) \varphi(t,s) f_j(s) ds\, dt$$

$$\textit{subtracting} \Rightarrow (\lambda_k - \lambda_j) \int_0^T f_k(t) f_j(t) dt = 0$$

Once demonstrated orthogonality, orthonormality can be accomodated by a multiplying factor for each function. These particular functions constitute the **Karhunen-Loève** basis: expanding $v(t)$ with this basis implies uncorrelated samples v_k, each with variance λ_k

$$\phi_{jk} = \lambda_k \int_0^T f_j(t) f_k(t) dt = \begin{cases} \lambda_k & j = k \\ 0 & j \neq k \end{cases}$$

and, since the samples are uncorrelated and Gaussian, they are also independent.

It is now simple to calculate the likelihood ratio. Suppose, for the sake of simplicity, that we base our decision on the first n samples only: since v_k are independent Gaussian variables with variance λ_k, under both hypotheses, zero mean under H_0 and mean value

$$E[v_k | H_1] = \int_0^T s(t) f_k(t) dt = s_k$$

under H_1, where s_k is the k-th coefficient of $s(t)$ expanded with Karhunen-Loève basis, the joint conditional pdfs of the samples will be

$$p_0(v_1, v_2 \ldots v_n) = \prod_{k=1}^{n} \frac{1}{\sqrt{2\pi\lambda_k}} e^{-\frac{(v_k)^2}{2\lambda_k}} \qquad p_1(v_1, v_2 \ldots v_n) = \prod_{k=1}^{n} \frac{1}{\sqrt{2\pi\lambda_k}} e^{-\frac{(v_k - s_k)^2}{2\lambda_k}}$$

from which it is easy to form the likelihood ratio $\Lambda_n(\underline{v})$. Letting $n \to \infty$, $\Lambda_n(\underline{v})$ becomes a sufficient statistics, the linear and log-likelihood ratios then become

$$\Lambda(\underline{v}) = e^{\sum_{k=1}^{\infty}\left[\frac{s_k v_k}{\lambda_k} - \frac{s_k^2}{2\lambda_k}\right]} \qquad G = \sum_{k=1}^{\infty} \frac{s_k v_k}{\lambda_k}$$

The log-likelihood ratio G is a Gaussian random variable, that can be practically computed avoiding the calculation of the coefficients v_k, s_k, λ_k. Resorting to the definition of v_k

$$G = \sum_{k=1}^{\infty} \frac{s_k}{\lambda_k} \int_0^T f_k(t)v(t)dt = \int_0^T \left[\sum_{k=1}^{\infty} \frac{s_k}{\lambda_k} f_k(t)\right] v(t)dt \qquad (2)$$

and the term in square brackets is the expansion, on the Karhunen-Loève basis, of a function $q(t)$ that is seen to solve the **inhomogeneous Fredholm equation**

$$\int_0^T \phi(t,u)q(u)du = \sum_{k=1}^{\infty} \frac{s_k}{\lambda_k} \lambda_k f_k(t) = s(t)$$

in which the useful signal $s(t)$ appears on the right hand side. Therefore, G can be computed by integrating the product $q(t)v(t)$ on [0,T]. Its statistics, conditional mean and variance, are

$$Var[G] = \int_0^T \int_0^T q(t)E[n(t)n(u)]q(u)dudt = \int_0^T q(t)s(t)dt \stackrel{\Delta}{=} d^2$$

$$E[G/H_0] = \int_0^T q(t)E[n(t)]dt = 0 \qquad E[G/H_1] = \int_0^T q(t)E[s(t)+n(t)]dt = d^2$$

where the symbol d^2 has been introduced: we will further exploit the meaning of this parameter. Through the Gaussian conditional pdfs of G, its likelihood ratio, which is referred to as the **likelihood functional**, can be computed

$$\Lambda(G) = \frac{P_1^G(G)}{P_o^G(G)} = \exp\left[\int_0^T q(t)v(t)dt - \frac{1}{2}\int_0^T s(t)q(t)dt\right]$$

where the first integral coincides with G and the second with the definition of d^2. The sufficient statistics G is suitable for the detection problem under both decision criteria: with Bayes detection it must be compared with the threshold G_0

$$G \underset{H_0}{\overset{H_1}{\gtrless}} G_0 = \frac{1}{2}d^2 + \ln \Lambda_0$$

while with Neyman-Pearson detection G_0 must be calculated from

$$Q_0 = \frac{1}{\sqrt{2\pi d^2}} \int_{G_0}^{\infty} e^{-\frac{G^2}{2d^2}} dG = erfc\left(\frac{G_0}{d}\right)$$

$$Q_d = \frac{1}{\sqrt{2\pi d^2}} \int_{G_0}^{\infty} e^{-\frac{(G-d^2)^2}{2d^2}} dG = erfc\left(\frac{G_0}{d} - d\right)$$

Using the latter expressions, it is possible, for a given Q_0, to evaluate G_0 for any value of d and then compute Q_d, deriving the plot of Fig.8, which is reprinted from [1]. This plot evaluates the performance of the receiver, under Neyman-Pearson criterion, giving the attainable power Q_d once the size Q_0 is chosen: the parameter d depends, by definition, only on the signal $s(t)$ to be detected and the noise autocovariance $\phi(t, u)$.

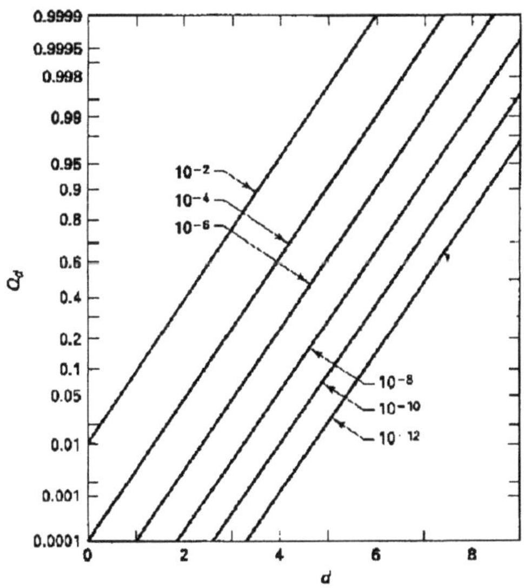

Figure 8. Probability of detection: curves indexed by the probability of false alarm Q_0.

Detection in stationary white noise

In the simple case of stationary white noise, assume the covariance spectrum is constant and equal to $\Phi(f) = N/2$, the autocovariance function is then $\phi(t,u) = (N/2)\delta(t-u)$ and the solution to the inhomogeneous Fredholm equation is simply

$$\int_0^T \frac{N}{2}\delta(t-u)q(u)du = s(t) \Rightarrow q(t) = \frac{2}{N}s(t)$$

provided that the signal $s(t)$ vanishes outside the observation interval [0,T]. In this case, the statistics G can be found through the integral

$$G = \int_0^T s(t)v(t)dt$$

referred to as the **correlation** between the useful signal and the observed process $v(t)$; the factor 2/N can be absorbed into the decision threshold G_0. The likelihood functional is simplified as well, and the term d^2 represents the signal energy to power spectral density ratio, briefly referred to as the **signal-to-noise ratio:**

$$d^2 = \frac{2}{N}\int_0^T s^2(t)dt = \frac{2E_s}{N}$$

Note that since d grows with T, performance also grows with the observation time, which is an intuitive result.

The optimality of the log-likelihood ratio G was first formalized by Granader in 1950, stating that the statistic G is optimum for the detection problem if $q(t)$ is square integrable on [0,T]; these results were further extended in 1967 by Kadota demonstrating that it suffices that d^2 is finite for G being optimum. If d is infinite, it is seen that Q_0=0 and Q_D=1: in this case a singular detection is said to occur.

The matched filter

In the general case of Gaussian noise, the correlation between the observed $v(t)$ and $q(t)$, defining G in accordance with (2), can be practically realized by constructing a filter with impulse response $k(t) = q(T-t)$, for $0 < t < T$, and zero elsewhere. This filter, first introduced in 1943, is called the **matched filter** since its impulse response is the "mirrored and shifted" version of $q(t)$. The output of the matched filter is found by convolution between $k(t)$ and the observed $v(t)$

$$y(t) = \int_0^t v(\tau)k(t-\tau)d\tau = \int_{t-T}^t v(\tau)q(T-t+\tau)d\tau$$

and can be sampled at time $t = T$ to yield the seeked value of G. Fig. 9 shows

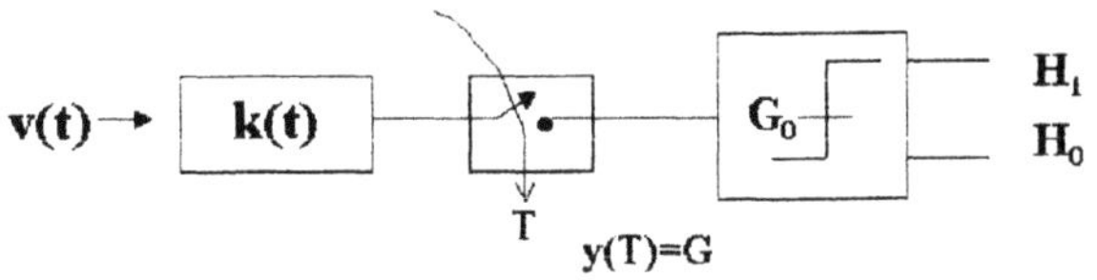

Figure 9. Optimal receiver scheme for binary detection in Gaussian noise.

the scheme of an optimal detector in this case: it is made up by a matched filter, followed by a sampler and a binary quantizer with the proper threshold G_0. It could be also demonstrated that the use of a matched filter minimizes the effect of noise on the random value of G that exits the sampler.

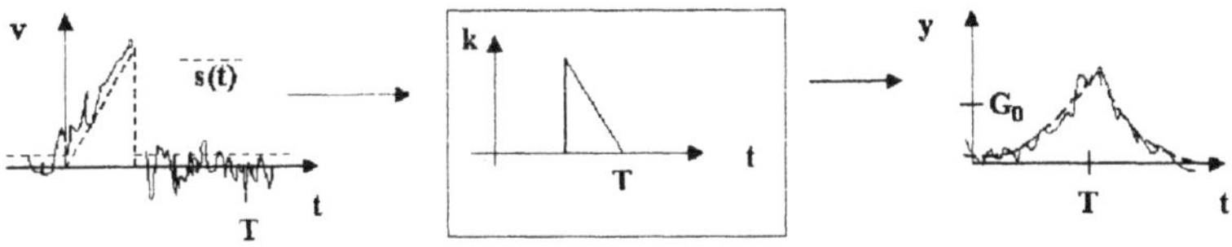

Figure 10. An example of the matched filter action, in the case of white noise.

In the simple case of white noise $q(t) = s(t)$, and thus the impulse response of the filter $k(t)$ will be matched to the useful signal $s(t)$. Fig. 10 depicts how the noisy signal $v(t)$, which is the sum of $s(t)$ (dashed line) and white noise, is filtered producing an output whose peak value would be on the sampling time T in the absence of noise (dashed line): random fluctuations, due to noise, may bring this value below the threshold thus causing a false dismissal. The use of a matched filter minimizes these fluctuations. The matched filter, previously introduced in communication theory, was first applied to a detection problem by Peterson, Birdsall and Fox in 1954. It must be noted that the signal to be detected has to be completely known in all its parameters, including the arrival time.

3 Elements of estimation theory

So far, we have been dealing with the problem of recognizing the presence of a finite number, one or more, of specified signals in the midst of noise; this problem is referred to as the **simple hypotheses testing.** Since these signals have been assumed completely known, and so is the statistical description of noise, the conditional pdfs $p_j(\underline{v})$ of the observed samples are also completely known. But signals to be detected are rarely univocally specified: more often it happens that these signals are specified only partially, since they depend on one or more unknown parameters (their amplitude, time of arrival, carrier frequency...). Thus the problem becomes to recognize if any among an entire class of signals has been detected (H_1) or only noise is present (H_0), the class $s(t;\underline{\theta})$ being specified by a vector $\underline{\theta} = (\theta_1, \theta_2, ..., \theta_m)$ of unknown parameters: in this case we talk of **composite hypotheses testing**. Even worse, it may be necessary, in the hypothesis H_1 that a useful signal is present, to evaluate those parameters, based on the observation of $v(t)$: this is the **parameter estimation** problem.

To introduce some terminology, the m unknown parameters are called the **estimanda**: these are, in general, random variables whose prior joint pdf $z(\theta_1, \theta_2, ..., \theta_m)$ may be known or not. Under hypothesis H_1, the observed signal is $v(t) = s(t;\underline{\theta}) + n(t)$ and the joint conditional pdf of data samples becomes dependent on the unknown parameters: it can be expressed as $p(\underline{v};\underline{\theta}) = p_0(\{v_k - s_k(\underline{\theta})\})$, where $p_0(\{n_k\})$ is the joint pdf of noise samples alone and $s_k(\underline{\theta})$ are samples extracted from the useful signal. This way, the estimanda become parameters of the distribution of the observations: our problem is then to seek those values of θ_i for which $p(\underline{v};\underline{\theta})$ best describes the observed data, in some sense. We must then seek a strategy that associates m numbers $\hat{\theta}_i$, called the estimates, to the estimanda, based on the data. This strategy will determine a vector of m functions of the data that performs this association: such a vector $\hat{\underline{\theta}} = (\hat{\theta}_1(\underline{v}), \hat{\theta}_2(\underline{v}), ..., \hat{\theta}_m(\underline{v}))$ is called the **estimator**.

We can think of the estimanda and the estimates to lie in a parameters space Θ: since the estimation is performed in the presence of noise, the estimation procedure repeated in two different experiments will produce two different values for the estimates, even if the estimanda are the same. Thus, the estimates will be always in error: estimates and estimanda are then, in general, represented by two different points in Θ. The aim of **estimation theory** is the design of receivers that minimize the errors due to noise and, possibly, predict how large the irreducible errors will be, on the average.

The key to the solution of the estimation problem is the knowledge of the joint conditional pdf of the data $p(\underline{v}|\underline{\theta})$, given the estimanda. In order to derive an estimation strategy, let us define some quantities that give indications about the goodness of a particular estimator: it is intuitive that a good estimator $\hat{\theta}_k$ is required to be close to the true value θ_k (the estimandum), on the average.

$$\bar{\hat{\theta}}_k \stackrel{\Delta}{=} E[\hat{\theta}_k(\underline{v}) \mid \underline{\theta}] \qquad\qquad E[\hat{\theta}_k(\underline{v}) - \theta_k) \mid \underline{\theta}] = \bar{\hat{\theta}}_k - \theta_k$$

The two above expressions define the estimator expected value and the mean deviation from the true value, also called the **bias**, respectively; if the bias is equal to zero, the estimator is said to be unbiased. Both expectations are conditioned on the true values of the estimanda; expected values are then evaluated with respect to the statistical variations of noise only. Another fundamental figure of merit for an estimator is the **Mean Square Error** (MSE)

$$\bar{E}_k \stackrel{\Delta}{=} E[(\hat{\theta}_k(\underline{v}) - \theta_k)^2 | \underline{\theta}] \qquad\qquad \Rightarrow \quad \bar{E}_k = Var[\hat{\theta}_k(\underline{v})] + (\bar{\hat{\theta}}_k - \theta_k)^2$$

that can also be computed through the variance of the estimator and the bias, as shown above. It is evident that a good estimator should have a low bias and a low variance but, in practice, this is often a compromise.

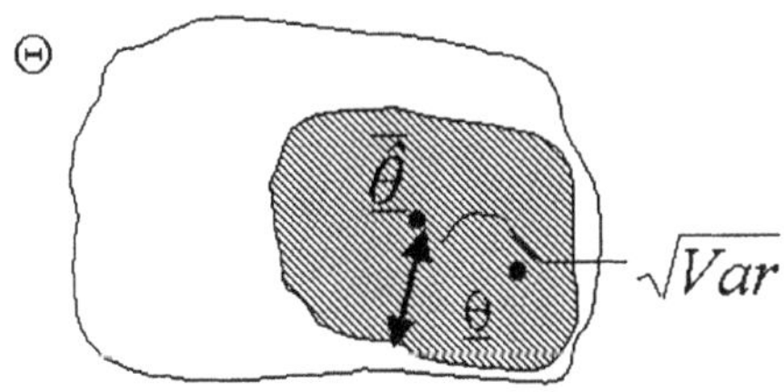

Figure 11. Estimator expected value and estimate, in the parameters space Θ.

Figure 11 shows the shaded region, whose *radius* is the estimator standard deviation, around the estimator mean value in which the actual estimator value will more often fall; the bias is the distance between the two points in the area. The calculation of bias and MSE requires the knowledge of the conditional pdf $q(\hat{\theta}_k | \underline{\theta})$ of every estimate: these are obtainable from the conditional pdf of data $p(\underline{v}|\underline{\theta})$, since the estimator elements $\hat{\theta}_k$ are deterministic functions of the data.

Figure 12. Partition of the parameters space: every region is associated with an hypothesis H_j.

3.1 *Maximum a posteriori (MAP) estimators*

Suppose to partition the parameters space in M disjoint regions, each called Δ_j and having a vector $\underline{\theta}_j$ in its "center", as depicted in Fig. 12. If we introduce M events H_j, each stating that "the vector of estimanda $\underline{\theta}$ lies in region Δ_j", we can choose one of these events, based on the observation of the data $\underline{v}$, and then associate the data to the i-th central point $\underline{\theta}_j$, so defining the estimator function: $\hat{\underline{\theta}}(\underline{v}) = \underline{\theta}_j$. This way, the estimation problem is reduced to a multiple hypotheses testing problem, whose solution can be found by applying the strategies devised in the previous chapter. In particular, following the Bayes's rule, one should choose the i-th hypothesis for which the posterior probability $\Pr(H_i—\underline{v})$ is largest. In the present case, these probabilities can be computed through the expression

$$P_r(H_i|\underline{v}) = \int_{\Delta_i} p(\underline{\theta}|\underline{v}) d^m\theta$$

which in turn requires the knowledge of the conditional pdf of parameters, given the data. This function is in general not given but it can be computed through the application of the Bayes's theorem in its continuous form

$$p(\underline{\theta}|\underline{v}) = \frac{z(\underline{\theta})p(\underline{v}|\underline{\theta})}{p(\underline{v})}$$

since we assumed to know the conditional probability of the data, given the estimates, appearing in the numerator.

It is evident that an estimator associating the data to the M points $\underline{\theta}_i$ is not the optimal choice but, passing to the limit for M$\rightarrow\infty$, the regions Δ_j become smaller and smaller thus making $\Pr(H_i—\underline{v}) \propto p(\underline{\theta}_i|\underline{v})$. The optimal estimator, according to the MAP strategy, implementing the Bayes's rule, is then the one selecting the estimates that maximize the conditional pdf of the estimates, given the data. Stated mathematically, the strategy is to choose that satisfies

$$p(\hat{\underline{\theta}}(\underline{v})|\underline{v}) \geq p(\underline{\theta}|\underline{v}) \qquad \forall \underline{\theta} \in \Theta$$

To gain a deeper insight on the MAP strategy, let us denote with $\underline{\theta}'$ one of the many possible estimates (not necessarily the selected one): the conditional pdf of the estimate $q(\underline{\theta}'|\underline{\theta})$, given the true estimanda, for $\underline{\theta}'$ to be a good estimate should be sharply peaked around the true value $\underline{\theta}'{=}\underline{\theta}$. While

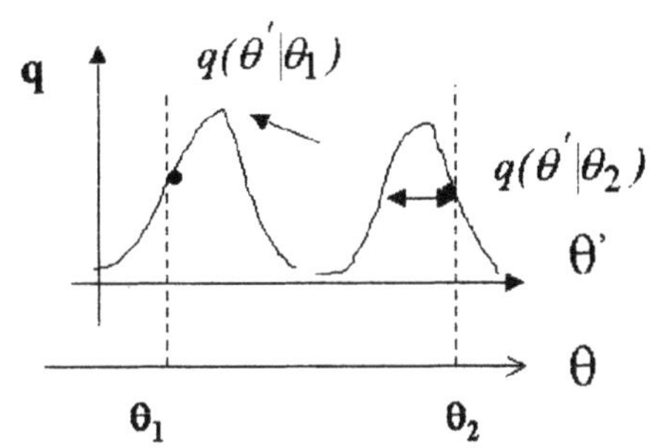

Figure 13. Distribution of the estimate θ', given two possible values for the estimandum.

in the estimation problem the estimanda are random by definition, also the selected estimates are random, because of noise, prior to the observation of data. In the pictorial representation of Fig. 13, where the estimandum is assumed a scalar θ for simplicity, even when the true estimandum is given, for example $\theta = \theta_2$, we do not know which $\underline{\theta}'$ will be the actual estimate before we observe the noisy data: $\underline{\theta}'$ a priori spans an entire interval, labelled $\leftrightarrow$ in the figure. The more $q(\underline{\theta}'|\underline{\theta})$ is peaked around $\theta = \theta_2$, the more accurate the estimate will be, on the average. Since we do not know what value the actual estimandum will have, it can be introduced a measure of the average peakedness for the pdf $q(\underline{\theta}'|\underline{\theta})$, according to the following expression

$$\bar{Q} = \int_{\Theta} z(\underline{\theta}) q(\underline{\theta}|\underline{\theta}) d^m \theta$$

that averages $q(.)$ with respect to the prior distribution of the estimanda. The meaning of $q(\underline{\theta}'|\underline{\theta})$ is the probability that the chosen estimate is $\underline{\theta}'$: it can be computed by summing the probability that data takes values on which the estimator $\hat{\underline{\theta}}(\underline{v})$ is equal to $\underline{\theta}'$, according to the following equation

$$q(\underline{\theta}'|\underline{\theta}) = \int_{R^n} \delta(\underline{\theta}' - \hat{\underline{\theta}}(\underline{v})) p(\underline{v}|\underline{\theta}) d^n v$$

Putting the latter expression into the previous integral gives a new expression for $\bar{Q}$, that can be further simplified using Bayes's theorem

$$\bar{Q} = \int_{\Theta} z(\underline{\theta}) \int_{R_n} \delta(\underline{\theta} - \underline{\hat{\theta}}(\underline{v}))p(\underline{v}|\underline{\theta})d^n v d^m \theta \quad \Rightarrow \quad \bar{Q} = \int_{R_n} p(\underline{v})p(\underline{\hat{\theta}}(\underline{v})|\underline{v})d^n v$$

Since the MAP strategy chooses the estimator maximing $p(\underline{\hat{\theta}}(\underline{v})|\underline{v})$, it also maximizes the last integral ($p(\underline{v}) \geq 0$) and thus $\bar{Q}$.

3.2 Maximum Likelihood (ML) estimates

Through Bayes's theorem it can also be said that MAP strategy seeks the maximization of the product $z(\underline{\theta})p(\underline{v}|\underline{\theta})$. Unfortunately, the first factor, $z(\underline{\theta})$, is often unknown or even with little meaning (for instance when the parameters are deterministic but unknown), thus making the applicability of MAP strategy impossible, since it requires the knowledge of the prior distribution of the parameters. In all those situations in which there is great uncertainty on the parameters, the prior distribution of these can be thought to be much broader than the conditional distribution of data samples $p(\underline{v}|\underline{\theta})$, which is only due to noise. Referring to Fig. 14, where scalars are employed for simplicity, suppose that the scalar datum v_0 is observed and that the conditional distribution of data $p(v|\theta)$ has a maximum on v_0 for $\theta = \theta_1$. Since the prior pdf of the parameter, $z(\theta)$, is increasing for $\theta > \theta_1$, we can expect that the optimal MAP estimate lies somewhere between θ_1 and θ_2, the latter point identifying the maximum $z(\theta_2)$ for the distribution of the parameter. If, as assumed, the function $z(\theta)$ is relatively flat, we can reasonably think that the MAP estimate, i.e. the one maximizing $z(\underline{\theta})p(\underline{v}|\underline{\theta})$, lies close to θ_1, which maximizes $p(\underline{v}|\underline{\theta})$ only.

The choice of an estimate for which the conditional distribution of data $p(\underline{v}|\underline{\theta})$ is maximum on the observed data $\underline{v}$ is called the **Maximum Likelihood** (ML) strategy: the estimator $\underline{\hat{\theta}}(\underline{v})$ is chosen as to satisfy the following property

$$p(\underline{v}|\underline{\hat{\theta}}(\underline{v})) \geq p(\underline{v}|\underline{\theta}) \qquad \forall \underline{\theta} \in \Theta$$

Thinking again of the product $z(\underline{\theta})p(\underline{v}|\underline{\theta})$, it is recognized that when the pdf of $z(\underline{\theta})$ parameters is very peaked, almost all the information on the estimanda is contained in it and, knowing $z(\underline{\theta})$, one can reliably guess an estimate without observing the data. If, on the other hand, the function $z(\underline{\theta})$ is really broad, it can be neglected in the comparison between different estimates, thus reducing the MAP strategy to the ML strategy. When $z(\underline{\theta})$ is not known, there is no choice: it must be neglected and the ML strategy is to be applied. A useful property of ML estimates is that, if in one problem it is easier to estimate not the parameter but a monotone function of it (e.g. $f(\theta) = \theta^3$) we can estimate $\hat{\theta^3}$ and then set $\hat{\theta} = \sqrt[3]{\hat{\theta^3}}$, using the inverse function of $f(.)$. This useful procedure cannot be applied with MAP estimates since a different weight is assigned, through $z(\theta)$, to corresponding ranges of θ and $f(\theta)$, in the transformation performed by $f(.)$.

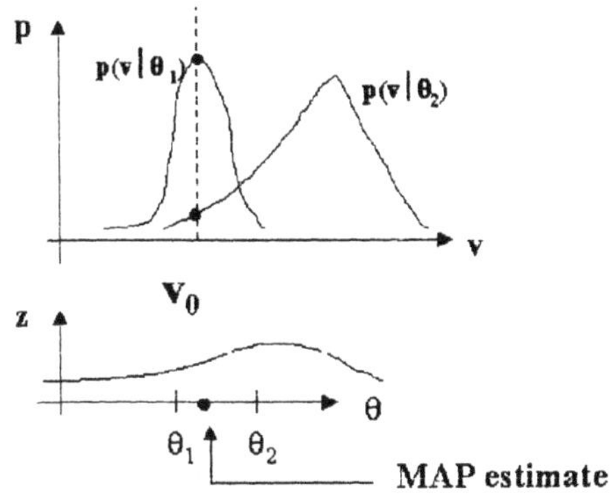

Figure 14. Correspondence between MAP estimate and ML estimate, for scalar parameter and data.

3.3 The likelihood functional and the Greater Likelihood Ratio Test

The strategies just described both require the comparison of the conditional pdfs of data $p(\underline{v}|\underline{\theta})$, for different values of $\underline{\theta}$, for deriving an estimator. To overcome the difficulty of having to deal with an infinity of samples, when the dimension n of the vector $\underline{v}$ grows, the latter pdf can be divided by the joint pdf $p_0(\underline{v})$ of pure noise samples, which does not alter the results of comparisons. Provided that $\underline{\theta}$ appears only in the expression of the useful signal $s(t;\underline{\theta})$, we can then resort to the function

$$\Lambda[v(t)|\underline{\theta}] = \lim_{n\to\infty} \frac{p(\underline{v}|\underline{\theta})}{p_0(\underline{v})}$$

which is referred to as the **likelihood functional**. Thus, the ML strategy requires finding the estimator $\hat{\underline{\theta}}[v(t)]$ for which $\Lambda[v(t)|\underline{\theta}]$ is the largest. In the case of white Gaussian noise, it is easy to see that:

$$\Lambda[v(t)|\underline{\theta}] = \exp\left[\frac{2}{N}\int_0^T v(t)s(t;\underline{\theta})dt - \frac{1}{2}\frac{2}{N}\int_0^T [s(t;\underline{\theta})]^2 dt\right] \qquad (3)$$

where the last integral coincides with the energy d^2 of the signal s with parameters $\underline{\theta}$.

How can we practically search for an estimate maximizing the likelihood functional: in the case of white Gaussian noise, a solution is to build a **bank of matched filters**, with impulse responses $k(t;\underline{\theta}_j)$, each filter being matched to a signal $s(t;\underline{\theta}_j)$ identified by a different set of parameters, as depicted in Fig. 15. As outlined in the figure, this approach corresponds to a *sampling of*

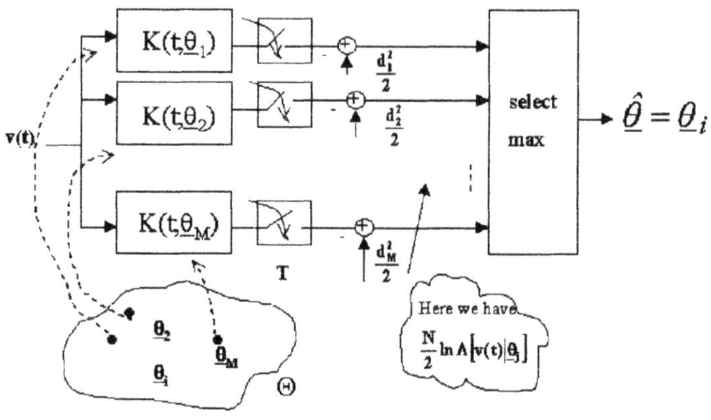

Figure 15. Receiver with a bank of matched filters.

the parameters space Θ: at the output of the j-th filter we have the result of the first integral appearing in (3), for $\underline{\theta} = \underline{\theta}_j$; through the subtraction of the corresponding energy d_j^2, a monotonic function of the likelihood functional can be maximized for choosing the estimate. Ideally, a continuum of filters would be needed, which is impossible, and thus the chosen estimate is always an approximation of the true estimanda. Some countermeasures to enhance the reliability of the estimate would be to provide parameter sets $\underline{\theta}_j$ that are densely spaced over the region of Θ where the true estimanda are expected to lie. Since it is not implicit that the observed signal $v(t)$ contains some useful signal in it (it can be only noise, in the hypothesis H_0) it is possible, after the estimation procedure, to switch back to the detection problem, deciding for a H_0 or H_1 based on the comparison of the selected likelihood functional with a threshold that can be chosen according to the guidelines of the previous chapter. This approach is called the **Greater Likelihood Ratio Test** (GLRT) and is formalized by the following decision rule

$$\max_{\underline{\theta}} \Lambda\left[v(t)|\underline{\theta}\right] = \Lambda\left[v(t)|\hat{\underline{\theta}}\right] \underset{H_0}{\overset{H_1}{\gtrless}} \Lambda_0$$

Estimation of arrival time

In general, the designer of the receiver should avoid building a bank of matched filters: however, this is possible only in some simple cases. To help understanding, suppose the problem is to determine if a certain signal $s(t)$ of finite duration T' is detected in the presence of white Gaussian noise. The signal can arrive at the receiver with a random time displacement θ, to be estimated, of which we don't know the prior distribution. The observed signal is thus $v(t) = s(t - \theta_0) + n(t)$, where θ_0 is the actual arrival time, as shown in Fig. 16. We assume that the observation extends in [0,T].

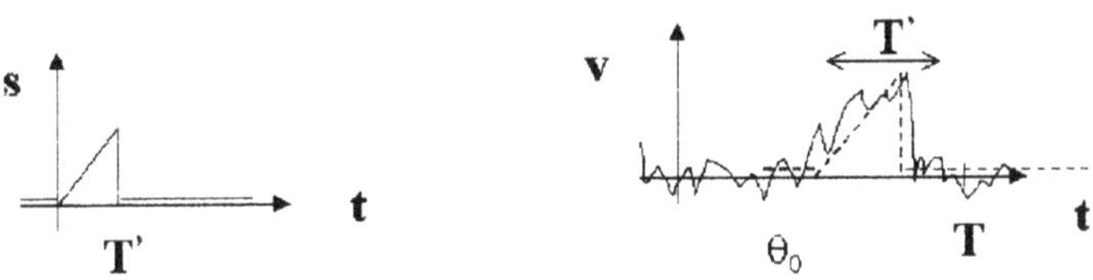

Figure 16. Signal with random arrival time corrupted by white Gaussian noise.

According to the ML estimation strategy, the receiver must maximize the likelihood functional or else its logarithm, the log-likelihood functional.

$$\max_{\theta} \Lambda\left[v(t)|\theta\right] \Rightarrow \max_{\theta}\left[\int_0^T v(t)s(t-\theta)dt\right]$$

This would require building a bank of filters, each with an impulse response $k(t;\underline{\theta}_j) = s(T - (t - \underline{\theta}_j))$. Practically, only one filter with $k(t) = s(-t)$ is needed: the output of the filter at time θ

$$G(\theta) = \int_0^T v(t)s(t-\theta)dt \approx \int_{-\infty}^{\infty} v(\tau)s(-(\theta-\tau))d\tau$$

coincides with the log-likelihood functional and thus the receiver will choose, as the estimate, the time t_m when the output reaches its maximum.

3.4 Performance of ML estimators

In this section we wish to derive an expression for the bias and the variance of a ML estimator, these parameters representing the figures for evaluating the performance of an estimator. For the sake of simplicity, we will develop a general approach for the case of a single parameter θ, with true value θ_0, using the natural logarithm of the likelihood functional, called **log-likelihood ratio** $g(v|\theta)$, where v briefly represents the observed signal. Since $p_0(v)$, appearing in the definition of $\Lambda[v|\theta]$, does not depend on θ, it will be omitted for the moment without loss of generality: the definition of the log-likelihood ratio is thus $g(v|\theta) = \ln p(v|\theta)$. In parallel, the approach will be applied to the practical case of estimation of the arrival time, introduced in the last subsection, using the complete definition for $g(v|\theta)$, that reads

$$g(v|\theta) = \ln\frac{p(v|\theta)}{p_0(v)} = \frac{2}{N}\int_0^T v(t)s(t-\theta)dt - \frac{1}{2}\frac{2}{N}\int_0^T [s(t-\theta)]^2 dt \qquad (4)$$

where the observed noisy signal is $v(t) = s(t-\theta_0) + n(t)$, in this case.

Since, in general the selected estimator $\hat{\theta}$ is the one maximizing g, the derivative of g with respect to θ must have a zero for $\theta = \hat{\theta}$ such derivative is denoted by g'_θ and is given by

$$g'_\theta(v|\theta) \stackrel{\Delta}{=} \frac{\partial g(v|\theta)}{\partial\theta} = \frac{p'_\theta(v|\theta)}{p(v|\theta)}$$

The value of g'_θ on the estimator value can be expressed through a Taylor series expansion in the neighbourhood of the true estimandum θ_0: assuming the estimator is sufficiently close to the estimandum, the terms above the first can be neglected

$$g'_\theta(v|\hat{\theta}) = g'_\theta(v|\theta_0) + (\hat{\theta}-\theta_0)g''_\theta(v|\theta_0) + \ldots = 0 \qquad (5)$$

In the expression (4), for the arrival time, the second integral is constant with θ, being the energy of $s(.)$, when the observation time T is sufficiently large; the first and second derivatives then read

$$g'_\theta(v|\theta) = -\frac{2}{N}\int_0^T v(t)s'(t-\theta)dt \qquad g''_\theta(v|\theta) = \frac{2}{N}\int_0^T v(t)s''(t-\theta)dt$$

Based on (5) we can then approximate the deviation of the estimator from the estimandum: in the general case and for the arrival time case its expressions is

$$\hat{\theta} - \theta_0 \approx -\frac{g'_\theta(v|\theta_0)}{g''_\theta(v|\theta_0)} \qquad \hat{\theta} - \theta_0 \approx -\frac{-\frac{2}{N}\int_0^T v(t)s'(t-\theta_0)dt}{\frac{2}{N}\int_0^T v(t)s''(t-\theta_0)dt} \tag{6}$$

respectively. The expected value of the numerator of the above formulas can be shown to be zero. For the general case, the expectation with respect to noise (the only stochastic entity appearing in the expression), denoted by $E_n[\cdot]$, is

$$E_n[g'_\theta(v|\theta_0)] = \int_{\mathbf{R}^n} g'_\theta(v|\theta_0)p(v|\theta_0)d^n v = \frac{\partial}{\partial_\theta}\left[\int_{\mathbf{R}^n} p(v|\theta)d^n v\right]_{\theta=\theta_0} = 0 \tag{7}$$

which follows from $p'_\theta(v|\theta) = p(v|\theta)g'_\theta(v|\theta)$ (see the definition of g'_θ) and the *normalization property* of a pdf. The same result is of course obtained for the specialized case of the arrival time, where the last equality in the following equation is again justified when T is long enough to contain the whole duration of $s(.)$.

$$E_n\left[\frac{-2}{N}\int_0^T [s(t-\theta_0)+n(t)]s'(t-\theta_0)dt\right] =$$
$$= \frac{-2}{N}\int_0^T s(t-\theta_0)s'(t-\theta_0)dt = \frac{-1}{N}[s^2(t-\theta_0)]_0^T = 0$$

An important formula can be derived for the expected value of the denominators in (6), by carrying on the calculations of the expectation in (7): through some algebra, we have, in the general case

$$\frac{\partial}{\partial_\theta}\left[\int_{\mathbf{R}^n} g'_\theta(v|\theta)p(v|\theta)d^n v\right]_{\theta=\theta_0} =$$

$$\left[\int_{\mathbf{R}^n} g''_\theta(v|\theta)p(v|\theta)d^n v + \int_{\mathbf{R}^n} g'_\theta(v|\theta)p'(v|\theta)d^n v\right]_{\theta=\theta_0} = \qquad (9)$$

$$\left[\int_{\mathbf{R}^n} g''_\theta(v|\theta)p(v|\theta)d^n v + \int_{\mathbf{R}^n} [g'_\theta(v|\theta)]^2 p(v|\theta)d^n v\right]_{\theta=\theta_0} = 0$$

$$\Rightarrow \quad E_n[g''_\theta(v|\theta_0)] = -E_n[[g'_\theta(v|\theta_0)]^2]$$

with the last equation relating the first and second derivatives of the log-likelihood ratio. Applying the last result to the arrival time problem, we get, for the expectation of the denominator in (6):

$$E_n\left[\frac{2}{N}\int_0^T [s(t-\theta_0)+n(t)]s''(t-\theta_0)dt\right] = \frac{2}{N}\int_0^T s(t-\theta_0)s''(t-\theta_0)dt =$$

$$= -E_n\left[\left(\frac{2}{N}\right)^2\int_0^T\int_0^T [s(t-\theta_0)+n(t)]s'(t-\theta_0)[s(u-\theta_0)+n(u)]s'(u-\theta_0)dudt\right] =$$

$$= \frac{-2}{N}\int_0^T [s'(t-\theta_0)]^2 dt \qquad (10)$$

where the autocovariance of white noise has been used and the result of the expected value in (8) has been used to simplify some terms. Rewriting the expression of the estimation error for the arrival time in a different form

$$\hat{\theta}-\theta_0 \approx -\frac{-\frac{2}{N}\int_0^T n(t)s'(t-\theta_0)dt}{-\frac{2}{N}\int_0^T [s'(t-\theta_0)]^2 dt + \frac{2}{N}\int_0^T n(t)s''(t-\theta_0)dt}$$

it could be demonstrated that, for high signal to-noise ratios (SNR), the second term of the denominator is negligible, compared to the first; the first term representing the denominator expected value (see (10)), and the second representing its statistical fluctuations. The latter is a general result: in the expression of the estimator error, for the general estimation problem, the numerator always contains noise terms, while the denominator consists of a nonzero expected value plus noise terms that can be neglected, for high SNR. With this result, that is not evident from (6), we can rewrite the general expression for the estimator error in (6), considering, in the denominator, only its expected value:

$$\hat{\theta}-\theta_0 \approx -\frac{g'_\theta(v|\theta_0)}{E_n[g''_\theta(v|\theta_0)]} \qquad (11)$$

now, computing the bias, only the numerator expectation must be evaluated: since the latter has been shown to be zero, the estimator is unbaiased, under our approximations. A more important result is obtained computing the estimator variance: using the expression (11), the result in (9) and some simplifications:

$$Var_n[\hat{\theta}(v)] = Var_n\left[-\frac{g'_\theta(v|\theta_0)}{E_n[g''_\theta(v|\theta_0)]}\right] = \frac{1}{E_n[[g'_\theta(v|\theta_0)]^2]}$$

The last expression, known as **Fisher's formula**, then relates the mean square error of an estimator to the variance of the first derivative of the log-likelihood ratio with respect to noise, this variance being computed on the true value of the estimandum.

Fisher's formula can be applied to the arrival time problem: having already computed $g'_\theta(v|\theta)$ and using (10) for its mean square value, we can write concisely

$$Var\left[\hat{\theta}(v)\right] = \frac{N}{2E\beta^2} \qquad d^2 \overset{\Delta}{=} \frac{2E}{N} \qquad \beta^2 \overset{\Delta}{=} \frac{\int_{-\infty}^{+\infty}[s'(t)]^2 dt}{\int_{-\infty}^{+\infty} s(t)^2 dt} = \frac{\int (2\pi f)^2 |S(f)|^2 df}{\int |S(f)|^2 df}$$

where T has been assumed sufficiently long, as to extend the integrals on the whole time axis, E is the energy of $s(t)$, d^2 is the signal-to-noise ratio, and the factor β^2 can be computed using well known properties of Fourier transforms. This latter factor is referred to as the **mean square bandwidth** of the signal $s(t)$: this name is justified by the last expression which gives an indication on the *dispersion* of the signal spectrum S(f) on the frequency axis.

As stated at the beginning of this section, we have used the conditional pdf of data $p(v|\theta)$ for our derivations; in general, the previously defined likelihood functional $\Lambda[v(t)|\theta]$ is appropriate, and all the results hold with $g[v(t)|\theta] = \ln\Lambda[v(t)|\theta]$ being the log-likelihood functional, computed on the observed signal.

The same approach followed here can be applied to the general case of multiple parameters estimation: we will omit the demonstrations but it can be shown that, for large signal-to-noise ratios, the ML estimates are unbiased and their covariance matrix can be computed, again, using the log-likelihood functional, according to

$$Cov\left[\hat{\theta}_j,\hat{\theta}_k\right]\stackrel{\Delta}{=}E\left[(\hat{\theta}_j-\theta_{j0})(\hat{\theta}_k-\theta_{k0})\right]=\left(\Gamma^{-1}\right)_{jk} \quad \textbf{(12)}$$

where the elements of the matrix Γ to be inverted are given by

$$\Gamma_{il}\stackrel{\Delta}{=}E_n\left[g'_{\theta_i}(v|\underline{\theta}_0)g'_{\theta_l}(v|\underline{\theta}_0)\right]$$

The matrix Γ is called the Fisher information matrix; we will further exploit its meaning for the case of white Gaussian noise, in the following section.

3.5 ML estimation in Gaussian noise: the ambiguity function

Recall that the likelihood functional, in the case of Gaussian noise, has the following expression

$$\Lambda[v(t)|\underline{\theta}] = \exp\left[\int_0^T v(t)q(t;\underline{\theta})dt - \frac{1}{2}\int_0^T s(t;\underline{\theta})q(t;\underline{\theta})dt\right]$$

where $q(t;\underline{\theta})$ is the function solving the Fredholm inhomogeneous equation

$$\int_0^T \phi(t;u)q(u;\underline{\theta})du = s(t;\underline{\theta})$$

Deriving the natural logarithm of Λ with respect to the i-th parameter, gives

$$g'_{\theta_i}[v(t)|\underline{\theta}] = \int_0^T n(t)\frac{\partial q(t;\underline{\theta})}{\partial\theta_i}dt + \text{constant terms}$$

where the constant terms are those not subject to stochastic variations. Thus, the covariance of the latter derivatives only depends on the noise process $n(t)$, and the elements of the Fisher information matrix can be computed by applying the definition of noise autocovariance function and the definition of q(.):

In the simple case of white noise, it has been seen that the function q(.) is proportional to the signal to be detected: for any value of the parameters set, $q(t;\underline{\theta}) = (2/N)s(t;\underline{\theta})$ The Fisher information matrix elements can then be written as

$$\Gamma_{il} = \frac{\partial^2 H(\underline{\theta}_1,\underline{\theta}_2)}{\partial\theta_{1i}\partial\theta_{2l}}\Big|_{\underline{\theta}_1=\underline{\theta}_2=\underline{\theta}_0} \quad (13)$$

$$\Gamma_{il} \stackrel{\Delta}{=} Cov\left[g'_{\theta_i}(v|\underline{\theta}_0)g'_{\theta_l}(v|\underline{\theta}_0)\right] = E_n\left[\int_0^T\int_0^T n(t)n(u)\frac{\partial q(u;\underline{\theta}_0)}{\partial\theta_i}\frac{\partial q(t;\underline{\theta}_0)}{\partial\theta_l}dudt\right] =$$

$$= \int_0^T \frac{\partial s(t,\underline{\theta}_0)}{\partial\theta_i}\frac{\partial q(t;\underline{\theta}_0)}{\partial\theta_l}dt$$

where the following function of the parameters, called the **ambiguity function**,

$$H(\underline{\theta}_1,\underline{\theta}_2) = \frac{2}{N}\int_0^T s(t-\underline{\theta}_1)s(t-\underline{\theta}_2)dt$$

has been introduced. The ambiguity function takes, besides the multiplying factor (2/N), the form of a correlation integral, over the interval [0,T], between the useful signal s(t), considered for two different values of the parameters set. Resorting to the interpretation of this integral, introduced in signal analysis, as a measure of the similitude between different signals, the ambiguity function gives us an indication on how much the signals $s(t;\theta_1)$ and $s(t;\theta_2)$ are similar on [0,T].

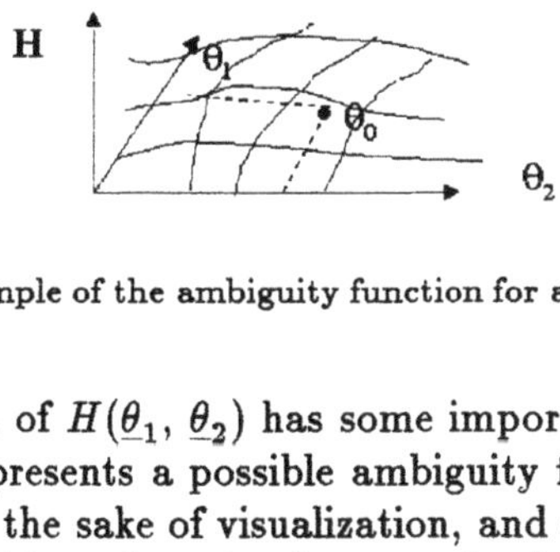

Figure 17. An example of the ambiguity function for a scalar parameter θ.

This interpretation of $H(\underline{\theta}_1, \underline{\theta}_2)$ has some important implications: referring to fig. 17, that represents a possible ambiguity function in the case of a scalar parameter θ, for the sake of visualization, and considering the function in the neighbourhood of $\underline{\theta}_1 = \underline{\theta}_2 = \underline{\theta}_0$, the more the function is flat around this point the less reliable the ML estimator will be. To demonstrate this result mathematically, consider that a flat function implies small partial derivatives, thus, according to (13), the Fisher information matrix elements are small; the inverse of this matrix has then large diagonal elements and, since, referring to (12), these represent the variance of the estimator elements, large variances result in a less reliable estimator, that is more subject to the fluctuations of its value induced by noise.

This important result could also be justified intuitively, again referring to the meaning of the correlation integral defining $H(\underline{\theta}_1, \underline{\theta}_2)$; if this function is relatively flat around the estimanda value $\underline{\theta}_0$, then significant deviations of the parameters from the true estimanda value do not change the useful signal shape significantly, since the value of the correlation integral remains nearly the same. The receiver has thus difficulty in estimating the true $\underline{\theta}_0$ because the useful received signal $s(t;\underline{\theta}_0)$ and another signal $s(t;\underline{\theta}_1)$, with $\underline{\theta}_1 \neq \underline{\theta}_0$ but close to it, are relatively similar, thus mistakeable, having a large correlation. If, on the other hand, the ambiguity function is sharply peaked around $\underline{\theta}_1 = \underline{\theta}_2 = \underline{\theta}_0$, the ML estimates are reliable, but, referring to fig. 15, a bank of many matched filters is needed. In other words, since the signal $s(t;\underline{\theta})$ is very sensitive to small variations of the parameters, a dense sampling of the parameters space (implemented by the matched filter bank) is required in order to have at least one filter whose impulse response is nearly matched to the true useful signal $s(t;\underline{\theta}_0)$ that is to be detected.

References

1. C. W. Helstrom, Elements of Signal Detection & Estimation (chaps. 1,2,6), Prentice-Hall, 1995.
2. C. W. Helstrom, Statistical Theory of Signal Detection, Pergamon Press, 1968 (2nd ed.).
3. H. L. Van Trees, Detection, Estimation and Modulation Theory, vol.1 – Malabar (Florida): Krieger; New York: Wiley, 1968 - 1971.

TIME-FREQUENCY ANALYSIS : AN INTRODUCTION

PATRICK FLANDRIN
Ecole Normale Supérieure de Lyon, Laboratoire de Physique (UMR 5672 CNRS), 46 allée d'Italie, 69364 Lyon Cedex 07, France
E-mail : flandrin@ens-lyon.fr

Time-frequency analysis is aimed at making spectrum analysis time-dependent, so as to permit a physically meaningful description of nonstationary signals. Basics of time-frequency tools are introduced, with emphasis on energetic ("Wigner-type") methods. The question of localizing a "chirp" signal on the time-frequency plane is addressed in some detail, with a potential application to the time-frequency detection of gravitational waves radiated by coalescing binaries.

1 Starting with Fourier

A signal $x(t)$ is generally given as a function of time, thus putting special emphasis on temporal events. Conversely, the Fourier representation $X(f)$ of the same signal, defined as[a]

$$X(f) := \int_{-\infty}^{+\infty} x(t)\, e^{-i2\pi ft}\, dt, \tag{1}$$

puts emphasis on oscillation properties, but without any localization in time, as well as without any necessary link with the physical structure of the signal.

Therefore, whereas the Fourier representation gives a mathematical description of particular interest in many cases (sustained oscillations, steady-state vibrations...), it is naturally faced with interpretation limitations as soon as the analyzed signals contain transient phenomena, abrupt changes, evolutions in their spectral content... in short, characteristic features which are qualitatively referred to as "non-stationarities". Rather than being satisfied with two descriptions of a same reality which are only partly adapted and mutually exclusive, it is tempting to look for a more general description which would encompass both aspects simultaneously, and concile mathematical description and physical interpretation : this is the purpose of time-frequency analysis.

The main difficulty faced by any attempt in this direction is that the two variables in play (time and frequency) are "canonically conjugated", i.e.,

[a] We will adopt the convention of using lower case symbols for time signals, and the corresponding upper case symbols for their Fourier transform.

associated with non-commuting operators[b]. A corollary is that localization of a signal with respect to one of the two variables is exclusive of localization with respect to the other. For any $x(t) \in L^2(\mathbb{R})$, this is evidenced, e.g., by the well-known inequality (also referred to as "uncertainty relation"[14]) :

$$\inf_{t_0,\omega_0} \|(t - t_0)\, x\|_2 \, \|(f - f_0)\, X\|_2 \geq \frac{1}{4\pi} \|x\|_2^2, \tag{2}$$

with $\|x\|_2^2 = \langle x, x \rangle$ and equality if and only if $x(t)$ is a Gaussian.

2 Mixed representations

2.1 From frequency to instantaneous frequency

Whereas Fourier analysis considers time and frequency in an exclusive fashion, a first improvement consists in considering one of the two variables as functionally dependent on the other. Such a point of view is clearly supported by intuition and the everyday experience of music, whistles... in brief, of frequency modulations. Formalizing the idea of an "instantaneous frequency" amounts therefore to generalizing the concept of a monochromatic wave $a \cos 2\pi f_0 t$ by allowing the amplitude a and the frequency f_0 to become time-dependent, so that it takes on the form $a(t) \cos \varphi(t)$, with $a(t) > 0$ and $\varphi(t)$ non linear in t. Given an observed signal, such a representation is of course not unique[23]. Following Gabor[15] and Ville[27], it is generally accepted to get rid of arbitrariness by considering a real-valued signal $x(t) \in \mathbb{R}$ as the real part of the complex-valued signal $z_x(t) := x(t) + i(\mathbf{H}x)(t)$, where $\mathbf{H}$ stands for the Hilbert transform. The rationale for introducing such an *analytic signal* $z_x(t)$ is that, when applied to a monochromatic wave, it simply reduces to a complex exponential, thus representing a "stationary" signal by a rotating vector whose modulus is constant and whose rotation is uniform. In the general case of an arbitrary signal, this leads naturally to define an *instantaneous amplitude* and an *instantaneous frequency* according to :

$$a_x(t) := |z_x(t)| \quad ; \quad f_x(t) := \frac{1}{2\pi} \frac{d}{dt} \arg z_x(t), \tag{3}$$

respectively. Other ways of defining an instantaneous frequency can be imagined[26] but, whatever the chosen definition, the principle of using a *one-dimensional curve* of the time-frequency plane has a natural limitation as soon as the analyzed signal is *multicomponent*, i.e., such that different frequency

[b]This situation is similar to that of Quantum Mechanics, with time and frequency in place of position and momentum[13].

contributions are allowed to be simultaneously present. In such a case, the representation can at best give some "average" description[c], and in no way the multivalued functions which would be necessary for a physically meaningful interpretation. Overcoming this limitation motivates the introduction of truely *mixed* (i.e., two-dimensional) representations.

2.2 *Atomic decompositions*

Generalizing upon the Fourier transform, a first approach is to *linearly* decompose a signal over a set of elementary *building blocks* with "good" localization properties in both time and frequency. More precisely, the value of a signal $x(t)$ at some time t_0 can be equivalently expressed as

$$x(t_0) = \int_{-\infty}^{+\infty} x(t)\, \delta_t(t_0)\, dt = \int_{-\infty}^{+\infty} X(f)\, e_f(t_0)\, df, \tag{4}$$

where $\delta_t(.)$ stands for the Dirac distribution at t and $e_f(t) := \exp(i2\pi ft)$. The idea is therefore to introduce some intermediate representation of the type :

$$x(t_0) = \iint_{-\infty}^{+\infty} \lambda_x(t, f)\, g_{tf}(t_0)\, dt\, df, \tag{5}$$

where the functions $g_{tf}(.)$ — which play in fact a role of *atoms*, in the sense that they are supposed to be well-localized elementary building blocks — allow for a transition between the two previous situations : perfect localization in time (resp., frequency) and no localization at all in frequency (resp., time) when $g_{tf}(.) \to \delta_t(.)$ (resp., $e_f(.)$).

For such a decomposition to make sense, it is naturally required to be *invertible* according to

$$\lambda_x(t, f) = \int_{-\infty}^{+\infty} x(t_0)\, g_{tf}^*(t_0)\, dt_0, \tag{6}$$

thus making of $\lambda_x(t, f)$ a (linear) time-frequency *representation* of $x(t)$[d].

There is of course much freedom in the selection of candidate atoms. A simple approach consists in generating the whole family $g_{tf}(.)$ *via* the action of a suitable transformation group, acting on a unique elementary element

[c]It is worth noting that this idea of "average" frequency cannot be followed up *stricto sensu* : for instance, the instantaneous frequency of a signal composed of two tones f_1 and f_2 generally contains contributions outside the interval $[f_1, f_2]$, whereas the signal itself is strictly limited to this frequency band[19].

[d]Generally speaking, the "synthesis" atom used in (5) does not necessarily coincide with the "analysis" atom involved in (6).

$g(.)$. Proceeding this way, the choice $g_{tf}(s) = g(s-t)\exp(i2\pi fs)$ leads to the family of *short-time Fourier transforms* (STFT's) with window $g(.)$:

$$F_x^{(g)}(t,f) := \int_{-\infty}^{+\infty} x(s)\, g^*(s-t)\, e^{-i2\pi fs}\, ds, \tag{7}$$

whereas imposing $g_{tf}(s) = (f/f_0)^{1/2}\, g((f/f_0)(s-t))$ (with $f > 0$, $f_0 > 0$ and $g(.)$ zero-mean) gives birth to the family of *wavelet transforms* (WT's)[20] :

$$T_x^{(g)}(t,f) := \sqrt{\frac{f}{f_0}} \int_{-\infty}^{+\infty} x(s)\, g^*\left(\frac{f}{f_0}(s-t)\right) ds. \tag{8}$$

2.3 Energy distributions

In many applications, relevant physical quantities are of *energy* type, suggesting to look for decompositions aimed no more at the signal itself, but rather at its energy. By definition, and from the isometry of the Fourier transformation, the energy $E_x := \|x\|_2^2$ of a signal $x(t) \in L^2(\mathbb{R})$ admits as equivalent expressions :

$$E_x = \int_{-\infty}^{+\infty} |x(t)|^2\, dt = \int_{-\infty}^{+\infty} |X(f)|^2\, df. \tag{9}$$

Reasoning as for linear decompositions, this suggests to consider joint *energy* distributions $\rho_x(t,f)$ such that

$$E_x = \int\int_{-\infty}^{+\infty} \rho_x(t,f)\, dt\, df. \tag{10}$$

The issue is therefore to define such a quantity $\rho_x(t,f)$, which leads to many possible approaches. A first possibility is to adopt an "operational" point of view, considering an energy distribution as potentially *measurable* by some device, *albeit* idealized. Here are some examples.

Spectrogram, sonagram, scalogram. The easiest solution amounts to using the "continuous basis" relationships (5)-(6), leading to the general definition

$$\rho_x(t,f) := |\lambda_x(t,f)|^2, \tag{11}$$

with the elementary atom $g(.)$ being normalized so that (10) be satisfied. In the case of windowed Fourier analyses, the resulting quantity admits two complementary interpretations, each attached to an effective apparatus. Considering first frequency as a function of time, one is faced at each time instant to a local energy spectrum (this is referred to as a *spectrogram*). Conversely, considering time as a function of frequency, what is observed is the output power of a battery of bandpass filters, as a function of time (*sonagram*). This latter interpretation can easily accomodate for the many variations in which the filterbank is no more uniform but, e.g., constant-Q (*scalogram*, i.e., energetic version of the wavelet transform), or for even more complex variations, as it may be the case when mimicking auditory models.

Page. One well-recognized weakness of classical Fourier analysis is to be time-independent. A possible way of re-introducing time in the picture is to make *causal* the computation of the spectrum, and to focus on the associated time variations. This is exactly what has been proposed by C.H. Page, who introduced the definition[10] :

$$P_x(t,f) = \frac{\partial}{\partial t}\left|\int_{-\infty}^{t} x(s)\, e^{-i2\pi fs}\, ds\right|^2 . \tag{12}$$

Rihaczek. One more different point of view consists in interpreting the local energy of a signal in a given time-frequency domain centered at (t, f) as the energy of interaction between a windowed version (centered at t) of this signal and a filtered version (centered at f) of the same signal. Making both windows go to zero simultaneously ("gedankenexperiment") led A.W. Rihaczek to define as a *complex energy density* the quantity[10] :

$$R_x(t,f) = \lim_{\delta T\,\delta B \to 0} \frac{1}{\delta T\,\delta B} \int_{t-\delta T/2}^{t+\delta T/2} x(s) \left[\int_{f-\delta B/2}^{f+\delta B/2} X(\xi)\, e^{i2\pi\xi s}\, d\xi\right]^* ds, \tag{13}$$

that can be easily checked to express as $R_x(t,f) = x(t)\, X^*(f)\, e^{-i2\pi ft}$.

These few examples do not exhaust in any way the totality of possible solutions. Their multiplicity (and diversity) raise however the issue of considering possible *links* between different solutions, i.e., the existence of *classes* of solutions. Two ways can be envisioned in this direction. The first one simply consists in *revealing* similarities between pre-existing definitions, whereas the second one would aim at *constructing* classes on the basis of a given set of requirements.

Unifications. Proceeding along the first line, a careful examination shows that all the definitions considered above are indeed *quadratic* transformations of the signal, and that they furthermore admit a common parameterization

$$C_x(t,f) := \iiint_{-\infty}^{+\infty} \varphi(\xi,\tau)\, x\left(s+\frac{\tau}{2}\right)\, x^*\left(s-\frac{\tau}{2}\right)\, e^{i2\pi[\xi(s-t)-f\tau]}\, ds\, d\xi\, d\tau, \tag{14}$$

based on the introduction of a suitable kernel function $\varphi(\xi,\tau)$ (for instance, Page's and Rihaczek's definitions correspond to $\exp(i\pi\xi|\tau|)$ and $\exp(i\pi\xi\tau)$, respectively). This approach was put forward by L. Cohen in the mid-sixties (in the context of Quantum Mechanics), and (14) is nowadays referred to as *Cohen's class*[9,10].

Such a unification represented a breakthrough, since it allows to easily get access to the properties of a given distribution by a simple inspection of its kernel function. Conversely, it also permits to create an infinite number of new distributions from some *a priori* specification of the kernel function. In particular, it turns out that the simplest possible choice, i.e., $\varphi(\xi,\tau)=1$, leads to the definition

$$W_x(t,f) := \int_{-\infty}^{+\infty} x\left(t+\frac{\tau}{2}\right)\, x^*\left(t-\frac{\tau}{2}\right)\, e^{-i2\pi f\tau}\, d\tau, \tag{15}$$

in which one can recognize the proposal made in 1932 by E.P. Wigner[29] (Quantum Mechanics) and in 1948 by J. Ville[27] (Signal Theory).

Covariances. The second way of constructing classes of solutions consists in imposing some (very general) *a priori* structure, and in deducing more restrictive parameterizations from the progressive imposition of further requirements considered as "natural". *Albeit* not stricly necessary, the usual framework is quadratic :

$$\rho_x(t,f) = \iint_{-\infty}^{+\infty} K(s,s';t,f)\, x(s)\, x^*(s')\, ds\, ds', \tag{16}$$

with the further constraint

$$\iint_{-\infty}^{+\infty} K(s,s';t,f)\, dt\, df = \delta(s-s') \tag{17}$$

applied to the arbitrary four-dimensional kernel $K(s,s';t,f)$ so as to guarantee that the resulting distribution be of energy-type, in the sense of (10). Given this framework, it then suffices to impose additional *covariance* constraints to appropriately reduce the space of admissible solutions. In brief, this amounts to imposing the commutative relation $\rho_{\mathbf{T}x}(t,f) = (\tilde{\mathbf{T}}\rho)_x(t,f)$, in which $\mathbf{T} : L^2(\mathbb{R}) \rightarrow L^2(\mathbb{R})$ stands for some transformation operator acting on signals (and $\tilde{\mathbf{T}} : L^2(\mathbb{R}^2) \rightarrow L^2(\mathbb{R}^2)$ for the corresponding operator acting on time-frequency distributions). In other words, the desired distribution is asked to "follow" a signal in the transformations that it undergoes.

The simplest example is that of *shifts*, in both time and frequency, for which the covariance principle leads the most general kernel to take on the simpler form

$$K(s, s'; t, f) = K_0(s - t, s' - t)\, e^{-i2\pi f(s-s')}, \tag{18}$$

where $K_0(s, s')$ is some arbitrary *two-dimensional* function. A remarkable result of this approach is that we then end up exactly with Cohen's class (14), provided that

$$\varphi(\xi, \tau) := \int_{-\infty}^{+\infty} K_0\left(t + \frac{\tau}{2}, t - \frac{\tau}{2}\right) e^{-i2\pi\xi t}\, dt. \tag{19}$$

The central role played by Cohen's class is therefore reinforced by the *constructive* argument according to which it is the class of *all* quadratic time-frequency distributions that are shift-covariant.

Generalizing the approach, a *deductive* construction of classes of distributions can be obtained on the basis of covariance requirements different from shifts. In particular, imposing covariance with respect to shifts in time and *dilations* leads — in the space of analytic signals, i.e., those signals whose spectrum identically vanishes on the real half-line of negative frequencies — to the so-called *affine* class. An admissible parameterization of this class reads[3]

$$B_X^{(k)}(t, f) = f \int_{-\infty}^{+\infty} X\left(f\lambda_k(u)\right) X^*\left(f\lambda_k(-u)\right) \mu_k(u)\, e^{i2\pi f t \zeta_k(u)}\, du, \tag{20}$$

with $\zeta_k(u) = \lambda_k(u) - \lambda_k(-u)$, and where $\lambda_k(u)$ is defined by

$$\lambda_k(u) = \left(k\, \frac{e^{-u} - 1}{e^{-ku} - 1}\right)^{1/(k-1)} \tag{21}$$

for $k \neq 0, 1$ (while being extended by continuity when $k = 0$ or 1). The weight $\mu_k(u)$ is *a priori* arbitrary, and it may be specified for controlling further properties of $B_X^{(k)}(t, f)$.

Other choices may be considered, at will. Covariance requirements with respect to *frequency-dependent* shifts (nonlinear group delays) lead, e.g., to *hyperbolic* and *power* classes[22].

3 Time-frequency analysis and "chirps"

3.1 Chirps

By convention a complex-valued signal will be here referred to as a "chirp" if its time-domain representation

$$x(t) = a(t)\, e^{i\varphi(t)} \tag{22}$$

is such that the oscillations controlled by the phase $\varphi(t)$ are slowly-varying as compared to the variations of the (positive) function $a(t)$. In the special case where such a signal is furthermore analytic, $a(t)$ stands for the instantaneous amplitude of the real-valued signal $\text{Re}\{x(t)\}$, whereas the phase derivative $\varphi'(t) := (d\varphi/dt)(t)$ coincides (up to a factor 2π) with its instantaneous frequency $f_x(t)$.

3.2 Time-frequency localization on chirps

Ideally, the time-frequency picture of a chirp should be an energy distribution whose support on the plane would reduce to the instantaneous frequency curve. In the case of windowed analyses (STFT or WT), it is intuitive that such a perfect localization cannot be obtained because of the introduction, in the measurement process, of some external quantity (the short-time window or the wavelet) whose time-frequency localization is jointly limited. The only exceptions to this situation correspond to pure tones and impulses, for which a perfectly localized window (in frequency or in time, respectively) yields a perfect localization on the plane too. These cases, while being of limited use by themselves, suggest to improve windowed analyses by *locally adapting* the window to the analyzed signal. In the STFT case, the easiest way of doing so is to mimick the well-known principle of *matched filtering* by choosing for window the time-reversed version of the signal itself, thus leading to :

$$F_x^{(x_-)}(t,f) = \frac{1}{2} W_x\left(\frac{t}{2}, \frac{f}{2}\right), \tag{23}$$

with $W_x(t,f)$ the Wigner-Ville distribution (WVD), defined as in (15).

The WVD admits therefore a new justification in terms of "matched" transform, with the localization property :

$$x(t) = \exp\{i2\pi(f_0 t + \alpha t^2/2)\} \Rightarrow W_x(t,f) = \delta\left(f - (f_0 + \alpha t)\right) \tag{24}$$

as a corollary[e]. This property can be viewed as a natural generalization, to

[e]Strictly speaking, the phase derivative $f_0 + \alpha t$ cannot be considered as an admissible instantaneous frequency, since the chirp signal in (24) is not analytic[23].

quadratic phases, of the classical localization property of the Fourier transform for *linear* phases :

$$x(t) = \exp\{i2\pi f_0 t\} \Rightarrow X(f) = \delta\,(f - f_0)\,. \tag{25}$$

One can give a very simple geometrical interpretation of this result, which gives some more insight on its origin and allows for deriving straightforward generalizations. When applied to a chirp as in (22), the WVD (15) reads :

$$W_x(t,f) = \int_{-\infty}^{+\infty} \left[b_t(\tau)\, e^{i\psi_t(\tau)}\right] e^{-i2\pi f\tau}\, d\tau, \tag{26}$$

with

$$b_t(\tau) := a\left(t+\frac{\tau}{2}\right) a\left(t-\frac{\tau}{2}\right) \quad ; \quad \psi_t(\tau) := \varphi\left(t+\frac{\tau}{2}\right) - \varphi\left(t-\frac{\tau}{2}\right). \tag{27}$$

Rewritten this way, the WVD is nothing but the Fourier transform of some modified chirp signal, whose local phase (in τ) is parameterized by the local time t and reads $\psi_t(\tau)$. As a consequence, if we assume that the analyzed chirp $x(t)$ is unimodular (i.e., that $a(t) = 1$), the resulting WVD will be perfectly localized — at each time instant t — as long as the partial derivative of the local phase, i.e.,

$$\frac{\partial}{\partial\tau}\psi_t(\tau) = \frac{1}{2}\left[\varphi'\left(t+\frac{\tau}{2}\right) + \varphi'\left(t-\frac{\tau}{2}\right)\right], \tag{28}$$

will be constant in τ. Considering the functional equation attached to this condition, the only solutions are given by phases $\varphi(t)$ which are *quadratic*, thus corresponding to a localization on straight lines of the plane. The localization property (24) admits therefore a very simple geometrical explanation in terms of *self-adaptation* of the analysis by a mechanism of phase compensation : an upgoing linear chirp exactly compensates a downgoing linear chirp for creating a pure tone[11].

This notion of phase compensation evidences the quadratic structure of the WVD. As a corollary, the analogy according to which this latter behaves, with respect to linear chirps, as does the Fourier transform with respect to pure tones cannot be followed in an exact fashion when several chirps are simultaneously present. Indeed, the WVD obeys a quadratic superposition principle, according to which $W_{ax+by} = |a|^2\, W_x + |b|^2\, W_y + I$, where I stands for an (oscillating) *interference* term, carrying information about the interaction between the two superimposed signals x et y. A detailed analysis of the interference mechanism[17] shows that any two components create on the plane an extra contribution located around the geometrical mid-point of the chord drawn between them. This result illustrates in fact that interference and localization cannot be disconnected, in the sense that it is the very same mechanism which underlines both phenomena. Indeed, any signal can always be segmented into arbitrary components : in the case of chirps, iterating segmentation *ad infinitum* leads to a pointwise application of the interference rule between any two points belonging to the (pseudo) instantaneous frequency curve $\varphi'(t)$. Therefore, localization is only possible if all interacting points belong themselves to the considered curve : one recovers this way the localization property of the WVD on straight lines of the plane (loosely speaking, linear chirps), a straight line being the only curve that can be defined as the locus of all of its mid-points.

This mechanism of localization *via* a constructive interference process is very general and it can be adapted to nonlinear situations, provided that interference construction rules are suitably modified[12]. For instance, we know[3] that specific Bertrand distributions (20) perfectly localize on *power-law* group delay curves. Specifically, it can be shown that the condition :

$$\mu_k(u) = (\lambda_k(u)\,\lambda_k(-u))^{1/2}\,\zeta_k'(u) \tag{29}$$

guarantees that (for any $k<0$, with a continuity solution for $k=0$) :

$$X_k(f) = \frac{1}{\sqrt{f}}\exp\left\{-i2\pi\left(t_0 f + \frac{c}{k}f^k\right)\right\} \Rightarrow B^{(k)}_{X_k}(t,f) = \frac{1}{f}\,\delta\left(t-\left(t_0 + cf^{k-1}\right)\right). \tag{30}$$

This result can be compared to (24), with straight lines replaced by generalized hyperbolae. As for the WVD, the localization properties of the (quadratic) Bertrand distributions can be viewed as a simple by-product of the pointwise application of their own interference rules. Using the same argument as above, claiming that a quadratic distribution perfectly localizes on a nonlinear curve of the plane is equivalent to saying that any two points belonging to the considered curve interact in such a way that their interfering contribution belongs itself to the same curve. In the WVD case, interference geometry is governed by a "mid-point" rule, associated to the usual idea of *arithmetic mean* : $(a,b)\mapsto A(a,b) = (a+b)/2$. One can then show[12] that the Bertrand case obeys a similar logic, with a construction rule being controlled by the (Galvani-Stolarsky) *generalized means* :

$$(a,b)\mapsto L_k(a,b) = \left(\frac{1}{k}\,\frac{a^k-b^k}{a-b}\right)^{1/(k-1)}. \tag{31}$$

One can remark that the arithmetic mean can be recovered from (31) by setting $k=2$, in accordance with the fact that the WVD is nothing but $B^{(2)}_X(t,f)$, when applied to analytic signals.

3.3 A time-frequency strategy for chirp detection

Localized distributions may be interesting in two respects, at least. First, *analysis* of some unknown chirp signal is made easier when a sharper picture is provided ; second, *decision* (i.e., detection, estimation, classification...) regarding a chirp signal can be simplified as well, in the sense that it may be built upon a natural and physically meaningful *signature* on the plane. We will briefly address in this section the problem of chirp detection, with a potential application to gravitational wave chirps.

Within the next years, several gravitational wave detectors (VIRGO, LIGO, GEO600) will be in operation. Their objective is to directly measure gravitational radiations, whose existence has been predicted long ago by general relativity. Many different gravitational wave sources can be considered, but it is almost universally accepted that the most promising one in term of detection is the coalescence of a massive binary system composed, e.g., of two neutron stars. In a first (Newtonian) approximation, when the two bodies are sufficiently far apart, the explicit expression of the waveform can be given as the real part of the complex signal[24,25] :

$$x(t; t_0, \mathcal{M}_\odot) = A\,(t_0 - t)^{-\alpha}\, e^{-i2\pi d (t_0 - t)^\beta}\, U(t_0 - t), \tag{32}$$

with $U(.)$ the unit step function, and where $\alpha = 1/4$ and $\beta = 5/8$. In this expression, t_0 is the coalescence time, and d and A are constants which depend upon some physical parameters characterizing the binary system. More precisely[25], the constant d is best expressed as $d \approx 241\,\mathcal{M}_\odot^{-5/8}$, by using the "chirp mass" $\mathcal{M} = \mu^{3/5}\, M^{2/5}$, function of the total mass $M = m_1 + m_2$ and the reduced mass $\mu^{-1} = m_1^{-1} + m_2^{-1}$, where m_1 and m_2 are the individual masses of the two bodies, and where the chirp mass $\mathcal{M}_\odot = \mathcal{M}/M_\odot$ is expressed in solar mass $M_\odot$ units. Looking at the phase in the exponential suggests that (32) is a chirp whose frequency evolves as a power-law with respect to time. In fact, a stationary phase approximation[7] of its Fourier transform shows that (32) is characterized by a group delay of the form : $t_X(f) = t_0 + ckf^{k-1}$, with a power index $k = -5/3$ and a modulation rate $c \approx 3.85 \times 10^5\, \mathcal{M}_\odot^{-5/3}$.

In practice, gravitational wave data will naturally be corrupted by many perturbations (including seismic noise at low frequencies and shot noise in the laser diodes at high frequencies). We will assume here that, in a first approximation[18] and within the measurement frequency bandwidth, these perturbations can be modeled as an additive stationary Gaussian noise of power spectral density $\Gamma_n(f) = \sigma^2 f^{-\epsilon}$, with $\epsilon \approx 1$ and $\sigma^2 = 0.7 \times 10^{-42}$/Hz. In this context, matched filtering is a natural candidate for detection, since it is known to be *optimal* (under the Gaussian assumption) in the sense of the "maximum likelihood ratio test"[28]. Matched filtering relies on a correlation measurement between the observed data $r(t)$ and the signal to be detected $x(t; t_0, \mathcal{M}_\odot)$, taken as a reference. When the absolute phase of the impinging signal is considered as a nuisance parameter, this correlation can be expressed in frequency as[28] :

$$\Lambda^w(r; t_0, \mathcal{M}_\odot) = \left| \int_0^{+\infty} \frac{R(f)\, X^*(f; t_0, \mathcal{M}_\odot)}{\Gamma_n(f)}\, df \right|^2, \tag{33}$$

but it also admits a similar expression in the time domain, thanks to the isometry of the Fourier transform. This naturally suggests a third possible approach, in which the inner product (33) would be written in both time and frequency, jointly, by means of a suitable time-frequency distribution $\rho_X(t, f)$ that we can think of as a well-adapted "signature" of the chirp signal to detect. Such an equivalence is obviously not guaranteed for all time-frequency distributions, but only for those having the *unitarity* property[10]. In this case, it becomes possible to write[7] :

$$\Lambda^w(r; t_0, \mathcal{M}_\odot) = \int\int_0^{+\infty} \rho_{X_1}(\tau, f)\, \rho_{X_2}(\tau, f)\, d\tau df, \tag{34}$$

where $X_1(f) = R(f)/\Gamma_n(f)$ and $X_2(f) = X(f; t_0, \mathcal{M}_\odot)$.

Within the family of unitary time-frequency distributions, choosing one such that $\rho_{X_2}(t, f)$ is perfectly localized along the group delay line $(t_X(f), f)$ permits to replace the double integral in (34) by a *path integration* in the time-frequency plane. In the context of the gravitational wave model (32), and in accordance with (30), localization amounts to using the *Bertrand distribution* of index $k = -5/3$. Strictly speaking, because of the power-law envelope of the chirp, the optimal time-frequency detector would require the computation of a filtered version of the Bertrand distribution of the data, thus implying a heavy computational burden. A more efficient solution, simpler yet precise, can be obtained by noting that the crucial characteristic of the Bertrand distribution is its perfect localization on the reference signal. This can indeed be achieved to an excellent approximation by using instead a so-called "reassigned spectrogram" $\check{S}_x(t, f)$, obtained from a conventional spectrogram by a suitable squeezing technique[1]. The amplitude of the reassigned spectrogram along the ridge has to be corrected too (so as to mimick the unitary property), ending up with the final detector structure :

$$\Lambda^w(r; t_0, \mathcal{M}_\odot) \propto \int_{\mathcal{L}(t_0, \mathcal{M}_\odot)} \check{S}_x(\tau, f) f^{-2/3}, \tag{35}$$

with

$$\mathcal{L}(t_0, \mathcal{M}_\odot) := \left\{ (\tau, f) \middle| t_0 - \tau = 3 \times 100^{8/3} \mathcal{M}_\odot^{-5/3} f^{-8/3} \right\}. \tag{36}$$

The proposed detection method should be understood as a preliminary step for further improved detection performance. The rationale for using a time-frequency approach is that, while being equivalent to conventional techniques in nominal conditions, it offers an appropriate language for encompassing more challenging situations[7,18]. In particular, *robustness* to inaccurate modelling (of either the chirp or the noise) is expected to be simply taken into account by adapting the strategy on the basis of suitable modifications of time-frequency signatures.

4 A reading guide

Basics of time-frequency analysis have only been sketched in this text, and the interested reader is referred to the vast existing literature for further details :

- Linear decompositions have been mentioned here, but with not much emphasis. These approaches are central in the monographs by Mallat[20] and Carmona-Hwang-Torrésani[5]. Moreover, these two books are supplemented by freeware libraries written in Matlab and S-plus, respectively.

- Quadratic energy distributions are central in the monographs by Cohen[9] or Flandrin[10], the latter being complemented by a freeware toolbox [2] written in Matlab and introduced by a tutorial booklet. From a more theoretical perspective, the monograph by Folland[13] gives a detailed account of harmonic analysis on phase space (pseudo-differential operators, symbols, quantization rules. . .).

- A number of other books or tutorial articles are worth citing too, such as those by Boudreaux-Bartels[4], Mecklenbräuker and Hlawatsch[21], or Papandreou-Suppapola *et al.*[22] and Chassande-Mottin *et al.*[8], where most recent developments can be found. Although it has been written some time ago, a special mention can be given to the survey by Hlawatsch and Boudreaux-Bartels[16], in which many applications in very different areas are discussed and illustrated.

- As far as chirp detection is concerned, and especially in the context of gravitational wave data analysis, the approach briefly described here is detailed in Chassande-Mottin and Flandrin[7].

Acknowledgments

Thanks to Eric Chassande-Mottin (AEI, Golm), with whom the work on the time-frequency detection of gravitational waves sketched here has been carried out.

References

1. Auger, F., Flandrin, P. (1995). Improving the readability of time-frequency and time-scale representations by using the reassignment method. *IEEE Trans. on Signal Proc.* **43** (5), 1068–1089.
2. Auger F., Flandrin, P. , Gonçalvès, P., Lemoine, O. (1997). *A Time-Frequency Toolbox, for Use with Matlab.* `http://crttsn.univ-nantes.fr/ auger/tftb.html`.
3. Bertrand, J., Bertrand, P. (1992). A class of affine Wigner functions with extended covariance properties. *J. Math. Phys.* **33** (7), 2515-2527.
4. Boudreaux-Bartels, G.F. (1996). Mixed time-frequency signal transformations. In *The Transforms and Applications Handbook* (A.D. Poularikas, *ed.*). CRC Press, Boca Raton (FL), 829–885.
5. Carmona, R., Hwang, W.L., Torrésani, B. (1998). *Practical Time-Frequency Analysis.* Academic Press, San Diego (CA).
6. Chassande-Mottin, E. (1998). Méthodes de réallocation dans le plan temps-fréquence pour l'analyse et le traitement de signaux non stationnaires. Thèse de Doctorat, Univ. de Cergy-Pontoise (F).
7. Chassande-Mottin, E., Flandrin, P. (1999). On the time-frequency detection of chirps. *Appl. Comp. Harm. Anal.* **6** (2), 252–281.
8. Chassande-Mottin, E., Auger, F., Flandrin, P. (1999). Time-frequency/time-scale reassignment. In *Wavelets and Time-Frequency Analysis* (L. Debnath, *ed.*). Birkhäuser, Boston (MA), to appear.
9. Cohen, L. (1995). *Time-Frequency Analysis.* Prentice-Hall, Englewoods Cliffs (NJ).
10. Flandrin, P. (1998). *Temps-Fréquence* (2ème éd.). Hermès, Paris (F). English translation as Flandrin, P. (1999). *Time-Frequency/Time-Scale Analysis.* Academic Press, San Diego (CA).
11. Flandrin, P., Escudié, B. (1984). An interpretation of the pseudo-Wigner-Ville distribution. *Signal Proc.* **6** (1), 27–36.
12. Flandrin, P., Gonçalvès, P. (1996). Geometry of affine time-frequency distributions. *Appl. Comp. Harm. Anal.* **3** (1), 10–39.
13. Folland, G.B. (1989). *Harmonic Analysis in Phase Space*, Ann. of Math. Studies **122**. Princeton Univ. Press, Princeton (NJ).

14. Folland, G.B., Sitaram, A. (1997). The uncertainty principle: a mathematical survey. *J. Fourier Anal. Appl.* **3** (3), 207–238.
15. Gabor, D. (1946). Theory of communication. *J. IEE* **93**, 429–457.
16. Hlawatsch, F., Boudreaux-Bartels, G.F. (1992). Linear and quadratic time-frequency signal representations. *IEEE Signal Proc. Magazine* **9**, 21–27.
17. Hlawatsch, F., Flandrin, P. (1998). The interference structure of the Wigner distribution and related time-frequency signal representations. In[21], 59–133.
18. Innocent, J.-M., Torrésani, B. (1997). Wavelets and binary coalescences detection. *Appl. Comp. Harm. Anal.* **4** (2), 113–116.
19. Loughlin, P., Tacer, B. (1996). On the amplitude and frequency-modulation decomposition of signals. *J. Acoust. Soc. Amer.* **100**, 1594–1601.
20. Mallat, S. (1997). *A Wavelet Tour of Signal Processing.* Academic Press, San Diego (CA).
21. Mecklenbräuker, W.F.G., Hlawatsch, F., *eds.* (1998). *The Wigner Distribution — Theory and Applications in Signal Processing.* Elsevier, Amsterdam (NL).
22. Papandreou-Suppappola, A., Hlawatsch, F., Boudreaux-Bartels, G.F. (1998). Quadratic time-frequency representations with scale covariance and generalized time-shift covariance : a unified framework for the affine, hyperbolic, and power classes. *Digital Signal Proc.* **8**, 3–48.
23. Picinbono, B. (1997). On instantaneous amplitude and phase of signal. *IEEE Trans. on Signal Proc.* **45** (3), 552–560.
24. Sathyaprakash, B.S., Dhurandhar, D.V. (1991). Choice of filters for the detection of gravitational waves from coalescing binaries. *Phys. Rev. D* **44** (12), 3819–3834.
25. Thorne, K.S. (1987). Gravitational radiation. In *300 Years of Gravitation* (S.W. Hawking, W. Israel, *eds.*). Cambridge Univ. Press, Cambridge (UK), 330–458.
26. Vakman, D. (1996). On the analytic signal, the Teager-Kaiser energy algorithm, and other methods for defining amplitude and frequency. *IEEE Trans. on Signal Proc.* **44**, 791–797.
27. Ville, J. (1948). Théorie et applications de la notion de signal analytique. *Câbles et Transm.* **2ème A.** (1), 61–74.
28. Whalen, A.D. (1971). *Detection of Signals in Noise.* Academic Press, San Diego (CA).
29. Wigner, E.P. (1932). On the quantum correction for thermodynamic equilibrium. *Phys. Rev.* **40**, 749–759.

INTRODUCTION TO THE DATA ANALYSIS IN INTERFEROMETRIC GRAVITATIONAL WAVE EXPERIMENTS

ANDREA VICERÉ

INFN, Sezione di Pisa, Via Vecchia Livornese 1291, I-56125 Pisa, Italy
E-mail: Andrea.Vicere@pi.infn.it

A concise introduction to the problems of data analysis for gravitational wave experiment is given. The aim was to offer a broad overview, which could help the newcomer in the field to approach the vast existing literature, while also giving a few computational details so as to keep the treatment self-contained, and appropriate for a Physics School. To this end the several simplifications have been made, trying to stick to the most crucial aspects of physics or methods.

1 Generalities on the signal

An interferometric detector is a very simple instrument, when looked at disregarding the complexity of the apparatus but focusing just on its output.

It produces a single time series $s(t)$ composed of a signal $h(t)$ added to a noise $n(t)$. We assume there is no multiplicative noise, and that it has been already compensated the effect of the instrument transfer function, so that $h(t)$ is related linearly to the metric change $h_{ab}(t)$: if the detector arms are oriented along the versors $\hat{X}$, $\hat{Y}$, then the output can be normalized to have

$$h(t) \equiv h_{ab}(t)\frac{1}{2}\left(\hat{X}^a\hat{X}^b - \hat{Y}^a\hat{Y}^b\right) ; \tag{1}$$

however the detector moves in space, and the versors $\hat{X}, \hat{Y}$ with it, so the metric h_{ab} is modulated.

Expanding the signal in series of plane waves, one can write down

$$h_{ab}(t) = \sum_{A=+,\times} \int_{-\infty}^{+\infty} df \int_{S^2} d\hat{\Omega}\, h_A(f,\hat{\Omega}) e^{i2\pi f(t-\hat{\Omega}\cdot\vec{x}/c)} \varepsilon^A_{ab}(\hat{\Omega}) \tag{2}$$

as a sum over frequencies, directions and polarizations, and therefore

$$h(t) = \sum_{A=+,\times} \int_{-\infty}^{+\infty} df \int_{S^2} d\hat{\Omega}\, h_A(f,\hat{\Omega}) e^{i2\pi f(t-\hat{\Omega}\cdot\vec{x}/c)} F^A(\hat{\Omega}) \tag{3}$$

where the antenna patterns $F_{+,\times}$ are introduced:

$$F^A(\hat{\Omega}) \equiv \sum_{ab} \varepsilon^A_{ab}(\hat{\Omega})\frac{1}{2}\left(\hat{X}^a\hat{X}^b - \hat{Y}^a\hat{Y}^b\right) , \tag{4}$$

containing the informations on the way the GW signal is translated into the output of the detector. The response depends on the position in the sky of the source, and is Doppler shifted by the motion $\vec{x}(t)$ of the detector. Notice that if the output of the detector is normalized (a procedure called h-reconstruction) so as to have the relation in Eq. (1) between the metric and the output, then different detectors can compare their data directly, without knowing instrument details, apart position and orientation of the arms.

In defining the framework of the analysis it is crucial the statistic of the noise $n(t)$: at the present status of development of the instruments one hopes it will be gaussian, with all the statistical properties known once it is known the two point correlation function

$$C(t,t') = E[n(t)n(t')]\,. \tag{5}$$

In the following we shall ignore any non-linear or non-stationary effect, and assume therefore also that $C(t,t') = C(t-t')$.

2 Noise characterization

In other courses the basics of spectral estimation and of techniques like the Kaharounen-Loève expansion, have already been laid out, hence we recall them briefly and give examples based on real data from the Caltech 40m detector, kindly made available by the LIGO collaboration [1].

2.1 *Spectral analysis*

The bread and butter of noise analysis is the evaluation of the noise spectrum, which is defined as the Fourier transform of the correlation function as [a]

$$S(f) = \int E[n(t)n(t+\tau)e^{i2\pi f\tau}\,d\tau \tag{7}$$

The estimation of the spectrum is of course affected by statistical and systematical errors: the former due to the variance of the samples, the latter due to the finite size of the sample, which corresponds to convolving the data with a

[a] This definition gives a bilateral spectrum, favored in signal processing, defined also in terms of negative frequencies: for instance the Percival relation is

$$E[n^2] = \int_{-\infty}^{+\infty} S(f)\,df \tag{6}$$

integrating both on positive and negative frequencies. In physics one uses a on-sided spectrum, which is twice as large: however in the following we shall use a bi-lateral one.

rectangular window. We are interested in strategies to control the estimate, that is balance between bias and variance: *control* and not *improve* because we did not define yet a criterium of goodness.

Most of the material is taken from Percival and Walden [14] which is the most complete reference on the subject we have found.

Process representation

A stochastic second order process can be written as a Riemann-Stiltjes integral

$$x_t = \int_{-1/2}^{1/2} e^{j2\pi f t} dZ(f) \tag{8}$$

where $dZ(f)$ are orthogonal increments, that is

$$E\left[dZ(f)dZ^*(f')\right] = \delta(f-f')S(f)df\,df' , \tag{9}$$

where $S(f)$ is the "true" spectrum.

Naive periodogram

The periodogram estimate of the spectrum is defined from the Fourier transform of the finite sample $\tilde{x}_N(f) \equiv \frac{1}{\sqrt{N}} \sum_{t=0}^{N-1} e^{-j2\pi f t} x_t$ as $S_N(f) \equiv \tilde{x}_N(f)\tilde{x}_N(-f)$. In terms of the orthogonal process $dZ(f)$

$$\begin{aligned} \tilde{x}_N(f) &= \frac{1}{\sqrt{N}} \int_{-1/2}^{1/2} \sum_{t=0}^{N-1} e^{j2\pi(f'-f)t} dZ(f') \\ &= \int_{-1/2}^{1/2} \sqrt{N} D_N(f'-f) e^{j(N-1)\pi(f'-f)} dZ(f') \end{aligned} \tag{10}$$

where the Dirichlet kernel D_N is defined as

$$D_N(f) \equiv \frac{\sin(N\pi f)}{N\sin(\pi f)} . \tag{11}$$

One therefore obtains

$$S_N(f) = N \int_{-1/2}^{+1/2} \int_{-1/2}^{+1/2} D_N(f'-f) D_N(f''+f) e^{j(N-1)\pi(f'+f'')} dZ(f') dZ(f'') \tag{12}$$

and for the expectation value one obtains immediately

$$E\left[S_N(f)\right] = \int_{-1/2}^{+1/2} N D_N^2(f'-f) S(f') df' \tag{13}$$

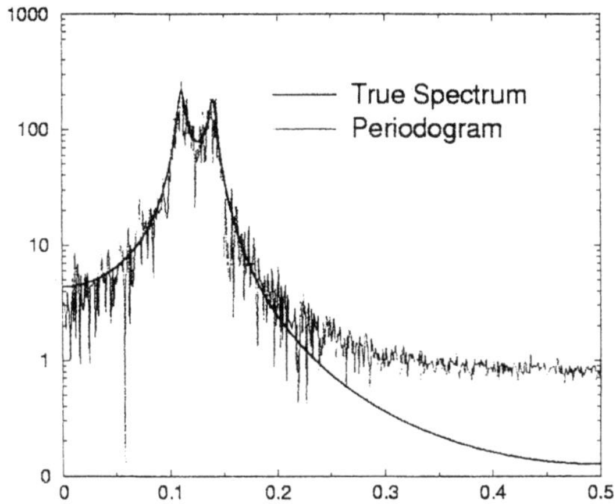

Figure 1. Comparison of true spectrum and naive periodogram.

where the Fejer kernel links the average sample spectrum with the true one

$$\mathcal{F}(f) \equiv N D_N^2(f) = \frac{1}{N}\sum_{t=1}^{N}\sum_{u=1}^{N} e^{j2\pi f(t-u)} = \frac{1}{N}\frac{\sin^2(N\pi f)}{\sin^2(\pi f)} \; ; \qquad (14)$$

this kernel leads to the *spectral leakage*, that is to a bias effect especially serious for spectra with large dynamic ranges, as in Fig. (1) for an AR(4) process. There are two ways to reduce this bias:

- altering the data sequence by *tapering* so as to modify the Fejer kernel.
- Reduce the dynamic range before the estimate (*pre-whiten* the data) using an approximate estimate of the spectrum. This strategy is particularly useful in connection with a parametric estimate of the spectrum [14].

Tapering (or windowing)

The idea of the tapering is elementary: replace the original sequence $x_0, \ldots x_{N-1}$ with a new sequence $h_0 x_0, \ldots h_{N-1} x_{N-1}$ and get a new relation between sample and true spectrum

$$E\left[\hat{S}_N(f)\right] = \int_{-1/2}^{1/2} \mathcal{H}(f-f')S(f')df, \; !\mathcal{H}(f) \equiv \left|\sum_{t=0}^{N-1} h_t e^{-j2\pi(f-f')}\right|^2 \qquad (15)$$

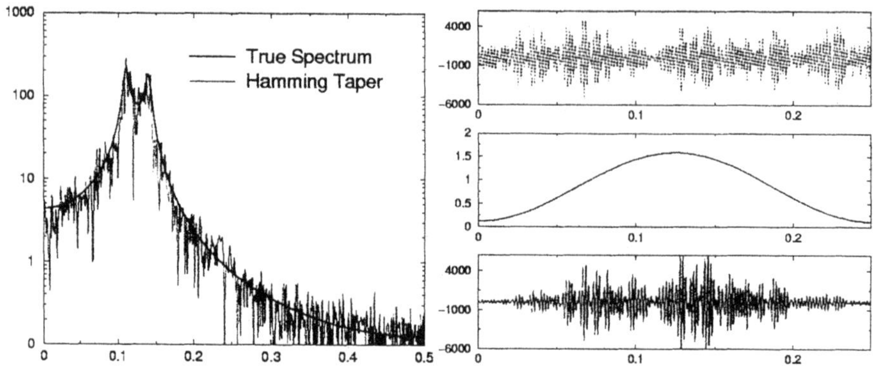

Figure 2. Left: the true spectrum and the periodogram with Hamming window; right, the data sequence, the window and the windowed data.

where for $h_t = 1/\sqrt{N}$ one recovers the periodogram estimate. The kernel $\mathcal{H}$ is linked to the Fourier transform of the window, which actually is zero for $t < 1$ and $t > N$, thus leading to the high frequency side-lobe effects.

The taper has to regularize the window at the extrema, weighting differently the data, and reducing the effect of the sharp transition, as in Fig. (2) where we show the effect of an Hamming taper[b]. The price is the increase of variance in the estimate: indeed if we look at the data sequence before and after the application of the window, as in Fig. (2), we see that values at the extrema are underweighted with respect to the data at the center of the sequence, and contribute less to reducing the variance.

Discrete Prolate Spheroidal tapers

A systematic way of choosing the taper(s) comes by classifying them according to their "energy concentration", that is by the ratio

$$\beta^2(W) \equiv \frac{\int_{-W}^{W} \mathcal{H}(f)df}{\int_{-1/2}^{1/2} \mathcal{H}(f)df} = \frac{1}{\sum_t |h_t|^2} \sum_{t,t'} h_t \frac{\sin\left[2\pi W(t-t')\right]}{\pi(t-t')} h_{t'}^* \,. \tag{17}$$

[b]The Hamming window is defined as

$$h_i = 0.54 - 0.46 \cos\left(\frac{2\pi i}{N-1}\right) \tag{16}$$

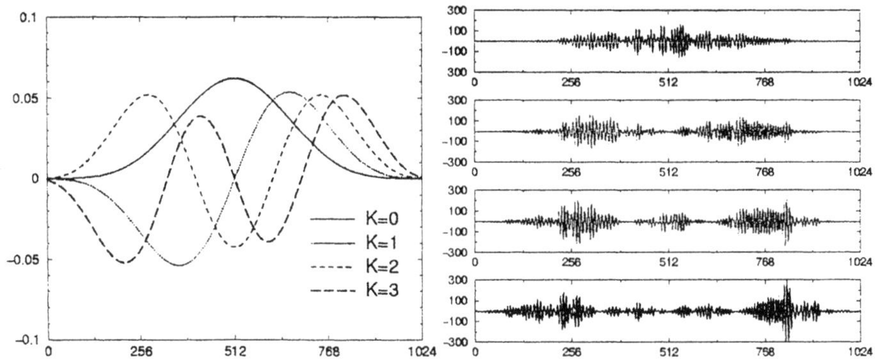

Figure 3. Left: first four DPSS, for $2NW = 4$; right, the effect of windowing a data sequence with these DPSS.

The sequences can be therefore classified as eigenvectors λ_i of the kernel

$$K(t-t') = \frac{\sin 2\pi W(t-t')}{\pi(t-t')} \tag{18}$$

and ordered in size $\lambda_0 > \lambda_1 > \ldots \lambda_N$.

The eigenvectors are called Discrete Prolate Spheroidal Sequences (DPSS) and at fixed product NW the zeroth order prolate gives of course the highest energy concentration.

We show in Fig. (3) the first four eigenvectors for $N = 1024, 2NW = 4$

The distribution of the eigenvalues is governed by the Shannon number $2NW$, in the sense that $\lambda_k \simeq 1$ for $k < 2NW$, corresponding to a good concentration, and $\lambda_k \ll 1$ for $k > 2NW$. The idea of multi-tapering is that once one has chosen the desidered spectral band resolution $2W$ one may want to reduce the variance of the spectral estimate by combining the estimates $\hat{S}_N^{(k)}(f)$ from different tapers $h_{t,k}$.

In Fig. (3) we show also the effect of applying different tapers to the same data: one evidences different data regions.

As the different DPSS are orthonormal $\sum_t h_{t,k} h_{t,l} = \delta_{k.l}$ the different spectral estimates are *approximately* uncorrelated [14] and can be combined. The most naive approach is to combine linearly with the same weight the estimates coming from K prolates with $K < 2NW$

$$\hat{S}_{N,W}^{(mt)}(f) \equiv \frac{1}{K} \sum_{k=0}^{K-1} \hat{S}_{N,W}^{(k)}(f) \; . \tag{19}$$

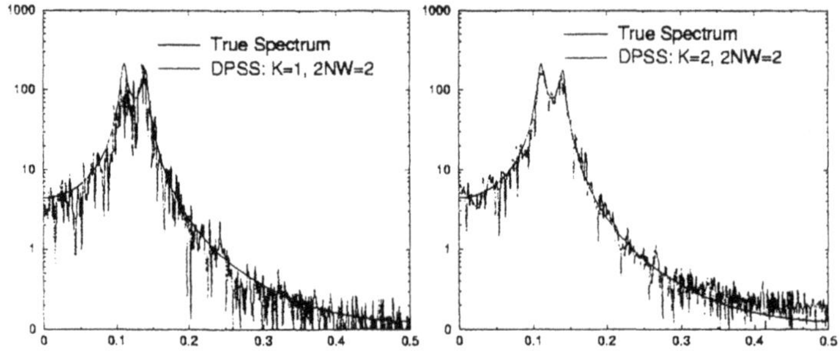

Figure 4. Estimates with 1 and 2 tapers.

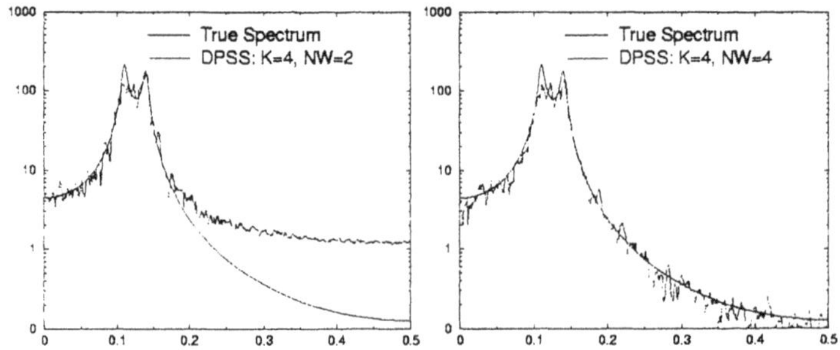

Figure 5. Estimates with 4 tapers, with $2NW = 2$ or 4

A refined approach weights the different estimates with the eigenvalues λ_k

$$\hat{S}^{(mt)}_{N,W}(f) \equiv \frac{1}{\sum_{k=0}^{N-1} \lambda_k} \sum_{k=0}^{N-1} \lambda_k \hat{S}^{(k)}_{N,W}(f) \ , \tag{20}$$

a combination optimal for gaussian white noise, an uninteresting case however because it does not have any leakage problem.

For instance, choosing $2NW = 2$ we can compare the estimates using 1 or 2 tapers, as in Fig. (4) where it is apparent a reduction of variance, especially around the peaks, at the price of a bias increase at higher frequencies.

We can further reduce the variance using more tapers, but then to avoid reintroducing bias we have to reduce the frequency resolution: we compare in Fig. (5) estimates made with 4 tapers, with different choices of W.

They exist "automatic" strategies to choose, at fixed resolution, the largest number of tapers which do not reintroduce bias: for an adaptive multi-

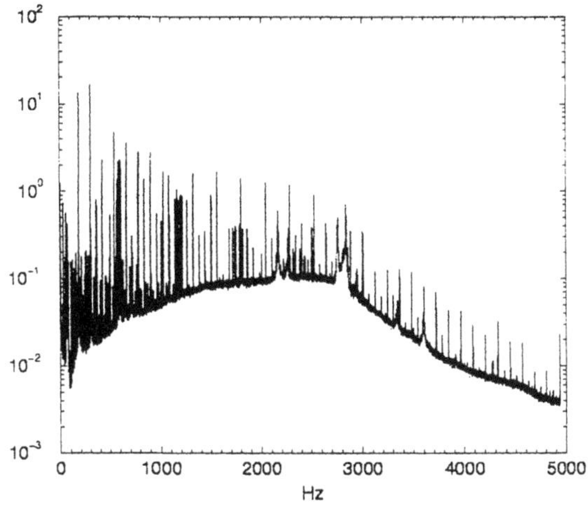

Figure 6. Typical LIGO 40m spectrum, arbitrary units. The frequency resolution is 0.3 Hz. The variance has been reduced with 128 spectral averages. Each sample was tapered with the Hamming window to reduce the bias.

tapering see for instance [14].

LIGO 40m data examples

We have tried a few experiments using data from the 19 November 1994 run of the LIGO 40m interferometer, selecting only data segments in which the interferometer was "locked". The typical spectrum in the locked state is shown in Fig. (6); it is populated by several lines, which are the main source of dynamics: the variation of the "base" value is not very large. Therefore we expect bias effects around the lines, but we do not expect "wideband" bias effects as those displayed in Fig. (1) . We first show in Fig. (7, left) an example of application of different windowing procedures to a short, uncalibrated, data segment: 8192 samples corresponding to 0.83012 seconds of data, displaying the region of frequencies around the 180 and 300 Hz resonances. The solid line displays the result of the naive periodogram, while the dotted line is obtained applying an Hamming window which is effective in reducing the leakage due to the narrow lines (f.i. at the 180 Hz peak); also evident is the increase in variance due to the underweighting effect of the window at the extrema of the interval. We also choose $NW = 2$ and consequently we used 4 tapers to estimate the multi-tapered periodogram, displayed with the dashed line, while

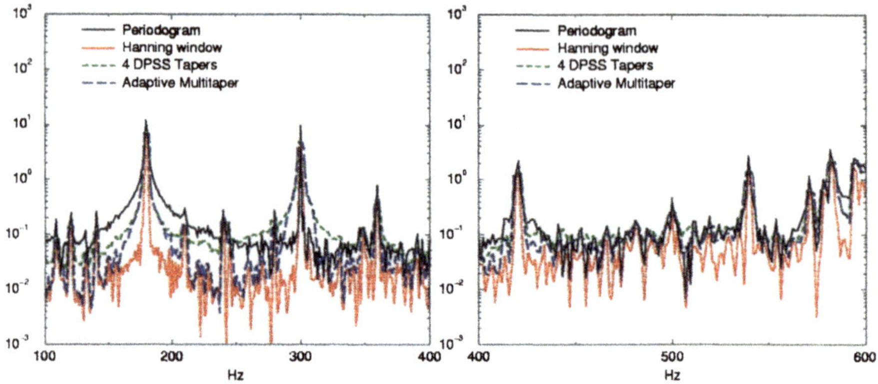

Figure 7. LIGO 40m data spectrum from 0.83012 seconds of data; left, around the 180 Hz resonance, right, around the 300 Hz resonance.

the long dashed line is the result of an adaptive multi-tapering procedure.

Concentrating on the 180 Hz resonance we see that the multi-tapering gives a lower variance, at a price of a higher bias than the Hamming window in reducing the bias, an effect compensated in part by the adaptive procedure [14], which f.i. recovers the level between the resonances.

Around the 300 Hz resonance (the 5th harmonic of the 60 Hz line) the multi-tapers appear to reduce the frequency resolution: this is not a surprise ($NW = 2$). The same does not happen at the 180 Hz resonance, because the chosen sample it corresponds roughly to 149 cycles of oscillation, and the 180 Hz line stays all in a single bin (I thank B.Allen for pointing out this).

sa

2.2 Karhunen-Loève expansion

The spectral representation corresponds to the idea that different frequency bins are uncorrelated, or in other words, that projecting the data over a basis of sine and cosine functions, the resulting coefficients are statistically uncorrelated.

However for fixed sample size this is not true, as suggested by the leakage: even for a stationary process, the correlation of two frequency bins is

$$E\left[\tilde{x}(f)\tilde{x}^*(f')\right] = \int_{-1/2}^{1/2} N D_N(f''-f) D_N(f''-f') S(f'') df'' \qquad (21)$$

where D_N is the Dirichlet kernel defined in Eq. 11; the kernel tends to $\delta(f-$

$f')\mathcal{F}(f''-f')$ only in the $N \to \infty$ limit.

Nevertheless even for fixed sample size it is still possible to define a basis of functions which lead to coefficients orthogonal in the mean-square sense.

Given a process in the continuum, even non-stationary, with a correlation function $R(t_1, t_2)$, it can be proved that the solutions of the integral equation

$$\int_0^T R(t_1, t_2)\phi(t_2)dt_2 = \lambda\phi(t_1) \tag{22}$$

form a infinite basis $\{\phi_n(t)\}$, on the interval $[0, T]$, with eigenvalues λ_n. A realization of the process can therefore be expanded as

$$x(t) = \sum_{n=0}^{\infty} c_n\phi(t), \text{ with coefficients } c_n = \frac{1}{T}\int_0^T x(t)\phi^*(t)dt \tag{23}$$

which are orthogonal in the MS sense

$$E\left[c_n c_m^*\right] = \lambda\delta_{n,m} \ . \tag{24}$$

In the discrete case we have correspondingly to solve the eigenvalue problem

$$\sum_{j=1}^{N} R_{i,j}\phi_j^n = \lambda_n\phi_i^n \tag{25}$$

which defines a basis of N vectors $\vec{\phi}$ over which any realization $\vec{x}$ of the stochastic process can be expanded as $x_t = \sum_{i=1}^{N} c_i\phi_t$. Besides the statistical independence of the coefficients c_i, it is important to notice that the size of the eigenvector λ_n provides a measure of its relevance in the decomposition (see Therrien [15]). If we order the basis elements to have $\lambda_i > \lambda_j \forall i < j$, and we decompose over a reduced basis

$$\mathbf{x} = \sum_{i=1}^{M} c_i\phi_i + \sum_{i=M+1}^{N} c_i\phi_i = \hat{\mathbf{x}} + \varepsilon \tag{26}$$

the error has the property that in average $E\left[\varepsilon^2\right] = \sum_{i=M+1}^{N} \lambda_i$. This suggests that one can choose a basis with optimal representation properties, to achieve a compression of the information (in the MS sense), representing the noise content with as little a number of coefficients as possible.

It is important to notice [15] that for large N and/or for a periodic random process the DKL basis converges to the DFT basis: therefore we shall always take in mind that the interest for using the DKLT always stems from the will to monitor the behavior of the system over *short* time scales.

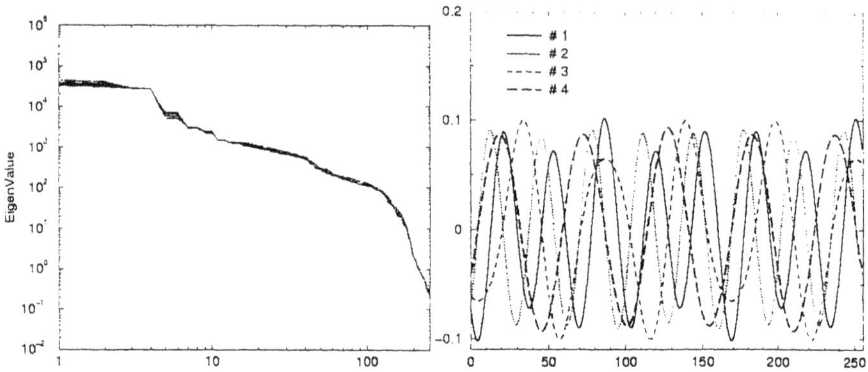

Figure 8. Left: eigenvalues relative to the basis for LIGO 40m noise; right, the first 4 basis elements.

Examples based on LIGO 40m data

The amount of compression depends actually on how "colored" is the noise: it is intuitive that no compression is possible when the noise is gaussian and white. In that case the eigenvalues would be all of the same size, and the eigenvectors could be chosen as the familiar sine and cosine functions.

On the other hand we display in Fig. (8), the eigenvalues obtained solving Eq. 22 for a 256×256 correlation matrix. The correlation matrix R was in turn estimated from a block of 32768 data, taken from a locked run of the LIGO 40m interferometer.

Several such estimates where taken, and are displayed with different colors, to show the stability of the eigenvalues in time. The basis functions are not greatly different from oscillatory functions, as it is shown in the same figure for the first 4 basis elements. It is clear that most of the RMS can be recovered by just using the first few eigenvalues (say, the first 32). We can compare in Fig. (9) the spectra of the original and "subtracted" data, subtracting $4, 8, 12, 16, 32$ elements of the Karhunen-Loève basis. The subtraction procedure actually selects definite spectral features, chosen in the order which contributes most to the RMS, while leaving essentially unaltered the rest of the spectrum.

2.3 Monitoring the coefficients

After having selected a basis, one can monitor the coefficients and understand from their behavior what is going on to the instrument, at least in

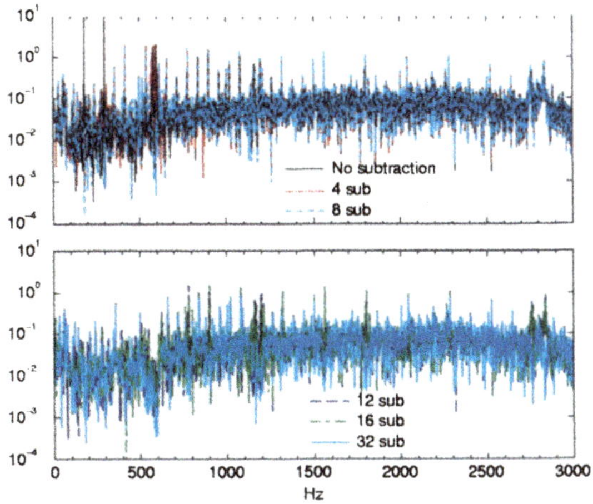

Figure 9. Comparison of spectra with different subtractions of Karhunen-Loève components.

correspondence to the most important (from the RMS point of view) spectral components. Given the basis consecutive chunks of data are projected on it and the coefficients give rise to new time-series. Pairs of coefficients with similar eigenvalues should be analyzed together, as they correspond roughly to the two quadratures of a certain spectral feature.

We used again the LIGO 40m data, divided in chunks of 512 elements (roughly 50 ms of data each), and we have decomposed the data over the first 128 elements of the basis, obtaining new time-series $c_n(t)$ which should be independent and gaussianly distributed as in Eq. 24. We display in Fig. (10) as an example an analysis of the behavior of the coefficients c_{32}, c_{33}; in the upper part we compare the histogram of the values with the gaussian distributions expected on the basis of the eigenvalues, and fitted from the data. In the lower part we display on the left side the scatter plot, which does not show any correlation. The coefficients relative to the same eigenvalue correspond to two quadratures of a KL basis element: we define the phase

$$\phi_n(t) = \arctan \frac{c_n(t)}{c_{n-1}(t)} ; \tag{27}$$

however the length of the samples is not in a simple relation with the period of the KL element, hence it is not convenient to plot $\phi(t)$ because its value

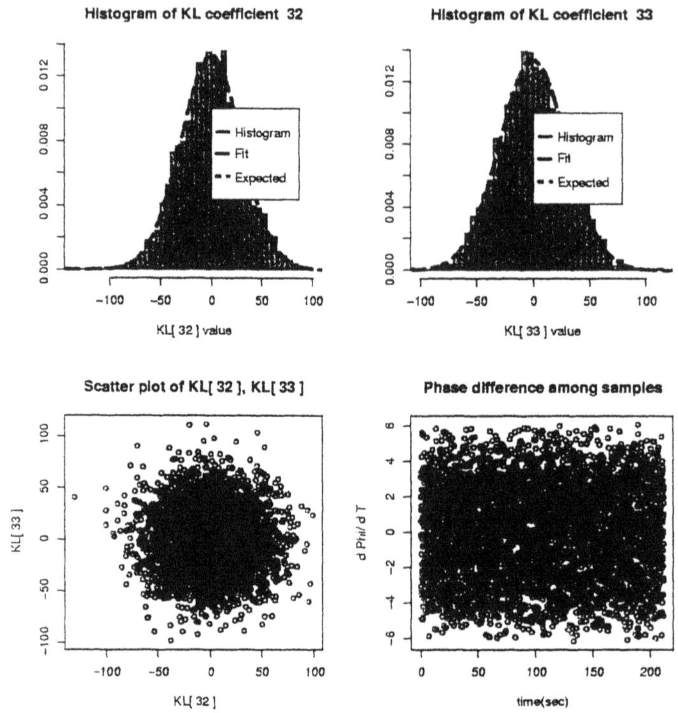

Figure 10. KL decomposition over basis elements 32, 33.

acquires a constant drift from sample to sample. In its stead we plot

$$\frac{d\phi}{dt} \equiv \frac{\arctan \frac{c_n}{c_{n-1}}(t) - \arctan \frac{c_n}{c_{n-1}}(t-1)}{dt} ; \tag{28}$$

where this shift cancels.

The coefficients obey the expected distribution, and the phase between the components is random. A completely different situation is displayed in Fig. (11), where we have decomposed over the first 2 elements of the basis. Here the distribution is far from gaussian, and there is almost perfect correlation between the coefficients, behaving as a vector of fixed amplitude, with a phase rotating from a sample to the next one due to the incommensurability of the two periods: in fact the plot of $d\phi/dt$ displays the long range correlation of the phase, giving a constant value modulo 2π. This means that the 1, 2

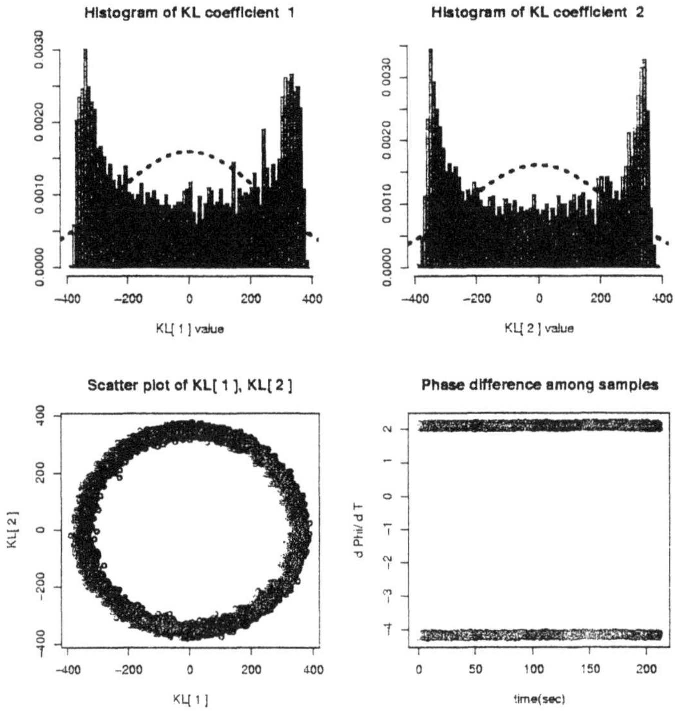

Figure 11. KL decomposition over basis elements 1, 2.

elements of the basis identify clearly a *discrete* component of the spectrum, in other words a deterministic signal, which is in this specific example the 3rd harmonic of the 60 Hz line.

Having understood the statistics of the KL coefficients, a possible noise monitor could track in time their value and detect changes in their distributions: examples could be made on the basis of the LIGO data, but it would bring us too far.

3 Coalescing binaries

Coalescing binaries are considered very promising sources, because the signal waveform can be predicted with a good accuracy, and a broad band detector can detect its frequency evolution over time periods ranging from seconds to

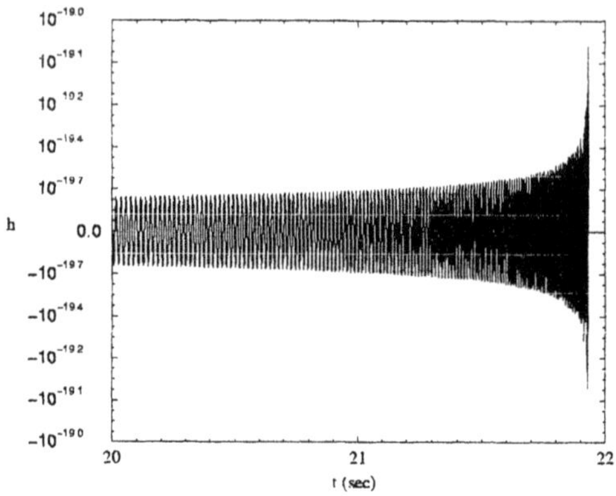

Figure 12. A "typical" chirping waveform, in the last seconds before the coalescence.

tens of minutes.

For what concerns signal detection in gaussian stationary noise, the data analysis algorithm is based on the standard matched filter, plus some more ingredients which we shall detail in the following.

3.1 The signal waveform

We know that the interferometer combines the two polarizations

$$h(t) = h_\times(t)F_\times(\theta(t),\phi(t)) + h_+(t)F_+(\theta(t),\phi(t)) \,. \tag{29}$$

The time variation of the response $F_{\times +}$ of the interferometer in general introduces an extra modulation, apart from the one due to the waveform $h_{\times,+}(t)$, but the observation time for interferometers on Earth is relatively short (compared with the variation of $F_{\times +}$ during the day) and we shall forget this effect.

Therefore for the purpose of data analysis the signal has the form

$$h(t) = A(\nu(t))^{2/3}\cos\left(\phi(t) + \phi_0\right) \tag{30}$$

where the A depends on the distance of the source and the relative source-detector orientation, but is otherwise constant in time, while we have made explicit the initial phase shift ϕ_0. We plot in Fig. (12) the last seconds of a coalescence, with the characteristic increase of amplitude and instantaneous frequency: $\nu(t)$ in the amplitude can be left to the Newtonian approximation

$$\nu(t) = \nu_0 \left(1 - t/\tau\right)^{8/3} \tag{31}$$

where ν_0 is the frequency at which we start observing (and start our clock), while τ is the time remaining before the coalescence, a function of the binary pair parameters, that is the masses and the spins.

The phase $\phi(t)$ has in general to be expressed as a series of contributions in a post-newtonian expansion and τ itself is a series $\tau = \tau_0 + \tau_1 + \ldots$ with $\tau_{0,1,\ldots}$ calculable functions of the binary's parameters.

The Fourier transform of this signal[c] can be well approximated by a stationary phase approximation: one has

$$\tilde{h}(f) = \int dt\, h(t) \cos\phi(t) e^{-2\pi i f t} \cong \frac{1}{2} h(\hat{t}) e^{i(-2\pi f \hat{t} + \phi(\hat{t}))} \sqrt{\frac{2\pi}{i\ddot{\phi}(\hat{t})}} \tag{33}$$

where $\hat{t}$ is related to f by

$$2\pi f = \dot{\phi}(\hat{t}) \quad \Rightarrow \quad \left|\tilde{h}(f)\right|^2 = \frac{h(\hat{t})^2}{4\dot{\nu}(\hat{t})}. \tag{34}$$

Inverting the relation in Eq. (34) allows to express the phase in the Fourier transform $\tilde{h}(f) \equiv \sqrt{\left|\tilde{h}(f)\right|^2} e^{i\Phi(f)}$: an explicit computation would give for instance the result in the PN1 approximation:

$$\begin{aligned} \tilde{h}(f) &= N\, f^{-7/6} e^{i\Phi(f)} \\ \Phi(f) &= \Phi_0 + 2\pi f t_a + \\ &+ 2\pi\nu_0\tau_0 \left[1 - \frac{f}{\nu_0} + \frac{3}{5}\left(1 - \left(\frac{f}{\nu_0}\right)^{-5/3}\right)\right] \\ &+ 2\pi\nu_0\tau_1 \left[1 - \frac{f}{\nu_0} + 1 - \left(\frac{f}{\nu_0}\right)^{-1}\right] \end{aligned} \tag{35}$$

where t_a is the time at which the signal has a instantaneous frequency ν_0, and N is a scale relevant for the SNR, but not for the search algorithm. At this order there are 4 parameters: the initial phase $\Phi_0 \in [0, 2\pi]$, the two PN

[c]at least in the so-called in-spiral phase, that is as long as the PN approximation remains accurate, and the rate of variation of the frequency over a period is much smaller that the frequency itself

$$\dot{\nu}(t)/\nu(t) \ll \nu(t) \tag{32}$$

parameters τ_0, τ_1, the arrival time t_a : detecting a signal in a specified train of data will mean. We have then to define the SNR for coalescing binaries.

3.2 Wiener filtering for coalescing binaries

We know that once the signal h_θ is known, in dependence of some parameters θ, the algorithm for assessing if it is present in a data series s is to compute the Wiener inner product $\langle h_\theta | s \rangle$, defined as follows:

$$\langle h|g \rangle \equiv \int_{-\infty}^{+\infty} \frac{\tilde{h}(-f)\,\tilde{g}(f)}{S_n(f)}\, df \tag{36}$$

and S_n is the two-sided noise spectrum . In absence of signal the result is a gaussian zero-mean random variable, with a variance

$$E\left[\langle h_\theta|n\rangle \langle n|h_\theta\rangle\right] = \langle h_\theta|h_\theta\rangle \quad \Rightarrow \quad SNR = \langle h_\theta|s\rangle \,/ \sqrt{\langle h_\theta|h_\theta\rangle}\,; \tag{37}$$

with this definition the SNR is proportional to the amplitude of the signal possibly present, while the overall normalization of the *template* h_θ cancels.

Now if we look at the parameters which need to be searched for, the situation may seem desperate: we have to search over all possible arrival parameters, all possible phases, all possible $\tau_{0,1,\dots}$. But actually some of these parameters come almost for free!

Effect of the initial phase

First of all, we should apparently use as template

$$h_\theta = h_0 \cos\Phi_0 + h_{\pi/2} \sin\Phi_0 \tag{38}$$

and vary the values of Φ_0 to cover the possible different phases: but actually we are interested in the value of the SNR, not of Φ_0! And we know that at Φ_0 the SNR will be maximum. Then we need only to maximize the SNR over all possible values of Φ_0 to get the maximum: we have

$$\begin{aligned} SNR &= \frac{\langle h_0|s\rangle \cos\Phi_0 + \langle h_{\pi/2}|s\rangle \sin\Phi_0}{\sqrt{\langle h_0|h_0\rangle \cos^2\Phi_0 + \langle h_{\pi/2}|h_{\pi/2}\rangle \sin^2\Phi_0}} \\ &= \frac{\langle h_0|s\rangle}{\sqrt{\langle h_0|h_0\rangle}} \cos\Phi_0 + \frac{\langle h_{\pi/2}|s\rangle}{\sqrt{\langle h_{\pi/2}|h_{\pi/2}\rangle}} \sin\Phi_0 \end{aligned} \tag{39}$$

and one has immediately that

$$\max_{\Phi_0} SNR = \sqrt{\langle q_0|s\rangle^2 + \langle q_{\pi/2}|s\rangle^2} \tag{40}$$

where we have also introduced the normalized template

$$q_\theta \equiv h_\theta / \sqrt{\langle h_\theta | h_\theta \rangle}\,; \tag{41}$$

this means that the SNR, in absence of an a priori knowledge of the phase, is given by the sum in quadrature of the response of two filters[d].

Arrival time

We have eliminated Φ_0 from the parameters of the signal, but the GW signal can start at any time, which means that we have to keep t_a as a variable, and consider all the possible normalized templates q_θ which start at different times t_a. For each of them (or better, for each pair) we have compute the SNR, and compare the values obtained.

Let us call $C(t_a)$ the result, and forget the two quadratures, as inessential for the reasoning that follows: one should compute

$$C(t_a) = \int_{-\infty}^{+\infty} \frac{\tilde{q}(t_a, -f)\tilde{s}(f)}{S_n(f)}\, df\,; \tag{43}$$

but templates differing for the arrival time differ in the Fourier space by a phase shift, that is

$$C(t_a) = \int_{-\infty}^{+\infty} \left[\frac{\tilde{q}(0, -f)\tilde{s}(f)}{S_n(f)}\right] e^{+2\pi i f t_a}\, df\,; \tag{44}$$

which is noting but the inverse Fourier transform, computed at point t_a, of the function in square parentheses.

This is of great computational importance: suppose one is given a time series with N samples: then computing the integral in Eq. (43) would cost $O(N)$ operations, and the integral should be repeated for all the possible arrival times, that is $O(N)$ points, with a total cost $O(N^2)$.

But actually this is not necessary: one has only to evaluate first just the *integrand* of Eq. 43 over the discrete $O(N)$ frequency values, and then perform $O(N \ln N)$ operation to compute the inverse Fourier transform of Eq. 44: the operation count remains $O(N \ln N)$, which is an enormous advantage.

[d]Notice that with the given normalizations, the two filters output normal gaussian variables with unit variance, and therefore the resulting SNR in Eq. 40 is a Rayleigh distributed variable: we recall that the Rayleigh distribution is

$$P_{Rayleigh}(\rho) \equiv \rho e^{-\rho^2/2} \tag{42}$$

where ρ is defined in $[0, \infty]$.

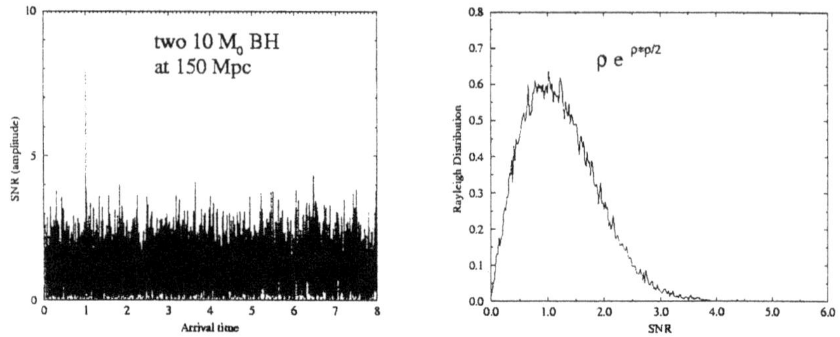

Figure 13. Matched filter output, signal arriving at time $t_a = 1$ s, and output distribution.

Summarizing, at fixed physical parameters each Wiener filter works transforming the input sequence $h(t)$ into a new output sequence $C_\theta(t)$: if there is a signal, in correspondence to its arrival time (that is, the time at which the template and the signal have best overlap) one sees a peak. We display in Fig. (13) an example of matched filter output, computed for two $10M_\odot$ black holes in coalescence: one sees a peak only in correspondence to the arrival time. Having summed in quadrature the two filters for the 0 and $\pi/2$ phases, the output is Rayleigh distributed, as shown in the second plot. We have now to consider different values of the physical parameters: but how different?

3.3 Mis-matched filters

Suppose we receive a signal with a certain set of parameters θ, but we use a template with slightly different θ'. Hence our signal is of the form

$$h(t) = A\, q_{\theta,t_a+\Delta t_a}(t) + n(t) \tag{45}$$

and therefore our filter outputs the correlation

$$C(\theta\,\theta', \Delta t_a) = A\,\langle q_{\theta,t_a} |\, q_{\theta',t_a+\Delta t_a}\rangle + \langle q_{\theta,t_a} | n\rangle\ . \tag{46}$$

Clearly the noise term is in average zero, so it is the first term which matters: now for the coalescing binary signal, a variation in the parameters can be partly compensated by a shift of the arrival times, in other words if we put in evidence the difference of arrival times of the two templates, and maximize over this parameter, we can obtain a larger correlation. Putting in evidence the arrival times, we can define the *ambiguity function*

$$C(\theta,\theta') = \max_{\Delta t_a} E\left[C(\theta,\theta',\Delta t_a)\right] = \max_{\Delta t_a} \langle q_{\theta,t_a} | q_{\theta',t_a+\Delta t_a}\rangle \,. \tag{47}$$

So the scalar product $\langle q_\theta | q_{\theta'}\rangle$, maximized over arrival time shifts, corresponds to a distance between the templates which tells us which fraction of the maximal SNR A the template can recover a signal which is not perfectly matched.

Note that if we take into account the two quadratures of the signal, we have only to slightly modify the definition, setting

$$C(\theta,\theta') = \max_{\Delta t_a} \sqrt{\langle q_{\theta,0,t_a} | q_{\theta',0,t_a+\Delta t_a}\rangle^2 + \langle q_{\theta,\pi/2,t_a} | q_{\theta',\pi/2,t_a+\Delta t_a}\rangle^2} \tag{48}$$

Now we can answer the question: how many templates are needed? It depends on the SNR fraction we are willing to recover, and the size of the physical parameter space.

3.4 Newtonian approximation example

What we have said is particularly clear in the Newtonian approximation: fixing the attention just on one quadrature, we have that

$$C(\theta_1,\theta_2) = \int_{-\infty}^{\infty} d\nu \frac{\tilde{q}(\nu,\theta_1)\tilde{q}^*(\nu,\theta_2)}{S_n(\nu)} = 2\int_0^{\infty} d\nu \frac{|\tilde{q}(\nu,\theta_1)\tilde{q}^*(\nu,\theta_2)|^2}{S_n(\nu)} \,. \tag{49}$$

which gives immediately

$$C(\Delta\theta) = \left[\int_{\nu_0}^{\infty} d\nu \frac{\nu^{-7/3}}{S_n(\nu)}\right] \int_{\nu_0}^{\infty} d\nu \frac{\nu^{-7/3}}{S_n(\nu)} \times \tag{50}$$

$$\times \cos\left\{2\pi\nu_0\Delta\tau\left[1-\left(\frac{\nu}{\nu_0}\right)+\frac{3}{5}\left[1-\left(\frac{\nu}{\nu_0}\right)^{-5/3}\right]\right]+\Delta\Phi+2\pi\nu\Delta t\right\}$$

This expression, evaluated for $\Delta\Phi = 0, \pi/2$, gives the two quadratures and remains to be maximized over Δt. Note also that the final result will in any case depend only on the difference $\Delta\tau$ of the times to coalescence, a fact which allows us to define a one parameter ambiguity function $C(\Delta\tau)$.

We show in Fig. (14) a plot of the ambiguity function, for different values of the lower frequency cutoff ν_0. Suppose we choose as lower cutoff 40 Hz, then we can set the spacing between templates at (say) 20ms with a SNR loss of the order of 15%. Of course, how good or bad is this result depends on the physical space: one has indeed that

$$\tau = 3\left[\frac{\mathcal{M}}{M_\odot}\right]^{-5/3}\left[\frac{\nu_0}{100\text{Hz}}\right]^{-8/3} \tag{51}$$

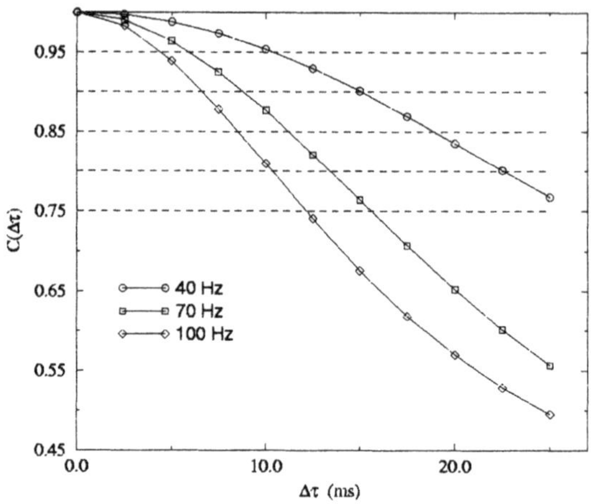

Figure 14. Decay of the correlation function

which means that allowing $\mathcal{M}$ masses as low as the Chandrasekar limit $\mathcal{M} = 0.25M_{\odot}$, one has $\tau_{max} \simeq 350$s, and therefore in VIRGO about 17×10^3 independent filters are needed.

3.5 First post-newtonian approximation

While in the Newtonian approximation the parameter space is one-dimensional, at the first post-newtonian it becomes two-dimensional: also at higher orders, neglecting spin effects, it would remain two dimensional, therefore this case is of general importance.

However the procedure is completely parallel to that of the Newtonian approximation: define a ambiguity function which depends on two parameters $C(\Delta\tau_0, \Delta\tau_1)$ where we are again blessed by the fact that it depends only on the difference of the parameter values. As before, the detailed form of the ambiguity function depends on the noise spectrum, but we can display in Fig. (15) an example for a VIRGO spectrum, fixing the lower frequency limit at 40Hz A complex structure, which at low SNR recovery becomes extended and branched. At high SNR recovery, on the contrary, the peak is quite similar to an inverted paraboloid, and therefore the levels become ellipses, which can be used to "tile" the parameter space.

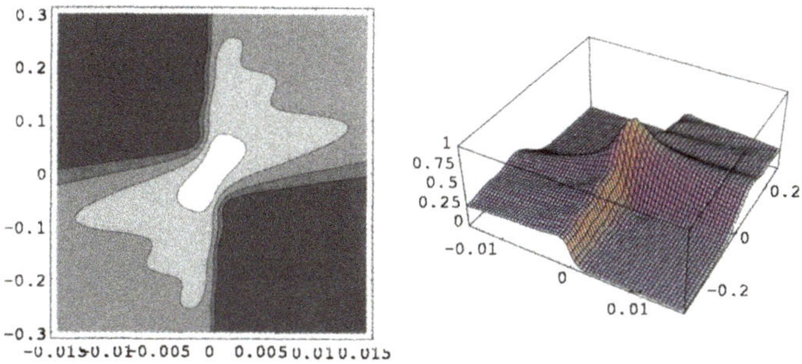

Figure 15. Two dimensional plot of the ambiguity function $C(\Delta\tau_0, \Delta\tau_1)$: at left a contour plot, at right, a 3D plot evidencing the "peak" structure. The $\Delta\tau_{0,1}$ axes units are seconds.

3.6 The parameter space to explore

The parameter space to explore is in our case the translation in the τ_0, τ_1 plane of the physical limits on the masses $m_{1,2}$ of the binary components, taking into account the lower frequency cutoff. At PN1 order it is a matter of a tedious but direct exercise of algebra, choosing a range of values for the chirp mass $\mathcal{M}$. As the chirp mass is an increasing function of both $m_{1,2}$, it follows that at the extrema of the range one may assume that the two masses are degenerate: this assumption leads to the identities

$$m_1 \geq 2^{1/5}\mathcal{M}, \;\; m_2 \leq 2^{1/5}\mathcal{M}, \;\; \mu \leq 2^{4/5}\mathcal{M}, \;\; \eta \equiv \mu/(m_1 + m_2) \leq 1/4 \,. \quad (52)$$

Armed with these identities the interested reader can easily compute by her/himself the kinematical limits: we plot in Fig. (16) an example obtained setting the frequency cutoff $\nu_0 = 40$Hz. Tiling the parameter space with the recovery ellipses is the procedure which allows to count how many "independent" filters are needed to recover the signal.

4 Stochastic background

At the other extremum of the possible signals is the stochastic background. We expect that, similarly to the electromagnetic background first discovered by Penzias and Wilson, and recently beautifully studied by the COBE experiment, there shall exist a background of gravitational waves. This can be due to sources of cosmological origin, or to the un-coherent signal of many sources at smaller distance: actually the nature of the stochastic radiation is

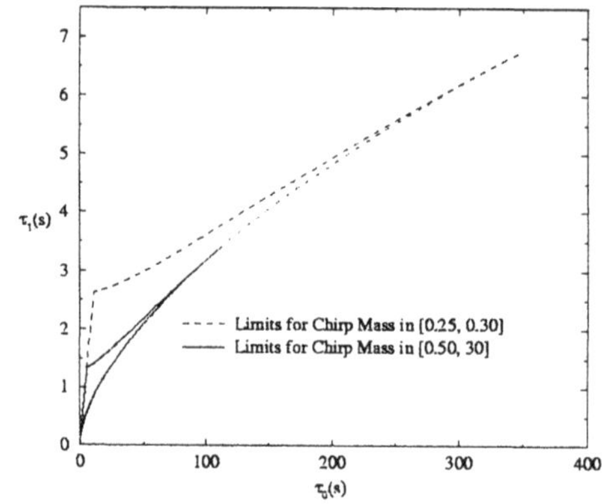

Figure 16. Kinematical range in the $\tau_{0,1}$ plane, obtained setting $\nu_0 = 40\text{Hz}$ and for two ranges of the chirp mass $\mathcal{M}$.

important in determining the correct detection strategy, because it depends both on the *spatial* and *frequency* distribution of the radiation.

In this treatment we shall closely follow the paper by Allen and Romano [6].

4.1 The stochastic signal

Suppose we are given two detectors i, j, at different locations on the Earth surface (like the LIGO pair, or VIRGO and GEO, or any combination of them). Forgetting the coherent signals, that is those originating for instance by particular sources, we can expect that the output $s_i(t)$ of each detector will be a combination of an instrumental noise $n_i(t)$ plus hopefully an un-coherent background of "gravitational" noise, that is

$$s_i(t) = n_i(t) + h_i(t) \tag{53}$$

where $h_i(t)$ will be the response of the interferometer to an un-coherent superposition of plane waves coming from every direction in the sky.

If it is so, the signal $h_i(t)$ at detector i can be written as follows

$$h_i(t) = \sum_A \int d\Omega\, F_i^A(\hat{\Omega}) \int_{-\infty}^{+\infty} df\, e^{i\,2\pi f\left(t - \frac{\vec{x}_i \cdot \hat{\Omega}}{c}\right)} \tilde{h}_A(f, \hat{\Omega}) \tag{54}$$

that is, as a superposition of plane waves, with the two possible polarizations $A = +, \times$, with random (complex) amplitude $\tilde{h}_A$. This signal is of stochastic nature in the sense that we are unable of predict its time evolution: we can only make assumptions on the expectation values of some moments of its distribution, for instance on the spectral properties.

The antenna pattern F_i^A introduces a slight complication, and is not essential for the arguments that follows, so we shall forget the spin 2 nature of the radiation, and go on as the GW radiation were of scalar nature! With this bold simplifications the signal becomes simply

$$h_i(t) = \int d\Omega \int_{-\infty}^{+\infty} df\, e^{i\,2\pi f\left(t - \frac{\vec{x}_i(t)\cdot\hat{\Omega}}{c}\right)} \tilde{h}(f, \hat{\Omega})\,; \tag{55}$$

we shall re-introduce the antenna pattern only at the end.

Let us make some assumptions on this stochastic signal, which further simplify the work.

Stationarity Aany correlation between variables $h_i(t)$, $h_j(t')$ depends only on the difference $t - t'$.

Gaussianity The "signal" $h_i(t)$ can be entirely described in terms of its spectral properties, that is just the second order: given the nature of the signal as a superposition of many random signals, also this is reasonable.

Homogeneity That is direction independence: this assumption allows to obtain "simple" results but it is not justified if the signal is not of cosmological origin: for instance a distribution of sources in our galaxy would be inevitably not homogeneous[e]

If these assumptions are correct, we can assume for the expectation values of the $\tilde{h}$ correlations:

$$E\left[\tilde{h}(f, \hat{\Omega})\tilde{h}^*(f', \hat{\Omega}')\right] = \frac{1}{2\pi} H(f)\, \delta(f - f')\delta(\hat{\Omega} - \hat{\Omega}') \tag{56}$$

where $H(f)$ embodies the spectral distribution of the radiation background, and the normalization will come handy later.

4.2 Correlation of two detectors

Suppose we decide to detect the background by correlating the output of two detectors: we shall start building the cross-spectrum, that is equivalently the

[e]Note that due to the explicit $x_i(t)$ motion of the detector, even in the scalar case an inhomogeneity would make effectively non-stationary the correlations!

correlation

$$C_{ij}(t,t') \equiv E\left[s_i(t)s_j(t')\right] \tag{57}$$

if we assume that the noises n_i of the two detectors are uncorrelated, we can expect that to this correlation contribute only the GW signal h_i and as we know the distribution of $\tilde{h}(f,\hat{\Omega})$, we can immediately compute it

$$E\left[h_i(t)h_j(t')\right] = \frac{1}{2\pi}\int d\Omega \int_{-\infty}^{+\infty} df\, e^{i2\pi f(t-t')} e^{-i2\pi f\frac{(\vec{x}_i(t)-\vec{x}_j(t'))\cdot\hat{\Omega}}{c}} H(f)\,; \tag{58}$$

now let us compute the integral in $d\Omega = d\phi\, d\cos\theta$, orienting the axis $\hat{z}$ as the difference $\vec{x}_i(t) - \vec{x}_j(t')$: we obtain immediately

$$C_{ij}(t,t') = \int_{-\infty}^{+\infty} df\, e^{i2\pi f(t-t')} \frac{\sin\left(2\pi f \Delta x_{ij}(t,t')/c\right)}{2\pi f \Delta x_{ij}(t,t')/c} H(f) \tag{59}$$

where $\Delta x_{ij}(t,t') = |\vec{x}_i(t) - \vec{x}_j(t')|$.

For coincident points the factor in square brackets goes to unity

$$C_{ii}(t-t') = \int_{-\infty}^{+\infty} df\, e^{i2\pi f(t-t')} H(f) \tag{60}$$

hence $H(f)$ is nothing but the spectrum of the GW noise at one detector.

For not coincident points, due to the two different times in Δx, in general we do not have that C_{ij} depends just on the difference $t - t'$; indeed it does not matter the distance between the two detectors, but the distance between the points where the two detectors are at times t and t', which changes as $t - t'$ gets large, due to the motion of the Earh.

However here it is possible to make a further simplification, namely assume that the spectrum $H(f)$ is broad-band, without dominating spectral features. If it is so, then the correlation $C_{ij}(t,t')$ decays rapidly as $|t-t'|$ grows, because of the oscillating factor $e^{i2\pi f(t-t')}$, and therefore we can limit ourselves to time intervals where the approximation $\Delta x_{ij}(t,t') = r_{ij}$ constant holds. In this case we can write the correlation as

$$C_{ij}(t-t') = \int_{-\infty}^{+\infty} df\, e^{i2\pi f(t-t')} \gamma(r_{ij}, f)\, H(f) \qquad \gamma(r,f) \equiv \text{sinc}\left[2\pi f r\right] \tag{61}$$

where the overlap function γ goes to 1 as $r \to 0$.[f] We can plot in Fig. (17) the simplified overlap function, which can be used in several ways. For instance,

[f] The overlap function depends only on the product $r\, f$ because of the simplification made, namely of scalar waves: in the real case, the γ depends also on the relative orientation of the detectors.

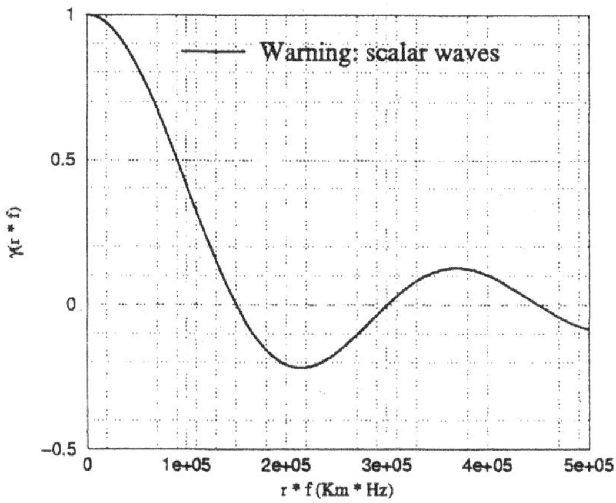

Figure 17. The overlap function for scalar waves, as a function of the product $r \times f$, where r is measured in Km and f in Hz.

considering the two LIGO detectors, separated by $r = 3001$Km, one finds the first zero at 50Hz. We remind the reader also that this is for *scalar* waves: for real GW waves the relative orientation of the detectors is also important, and can reduce below 1 the correlation at zero frequency, and also give anti-correlation ($\gamma(0) < 0$) when (for instance, as in the case of the two LIGOs) one of the arm is oriented oppositely.

4.3 The detection algorithm

One could at this point say that the correlation just computed, or better its Fourier transform, can be considered as the measurement, and then we could proceed to compute its noise level. However we can maximize the SNR by choosing carefully the way we correlate the two outputs: define as observable a general linear combination of the correlation at times t, t':

$$x = \int_{-\infty}^{+\infty} dt \int_{-T/2}^{+T/2} dt'\, h_i(t) h_j(t')\, Q(t - t') \tag{62}$$

with the purpose to find an optimal Q maximizing the SNR. Note that we are letting one of the limits of the integration run in $[-T/2, T/2]$, while the other is left running in $[-\infty + \infty]$, because its range is actually determined by the correlation time of Q; this choice simplifies the computations later.

Let us therefore first compute the expectation value:

$$\mu = \int_{-\infty}^{+\infty} dt \int_{-T/2}^{+T/2} dt'\, E[h_i(t)h_j(t')]\, Q(t-t') \tag{63}$$

$$= \int_{-\infty}^{+\infty} dt \int_{-T/2}^{+T/2} dt' \int_{-\infty}^{+\infty} df\, e^{i2\pi f(t-t')} H(f)\gamma(rf)\, Q(t-t')\,;$$

the integral in t gives the Fourier transform $\tilde{Q}(f)$ of the sought for filter function, while the remaining integral in t' gives a T, factor: one obtains

$$\mu = T \int_{-\infty}^{+\infty} df\, H(f)\gamma(rf)\tilde{Q}(f)\,. \tag{64}$$

To compute the variance we consider the value of μ when there is no signal:

$$x = \int_{-\infty}^{+\infty} dt \int_{-T/2}^{+T/2} dt' n_i(t)n_j(t')\, Q(t-t') \tag{65}$$

which has zero mean, therefore the variance is given by $E[x^2]$:

$$\sigma^2 = \int_{-\infty}^{+\infty} dt_i \int_{-\infty}^{+\infty} dt_j \int_{-T/2}^{+T/2} dt'_i \int_{-T/2}^{+T/2} dt'_j E[n_i(t_i)n_i(dt'_i)]\, E[n_j(t_j)n_j(t'_j)]\, Q(t_i - t'_j)\, Q(t'_i - t_j)$$

$$= \int\int\int\int \int_{-\infty}^{+\infty} df \int_{-\infty}^{+\infty} df' S_i(f)S_j(f') e^{i2\pi f(t_i - t'_i)} e^{i2\pi f'(t_j - t'_j)}\, Q(t_i - t'_j)\, Q(t'_i - t_j)$$

shifting $t_i \to t_i + t'_j$, $t_j \to t_j + t'_i$, and performing the integral in t_i, t_j

$$\sigma^2 = \int_{-T/2}^{+T/2} dt'_i \int_{-T/2}^{+T/2} dt'_j \int_{-\infty}^{+\infty} df \int_{-\infty}^{+\infty} df' e^{i2\pi f(t'_j - t'_i)} e^{i2\pi f'(t'_i - t'_j)}\, S_i(f)S_j(f')\, \tilde{Q}(f)\tilde{Q}(-f')\,. \tag{66}$$

For large T, we have that approximately

$$\int_{-T/2}^{+T/2} dt'_i e^{i2\pi(f'-f)t'_i} \simeq \delta(f'-f)\,; \tag{67}$$

it follows

$$\int_{-T/2}^{+T/2} dt'_j \int df\, S_i(f)S_j(f)\, \left|\tilde{Q}(f)\right|^2 = T \int df\, S_i(f)S_j(f) \left|\tilde{Q}(f)\right|^2. \tag{68}$$

This gives an SNR

$$SNR = \sqrt{T} \frac{\int_{-\infty}^{+\infty} df\, H(f)\gamma(rf)\tilde{Q}(f)}{\sqrt{\int_{-\infty}^{+\infty} df\, S_i(f)S_j(f) \left|\tilde{Q}(f)\right|^2}} \tag{69}$$

which grows as $\sqrt{T}$. Now we ask which $\tilde{Q}$ function gives the maximal SNR, or equivalently SNR^2: the problem is to maximize the ratio

$$R = \frac{\left[\int A(f)\tilde{Q}(f)\,df\right]^2}{\int B(f)\left|\tilde{Q}(f)\right|^2 df} \tag{70}$$

equating to zero the variation $\delta R/\delta\tilde{Q}(f')$ with respect to $\tilde{Q}(f')$

$$\frac{2A(f')\left[\int A(f)\tilde{Q}(f)\,df\right]}{\int B(f)\left|\tilde{Q}(f)\right|^2 df} - \frac{2B(f')\tilde{Q}(-f')\left[\int A(f)\tilde{Q}(f)\,df\right]^2}{\left[\int B(f)\left|\tilde{Q}(f)\right|^2 df\right]^2} = 0 \tag{71}$$

thus obtaining

$$\tilde{Q}(f) = \frac{A(f)}{B(f)} \quad \Rightarrow \quad \tilde{Q}(f) = \frac{H(f)\gamma(rf)}{S_i(f)S_j(f)} \tag{72}$$

the filter weights more the frequencies at which the overlap function γ is large, while depressing those where the product of noises $S_i(f)S_j(f)$ is large. Moreover, it contains as a factor the spectrum $H(f)$ expected for the GW background, which means that we shall have to set up several filters for the different kinds of spectra we want to statistically test for.

For given $H(f)$ the SNR recovered by this filtering method is given by

$$SNR_{max} = \sqrt{T}\left[\int_{-\infty}^{+\infty} df\, \frac{H^2(f)\gamma^2(r_{ij}f)}{S_i(f)S_j(f)}\right]^{1/2} \tag{73}$$

which is the quantity to measure, if we have any hint on the spectrum $H(f)$.

4.4 Actual measurement

We should stress that the time T is the one used to estimate one sample: however in general we shall repeat the measurement n times over a total observation time $T_{tot} = nT$, and average all the results. In this way we shall get a sample mean $\hat{\mu}_n$ which shall be an estimate of the "mean".

Actually the measurement will therefore be done as follows:

1. preliminarily measure the noise spectra S_i, S_j;
2. define $\tilde{Q}(f)$ for each spectrum $H(f)$ we are interested in
3. for each time interval, FFT the data $s_i(t), s_j(t)$ of each detector.

4. build the estimate x_k of the optimal correlation[g]

$$x_k \equiv \int_{-\infty}^{+\infty} df\, \tilde{s}_i(f)\tilde{s}_j^*(f)\,\tilde{Q}(f) \tag{74}$$

5. average the values, building the sample mean and the sample variance

$$\hat{\mu}_n = \frac{1}{n}\sum_k x_k, \qquad \hat{\sigma}_n^2 = \frac{1}{n-1}\sum (x_k - \hat{\mu}_n)^2 \ ; \tag{75}$$

   note that the sample mean will have variance $\hat{\sigma}_n/\sqrt{n}$.

6. Use a statistical criterium to decide if the observed value $\hat{\mu}_n$ is consistent or not with the hypothesis that the average μ is not zero.

There is a point however that we have not discussed: given a total time T_{tot}, how we decide the length of the sub-periods T to be averaged in order to build the estimate $\hat{\mu}$? As long as T is sufficiently large, so that subsequent estimates x_k are statistically uncorrelated, the choice does not matter. In fact $\frac{\hat{\mu}_n}{\hat{\sigma}_n}$ is an estimate of the quantity in Eq. (73), proportional to $\sqrt{T}$, therefore

$$\sqrt{n}\frac{\hat{\mu}_n}{\hat{\sigma}_n} \propto \sqrt{T_{tot}} \tag{76}$$

and the proportionality factor depends only on spectral quantities, which do not vary (in average) with T.

4.5 Modifications when dealing with real GW

We have discussed a scalar case: in the real experiment however we shall have a tensor field like in Eq. (54), and some formulas will be changed: we refer to Allen and Romano [6] and references therein for a detailed treatment, while we just quote here the results.

First of all we have that the overlap function γ is different: one has

$$\gamma_{ij}(f) = \frac{5}{8\pi}\sum_{A=+,\times}\int d\Omega\, e^{i2\pi f\hat{\Omega}\cdot\Delta\vec{x}/c} F_i^A(\hat{\Omega})F_j^A(\hat{\Omega}) \tag{77}$$

where $F_i^A(\hat{\Omega})$ is the antenna pattern, for polarization A, of antenna i. This is the only modification needed to the detection formulas.

---

[g] this integral should be of course replaced by the discrete time, finite interval version, that is by a finite sum over frequencies, ranging from $-f_{Nyquist}$to $+f_{Nyquist} = 1/(2\Delta t)$ in steps of Δt, where Δt is the sampling interval.

Then the quantity $H(f)$ defined in Eq. (56) is not what the theoreticians use when characterizing the spectral background: they introduce

$$\Omega_{gw}(f) \equiv \frac{1}{\rho_{critical}} \frac{d\rho_{gw}}{d\ln f} \tag{78}$$

where ρ is the energy density of the GW waves, connected in a direct way to the $H(f)$, and $\rho_{critical} = 3c^2 H_0^2/(8\pi G)$ is the density needed to close the universe.

We can here just quote the result [6], namely one should pose in all the previous formulas, and particularly in Eqs. (73,72)

$$H(f) = \frac{3H_0^2}{16\pi^2} |f|^{-3} \Omega_{gw}(|f|) . \tag{79}$$

Apart from these modifications, all the reasoning goes unchanged.

5 Continuous signals

While there is ambiguity on the spectrum of the SGW, and also we cannot be certain of the shape of the coalescing binary signal, when it is close to the coalescence, there is on the contrary perfect understanding on how they will be the signals from the pulsars: continuous, weak, modulated by the Earth motion. Despite all this, they are probably the most difficult to detect, for lack of an adequate algorithm (see in Schutz [8]).

5.1 The pulsar signal

We refer to the paper by Zimmermann [7] for a derivation of the signal emitted by a spinning neutron star: we have

$$h(t) = F_+(\hat{n})h_+(t) + F_\times(\hat{n})h_\times(t) \tag{80}$$

where $\hat{n}$ is the direction of the source in the sky, and

$$\begin{aligned} h_+ &= h_0 \sin\alpha \left[\frac{1}{2}\cos\alpha\sin i\cos i\cos\omega\tau(t) - \sin\alpha\frac{1+\cos^2 i}{2}\cos 2\omega\tau(t)\right] \\ h_\times &= h_0 \sin\alpha \left[\frac{1}{2}\cos\alpha\sin i\sin\omega\tau(t) - 7\sin\alpha\cos i\sin 2\omega\tau(t)\right] \end{aligned} \tag{81}$$

where α is the angle between the rotation axis and the distortion axis, i is the angle between the line of sight of the instrument and the rotation axis, and ω is the rotation pulsation. We forget the pulsar spin-down or Doppler effects related to the orbital motion of the source, but we do not forget the receiver's motion, accounted for by the time dependence of $F_{+,\times}$, and by

$$\tau(t) = t + (\hat{n} \cdot \vec{x}(t))\,/c \tag{82}$$

with $\vec{x}(t)$ the receiver's position at time t.

Notice that each spinning neutron star emits at two distinct pulsations, both ω and 2ω, with an amplitude depending on its distortion and rotation parameters, and on the angle under which we see it.

Second, we have that the signal around each characteristic frequency is almost monochromatic: indeed the fastest time dependence is in the phase(s)

$$\phi(t) = \omega(t + \frac{\hat{n} \cdot \vec{x}(t)}{c}), \quad \Rightarrow \quad \frac{d\phi(t)}{dt} = \omega\left(1 + \frac{\hat{n} \cdot \vec{v}(t)}{c}\right); \tag{83}$$

as $\vec{v}/c \ll 1$ one would be tempted to drop the last term, but this is wrong because the accumulated phase is large. Think that the diameter of the Earth orbit is about 15 light minutes, which means that in half a year the variation of $\hat{n} \cdot \vec{x}/c$ can be of the order of 1000, completely altering the phase value.

5.2 *Relevance of the Doppler effect*

If the detector were fixed, we should just evaluate the spectrum of the data acquired: in presence of a sinusoidal signal

$$h(t) = A\cos(\omega t + \phi) \tag{84}$$

one would have a contribution to the spectrum of the form

$$S_{signal}(f) = \frac{T}{4}A^2\left[\frac{\sin(\omega + 2\pi f)T/2}{(\omega + 2\pi f)T/2} + \frac{\sin(\omega - 2\pi f)T/2}{(\omega - 2\pi f)T/2}\right] \tag{85}$$

which peaks at the frequencies $f = \pm\frac{\omega}{2\pi}$, giving for large T an effective SNR

$$SNR = \frac{\sqrt{T}A}{2\sqrt{S_{noise}(f)}} = \frac{\sqrt{T}A}{\sqrt{2}\tilde{h}_{noise}(f)}. \tag{86}$$

where we have introduced the sensitivity level $\tilde{h}_{noise}(f) \equiv \sqrt{S_{noise}(f) + S_{noise}(-f)}$. All this is well known, but it is completely changed when the Doppler effect comes in the game, because the signal does not contribute anymore to a single "bin" of the spectrum.

Suppose for instance that the detector motion is just a rotation motion:

$$\hat{n} \cdot \vec{x}/c = \beta\cos(\omega_m t) \tag{87}$$

then the originally monochromatic signal becomes

$$\cos(\omega(t+\beta\cos(\omega_m t))) = \sum_{k=-\infty}^{+\infty} (-1)^k J_{2k}(\omega\beta)\cos(\omega+2k\omega_m)t \qquad (88)$$
$$- \sum_{k=-\infty}^{+\infty} (-1)^k J_{2k+1}(\omega\beta)\sin(\omega+(2k+1)\omega_m)t$$

that is the signal of the carrier ω is split over an infinite number of sidebands $\omega \pm k\omega_m$. The amplitude of the different sidebands depends on the argument $\omega\beta$ of the Bessel functions: if $\omega\beta \ll \pi/2$, the most relevant sidebands are just the two closest ones, but if the product grows the energy spreads.

An useful identity is the following [9]:

$$\frac{\sum_k k^2 J_k^2(x)}{\sum_k J_k^2(x)} = \frac{x^2}{2} \qquad (89)$$

which gives the "variance" of the k index, that is the spread of it under the distribution $J_k^2(x)$. It is clear that for large x, large values of k are important.

To get an idea of the effect we estimate the modulation index β for the Earth revolution: one has $\beta \simeq 3 \times 10^3$ and a signal at $\omega = 2\pi\,100\text{Hz}$ would be spread over $O(10^6)$ sidebands, each smaller of a factor $1/\sqrt{\omega\beta} \simeq O(10^{-3})$.

More generally the motion of the detector entails both Earth rotation and revolution: overall the signal is spread over a large number of sidebands, a number which grows with the observation time T.

5.3 The matched filter and its problems

If the position of the source is known, actually this spread is not a problem and the signal can be "re-compacted": actually we compute the spectrum over not-equispaced time interval, taking into account that equal steps in $\tau(t)$ do not correspond to equal steps in t.

Indeed if we know $\tau(t)$, from the position of the source, then a general method exist, known in literature (see in Scargle [10]) as the Lomb periodogram, which we shall derive for exercise.

The data model is $x_i = A\cos\omega\tau_i + B\cos\omega\tau_i + \sigma\xi_i$, that is a combination of signal, with unknown amplitudes A, B, and noise with known variance σ: we consider the data sampled with interval Δt, that is $\tau_i = \tau(i\,\Delta t)$.

We restrict to gaussian noise for simplicity: then the probability of observing a certain sample $\mathbf{x}$, of length $N = T/\Delta t$ is given by

$$P(\mathbf{x}) \propto \frac{1}{\sigma^N} \exp\left[-\frac{1}{2\sigma^2}\sum_i (x_i - A\cos\omega\tau_i - B\cos\omega\tau_i)\right]$$
$$= \frac{1}{\sigma^N} e^{-\frac{N}{2\sigma^2}E[x^2]} e^{-\frac{1}{2\sigma^2}\left[A^2\sum_i \cos^2\omega\tau_i + B^2\sum_i \sin^2\omega\tau_i\right]}$$
$$\times\, e^{\frac{A}{\sigma^2}\sum_i s_i\cos\omega\tau_i + \frac{B}{\sigma^2}\sum_i s_i \cos\omega\tau_i} \times e^{-\frac{AB}{\sigma^2}\sum_i \cos\omega\tau_i \sim \omega\tau_i}$$
$$= \frac{1}{\sigma^N} e^{-\frac{N}{2\sigma^2}(\mathbf{x}\cdot\mathbf{x})} e^{-\frac{1}{2\sigma^2}\left[A^2(\mathbf{c}\cdot\mathbf{c}) + B^2(\mathbf{s}\cdot\mathbf{s})\right]}$$
$$\times\, e^{\frac{A^2}{\sigma^2}(\mathbf{x}\cdot\mathbf{c}) + \frac{B^2}{\sigma^2}(\mathbf{x}\cdot\mathbf{s}) - \frac{AB}{\sigma^2}(\mathbf{c}\cdot\mathbf{s})} \tag{90}$$

where we have introduced a short hand notation for the sums.

Then a standard trick (see in Press [11]) is to redefine the scale of the time in order to cancel the cross term $(\mathbf{c}\cdot\mathbf{s})$ [h] so that the expression simplifies to

$$P_N(\omega, A, B) = \frac{1}{\sigma^N} e^{-\frac{N}{2\sigma^2}(\mathbf{x}\cdot\mathbf{x})} e^{-\frac{A^2}{2\sigma^2}(\mathbf{c}\cdot\mathbf{c}) - \frac{B^2}{2\sigma^2}(\mathbf{s}\cdot\mathbf{s})} e^{\frac{A}{\sigma^2}(\mathbf{x}\cdot\mathbf{c}) + \frac{B}{2\sigma^2}(\mathbf{x}\cdot\mathbf{s})} \tag{94}$$

where we have used the Bayes theorem to re-interpret the probability of the data at fixed signal as a probability of signal parameters at fixed data.

Now if we integrate over the nuisance parameters A, B, namely if we are interested in locating what is the most probable frequency, for any value of the signal amplitude, then the expression further simplifies and we obtain

$$P_N(\omega) = \frac{1}{\sigma^{N-2}} \frac{1}{\sqrt{(\mathbf{c}\cdot\mathbf{c})(\mathbf{s}\cdot\mathbf{s})}} e^{-\frac{1}{2\sigma^2}\left[\frac{(\mathbf{x}\cdot\mathbf{c})^2}{(\mathbf{c}\cdot\mathbf{c})} + \frac{(\mathbf{x}\cdot\mathbf{s})^2}{(\mathbf{s}\cdot\mathbf{s})}\right]} \tag{95}$$

where we recall that $(\mathbf{x}\cdot\mathbf{c}) \equiv \sum_i x_i \cos\omega\tau_i$. Hence apart correction due to the

---

[h] indeed one has

$$(\mathbf{c}\cdot\mathbf{s}) \equiv \sum_i \cos\omega\tau_i \sin\omega\tau_i = \frac{1}{2}\sum_i \sin 2\omega\tau_i \tag{91}$$

and shifting $\tau_i = \tau_i - \theta$ one has

$$(\mathbf{c}\cdot\mathbf{s}) = \frac{1}{2}\sum_i \sin 2\omega\tau_i \cos 2\omega\theta - \cos 2\omega\tau_i \sin 2\omega\theta \tag{92}$$

so that setting

$$\tan(2\omega\theta) = \frac{\sum_i \sin 2\omega\tau_i}{\sum_i \cos 2\omega\tau_i} \tag{93}$$

one has cancellation.

front factor, the most probable frequency is the peak of the Lomb periodogram

$$L_N(\omega) = \frac{1}{2}\left[\frac{\left(\sum_i x_i \cos\omega(\tau_i - \theta)\right]^2}{\sum_i \cos^2\omega(\tau_i - \theta)} + \cos \to \sin\right] \tag{96}$$

The result is clear: given the corrected time $\tau(t)$, one can construct a modified periodogram which does not exhibit the many side-lobes of the standard one, and which will peak at the right frequency value: note also (see *Numerical Recipes*) that efficient FFT based algorithms for evaluating $L_N(\omega)$.

If we do not know the position of the source a problem similar to that of the ambiguity function studied for the coalescing binaries occurs.

Namely, if we choose the wrong direction $\hat{n}'$, we have an inefficiency which we can quantify exactly as we did before: only we forget the noise, or just assume that it is white. We also forget all the other effects, for instance the one (very important actually) of the modulation induced by the time dependence of the detector antenna pattern. With so drastic simplifications we have [12]

$$C(\hat{n}, \omega, \hat{n}', \omega') = \frac{1}{T}\sqrt{I_c^2 + I_s^2} \tag{97}$$

having introduced the superposition integral for the two signal components:

$$I_c = \int_0^T \cos\omega\tau(t)\cos\omega'\tau'(t)\,dt \tag{98}$$

and similar for I_s. For large T only the term with the difference is relevant

$$I_c = \frac{1}{2}\int_0^T \cos\left(\omega\tau(t) - \omega'\tau(t)\right)\,dt\,. \tag{99}$$

One should now write an explicit parameterization for $\tau(t)$ in terms of the source position, and start a numerical analysis to tile, as done before, the parameter space (frequency $\times$sky): we have no time to do that, so we give a further simplified example, forgetting the effect of the daily motion and just considering the revolution motion of the Earth. In this case we have, for signals at the same frequency, [12]

$$C(\hat{h}, \hat{n}') = \frac{1}{T}\left|\int_0^T \cos\left[\omega\kappa\sin\left(\omega_m t\right)\right]\right| = |J_0(\omega\kappa)| \tag{100}$$

where

$$\kappa = \frac{R}{c}\sqrt{\Delta_1^2 + \left(\Delta_2\cos\varepsilon + \Delta_3\sin\varepsilon\right)^2}$$

$$\Delta_1 \equiv \cos\beta\cos\alpha - \cos\beta'\cos\alpha'$$

$$\Delta_2 \equiv \sin\beta \sin\alpha - \sin\beta' \sin\alpha'$$
$$\Delta_3 \equiv \cos\alpha - \cos\alpha' \tag{101}$$

here $\alpha\,\beta$ are the two angular coordinates of each pulsar, while ε is the angle between the ecliptic and the equatorial plane, and finally R is the average radius of the Earth orbit.

For good efficiencies, the condition $C(\hat{n}, \hat{n}') > \eta$ where η is a predefined efficiency close to 1 translates into a limit o on the argument of the Bessel function, namely $\omega\kappa < l_\eta$, which in turn defines a solid angle in the sky (around the $\alpha\,\beta$ position)

$$d\Omega \simeq \frac{\pi\, l_\eta^2 c^2}{\omega^2 R^2} \frac{1}{|\sin\alpha \sin\beta \sin\varepsilon - \cos\alpha \cos\varepsilon|} \tag{102}$$

as an order of magnitude, one obtains from this equation that

$$d\Omega \simeq \frac{10^{-8}}{(\omega/\mathrm{Hz})^2} \tag{103}$$

which gives an idea of the problem: to search for pulsars at 100Hz one would be forced to cover the sky with 10^{13} angular patches.

We have no time to further elaborate on this: the interested reader can find much more for instance in the papers by Krolak [13] and references therein.

References

1. Barry C. Barish, *The Detection of Gravitational Waves with LIGO*, submitted to DPF'99.
2. Peter R. Saulson, *Fundamentals of interferometric gravitational wave detectors*, (World Scientific, Singapore, 1994).
3. W. Misner, K. S. Thorne, and J. A. Wheeler, *GRAVITATION*, Freeman and Company, New York, 1970.
4. L. Blanchet, B.R.Iyer, C.M.Will and A.G.Wiseman, *Class.Quant.Grav.* **13**, 575 (1996).
5. C.Cutler and E.E.Flanagan, *Phys. Rev.* D **49**, 2658 (1994).
6. B. Allen and J.Romano, *Phys. Rev.* D **59**, 102 (1999).
7. M. Zimmermann, *Phys. Rev.* D **21**, 891 (1980).
8. B. Schutz in *The detection of gravitational waves*, eds. D.Blair et al. (Cambridge University Press, 1991).
9. S.Gradshteyn and I.M.Ryzhik, *Tables of integrals, Series and Products*, Eq. 8.563.2, (Academic Press).
10. J.D.Scargle, *Astrophys. J.* **263**, 835 (1982).

11. W.H.Press and G.B.Rybicki, *Astrophys. J.* **338**, 277 (1989).
12. M.Beccaria, G.Cella, A.Ciampa, E.Cuoco, G.Curci, A.Viceré, Gravitational wave search: real time data analysis strategies on parallel computers, VIRGO note NTS96 24 (1996)
13. P.Jaranowski, A.Krolak, B.F.Schutz, *Phys. Rev.* D **58**, 63 (1998).
14. D.B.Percival and A.T.Walden, *Spectral analysis for physical applications*, Cambridge University Press, 1993
15. C.W.Therrien. *Discrete random signals and statistical signal processing*, Prentice-Hall, Englewood Cliffs, 1992.
16. G.H.Golub and C.F.Van˜Loan, *Matrix Computations, second edition.* The Johns Hopkins University Press, 1989.
17. A.Papoulis, *Probability, Random Variables, and Stochastic Processes*, McGraw-Hill, Electrical & Electronic Engineering Series, New Yortk, 1991.

R&D FOR INTERFEROMETRIC GW DETECTORS

ALAIN BRILLET
Observatoire de la Côte d'Azur,
06304 Nice Cedex 4, France
E-mail: brillet@obs-nice.fr

A short review is presented on the history and development of the idea of "gravitational wave", the present status of art and foreseen R&D for gravitational antennae, with particular emphasis on VIRGO detector.

1 Introduction

Gravitational waves (GW's) were already present in Einstein's theory, in 1917, but they did not attract much interest in the beginning. They were considered for a while as mathematical artefacts, "propagating at the speed of thought", according to Dirac. In the 1950's the British physicist Pirani tried to answer the question: what happens in my laboratory when a gravitational wave comes through? This was the beginning of a whole new research field, which has now been spreading in many directions, from pure theory to many domains of applied physics, that we will try to review. Since this book contains high quality technical papers about all the important topics, we will not repeat here the formulas and references that they already contain.

For the specialists in relativity and gravitation, Pirani's question was not so trivial: it resulted in an intense activity until the 1980's, when a reasonable consensus was reached as concerns the effect of gravitational waves on elastic classical bodies (resonant bars), and on free masses (interferometers). This understanding is now extending to their interaction with quantum objects, in particular with atom interferometers. Today, most of the theoretical work concerns the study of astrophysical GW sources (signal characteristics, and event rate, for black holes and other condensed objects).

The weakness of the interaction of GW's is such that it became useful to question the possibility that they could be detected without violating fundamental incertitude principles. Many aspects of the quantum theory of measurement were revisited, to end up with positive conclusions. Not only we can design detection instruments with low enough quantum limits, but also we found that the so-called "standard quantum limit" is usually not a hard limit: it can be beaten with appropriate measurement techniques. For instance, the quantum limits in interferometers are:

- The shot noise, and the radiation pressure noise which depend on photon statistics.
- Shot noise is usually dominant. It can be lowered by increasing the light power, and beaten using "squeezed states" of the light.
- In the case of a position measurement, the shot noise decreases with increasing photon number, while the radiation pressure noise increases. Thus, when the photons are in a coherent state, there is a minimal incertitude, called the "standard quantum limit", which depends only on the mirrors masses, and corresponds to the Heisenberg incertitude principle. It can be improved by increasing the mass of the mirrors, and may be beaten by using "back action evasion" measurement techniques.
- Such techniques have been demonstrated experimentally on simple systems. They are not yet fully applicable to GW detectors, difficult theoretical and experimental studies remain to be done, but they do open a road for future upgrades of the detectors.

Before this becomes possible or even useful, the first challenge consists in reaching the sensitivity limits of the real instruments: this is already an enormous difficulty, which requires a large R&D effort in many domains. We will try to give a flavour of each of these research activities, insisting particularly on the metrology and technology aspects, which are the key for success.

It is important to note, at this point, that the GW research community is still rather small, even though it is growing rapidly. There is little redundancy and competition: ideas and results are rapidly exchanged throughout all the projects in the world, and most of the R&D is performed under international collaborative agreements. In addition, the fact that a number of large interferometers are presently being built generates opportunities for young researchers at the thesis and post-doc levels, throughout the world, and there are real possibilities for international exchanges of physicists. Up to now, the international collaboration relies principally on personal initiatives, and is not yet very much "institutionalised", but the initiatives in this direction are well supported by the institutions.

2 Interferometer design, and technical challenges

Let us first remind some orders of magnitude concerning an interferometer of the first generation: the mirrors in both arms of the Michelson interferometer being separated by a distance of 3-4 km, the events to be detected correspond roughly to a differential length variation of about 10^{-20} m, over a timescale of a tenth of a millisecond to a tenth of a second. In addition, the events are expected to be rare (a few per year, with the initial sensitivity of the large interferometers), so that the stability and reliability problems are very acute.

The first step is to reach the required measurement sensitivity by lowering the shot noise. This requires a high power laser, the use of light recycling techniques, and the general use of ultra high quality optical components (materials and coatings), with very low losses in order to allow for a good recycling level, and with a very good power handling capacity, because the light power which circulates in the interferometers reaches some tens of kilowatts. All the technology had to be developed on purpose, because the requirements are specific and/or exceed, by far, those coming for all other applications.

The second step consists in minimising all the perturbations of the mirrors position and of the light path, which could simulate a real event (the first goal is to reach the quantum limit). To estimate and to quantify them constitutes a research field in itself. It will have required years of physical modelling and simulation studies, in order to design properly each one of the instrument subsystems. These studies result in very strict, and often unusual, constraints on most subsystems: each one of them requires a very specific design, and a specific metrology, in order to be characterised. We have made very large improvements in all these fields, since the beginning of the 1980's. Yet, it is not possible, with the present best technologies, to ensure that the sensitivity will remain adequate at all times and over all the measurement frequency range. Seismic noise, Newtonian noise, and thermal noise will initially bring some limitations that we will have to keep improving during the coming years, through research on the physical models and the technology.

Finally, whatever the noise level of the interferometer, one will always try to operate it at the limit of the achievable sensitivity, in order to detect weak, or remote, events. This means that one should permanently expect to detect more "false events" than real ones, and that signal analysis is to become an important research activity in order to interpret properly each event. This also means that the confidence in the first detections will be strongly enhanced if the same signal is detected in coincidence by different instruments, in different places of the world.

3 Theory of GW's

As we wrote above, one can consider that the properties of GW's, and their interaction with the detectors, are now well understood. This is comforted by the beautiful agreement that has been established between the observation and the physical model of PSR 1913+15. The present research in this field concerns mainly the following questions:

- Once we detect some GW events, to what extent can we use them to test the theoretical predictions? How can we use these events in order to discriminate between the different possible theories?
- In this respect, how to make the best use of an array of detectors?
- Are not there other and/or better detectors than bars and interferometers?

4 Astrophysics and GW sources

The following fields of astrophysics are particularly stimulated by the construction of sensitive detectors:

- Search for new kinds of GW emitting sources
- Statistics of known sources
- Physics of known sources and correlation between GW's and different signals (γ rays, X rays, neutrinos, visible light, ...)
- Accurate modelling of star collapse, binary stars, black hole vibrations and collisions, etc... (numerical relativity)

5 High vacuum technology and contamination studies

The vacuum systems which house the interferometers have some specific requirements:

- A very high vacuum (10^{-8} mbar) in order to minimize the possibility that statistical pressure fluctuations will phase modulate the light beams.
- An ultra low level of contaminants (hydrocarbons) must be maintained in order to avoid the risk of polluting the mirror surfaces, even after years of operation.
- Dust particles should be avoided, because they can pollute the mirrors, and also because dust particles falling through the laser beam would simulate a GW signal.

- The pumping system should not produce vibrations, nor pressure fluctuations.

In addition, economical reasons call for a minimal maintenance, which means a very low outgassing rate of the whole equipment, and for a high safety level. The corresponding studies, which lasted for about ten years, have been surprisingly successful: in particular we have understood the reason for the usual outgassing rate of stainless steel, and reduced it by two to three orders of magnitude, as compared with previous best technologies. The only remaining problem seems to be quality control.

Yet it looks very important, in case of accidental pollution of the optics, to be able to clean them in situ, without breaking the vacuum. We have started experimenting a pulsed laser cleaning technique, which looks very promising.

In addition to these requirements, there is a very specific constraint: all internal reflections in the vacuum system, which could mix scattered light with the main beam, must be prevented. Otherwise the modulation of the scattered light by the vibrations of the vacuum vessels would generate a very high noise level. The quantification of this effect has been very difficult theoretically and numerically, and has required years of collaboration between Virgo and LIGO. It has important consequences on the design and the choice of materials for the vacuum vessels and tubes, which will only be confirmed once the first interferometers are running.

6 Seismic isolation

It has long been recognised that it is important to be able to operate the interferometers in the low frequency range (below 10 Hz). The difficulty to get a sufficient seismic isolation is such that only Virgo accepted the challenge for the first generation detector. Other groups have chosen to operate only at higher frequencies, so missing a number of possible signals. It looks clear that the solution is a complex mixing of mechanics and electronics, which consists in some active isolation at very low frequencies, based on very low noise accelerometers and on a sophisticated multidimensional feedback system, followed by a few stages of passive isolation, which can provide very high gain at high frequency. Different variations on this scheme are being studied. Full size test facilities are being developed by Virgo, LIGO, and GEO.

A new field of research is starting to develop, in order to understand the so-called Newtonian noise: the ground and underground motions which create seismic noise are also the cause of variable gravity gradients which directly affect the mirrors motion and cannot be shielded. We begin to understand their effect, and suspect that they will contribute to the low frequency noise of the interferometers, once seismic isolation will be good enough. We also foresee that this Newtonian noise could be evaluated from the data given by a network of accelerometers around the site, and possibly cancelled, or subtracted from the signal. This will be an important research field in the near future.

7 Low dissipation materials

Thermal noise is slightly less critical in interferometers than it was for the Weber bars: these need to operate at very low temperatures, while an appropriate choice of low damping materials could already be sufficient for the first few generations of interferometers. R&D on thermal noise require improvements in the theory, in material science and in bonding technologies, plus a lot of difficult experimentation.

Presently, Virgo and GEO are validating together a new design of a monolithic silica pendulum for the first generation of the Virgo interferometer, which is an evolution of the first GEO suspension, presently being installed in Hannover.

Virgo, LIGO, and the Australian team, have started a common feasibility study of sapphire as a mirror material. Other crystalline materials are also considered (YAG, CaF2).

Apart from the choice of appropriate materials, the obvious long-term solution consists in cooling the mirrors and the last suspension stage to very low temperatures. This is technically very difficult, and a priori not compatible with the very high light intensities, but nevertheless the Japanese team is working at it and plans to start the construction of a large cryogenic interferometer in the coming years. The European groups also plan to study this idea, later.

Yet another solution would be to measure the thermal noise of each critical component, and to subtract it, or even better to cancel it with some feedback loop, as we have learned to do with many other kinds of noise. This is just more difficult here, because the motion is very small and, mainly, because it is not uniform across the surface of the mirror: the motion of each point of the mirror has to be measured and weighted appropriately.

8 Ultrastable high power lasers

The choice of an appropriate laser system was one of the initial problems, because the requirements of high power, ultrahigh stability, and reliability were rather incompatible. The appropriate techniques: diode-pumped YAG lasers (which provide power and reliability), injection locking (which allows for the amplification and the coherent addition of laser beams), and the MHz bandwidth frequency and power control systems (which provide the stability), were first designed and proposed during the Virgo feasibility studies, and are now in use in all the groups.

The frequency and amplitude control loops are examples of a situation where the fluctuations can be measured with a sensitivity that is better than the "fundamental" fluctuations, as determined by stimulated and spontaneous emission, and the photon statistics in the laser. The stabilised laser does fluctuate less than an ideal, quantum limited, laser.

For the first generation detectors, which require a laser power of 10-25 W, the problem is now practically solved. But R&D continues toward higher power (> 100W) and still better stabilities. In particular, the following questions will have to be addressed:

- Improvement of the laser power stability, in the signal range (10 Hz to 10 kHz), and in the modulation frequency range (10 to 30 MHz)

- New solid state lasers (improved efficiency, lower thermal effects), and frequency doubling.

- Improvement of the geometry of the light beam (active mirrors, holographic techniques)

- Improvements and new designs for the laser control components (PZT, electro-optic components, Faraday isolators, mode-cleaning cavities).

9 Optics and metrology

In 1990 it was still not clear whether the mirrors could ever reach the required low level of losses (scatter and absorption). The problem was twofold:

- Economically, it was not interesting for any private company to start the R&D for it, because GW interferometers constitute a small market, and there was no reason to think that other customers could need similar performances.

- Technically, the specifications were so strict that they could not even be checked with the existing metrology tools.

LIGO chose to rely on some hi-tech US companies (which were more advanced than their European competitors), and Virgo chose to develop the metrology and the mirror coatings inside CNRS research laboratories. Today, it looks like both approaches will succeed, partly because the collaboration has been very good between US companies and French research labs. But the choice made by Virgo may even result in significant future improvements, which could be difficult to obtain from the private companies.

The overall progress is quite remarkable, and reveals (if needed) the importance of having a very good metrology:

- Coating absorption: no private company knew how to measure absorption at the ppm level (10^{-6}) in 1990. We have developed in Paris a tool, which can measure down to 10 ppb (10^{-8}). This allowed for systematic studies at the ppm level, which improved our understandings, and the goal has finally been achieved.

- Substrate absorption: we needed 1 ppm/cm, while the previous available metrology was about 100 times worse. Once we became able to measure at the level of 0.1 ppm, we could interact with a German producer, which is now manufacturing for us a special, low absorption, silica.

- Mirror shape: the precision with which our mirrors approximate their perfect shape (either flat or spherical) is usually well measured during the polishing, but it is changed in an unknown way by the stress due to the coating layers and by the thickness inhomogeneities of the coating layers. Coating factories use to anneal the mirrors after coating in order to reduce the effect, but they don't control it.

Having brought the right metrology tools to our coating factory, we became able not only to measure, but also to understand and to correct for these effects. We have learned how to add a "corrective layer", which improves very much the quality of the reflected wavefront. In consequence, a whole new laboratory has been built in Lyon for the coatings manufacture and R&D. Work will continue in the following areas:

- Lower absorption and scattering losses

- Better wavefront metrology

- New components, for instance low loss, all reflective optics.

10 Electronics and Control

In Virgo, all the mirrors and main optical benches are suspended from a chain of pendulums. All the degrees of freedom of each mirror, and a number of points in each suspension chain, have to be monitored, and controlled by feedback systems. In addition, there are many interactions between these servo loops, and the mechanical transfer functions are very complex (many sharp resonances).

This was an ensemble of good reasons for us to decide developing a digital control system. The first generation system uses an ensemble of DSP's (digital signal processors), and microprocessors. R&D will go on in this domain, in order to be able to follow the evolution of the technologies, and probably to design an expert system in the future.

11 Modelling and simulation

11.1 Physics of the detector

GW interferometers are much more complex than a simple Michelson interferometer: they contain in addition some resonant cavities. Moreover, for control reasons, and also for signal detection, the light is phase modulated, so that a few different frequencies (carrier and sidebands) are propagating therein. Their transfer functions, for the various noises and for the signal, are thus complex, and they have to be estimated dynamically, because they depend on the status of the instrument. This study is still very active.

Each one of the subsystems which constitute Virgo has to be modelled and understood in detail, to ensure that it will not be a cause of extra noise. Much work has been done, but the tests of the Virgo central interferometer, of the GEO and TAMA interferometers, and of the first generation LIGO detectors will soon tell us were did we make mistakes, and which studies should be restarted.

11.2 Global noise studies

In addition to the individual contribution of each subsystem, some non-linear interactions have to be expected, which will introduce couplings between a noise source and some static defect or offset in the interferometer. For instance, laser power fluctuations and an offset relative to the dark fringe tuning, or laser beam jitters and a static misalignment, thermal noise of the output mode-cleaner and a frequency offset, and probably a few more that remain to be discovered. The simulation programs, which are developed for Virgo

(SIESTA) and LIGO (end-to-end model), help understanding these effects. The analysis of the first interferometers should also teach us a lot, and may show the necessity to start new studies.

11.3 Locking and signal reconstruction

Locking consists in aligning each component relative to others (to nanoradians) and fixing the distances between each pair of suspended optical components (to picometers), in terms of the laser wavelength, which is, itself, stabilised on a mechanical length. This represents a few tens of servo-loops, which have to be closed one after the other, respecting some hierarchy rules. Different strategies can be considered, which are presently evaluated using simulations, and will soon be tested in reality. An important question is how can we use the signals delivered by the interferometer in order to reconstruct the status of the interferometer, and the signal itself.

Progresses in our understanding and in the computer technology should soon make it possible to design a real time expert system.

12 Data Analysis

Once the interferometers operate properly, signal analysis is the last challenge. Since the detectors will be running at all times, it is necessary that the first analysis (detection of possible events) be able to run in real time. This means that tens of thousands of matched filters will have to run in parallel, triggering typically each millisecond, which may require large parallel computers. The present tasks consist in developing algorithms, optimising them, and evaluating precisely which computing power will be needed. This depends on some realistic astrophysical assumptions, and requires collaboration with astrophysicists.

The technical computer aspects will reveal very important, as it is likely that the computing requirements will keep asking for very large and expensive parallel machines, in spite of the very fast technological progress which gives to today's PC's (G4) the power of last year's workstations.

The organisation of coincidence detection with other detectors (GW detectors and others), is another priority.

13 Long term R&D

We have given above an idea of the many research fields which are being explored in order to improve (upgrade) the subsystems of the present detectors. **It is important to stress, again, that the key for experimental progress is metrology: any noise and any imperfection that can be measured, can be understood and eliminated by some kind of feedback system, or subtracted by calculation.**

It is also important to start working at some more futuristic ideas (new kinds of detectors, very "advanced" subsystems for the present detectors, or just new technologies, which may be useful in the future. This is difficult, in practice, because all the teams involved in GW research are presently busy and understaffed, and their budget is not unlimited. But these activities can also be performed by other physics groups, which are not inside, but "around" the GW community (particularly in astrophysics labs, in quantum optics, cryogenics, signal analysis and computer science). It is important to maintain links and mutual interest within this enlarged community, so that GW groups remain able to follow the external developments, and peripheral groups tend to explore the ideas which can be useful for us.

In France, a larger community (Experimental Gravitation) meets yearly at the GREX meetings.

At the international level this is partially the role of the Aspen-Moriond conference, which takes place each year in January or February, either in Aspen (Colorado) or in Les Arcs (France).

14 Conclusions

The fact that large and expensive GW detectors are being built today is the consequence of a wide R&D activity, in many different domains, which had started more than 20 years ago. The main effort has presently turned to construction activity, but R&D is restarting and must come back to its highest level, in order for us to become able to make the best use of the detectors, and to upgrade them in a few years. This is true for all the projects in the world: we have understood together that this R&D should be made in a collaborative way, and our institutions start to react positively to this idea.

This means that GW research should become a very attractive field for students and post-docs, from pure theory to very applied physics, not only because the challenge is fascinating, but also because it is offering more and more opportunities for working at the international level in all these domains.

15 Useful addresses

The interested reader could explore the following web pages,

GEO : http://www.geo600.uni-hannover.de/

LIGO: http://www.ligo.caltech.edu/
Virgo: http://www.virgo.infn.it/
Others: http://www.virgo.infn.it/facilities/othergw.htm

And contact the following persons, for technical questions or job opportunities:

- Theory:
  Thibault Damour, damour@ihes.fr
  Philippe Tourrenc, pht@ccr.jussieu.fr

- Quantum effects:
  Christian Bord, chbo@ccr.jussieu.fr
  Antoine Heidmann, heidmann@spectro.jussieu.fr

- Astrophysics:
  Jean Paul Scheidecker, jpaul@obs-nice.fr

- Vacuum and contamination:
  Carlo Bradaschia, bradaschia@pi.infn.it

- Seismic isolation:
  Stefano Braccini, braccini@pi.infn.it

- Thermal noise:
  Luca Gammaitoni, gammaitoni@pg.infn.it

- Newtonian noise:
  Flavio Vetrano, vetrano@fis.uniurb.it

- Lasers:
  Nary Man, man@obs-nice.fr

- Optics:
  J.M.Mackowski, mackowsk@ipnl.in2p3.fr

- Metrology:
  Claude Boccara, boccara@optique.espci.fr

- Electronics:
  Benoit Mours, mours@lapp.in2p3.fr

- Control:
  Gianfranco Giordano, gianfranco.giordano@lnf.infn.it

- Modelling:
  J.Y.Vinet, vinet@obs-nice.fr
- Simulation:
  Andrea Vicere, vicere@pi.infn.it
- Data analysis:
  Alain Brillet, direction@virgo.infn.it

www.ingramcontent.com/pod-product-compliance
Lightning Source LLC
LaVergne TN
LVHW051057210926
848486LV00002B/139

* 9 7 8 9 8 1 0 2 4 3 0 6 7 *